AF888159

Springer Geography

Advisory Editors

Mitja Brilly, Fac. Civil & Geodetic Engineering, University of Ljubljana, Ljubljana, Slovenia
Richard A. Davis, School of Geosciences, dept. of Geology,
University of South Florida, Tampa, FL, USA
Nancy Hoalst-Pullen, Dept. Geography & Anthropology, Kennesaw State University, Kennesaw, GA, USA
Michael Leitner, Department of Geography & Anthropology, Louisiana State University, Baton Rouge, LA, USA
Mark W. Patterson, Dept. Geography & Anthropology, Kennesaw State University, Kennesaw, GA, USA
Márton Veress, Department of Physical Geography, University of West Hungary, Szombathely, Hungary

The Springer Geography series seeks to publish a broad portfolio of scientific books, aiming at researchers, students, and everyone interested in geographical research.

The series includes peer-reviewed monographs, edited volumes, textbooks, and conference proceedings. It covers the major topics in geography and geographical sciences including, but not limited to; Economic Geography, Landscape and Urban Planning, Urban Geography, Physical Geography and Environmental Geography.

Springer Geography—now indexed in Scopus

More information about this series at http://www.springer.com/series/10180

Mainak Ghosh
Editor

Perception, Design and Ecology of the Built Environment

A Focus on the Global South

Springer

Editor
Mainak Ghosh
Department of Architecture
Jadavpur University
Kolkata, India

ISSN 2194-315X ISSN 2194-3168 (electronic)
Springer Geography
ISBN 978-3-030-25878-8 ISBN 978-3-030-25879-5 (eBook)
https://doi.org/10.1007/978-3-030-25879-5

© Springer Nature Switzerland AG 2020, corrected publication 2020
This work is subject to copyright. All rights are reserved by the Publisher, whether the whole or part of the material is concerned, specifically the rights of translation, reprinting, reuse of illustrations, recitation, broadcasting, reproduction on microfilms or in any other physical way, and transmission or information storage and retrieval, electronic adaptation, computer software, or by similar or dissimilar methodology now known or hereafter developed.
The use of general descriptive names, registered names, trademarks, service marks, etc. in this publication does not imply, even in the absence of a specific statement, that such names are exempt from the relevant protective laws and regulations and therefore free for general use.
The publisher, the authors, and the editors are safe to assume that the advice and information in this book are believed to be true and accurate at the date of publication. Neither the publisher nor the authors or the editors give a warranty, expressed or implied, with respect to the material contained herein or for any errors or omissions that may have been made. The publisher remains neutral with regard to jurisdictional claims in published maps and institutional affiliations.

This Springer imprint is published by the registered company Springer Nature Switzerland AG.
The registered company address is: Gewerbestrasse 11, 6330 Cham, Switzerland

Foreword

The particular issues of the Global South need urgent attention for a pluralistic multifaceted vision of improvement and sustainable development. This book is an attempt to address this difficult and interdisciplinary topic. Why do the architecture and built environment of the Global South remain so vastly different from those of the North, often despite the rich heritage values and significant spatial inheritance of the past? The arc of progress seems to have dampened with time and accelerated during the colonization and after the industrial revolution. Today, globalization, aided by scientific and technological worldviews, has made the world a smaller, more known, and accessible place. However, as the chapters in this book show, the differentiators need careful appraisal. The various divides that exist among territories marked by political, cultural, and socioeconomic boundaries have impacted everything that human civilization had ever produced. With the growing intellectual exchange and media access, there is every reason to assume that these differentiators can be better understood.

The growing usage of the terms Global North and Global South, today, is largely an attempt to bridge the gaps rather than reinforcing and accepting modes of differentiation in the world. There is a need to address a multiplicity of views, and cross-sectional knowledge needs to be dissipated about the characteristics and criteria of the Global South, so that, first, the awareness is raised, and, second, forces and parties join hands to solve problems to make the world a better place to live and grow in the future. Many of the chapters of this book have been framed in the form of stories from different parts of the world. These depictions are specific but are ubiquitous in the developing countries of the world. The presentation of the book is vivid and catches the attention of the reader, each case study painting a conceptual picture of typical Global South scenario, irrespective of place specificity. Having been born and brought up in a developing country and now living in a Global North country in the Southern Hemisphere for many years, I could relate to this book in a manner of dual readership, experiencing both dichotomy and unison.

This book is an excellent resource for architects, designers, planners, environmentalists, sociologists, and policy-makers, for it projects the setting of built environment in the Global South. Though the book is academic- and research-driven, an overall read would not be too strenuous for the interested reader or a student curious to learn about the built characteristics of the developing world. With my research area encircling around computational techniques in architecture and urbanism, I have been fascinated by the simplicity with which the key features of built environment has been postulated in this book, namely, the environment (ecology), perception, and design. Overall, the book has harnessed on one or more of these themes in different chapters of the book fashioning a subtle integration. The interwoven characteristics of these themes, with multitude of places and different styles of writing by the different chapter authors, have given it a form which resonates with the essence of how the Global South actually exists – diverse, rustic, vibrant, and perhaps chaotic.

I would contend that this book is but a tiny grain of sand compared to the vastness of the Global South. Large and overwhelmingly diverse, difficult to research and document, the book is a timely compilation on this important subject matter. I am certain that this is a ripple which would ambitiously multiply over the coming years.

Professor, Curtin University
Perth, WA, Australia

Sambit Datta

Acknowledgment

The countless anonymous people around me, passer-by encountered while living my day-to-day life in a developing country, gave me the impetus to frame the book. Observing the habits and habitat of my fellow inhabitants in the Global South charged me with the thought that more needed to be explored and documented on this front. Hence, this is to acknowledge those whom I do not know by name but are part of this same world, living a life of difference in a substantially dissimilar environment than that of the developing nations.

We all reside in a zone of interface of place and people; without place, we do not exist, and places get their meanings because of people. And this place is to convey my special humble acknowledgment to the important people in my life supporting this work.

My sincere expression of gratefulness to all the chapter contributors of this book from different countries of the world. Without the support and cooperation from all the authors, it would have been impossible to give a shape to this thought which is vast and obscured. Discussion with colleagues and friends about the idea of the book helped me get clearer perspectives: Professor Souvanic Roy had played an important role in formulating the title of the book and Professor Keya Mitra, Neeta Das, and Venance in connecting up with the authors from other countries. Thanks are also due to Professor Sanjib Nag for his support.

I have been fortunate to receive the foreword from Professor Sambit Datta, Curtin University, who is fondly keen on the developments of the Global South. With his extended travel experience and expertise on the subject, living in Global North but originally from Global South, he is perhaps the best person who could review this book.

My students have been the constant source of inspiration, and I find it interesting to learn from them. I especially thank my research scholars, Farha Shermin and Shreeja Ganguly, for helping me with some of the copy editing part.

I am obliged to the infrastructural support provided by Jadavpur University, Kolkata, India, where I am attached professionally at present to teach and research. Also, thanks to my university colleagues and staff. Professor Samantak

Das from the Department of Comparative Literature has been a source of great encouragement.

I hereby express my humble gratitude and regards to my parents who have been the silent supporters in all my pursuits. I have lost my mother, Mohua Ghosh, during the book project. I am grateful to the publisher and authors for staying by my side patiently, despite the delay in the book project. I dedicate this book to my mother.

I thank my family, Sudipta and Anthea, for their cooperation.

Contents

Chapter 1
Built Environment in Response to the Ecology, Design, and Perception of the Global South

Mainak Ghosh

Abstract The need is highlighted never than before, that the urbanization and proliferation of built environment of the Global South require "new theoretic approaches" and in-depth analysis based on case studies, evidences, and scenarios since the causes of urbanization and its outcomes are significantly different from those in Global North. Hence it is evident today the Global South indicator is not based on geographical parameters but relies on various other considerations related to urbanization, development, and built environment. This particular chapter compiles some of such individual studies of the Global South which are very specific; however when looked into the analytical sublimity of these cases, its specificity gives way to the generic characteristics of these cases, merging with the common issues of the urbanized built environment and culture of the Global South. Hence, seemingly the built environment is to deal with buildings, urban regime, and human settlements; however a deeper introspection here extends its significance into much larger areas and forms. Thus when the design of the built is considered holistically against perception of the built and its environment, it gets truly understood. This chapter gives an overview to this understanding, which is discussed and elaborated substantially in the next chapters. The aim is to propound that it is very crucial to know that every component of the built environment is defined and shaped by context and each and every element impacts either positively or negatively to the overall quality of environments for built and natural and to human-environment relationships. Added with political and social complications, the urbanization of the Global South has been steadily and sporadically increasing with degradation of quality of urban environment and living. The skewed design of the built environment for these urbanized worlds, the conflicting perception of different stakeholders towards it, and the degraded ecology of the development remain ubiquitous throughout these countries or rather regions; hence a summarized view is presented in this chapter for a comprehensive understanding of the overall compilation of the book.

M. Ghosh (✉)
Department of Architecture, Jadavpur University, Kolkata, West Bengal, India
e-mail: mainak.ghosh@jadavpuruniversity.in

© Springer Nature Switzerland AG 2020

M. Ghosh (ed.), *Perception, Design and Ecology of the Built Environment*,
Springer Geography, https://doi.org/10.1007/978-3-030-25879-5_1

Keywords Global South · Developing country · Sustainability · Ecology · Design · Perception

Global South

It is quite a challenging task to define "Global South" more so when the world has become a melting pot of every aspects to do with human civilization. Aided with technological advancements, ever since industrial revolution, the world has stepped towards regionalism in terms of economics and associated politics; at the same time, it has fallen flat in terms of culture, living, and lifestyle and societal aspects at large. Post-1990s, marking the end of Cold War expression, the categorization of First, Second, and Third World has become outdated. Even the prominence of the terms which have been in abundant use, namely, "development" and "developing countries" has diminished considerably. Though they are still used eloquently and ubiquitously in a colloquial perspective, it has questionable validity and vitality, as the center of global economic growth has been shifting to places outside the so-called West (Europe, North America). These changes triggered debates and dialogues on varied aspects of postcolonial and developmental scenarios, which though started in a discursive manner soon gathered momentum. And one of the rather stable terminologies which surfaced out resulted into the origination of the divisions of "Global North" and "Global South" (Levander and Walter 2011). Though these terms have been coined by and predominantly used by social science and the humanities, in the initial stages, today natural sciences, medicine, and other sciences have been increasingly using the term. Business, management, law, and other disciplines are also finding the use of North-South dichotomy to be emergent. Research has shown that there has been a steady and steep growth in the usage of the term Global South in publications since 1996 (Sivertsen 2014). This is just a small indication that Global South is not only gaining importance as a terminology but also is being rendered intellectually and scholastically by different researchers in different disciplines.

Apart from this quantitative predominance of the Global South, it is important to note the qualitative notion of Global South. It is associated with the negative impacts of capitalism, mostly by Western actors, and served by the weak regulations, acting on the inhabitants of the countries or regions of the Global South. In contrast, Global North is often linked to strong governance and public sector. Another dichotomy noted for these terms relates to urban and rural. Global North is symbolized as having accomplished the complete urbanization process in an organized and orderly fashion, in the postindustrial revolution world, whereas Global South is visualized as predominantly rural, backward, and underdeveloped. It could be observed; there is always a lot of scope to describe the inadequacies and deficiencies in Global South compared to the urbanization standards set by the Global North. The urbanization process of Global South is rather termed as "chaotic" and "roblematic"

(Pagel et al. 2014). These researchers have highlighted that the urbanization and proliferation of built environment require "new theoretic approaches" and in-depth analysis based on case studies, evidences, and scenarios since the causes of urbanization and its outcomes are significantly different from those in Global North. Hence it is evident today the Global South indicator is not based on geographical parameters but relies on various other considerations related to urbanization, development, and built environment. Methodological national boundaries could be avoided; rather sub-national regions or localities could be considered, since the "internal conflict and contradictions" could lie anywhere in the globe (Dirlik 2007). With narrower and idiomatic usage of "North" as equivalent to the capitalist elite class and "South" as equivalent to marginalized segments, today indeed Global South could be anywhere and everywhere (Sparke 2007).

It is also observed that within Global South while there are similarities of pattern, diverse variations and heterogeneity exist which requires in-depth individual studies. A gross generalization of issues of Global South is almost impossible. This particular book compiles some of such individual studies of Global South which are very specific; however when looked into the analytical sublimity of these cases, its specificity gives way to the generic characteristics of these cases, merging with the common issues of the urbanized built environment and culture of the Global South.

Built Environment

Man like animals have looked at environment as a source for mitigating needs of self. The immediate environment has provided them with basic resources. And in many cases, if the immediate environment has been exhausted or not sufficient, man has resorted to migration. However unlike animals, man started looking at environment not just as a resource but as an inspiration that could be changed. The very basic nature of human exploration is visible if one observes how the children interact with the environment. Children with no particular experience of how to use a natural element come up with a very innovative approach when encountering a natural form, for example, a twig of a tree. In contemporary day-to-day living, a child would not see much use of a twig of a tree except that of observing twigs to be pruned at times. However when they are handed with a twig, they would find multiple uses; the twig then starts acquiring a greater importance than it could be thought of; hence the twig starts acting as an inspiration to the children. Children try to change their surrounding every time, with different materials and tools available to them. One would surely recall the experience of the child doodling on the clean walls. It is certainly a very simple example of changes being brought to the environment. Hence human being by very nature is a changemaker and brought changes to immediate or remote environment. As a part of this endeavor, man acted on the environment to build. Now the dictionary definition of "build" says "Construct (something) by putting parts or material together" (Oxford Dictionary 2018), whereas the definition hinted for the "built" suggests that something is "made, formed, or shaped in a specified way" (Merriam-Webster Dictionaries 2018). It is understood from these definitions that both the words related to tangible form which are meant to exist in the environment and be part of it. Hence there is a strong connection of the verb "build" and the noun or adjective "built" to the environment.

Historically speaking the built environment is one of the most crucial parameters to mark human civilization. Today the built environment and forms are almost "natural" and elemental to us. People find them obvious. The built and the natural go hand in hand. However this is not to judge its impacts. From a theoretical discourse, every bit of earth has been transformed into built environment. One may argue what about pristine oceans, impenetrable forests, or frozen polar regions? The oceans are surfaced with countless voyages. Diving explorations, dumping of wastes, floating plastic bottles, sinking carcasses of junk machines, weapons, and the like have extended the "built environment" into the unimaginable spreads and into the depths of the ocean. The trackers fitted on wild animals, the satellites sending pictures from the deepest cores, and the aircrafts hovering over the densest foliage of the world have turned the scariest of the forests, heard in stories into the "built," the "built" that one experiences every day while sitting in an office room or strolling in an urban park. Needless to talk about the sinking ice, greenhouse effect, global warming, etc. and their impacts on the poles, the effects are not just restricted to the poles but percolate to every continent. In a broader sense, it deals with water. Which is essential for life, too much or too less of it, both are of great threat to human civilization. Hence, seemingly the built environment is to deal with buildings and urban regime; however a deeper introspection extends its significance into much larger areas and forms.

The most dominant impacts and effects of built environment are realized in urban areas, not to neglect the rural and other forms of settlements though. It is just that urban areas have higher density and large number of human interventions as seen against the natural environment. The rise of urbanism is a significant cause of concern. The United Nations data set predicts that 68% of the world's population would live in urban areas by 2050, which in 2018 was summed up to 55%. This increase in urban population resulting gradual shift in residential settlements from rural to urban areas and also taking into account the overall growth in global population by 2050 would add another 2.5 billion people to urban areas. It is also notable that UN report states that close to 90% of this increase take place in Asia and Africa (United Nations 2018). And because of rapid urbanization, there are other issues emerging. It is obvious that this exuberant urbanization is giving rise to growth of urban built environment which was never seen before. The built environment from its very inception has been seen as a place of recess and safety, a resort to protect from the turmoil prevalent in nature, such as storms, rains, harshness of temperatures, and the predictable and periodical ones, and also from the unpredictable ones such as attacks of wild animals or landslides. It still continues to serve these purposes of protection and harnessing. But with the growth of enormous built-up forms in contrast to its surroundings, it has started posing newer challenges to mankind. Pollution is one of the examples. The by-product of urbanization has led to a threat to the human habitation again. The same notion of recess and safety is now contested. Hence as man starts altering his surroundings and his immediate environment, he also owns the responsibility of a holistic betterment and not a skewed one.

Environment, Perception, and Design of the Built

Built environment or man-made environment has various dimensions and aspects related to it. There are multiple disciplines which take different perspectives and discourses. This particular book attempts to look at the built from three distinct perspectives; these are based on aspects of environment, perception, and design. These are very broad areas of discussion and embrace interdisciplinary and varied modes. In most cases these perspectives seamlessly merge, one into another. However for a holistic understanding of built environment, one should be aware of these three perspectives and analyze its interrelationship while considering a built environment. Let us consider the example of peri-urban or urban areas. The growth of fringe areas of a city is inevitable. In many developing countries, the cities merge with the rural surroundings. The urbanism sprawls into the nonurban settings. This gives rise to unique characteristics not only to the built environment but also to the natural and sociocultural ambience. One would come up with an interesting analysis if the residential houses are considered in these areas. Many of the residential buildings use a combination of masonry, concrete along with natural materials such as timber structural members, and thatch or burnt clay tiles. Now, this perspective of the built form is a unique design endeavor by the local masons and housebuilders. Often the residents themselves build up their shelter using this unique hybrid technique developed indigenously. From a design standpoint, this is novel and technologically innovative. However these innovative frugal design attempts could be tested for its technological soundness, efficiency, and safety. Some of these stand very sound over prolonged trial and error. For example, the way bamboo roof framework is inserted and supported on masonry walls, concrete lintels or ring beams, and concrete columns or pillars is robust and effective. These houses are often considered to be "kuccha" or temporary. Many of the authorities and governing bodies do not even acknowledge these to be legitimate buildings. And compared to the buildings which are purely made out of masonry, stone, and concrete, these buildings are recognized to be inferior. They are socioculturally portrayed as of lower class, representing poorness and inadequacy. It is neglected and not praiseworthy. The perception towards these designed forms of the built environment starts coming into picture. Hence the design of the built could be analyzed based on the perception of the built. A simple place of living starts making a mark on social discrimination and on cultural mindsets. In the same neighborhood, a person having such "kuccha" house would receive social bias compared to a person living in a concrete-clad residence. However ecologically and environmentally speaking, it is found that these hybrid models often turn out to be climatologically and economically very effective. Thus these are perhaps more sustainable scientifically compared to the concrete buildings. Thus when the design of the built is considered holistically against perception of the built and its environment, it gets truly understood. Just a partial understanding of the built environment is therefore always very limited and narrow. Until analyzed against these three perspectives, a discourse on the built environment remains incomplete and incorrigibly deficient.

Environment and the Built: Earth as an Inspiration

About 13 million years ago, man was a hunter and gatherer and lived in caves. With the advancement and invention of simple tools and discovery of fire, he advanced into the barbaric age. He then learnt that living in groups was safer and living closer to the water sources ensured a good supply of food and water. And this is when clusters and settlements started to form. With the discovery of agriculture, man learned to live on fruits and vegetables, and thus he did not have to move anymore in search of food; the settlements became more permanent. This journey from the caves to settlements had already taught humans the importance of the "built," which provided safety, security, and feasibility in all aspects.

If we talk about the first house that was built or the first ship that was built, it will be undoubtedly from the most local and natural materials which were found in vicinity, like timber, stone, thatch, etc. As technology advanced and new inventions happened, the built took different shapes and forms, like the monumental dome built by the Romans at the Pantheon and the gigantic Greek orders at the Parthenon.

Moving centuries later to a very important and life-changing phase in human history, the industrial phase, we start to see large-scale production of iron. The ready availability of iron contributed to the development of machinery, notably James Watt's double-acting steam engine of 1769. Although application of iron took place centuries before the industrial revolution, the first large cast-iron structure was the bridge over the River Severn at Ironbridge, a town in England. Built by the iron founder Abraham Darby III between 1777 and 1779, it has a span of 30 m. An extravagant sequence of iron and glass structures continued to the end of the century. The most important of these was the Crystal Palace, built in London's Hyde Park with standardized parts. This was to accommodate the Great Exhibition of 1851 (César 2017).

The built has come a long way since then, and in every stage, it has always had an intricate relationship with the environment. Firstly, by using the materials that the environment provided, the built took inspiration from natural forms only to exploit the environment at some point of time.

The environment has not only provided the materials but has also shaped the way we built in the past. A very good example would be a vernacular building of any area. Let us consider the vernacular stilt houses of Assam as an example. Being in a very heavy rainfall zone, the high plinth not only helps during floods but is also used to shelter cattle. The very convenient sloping roofs make it easy to drain off rain water. The materials used are locally available ones like bamboo, mud, and thatch which are easily replaceable when needed. Moreover, these materials have insulating properties which keep the interiors warm in the harsh winters and cool in the summers. Being lightweight and constructed with flexible connections, such structures are very efficient during earthquakes which this zone is extremely prone to (Kaushik and Babu 2009).

Today, we often talk about the terms "green" and "environment-friendly" and we try building with materials which have lesser impacts on the environment. In this context the methods and materials used to construct a building that affects the environment are as important as to the knowledge on how the built is operated. For example, use of heating, ventilation, and air conditioning (HVAC) system is not efficient or sustainable having a long-term negative impact. Several other factors make huge impacts but are not immediately noticeable, like the energy used for building a shelter in the form of electricity and fuel consumption. The fossil fuel usage for transportation related to construction of contemporary buildings, unlike vernacular, is higher, as they do not tend to use locally available materials. Added is the transport of waste, produced from construction and demolition. Today perhaps we have reached the threshold, and thus, now it is absolutely important to understand and respect the relationship of the planet with the built and act accordingly.

So far man has created tools, harnessed fire, manipulated spaces, and modified the environment to make his life safer, comfortable, productive, and enjoyable. Now, times have changed, and man-made changes have increased, and populations have blown up, but the basic reasons for creating the built environment remain vitally the same; however the means have changed.

It is very crucial to know that every component of the built environment is defined and shaped by context; each and every element impacts either positively or negatively to the overall quality of environments both built and natural and also to human-environment relationships. These impacts are sometimes local but are experienced at every scale, including global and even planetary.

Perception and the Built: Eyes of the Mind

The Oxford Dictionary defines perception as "the way you notice things, especially with the senses." Thus, the way we realize, understand, or experience the surroundings depends on how we perceive it.

The built around us is guided by several factors, starting from basic ones like the site, climate, etc. to complicated ones like issues of territory and privacy, as well as the question of who determines and designs the physical spaces. Such factors often neglect the individual as well as the collective perception. Also, the implicit nature of our daily surroundings often changes individual perception.

Architecture as a crucial part of the built environment acts as an identity for the given region, community, or culture. It reflects culture closely with structural, historical, political, economic, and social features of society. The changing cultural and social attitudes in communities have the greatest impact on the built (architecture), and the key actor and factor is perception.

Without knowledge our eyes are nothing but mere instruments to identify and measure, and we tend to develop incapacity to determine meanings in what we see (Youssef 2015). This knowledge comes from the cultural and social aspects. It is interesting how the cultural and social perspectives can change the value of a building. Let us consider the case of Mawlynnong, a historic settlement in Meghalaya, as an example.

Mawlynnong is a small hamlet of about 500 residents located in the East Khasi Hills, about 90 km from the capital city of Shillong. Nearly 130 years ago, there was an outbreak of cholera in this remote settlement. Without the availability of medical facilities, the Christian missionaries suggested that cleanliness was the only solution. Since then, cleanliness has become a daily ritual for the local inhabitants.

Until 2003 it was like any other remote settlement in the state. No tourists visited the settlement as it had no proper roads and was accessible only by foot. Later in 2003 it was voted as the cleanest village in Asia by *Discovery India* magazine. It was reinforced by BBC and National Geographic in 2004, and UNESCO endorsed it in 2006, and thus the village won fame in the national and global scenario.

The architecture in this settlement is very traditional with typical features like sloping roofs with palm leaves, bamboo walls, and stilt houses with wooden columns. All the materials and techniques are sustainable and locally available, and the settlement has survived and flourished over the years in the same way.

Now with the inflow of tourist and increased interaction and mixing of cultures, the settlement suddenly has an outburst of concrete buildings. To cater to the tourists, modern sanitary facilities are provided, several houses have been converted to restaurants, and homestays, and a complete change in perception towards the local architecture is observed. The inhabitants who were once very proud of their traditional bamboo houses and showcased the same by building tourist lodges with bamboo are today willing to shift to concrete buildings at any cost. Concrete buildings are wrongly perceived as a symbol of status, and several traditional houses have been replaced.

It is also very interesting as how a neighboring object affects an object, with respect to perception. For example, one may find a sculpture simple because he is unaware of its intricacy; or he finds it confusing and complex because he is unaware of its intricacy; or he may be puzzled only because he is not accustomed to a new, modern style of shaping things.

To conclude, one can say that perception is one of the main factors which determine the success of the built in the long run. For a designer it is important to understand the perception of the user before putting forward his ideas. Secondly, we see that sociocultural aspects impact perception to a great extent, and in such cases knowledge in terms of awareness becomes crucial.

Design and the Built: Man the Place Maker

Design is a very broad concept. Generally speaking it is the process of envisioning and planning the creation of objects which in this context is the built. Designing is an intentional process. It is often misinterpreted as something that is done to make things look more aesthetically pleasing. Of course, no one likes to see anything unpleasant unless there is a specific reason to do so. But that is not what design is about. Most importantly, it is about making the user's interaction with the environment more natural and complete. It is a combination of both art and science; in brief it takes the expression of art and problem-solving aspect of science. And at the end, it is an attractive keyword, as it conveys to people the associated notions of ability and thought (Taura and Nagai n.d.).

Man has over the years mastered the several aspects of design in all possible ways. This process has developed with the new inventions, discoveries, identification of new materials, advancement of technology, and interaction with different cultures. Such knowledge has been passed down from generation to generation to have what we have today.

From the early civilization of Mesopotamia built in mud to the gigantic pyramids of the Egyptian civilization and to the more advanced brick houses and strategically planned settlements of the Indus Valley Civilization, man had it all sorted. We design and build our lives from one experience to another. Based on those experiences, components of the built emerge from human needs, thoughts, and actions.

Man has always been able to establish himself as a superior being on the planet. Our actions have built civilizations, unimaginable skyscrapers, bridges, roads, and dams. We have designed and built for the good of culture, religion, and society. But the irony is that sometimes, the essence of human actions is magnificent, and we design and plan quality life experiences for ourselves and others. At other times, the same human actions are shortsighted which ends up in creating uncomfortable situations that are not very feasible for healthy human activities and impacts the environments negatively.

Like mentioned before, design is art in terms of expressionism. A designer not only creates a space for a user but also expresses his thoughts and ideas with the same. Thus, there is always a cause-and-effect relationship between human purpose and the things we design and create. Human needs and values are manifested in the built environment. A persuasive historian and English Prime Minister, Sir Winston Churchill, expressed that "We shape our buildings; thereafter they shape us." Another noted historian, Arthur Cortell, conveys this same interdependency by claiming: "Tell me the landscape you grew up in and I will tell you about yourself."

When we look into the past, we know that most of society's knowledge of past civilizations, cultures, their habits, and everything for that matter is derived from remnants of the built environment. We would probably never have known the use of arches in a building if the Romans had never designed their temples using them. In the same way, our present cultures would be judged someday in the future by what have been created today. We should think about the results then as whether our present practices will be applauded to or will be condemned as being insensitive and brutal. Perhaps when we design with such questions in mind, the true sense of design is achieved, which is through understandings of the present, with the knowledge of the past and keeping in mind the values of the future.

Urbanism and Urban Transformations

It is easily understood that the divide of the Global North and Global South is based on urbanization factor in a very significant manner. Though there are various other factors relating to difference in economics, lifestyle, culture, technology, etc., it could be inferred easily that many of these are rather by-product or outcomes of urbanization process. Kaul (2013) has strikingly projected the growth and rise of Global South in his report on United Nations Development Program. He also mentions that countries of the Global South have paced forward at different speeds and have been able to join the group of major global policy-shapers. On the other hand, in a need of their own sovereignty concern, the countries of the Global North have responded to this increasing power of Global South. Political multipolarity is adding economic multipolarity and military multipolarity, assimilating the role of Global South into the world economy. Kaul (2013) even states ambitiously, "the world has been fortunate that increasing economic openness and deepening policy interdependence have coincided with the rise of the global South."

With all these the immediate direct impact which has been noticeable is the growth of urbanization both in a planned and unplanned manner. Many of the countries of Global South have been influenced by the effects and privileges of urban settlements during the colonial stages. However the postcolonial challenges have shifted these urban lands into cluttered settlements in most cases. Added with political and social complications, the urbanization of Global South has been steadily and sporadically increasing with degradation of quality of urban environment and living. Again with the gradual rise and importance of these countries in the Global scenario for the past decade, the urbanization process is under strict vigilance by most of the stakeholders. Hence the urban transformation of the Global South is ever-changing and continuous.

This particular book starts with chapters on urban transformations with stories of some of the countries of Global South. These transformations are rather predominant and common for most of the countries in Global South. The design of the built environment for these urbanized worlds, the perception of different stakeholders towards it, and the ecology of the development remain ubiquitous throughout these countries or rather regions.

Discussions on Built Environment of the Global South: Structure of the Book

In the first chapter, the current state of some of the towns has been elaborated, which developed into urban settlements during the time of colonial proliferation around the globe. This is comparable to most of the countries in Asia, Africa, and South America. These smaller settlements were brought into the surface of the earth based out of native emergent settlements. These settlements manifested substantial rural characteristics with being hubs of economic prominences dependent on trade or agriculture. In India interestingly enough, on the eastern fringe in the state of West Bengal, series of colonial settlements started emerging influenced by the colony established by different countries of Europe. Such a combination was unique in the history of colonial period, and this region was often colloquially referred to as "Little Europe," though later on only one town emerged to become the capital of the country, erstwhile Calcutta. The saga and grandeur of the other town faded with passage of time. Towns such as Bandel which was a Portuguese colony, while Chinsurah – a Dutch settlement–Chandernagore, and Serampore corresponding to French and Danish provinces, respectively, have turned into mundane native urban settlements with legacy of the past. The built environment went through gradual change based on the needs of the people and government, on top of the initial colonial footprints of the cities, with changes which are incongruent and haphazard. Today with a huge pressure on urban lands, and with the political and economic shift of attention to smaller towns and cities, these cities are experiencing immense change. The focus has shifted from few prominent, populous, powerful cities of the developing countries coined as megacities or metropolitans to transforming urbanization of less popular cities which have been dormant for years. These changes are occurring in an uncontrolled and sporadic fashion, creating undesired and disorderly urban environments. This chapter discusses some of these issues and highlights on the need of quick intercession, to check further damage and also to redress the harms previously thrust upon in these cases. Role of residential settlements in the urban transformation process has been investigated in some of these cases, along with the function and impact in relation to the built environment of the past.

While the first chapter explores the part of residential areas of an urbanized domain in Global South, the next probes into the commercial arena. Twenty-first century marks the increase of commercial centers across the world in response to economic development. The "West" has been distinguishable in such propagation. The influence of the West percolated to the Global South inadvertently and witnessed local variations and conditioning. These transformations like the earlier case are facing turbulence too and eventually causing an overall impact on the urban development of these cities at large. This particular chapter looks at the combination of commercial centers and urbanization with scrutiny of cases in Africa and Asia.

If a child is asked to draw a city, his view of urban environment would certainly house roads and roads with vehicles. Apart from dense residential setting and commercial projections, the scene of a city is incomplete without its transport. Many of the cities designed and developed in the USA were inclined towards strong bias on vehicular or transportation networks. It is indeed essential to run the functions and services of a city. The cities of the Global South experienced vehicular influx rather late in the timeline; however the growth rate is steep and, combined with lack of planning and designing which is oriented towards transportation from the inception, led to serious concern in the developing world.

The third chapter deals with issues and challenges for transit-oriented development (TOD) in a scenario of developing country. TOD ensures meaningful urban development and growth which is supported by public transits. This curtails the need to use private vehicles. The increasing number of private vehicles owing to the dense population and financial aid supportive policies, in the developing countries, causes enormous challenges and threats, such as traffic congestion, degrading urban environment, inequity in allocation of road space between vehicles and pedestrians, increasing road accidents, etc. to name a few. In India there has been an emergence in the rapid transit systems (RTS) in the last decade, and there is a strong need for appropriate TOD guidelines. Thus the entire fabric of the built environment today is governed by the transport modes and allied functions. Needless to say many of the developing countries are not really prepared for the outburst of vehicular load, related to changing governmental policies, and newer systems of transports such as RTS. On one hand, there is massive pressure towards mitigation of fast, effective, and comfortable transport needs of heavy population of these cities, and on the other hand, the aftermath confronted by the newer and advanced means of transport needs to be resolved. Hence the balancing act requires prudent decision-making, public support, and participation.

In continuation to the transit-oriented development, the next chapter looks into the possibility of conventional public transport as a solution towards the issues and challenges which have been raised in the previous chapter. While the governments are in constant strain to build more roads, flyovers, rail systems, and smart infrastructure, with immense investments, they are concerned with the returns and profitability of these ventures. At the same time, with all these developments towards betterment and ease of living and working, the quality of life ensured to the citizens is questionable. The fourth chapter focuses on the conventional bus transport which has existed in many of the developing world for years as an efficient and comfortable mode of transport. Though the newer and faster modes of transport are inevitable to flourish in future years, shaping the cities in a different form than that of present, the conventional means should not be neglected. Multimodal development needs to be explored in its true sense. These transport interventions have lasting impact on the shape and character of the city. The bus and similar public transit call for bus stops and shelter at intervals which again are designed at walkable limits of the neighborhoods and workplaces. Dedicated, safe bicycle ways are almost invisible in the developing countries, despite large number of users and strong urge to use this low-cost, eco-friendly mode, whereas the Global North has built environment responding to this need accordingly. The more the train tracks and fast tracks, the more the use of underground tunnels or overhead passageways, walk bridges, fly overs, complex intersections, and interchanges in the city landscape. Thus the built environment goes hand in hand with some of these infrastructural inputs essential to urbanization.

The transport and its adversities raised the concern long back. Thus landscape urbanism was instigated in the history of time as a reaction towards these undesirable impacts. Today the growing concerns of global warming, greenhouse gases, and vanishing fossil fuel impeach existing transport systems to a large extent in Global South. Added to these are the alarming rise of accidents and degradation of quality of urban life. The government policies play an important role in the control of these, which is rooted to the issues of transport systems, traffic, and vehicular density. For example, in the capital city of India, in order to regulate the number of private cars, which causes choking and enormous congestion every day, the authority had set the rule to use car numbers ending with even and odd digits on alternate days of the week. This is not to discuss the suitability and effectiveness of such policy, but it is important to note that rules and regulations are imbibed into urban regime in order to bring greater good to the society. Various rules and codes have been formulated to make habitable environments with the limitations and constraints of each city. Each house and building that is being seen in a city is impregnated with countless codes. In developing countries these rules are often a device to counter alteration, anomaly, and biased advantage. The focus is rather based on unpleasant and tacky experiences faced by the authority.

The next chapter raises an important question to rethink some of these rules and codes, which play a central role in shaping up the built environment. It is often a matter of fact that these rules are very stringently adhered to in urban areas whereas loosely dealt in peri-urban, rural, and urban fringes. This is a detrimental mode of development for developing countries stagnating growth and future prospects of planned urbanization. Sustainable growth strategy and related modifications to the rules and codes are thus the focus of this chapter. An example from Bangladesh is cited to explain the pitfalls of textual document simply stating series of dos and don'ts. The Chittagong city of Bangladesh is the second largest city in Bangladesh blessed with ecological setting of mountains, forests, water bodies, and coastline. Urbanization has eventually desolated many of the natural richness. The "gray" patches have deluded the "greens" aided with the thoughtless rules such as – the setbacks along the perimeter of the site where no building could be built or the maximum ground coverage, which limits the amount of land in the site which has to be left unbuilt. Most of the city authorities resort to these rules for urban development and built environment; however a minor slip could be interpreted differently and unfairly, or taken undue benefit of, which ultimately leads to disastrous consequences. Hence a holistic and sustainable thought process is essential to be adopted for formulating these guidelines and documents, present in most of the cities of Global South.

From Asian context of discussion, the sixth chapter takes one to the Caribbean Region where the Americas merge in the realm of Global South in the Gulf of Mexico. Here the policies and guidelines are adjudged for natural disaster. These regions like many other regions of Global South lying in the equatorial to tropical zones of the earth face the risk of storms, hurricane, cyclone, tsunami, and the like.

Also with climate change, indicators such as soaring high menace of rising sea and sinking coastlines are not to be ignored anymore. The role of regional government and relevant policies protecting the inhabitants of the coastal regions, be urban or rural, is crucial. These require prudent analysis of the risk, benefit, and activities or uses of the region for climate resilience and preparedness. This particular chapter takes a comparative approach through the study of well-developed policies and rules related to coastal climate readiness and preparedness, of certain places like Florida, and comes up with an understanding of how these should be incorporated into the practices of the developing world which unfortunately encounter such coastal disasters. Loss of property and life and cost of mitigation of post-disaster damage, along with low profitability of land and property market, render these regions into territories of low and abandoned development. This chapter deep dives into how effective markets compensate for incorporated natural hazard risks and comes up with applicable models for Cuba and the Caribbean Region and other similar areas across the globe which have similar risk exposure and morphology but vary on national and sub-national institutions.

The sub-national institute in rural setting of the villages of India is termed as "Gram Panchayat." These village councils act as rural local governments with elected heads and act for local self-governance. This self-administrative structure is similar to many countries of the Global South such as the Majlis Daerah in Malaysia or Kelurahan, Kepala Desa in Indonesia, Gaunpalika in Nepal, Barangay Council in the Philippines, and Partidos in Paraguay. This particular chapter highlights the barriers of post-disaster recovery in rural and peri-urban dominions of a developing country based on the role of local governance. This is based on a post-tsunami case in India, in 2004, which had devastating impact not only in India but whole of the Southeast Asia. The villages had been worstly affected, and post tsunami the local village government or the Panchayat was key in decision-making for recovery. The effects have been flawed, delayed, and wasteful. From a built environment perspective, this is a good example which merges the boundaries of environment, perception, and design. An environmental impact of tsunami, perceived arguably by stakeholders – Panchayats, has complicated the design intervention and implementation of reconstruction. This is an undesirable circumstance. As discussed earlier, the role of built environment is to provide a safe haven and place for refuge for human beings. Its failure due to sociopolitical perception factors and design defects which were unsuccessful to encompass practical economic and management barriers, no matter how much technologically sound, is a glitch. This chapter tries to create awareness towards the same and at the same time researches on methods for future attentiveness.

Post-disaster recovery procedure is an aftermath, but the question remains "Can built environment be constructed sturdy and resilient enough to withstand natural anomalies to the maximum in the first place?"

Man has played various roles over course of time ever since he evolved superior than other animals, from hunter and gatherer to harvester, maker, and designer. The role of the primordial designer in response towards the needs has shaped innovations and civilizations. From family to society and from simple tools and weapons for scavenging to shelters for protection against nature and beasts, man has been the creator of the "optimum" which balances issues, with best possible resources available. It does not necessarily have implications to the "most functional" or "best looking" or "most economical," rather the designed world of traditional man looked seamless, effortless, and almost a continuum of nature. Man as a designer has acted empathetically, and designs reflected long synergy of personal or societal observations and functions of nature. This led to harmony in the built forms reflecting wisdom and poise. This particular chapter looks at how traditional designs ingrained with knowledge from earlier generations built resilient environments, in one of the most disaster-prone areas on earth, the Himalayas. Difficult terrain, extreme climate and frequent hazard of hail, landslide, and earthquake have marked the Himalayas as one of the most difficult and adverse sites for human habitation. However the traditional forms of buildings have withstood these diverse forms of hazards effectively. With the increase of usage of cheaper concrete, easy erection workability, these traditional buildings are fast vanishing even in remotest part of the Himalayas despite difficult transportation and reduced performance of nontraditional materials. Over several past earthquakes, it was evident that the newer, engineered concrete structures have failed to survive unlike the commonplace traditional construction using locally available materials and using indigenous knowledge. Dhajji-diwari and Taq in Kashmir, Ikra in Assam, and Shee-Khim in Sikkim are some of the examples. Though these are specialized construction suitable for hilly region prone to disasters, the essence of this chapter is to highlight design for resilience using traditional knowledge of the built environment, which is effective, low cost, and locally pruned. These make it suitable for settlements of Global South, especially places prone to disasters. Harnessing vernacular architecture is effective in formation of resilient environments in the developing countries rather than resorting to mindless concretization.

Discussing on the Himalayas, one gets reminded of the impacts of climate change and global warming. One of the popular Indian dailies reported in their science news with headline stating "Himalayas will melt by 2100," based on the report by 350 researchers and policy experts from 22 countries, crafted over 5 years. It sounds unimaginable, but bit by bit, over the course of time, the contributors of global warming would cause this unprecedented phenomenon if remains unchecked. Urbanization is directly linked to this global climate and environmental change (Grimmond 2007). It is evident since globally, the urban areas almost always remain warmer than adjacent rural areas (Oke 1973). One of the noticeable facts in urbanization is the issue of growing urban heat islands. This is perhaps one of the most challenging climatic issues of this century, related to built environment. With the rise of built environment and with greater use of concrete and road surfaces, coupled with dense building fabric, lower ventilation, and climatic permeability, the persistent rise of temperature has caused pockets of perpetually heated regions both during the day and night. The problem is more severe in equatorial, tropical, and subtropical portions of the earth; the major chunk of Global South is constituted of this geographical belt. Studies have shown that heat stress-related morbidity is on the rise in the Southeast Asia. Needless to say, if not loss of lives, the comfort thresholds have certainly crossed in the urban areas. And to mitigate this, people are heavily relying on electromechanical means of cooling and ventilation, which is an added burden to the energy consumption loads and economic sustenance. Both indoor and outdoor thermal comforts are heavily affected by urban heat island effect in the Global South. While indoor environment heavily resorts to conventional modes of comfort, outdoor environment still lacks a remedial approach. This particular paper raises a concern and provides a solution through informed and sustainable planning intervention in urban areas for achieving desired outdoor thermal comfort. This improvement is extremely needed for enhancing quality of outdoor urban spaces for its better usability and utilization finally affecting the betterment of human health and living, as well as bringing vitality to economic development.

Heat island is one of the prominent and conspicuous negative impacts of urbanization, and initially the Global North was blamed for this, but rapid and often unplanned urbanization in Global South is of significant damage too. Another pessimistic outcome of urban succession is the rise of waste generation. As per the World Bank (2018) analytics, each person on the globe generates 0.74 kilograms of solid waste per day, amounting to a total of 2.1 billion tonnes, which is likely to shoot up by 70% in the next 30 years. The report also states that compared to developed nations, citizens of developing countries, especially the urban poor, suffer the consequences of unmanaged solid waste, leading to serious health, safety, and environmental hazards. It is found to act as breeding ground of diseases, pollution, and contaminants, promotes urban violence, and contributes to global climate change again, through methane generation. The melting mighty Himalayas has been brought into focus earlier, but another facet remained undisclosed. Himalayan glaciers feed ten of the most important river systems in the world. Oldest civilizations have been nourished by the water of these rivers, and the combined drainage basin is home to approximately 600 million people. The rivers have played an important part in shaping the built environment. Cities have been conceived and grown on the river banks; these settlements like Varanasi and Harappa were cradle of strong culture and civilization. However with the passage of time, the built environment encroached into the sanctity and entity of the river itself. The issue is more pronounced in case of Global South due to population density, poor water, and waste management systems owing to economic and sociopolitical dissonance. This particular chapter draws attention to the harmful by-product of the city and culture which continuously flow into the river, making it polluted and hazardous for other usage. Though the chapter is a thorough and in-depth study of a particular case, it would be interesting to note the details through which a river contamination needs to be examined which arises out of practices of the riverside habitation.

Some of the most important rivers and river system supporting chunk of population is situated in the Global South. While these rivers are the lifeline of these settlements and societies, it is polluted every minute. The rivers have become vulnerable. As a result they pose the civilization with newer challenges of health, hygiene, drainage, irrigation, and many more. The most ancient of the human civilizations emerged in the cradles of these fertile river basins, having similar characteristics. Thus these places have become ecologically and culturally potential for tourism.

Tourism is also relevant from economic perspective. However there lurks a danger of overexploitation and unsustainable patterns, which would again add on to the existing damages to these rivers and supporting systems. This particular chapter looks into a meaningful analysis of riverine Global South in terms of tourism development. This analysis is based on geospatial and built environment characteristics; therefore the river and subsidiary settlements are focused. This leads to sustainable approach for water-based development in the Global South, leading to economic boost but not at the cost of ecological depredation as a result of mindless tourism. The built has percolated to every facet of nature today and needs a prudent justification towards the departure from the intrinsic agreement.

Chapter 13 brings forward the story of Indonesian Archipelago of the Global South which reminds one of the innate agreement which was referred to. The picturesque description of the built environment enigmatically emerges out of the fabric of the natural setting comprising of mountains, volcanoes, reticulate rivers, saline sea, and coasts. This characterizes the generic ecology and built form of archipelago in the tropical belt. This particular chapter redirects us to the basic settlement structure which governed the built environment of human beings for centuries. Farming and farm produce are the key to give form to buildings and related structures. The grain transcends from meager object of need to a symbol and value shaping society, customs, culture, and practices. Very importantly it takes a holistic approach towards the built environment in harmony with ecology and perceptual factors of the inhabitants. The design and architecture were not a need-based imposition but rather a humble expression of continuity of the social practices and cultural values, blended with local materials, ecologically supple. Thus Wanuas – the villages of Indonesia – were formed. This pattern of development is common in most of the agrarian societies and settlements of the tropical regions of the Global South.

From abundance and rich Indonesian vernacular architecture, the next chapter shifts into the discourse of housing shortage to one of the most economically backward continents of the world, Africa. The chapter title "Reflection on Rhetorics, Appropriate Building Materials and Domestic Utilities Towards Reduction of Housing Costs in Africa" is self-explanatory. The authors have addressed the need for suitable supply towards the scarcity of housing resources. Global housing rhetorics such as shelter for all by the year 2000, homelessness, low-cost housing, affordable housing, national housing corporations, housing corporative, and housing providers and enablers' strategies have existed for decades, but little is tangibly achieved by the poorest population of the world. Most of them dwell in the countries of the Global South. It is almost two decades after UN habitat declaration to have shelter for all by the year 2000, but statistics show there is a shortage of millions of housing units in Africa alone till date. Overdependence of industrial building materials reduces the number of required total number of housing units; hence there is a need for alternative building materials and sustainable use of domestic utilities such as water and electricity. Besides, there is limited information on how to reduce construction cost of housing as an attempt to provide affordable housing. Lack of knowledge on the effective utilization of both appropriate building materials and utilities has led to unresolved economic burden to the poor population in Africa. Perception of the past housing policies and strategies at global, regional, and national levels provides self-assessment of how governments, professionals, and housing developers have contributed to the success or failure of the housing sector in different socioeconomical contexts. The chapter promotes the use of different materials and construction techniques, upholding principles of "3R," namely, "reducing," "reusing," and "recycling," citing examples from Tanzania. This has led to momentous cost reduction in the construction and utilization of housing resources. In developing countries the stakeholders of production and supply of housing should be made aware to use of appropriate and efficient building materials and apply 3Rs principles in housing, instead of conventional housing design to harmonize with ecology and make it cost-effective.

In the preceding chapters, the discussions expose that conventional and contemporary architecture and built forms seem to have posed issues in Global South and have worked towards inefficiency and ineffective modes compared to those of traditional forms. In the age of globalization and enculturation, in the developing world, the traditional buildings and houses are often perceived as financially nonviable and inefficient. The next chapter talks about settlement in a tiny island of the Aegean Sea. Part of Turkey but by all means part of the Global South, this island today is predominantly an urban piece with natural and archaeological abundance. Just like the case of Indonesian village dwellings in the archipelago, the houses of this island, in the far apart Mediterranean, were shaped in forms suitable for storing items produced in those houses such as grapes, wine, olives, and olive oil. And interesting to note, the building typology was honored and kept unchanged by the new owners of these houses after the exchange of populations. Documentation of these built forms is certainly precious and relevant from a perception point of view of the built environment.

In contrast to the discussions on residential settlements and houses, the next two chapters look at built environment corresponding to religious entities. Both argue analytically towards regionalization of the religious built form in context to local people and societal needs. The first one revolves around the design and construction of a mosque in West Sumatra. The icon of the Islamic architecture named as *Masjid Raya* was built in 2007. The regionalism has flowed into this building form through usage of local inspirations. People using the mosque and the immediate community have been taken into cognizance during the design process. The author also finds it important to comment on the role of critic in propounding the contrasts and harmony into the realm of social acceptance of the local community. The perception of inhabitants of the Global South towards unconventional built form is unique and requires conditioning owing to their variance in education, exposure, and diverse aesthetic appreciation. The other chapter discusses spatial characteristics of Hindu and Jain Temples. These temples have played a central role in building cities, forming sociocultural contexts, leveraging economy, trade, politics, and external relations spreading over space and time. From simple forms temple architecture construed complexity due to various factors. The evolution also influenced everything else directly or indirectly. Though it is debatable if the role of temple still holds an overwhelming dominance in shaping the built environment in present states, the Hindu religious influence never ceased its crucial role. The predominance of *Vaastu Shastra*, which is the traditional methodology and framework for the buildings and built environment based on religious ideologies of Hinduism, has shown powerful trends in contemporary India. The design of the new capital city in 2015, for the state of Andhra Pradesh in India, called Amaravati, and its important public buildings have been reported to be guided by *Vaastu Shastra* (Srivasthsan 2015). Thus religious impacts on the built environment as a cultural milieu have not only been predominant in the past but are crucial in present too. The effects are more prominent in the Global South. This is a good example where human perception towards abstract form of religion amalgamates with something as tangible and practical as a building, a square, or a neighborhood. Temple towns are common in India and in some of the Southeast Asian countries. Today the design of these built environments, edifice of frozen human imagination and perception, serves to attract pilgrimage and tourism, boosting the economy and encouraging the heritage value in the developing nations.

Cultural heritage is often associated with architecture of a place along with many other ingredients which blend with entities of ethnography and other natural assets. Many of the cultures and communities of Global South have embraced natural resources, flora, and fauna as a part of their living and culture. This has percolated to the character of the built environment as well. Contrary to so-called conflict with nature as have been the case, postindustrial revolution in the developed nations, the developing countries have learnt to coexist. The harmony resulted in resilient and sustainable solutions instead of aggressive natural devastation. Chapter 18 with the title "Co-existence – migrated settlement redefining cultural heritage" based on a case in Bangladesh elaborates this further. This chapter is unique and puts up a case where the concordance of human settlement and buildings with hills, harbors, and forests has been investigated, to the extent of considering interactions and encounters with the wild animals. The particular tropical region is home to large mammals like Asian elephants. These giant beasts often explore into human territories and built forms, causing unintentional damage to the buildings, crops, and livelihood. The residents of the settlement have combined their wisdom and cultural values and weaved them into the design of their mud houses, so that it can resist the elephant attack in a passive fashion. It is a matter of perception towards the situations and solutions through design of the built, in context to the ecology and environment, that leads to "existence" and "existence" which is primordial to sustainability.

Talking about perception the next chapter focuses on how mindless urbanization, often a drive by the "West" later transformed into cultural and economic overburden of the Global North, victimized the Global South in terms of loss of identity. Place identity is governed by "imageability" which is related to a two-way process: one, the user or the audience perceives built form based on the experience and cognitive capabilities, and the other, where the built spaces are designed to reflect the cultural values and make it congruent to the perceiver's cognitive choices. Both cases are very sensitive and fragile to externalities of cultural cross pollination and technological advancement such as advanced information and communication systems of telephone, the Internet, etc. Urbanization process has also brought into massive use of such systems, thus "altering the perceptions of culture" – the title of the Chap. 18 banks on this phrase. This chapter deep dives into architectural style that has balanced the indigenous knowledge system with the modern forms and in other words semantically differentiates and reunites the perceptual concepts in built environment, where there are certain things which are indigenous and others which are modern. In many of the Global South this dichotomy prevails, where postcolonial era, there has been introduction of the "modern" which was hitherto unfamiliar in contrast to indigenous practices and styles. Over generations there had been transition of these alternations into the population of the developing nations (Madeleine and Lee 2007).

A noticeable turning point in terms of demographic data of the Global South is related to gerontology, which deals with the aging population. Not too long ago, in 1975, the developed countries housed majority of the world's elderly people. Today this scenario has flipped making the Global South the home for the most elderlies. By 2020 this shift is predicted to reach 67% by the UN which is equivalent to double the population of the entire North America. The rate of this growth is faster than that of the rate of overall population increase; hence the proportion of aged persons in world population would sharply increase. And the trend of such growth is thrice larger for the Global South compared to Global North (Shrestha 2000).

With such alarming statistics, the obvious question posed today is if the developing nations around the world are ready for accommodating such demographic transferal. Rethinking existing built environment is certainly one of the concerns addressing this issue. Accessible environments are desired which is less accident prone and safe. Chapter 20 discusses the need for this inclusive design process of the built environment by analysis of various strategies adopted for creating urban public spaces responding to the needs of the elderlies in China. There is no other way than to bring gerontological perspective into the folds of urban built environment of the Global South. While these adaptations are essential based on the needs of the changing society, it is also important to look at the spatial limitations in developing countries. Many of these cities are densely nestled with cramped spaces, slums, alleys, and reticulate narrow road networks, with lesser public open spaces. The general populations in most cases do make-shift arrangements and adaptations which are feasible within the physical restrictions and not going too much beyond the legal limits. The story of alleys and living in these alleys of Indonesia in the next chapter would provide a vivid description of adaptive mellifluous simultaneous usage of restrictive physical built environment. The inhabitants have adjusted and accommodated the regular functions of living with the built form they are thrust into. There is a peaceful negotiation, though apparently seen as a urban blight; such settlements and patches are extremely common features of any country of the developing world.

The meandering alleys, with dense settlement around, prowling chaos, and puzzling clutter, often infringed with socioeconomically weaker strata of the city, are not the only symbolism to be related to the urban areas of a developing country. Often the alleys are lined up with havens of history and deep cultural icons. The walkways not only connect the buildings and the roads but also in a way connect the past with the present. A walk through some of the cities of the Global South is termed to be rich and vibrant owing to the proliferation of life of people and society into public spaces or spilling into spaces beyond the restriction of the walls and buildings. The reverence towards heritage and related tangible assets is rather low in core capitalist regime since there is always an economic pressure exerted to demolish low-rise heritage built form for converting into a lucrative taller moneymaking device. There is a growing need to bring economic sustenance to counter such unwelcome challenges through tourism, heritage management, and other strategies (Ebbe 2009). One such mechanism has been discussed in the Chap. 22 through the usage of heritage walks in the city of Hyderabad in India. This not only serves as a component for cultural tourism integrated with the tourism prospects of the region but also seeps a sense of heritage awareness ingrained in the built environment into the minds of younger generation.

The next three conclusive chapters independently delve into three prominent areas related to built environment, its design, and perception in context to the Global South. With focus on many of the challenges of urbanization in the developing countries, it is obvious that the questions arise in terms of what is the way forward. The changes have been ever persistent, and the dynamics will keep changing with passage of time, but the questions remain on the approach towards the future. The first in the series is to understand the rural as against the urban, in developing country context. This understanding clarifies some of the social and economic juxtaposition which leads to a misleading direction for the developing nations, such has been in the case of Nepal. The perception of the residents and their beliefs and aspirations are key to shaping a built environment and not vice versa. This thought has been analytically presented in Chap. 22, whereas the next takes an outlook to "lively urban spaces." The future urban spaces irrespective of any classification in terms of South or North need to be lively and livable. Without synergy of life into the built environment whether a city or a village, it is no grimmer than a necropolis. Using open spaces and ecological facets, one may create hub of miscellaneous activities and social interests merged into the fabric built forms. A case from Brazil is cited to give more justification to the argument. The third industrial revolution began at the end of the twentieth century, and the digital revolution continues to the twenty-first century. This century is also to witness the culmination of fourth industrial revolution involving cyber-physical systems. In this context, one of the buzzwords which have made strong entry in this century is "smart" linked to the "Internet of things" (Satell 2014). Many of the smart city missions have been strongly patronized in Southeast Asia, South America, and Africa. Hence the string of "smart" in this century binds the Global North and Global South alike, honoring the rise of the Global South. The chapter as a conclusion investigates the latest trends in smart city, as a project for a built environment. The examination has been done in three specific areas, namely, architecture, technology and communication, and sociology of space. The smart cities or rather smart territories are expected to emerge as technologically aided design solutions of the built environment which are more sustainable, in terms of ecology, economy, and society. Economic and social perspective towards the built is packaged into "perception" (socioeconomic), as an expression in this book.

Bibliography

César LI (2017) World expos a history of structures (Prieto MB (eds), 1st edn). Architect Publications, Barcelona, pp 29–32

Dirlik A (2007) Global south: predicament and promise. Glob S 1(1):12–23

Ebbe K (2009) Infrastructure and heritage conservation: opportunities for urban revitalization and economic development, directions in urban development, Report of Urban Development Unit, World Bank

English Oxford Living Dictionaries Definition of build in English. Accessed 10 Oct 2018. https://en.oxforddictionaries.com/definition/build

Grimmond S (2007) Urbanization and global environmental change: local effects of urban warming. Geogr J 173(1):83–88

Kaul I (2013) The rise of the global South: implications for the provisioning of global public goods, Occasional Paper 2013/08, United Nations Development Programme, Human Development Report Office

Kaushik H, Babu K (2009) Report #154: Assam-type House. Accessed 11 Aug 2018. http://www.world-housing.net/asia/report-154-assam-type-house

Levander C, Walter M (2011) Introduction: the global South and world dis/order. Glob S 5(1):1–11

Madeleine L, Lee B (2007) Lost in translation. Nature 445:454–455

Merriam-Webster Dictionaries. Definition of built, more definitions for built. Accessed 11 October 2018, https://www.merriam-webster.com/dictionary/built

Oke TR (1973) City size and the urban heat island. Atmos Environ 7:769–779

Pagel H, Ranke K, Hempel F, Köhler J (2014) The use of the concept "global south" in Social Science & Humanities, Globaler Süden/Global South: *Kritische Perspektiven*, Institut für Asien- & Afrikawissenschaften, Humboldt University, Berlin

Satell G (2014) Why the digital revolution is really just getting started, forbes. Accessed 5 Apr 2018. https://www.forbes.com/sites/gregsatell/2014/04/05/why-the-digital-revolution-is-really-just-getting-started/#5cd1ff4e3a72

Shrestha LB (2000) Population aging in developing countries. Health Aff 19(3):204–212

Sivertsen G (2014) Scholarly publication patterns in the social sciences and humanities and their coverage in Scopus and web of science. In: Noyons E (ed) Proceedings of the science and technology indicators conference 2014 Leiden. Universiteit Leiden, Leiden, pp 598–604

Solid Waste Management, The World Bank. www.worldbank.org/en/topic/urbandevelopment/brief/solid-waste-management, retrieved on 10 October 2018

Sparke M (2007) Everywhere but always somewhere: critical geographies of the global south. Global S 1(1):117–126

Srivathsan A (2015) Amaravati reveals how public projects in India remain dependent on whims of politicians, WIRE. www.google.com/amp/s/thewire.in/culture/amaravati-reveals-public-projects-india-remian-dependent-whims-politicians/amp/

Taura T, Nagai Y (n.d.) A definition of design and its creative features, Research.kobe-u.ac.jp. Accessed 21 July 2018. http://www.research.kobe-u.ac.jp/eng-mech-design/taura/img/taura_pub/72.pdf

United Nations (2018) World urbanization prospects: the 2018 revision, Department of Social and Economic Affairs, population Devision, No 2018/1

Youssef D (2015) Architecture: Perception and Concept, Researchgate. Accessed 10 Mar 2018. https://www.researchgate.net/publication/275971712_Architecture_Perception_and_Concept

Chapter 2
Urban Transformations of Residential Settlements in Colonial Towns: Case Study of Chandernagore and Serampore

Ruchira Das, Sanjib Nag, and Keya Mitra

Abstract Urban transformation is a social, economic and physical and ecological change that shapes a settlement. In this process of transformation, an urban area redefines itself within the existing city structure. Historically, several forces have triggered transformation processes in the world, and colonization was arguably one of the important ones. Settlements were established and developed, and they also flourished, often, as a direct result of the administrative agendas of the colonising powers. Within these settlements, the residential neighbourhoods were unique. Colonial architectural expressions and settlement planning ideals came into contact with the local social, cultural and climatic needs, resulting in a unique neighbourhood character, that yielded building typologies that were derived through this meeting of cultures and worldviews.

Soon after the end of the colonial era, many of these towns became small dormitory towns catering to the nearby metropolises, and they received scant attention till the next international phenomenon, that is, globalisation achieved centre stage in the narrative. Like other parts of the world, in the past few decades, the development pressure due to globalisation, around the adjacent areas to Kolkata, has increased dramatically leading to a real estate explosion in these towns too. As the residential neighbourhoods are privately owned, it is here that the process of transformation occurred rapidly, without any controls exclusively for retention of the unique character of these places.

In this research, an attempt has been initiated to understand the transformation and its implication in urban fabric of the residential areas of selected colonial towns, namely, Chandernagore and Serampore.

Keywords Urban transformation · Residential settlement · Colonial town

---

R. Das (✉)
Kolkata, India

S. Nag
Department of Architecture, Jadavpur University, Kolkata, West Bengal, India

K. Mitra
Department of Architecture, Town & Regional Planning, Indian Institute of Engineering Science & Technology, Shibpur, Howrah, India

© Springer Nature Switzerland AG 2020

M. Ghosh (ed.), *Perception, Design and Ecology of the Built Environment*, Springer Geography, https://doi.org/10.1007/978-3-030-25879-5_2

Introduction

Urban transformation is a change of urban pattern which is a major segment of urban development. It has been an ongoing process, shaping cities throughout human history. It therefore can be referred as the evolutionary process of modification or adaptation in urban areas, due to the change in political, economical and social pattern. This in turn generates diverse urban physical forms. The physical change of urban fabric can be termed as urban physical or spatial transformation. Urban physical or spatial transformation can be most exceedingly visible in residential settlements, because here the owner-specific properties can be converted into various building use typologies, according to the changing needs of the surroundings.

Residential settlements are areas where the predominant typology is residential rather than commercial, industrial or institutional. These are usually designated by the development plan of the city administrative authorities, including the amount of units and consequently, the approximate number of people living in that area, following the prescribed density. They are connected in some ways, with the adjacent activities, by their physical or spatial structures, functional attributes and user group preferences. The physical or spatial structures and functional attributes differ from one to another based on several factors like occupancy pattern, built form, adjacent activities, origin of establishment, etc. Thus, settlements formed by indigenous people always differ from that of the foreign people. Consequently, the spatial structures of residential settlements in colonial towns are unique in nature. This type of colonial propagation-based settlement pattern can be observed in Asia, Africa and South America.

Colonial towns are settlements which are resultant of colonial manifestation of production, consumption and sustenance. These settlements grew, survived and sustained on the basic whims of colonial powers and their administrative agendas. They were symbols of power, authority and administration. After independence, these towns became a part of larger administrative regions and started following the development agenda of the nearby metropolises, or they themselves became the metropolises generating development situations. Colonial towns, which remained as small towns within the close vicinity of the major metropolises and functioned as selfsufficient units, have started transforming themselves, due to tremendous development pressure, occurring in the nearby metropolises at present.

In today's world, the colonial towns can be divided into three categories, according to their socioeconomic development patterns, such as colonial towns of developed countries, developing countries and underdeveloped countries.

The developed countries have already formulated development guidelines, for all types of urban systems. The underdeveloped countries are not in a position for formulating guidelines, for their same, as they are not prepared for actual development because of economic constrains. In this scenario, colonial towns of the developing countries are undergoing considerable unhindered transformations due to constant social, economical and political reformations. These colonial towns are grappling with a colonial legacy vis-à-vis the driving forces of modern-day development.

One such country is India where several colonisers have imprinted their inheritance in the form of new settlements. Coastal zones of India such as Kerala, Tamil Nadu, Maharashtra and West Bengal were mostly affected area where colonial towns are found. Gangetic delta was the most promising place during that time apart from other areas where European traders were settling down their business as well as settlements rapidly.

Little Europe is one such stretch in the bank of Hooghly River where European traders and their settlements were flourishing. It describes the rise and fall of European powers beginning with the Portuguese, followed by the Dutch and the French, who were finally overpowered by the British, in the mid-eighteenth century. The colonial powers played out their traditional rivalries and alliances in the colonies along the Hooghly, bringing this stretch of river so far removed from Europe, into the centre stage of attention. All of them left their physical mark among these settlements in terms of sociocultural impact as well as architectural legacy.

The settlements of European countries on the West Bank of Hooghly were some of the first places where European trade started and turned them into the first places of penetration of colonization into South Asia. Colonial modernity generated into the subcontinent as a result of the cultural encounters of the East and West in these regions. As Britain sovereignty was cemented along the West Bank baring Chandernagore, it led to these areas developing as per a British mould. However one can't ignore the deep influence the Portuguese cast in Satgaon, Bandel and Hooghly, that of French in Chandernagore and the Danish in Serampore. Among all these settlements, colonial imprint in settlement structures is most evidently visible, in Chandernagore and Serampore. The lasting influence is evident by the European buildings, monuments and artefacts that survive in these areas. Time has not yet obliterated the legacy left by these European states along the West Bank of Hooghly. A cruise from Calcutta to Bandel will reveal to the eyes the truth of such a statement. Not only was the Bengal vocabulary and literature made fertile and rich, Bengal also prospered in sciences and other arts due to their exposure to a range of western knowledge. These settlements turned into nodal points of culture and politics. The need to preserve and reconstruct these influences should be strong.

Department of Tourism, Government of West Bengal is currently promoting 'Little Europe' – the four towns in west bank of river Hooghly – to establish the need of understanding the spirit of such areas and their history. Resultantly, it will give rise to the development of the fronts of tourism that will benefit the people local to the area economically. River transportation along this area is something that has been long neglected; its potential needs to be explored once more. Government has ambitions to develop the river transportation in this area, which used to be its lifeline in history. It will along with establishing better connectivity among the areas also rejuvenate the history of the big trader ships docking at these banks with dreams of making a fortune.

This research focuses on the role of residential settlements of colonial towns in urban transformation process and its impact on overall urban structure of these towns in general. Accordingly, an attempt has been made to understand urban transformations considering physical/spatial developments of residential settlements in selected colonial towns in developing countries for establishing a parametric relationship between these aspects. Based on this, an attempt would also be made to examine cases of similar transformations, in selected colonial towns of Little Europe in West Bengal, India, and to give a way forward for such further future transformations, in desired directions. Hence, here an endeavour has been made to understand the impact of the same in Chandernagore and Serampore.

Exploring the Two Towns

Chandernagore is a corporation city and former French colony located about 35 km (22 mi) north of Kolkata, in West Bengal, India. It is one of the seven municipal corporations in West Bengal. Located on the bank Hooghly River, the city has been able to maintain a unique identity different from all other cities and abide by its own characteristics. The total area is 19 square kilometres (7.3 sq. mi) and had a population of 166,867 as per 2011 Census. Chandernagore is connected to Kolkata by railway, roadways and Hooghly River, and it takes about an hour to reach there by car. The French created their trading post by amalgamated three noticeable villages, namely, Gondolpara, Boro Kishangang and Khalishani in the year 1673 by obtaining permission from Ibrahim Khan, the then Nawab of Bengal. The first director of French East India Company, Deslandes, got control over these lands in 1688 from Mughal Subahdar. In 1730 Joseph Francois Dupleix came as a governor of the city. During his reign, trade and commerce flourished. Chandernagore became the most prominent business hub of European traders and attracted people from all over India for considerable amount of mercantile activity (Fig. 2.1).

Serampore is a famous and historical city in the Indian state of West Bengal. It is a precolonial town (city) on the West Bank of the Hooghly River. The total area is 11.6 square kilometres (4.48 sq. mi) and had a population of 181,842 as per 2011 Census. Serampore is connected to Kolkata by railway, roadways and Hooghly River, and it takes about an hour to reach there by car. A few private bus services are also available to connect with Kolkata from Serampore. The colonial history of Serampore began in 1755 when the Danish East India Company was authorised by the local prince to set up their trade post along the Hooghly River, almost 25 km north of Kolkata. The settlement was named as Frederiksnagore but was usually called by its Indian name Serampore. It has a multilayered history with Indo-Danish-British built and cultural heritages through different periods of time (Fig. 2.1).

Colonisation had a profound influence on development of urban character around the globe, and most of the colonial towns in developing countries reflect the European influence in their urban pattern till date. The primary intentions of colonisers were trade and commerce, missionary activities and military purpose. The establishment of European enclaves in the colonised territory was planned to build up an urban network, a cultural domination, a social set-up and a physical indentation that would be easily recognisable right through the colony. Areas in and around the colonial core were

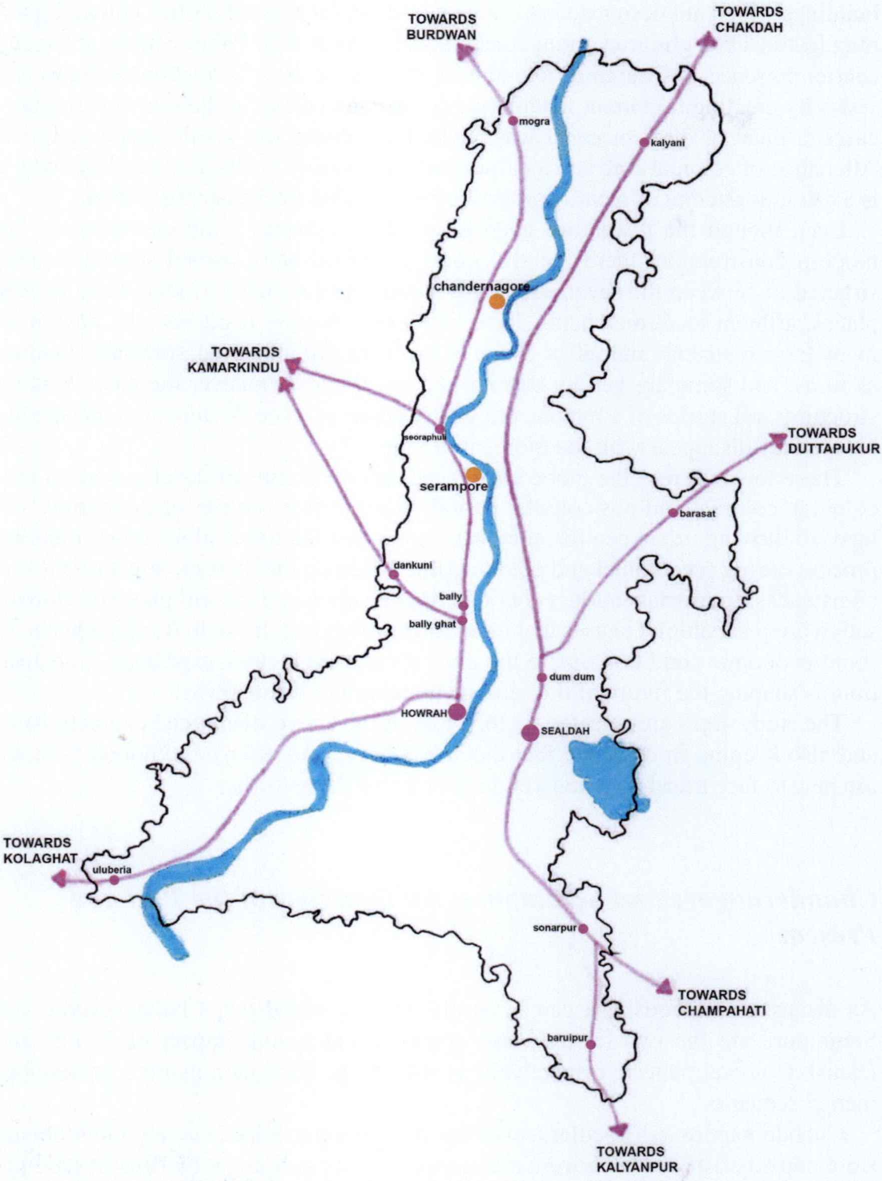

Fig. 2.1 Location of Chandernagore and Serampore within Kolkata Metropolitan Area

highly influenced by the urban pattern of the colonisers creating a unique juxtaposition. Both the towns under discussion in this research fall under the same category.

Further if we dig deeper, few more characteristics of these towns could be identified, such as area in and around the colonial core encompasses most of the colonial buildings which are designated as colonial heritages nowadays. A few colonial planning features like grid iron arrangement model, provision of Public Square at the city centre, presence of Church as a landmark around the core, formation of views and vistas by erecting important landmarks, etc. are noticeable in these towns. In many cases, a moat or ditch or canal was made, for defence and goods supply purpose. Alienation of colonial core and local settlement is mostly visible in every town which is a cautious attempt of racial segregation between the blacks and the whites.

Even though the towns are getting rapidly transformed and characterized by modern construction, there are still quite a few privately owned structures constructed in between the seventeenth and nineteenth centuries. Traders from nearby places, affluent local merchants, local landlords, wealthy residents, etc. have built many large mansions and villas. Some of them are still inhabited, some are standing as ruins, and some are getting demolished as rebuild. Whatever the state is, these structures tell stories of a bygone era which deserves to be documented thoroughly before they disappear with the moment in time.

These towns across the globe have faced several phases of development in precolonial, colonial and postcolonial period. Starting from cluster of flourishing villages to thriving urban centres, they have gone through tremendous transformation process during precolonial and colonial time based on the then socio-economic circumstances. Industrialisation, economic liberalisation and current phase of globalisation are postcolonial phases that these towns are going through. A rapid change in socio-economic condition due to the three mentioned factors, especially globalisation, is shaping the future of the settlement pattern of these towns.

The study areas are selected on the basis of the above-mentioned characteristics and also keeping in mind the fact that areas developed after postcolonial time are too new to face transformation yet, in the case of these towns.

Chandernagore and Serampore: An Overview of the Past and Present

As discussed previously, it can be easily acknowledged that Chandernagore and Serampore are the two towns where physical and spatial impact of French and Danish Colonial pattern, respectively, is still on the edge with its present developmental scenario.

Chandernagore was a settlement of few flourishing hamlets, namely, Gondolpara, Boro and Khalishani which were mentioned in Manasa Mangal by Bipradas around 1495 and Kabikankan Chandi of Mukundaram around 1533–1600. After the French took over the town, it has grown as remarkable urban centre of trade and commerce recognised during the late seventeenth century when the foundation of Kolkata was just getting laid. The town was fortified, an urban grid was laid, and many houses were established. During the war between British and French, the town was taken over by the British, and they demolished it extensively. The city was returned to the

Frenchmen in 1812, but by the time its commercial importance was gone, and it remained as a calm dormitory town of Kolkata. But it was always prominent for its clean wide thoroughfares with many elegant dwellings along the river bank. Chandernagore remained French establishment, and it got independence from French colonial rule on June 1952. Thus, this town had a better blend of French and local mix in terms of both tangible and intangible aspects of urbanism.

Serampore was a cluster of small villages which developed after the Danes took over the place. The Danish trade post was a fortified area consisting administrative and residential structures. Other migrant European communities were also there other than Danes who built villas along the main street and the riverfront. Indian merchants also were attracted and got settled in the Serampore, and palaces were built alongside with European houses. In 1845 Serampore was sold to the British East India Company by Danes. It is the town where Danish administration allowed the English Missionaries to work freely which had a great impact on the educational revolution, and thus their physical manifestations were the esteemed educational precinct and the residential settlement around the colonial core.

Despite tremendous development pressure, the towns are able to maintain its street network and urban structure. A large part of population comes to these towns from neighbouring areas for job opportunities. The riverside settlements also attract students from places for educational opportunities which generate economical vitality to localities in a large scale.

In this context, this two is considered for further investigation and chosen as case study sites for this research. The area under scrutiny has been zoned in and around the colonial core of the town, where the predominant building use is residential and a remarkable number of heritage buildings are available as shown in Figs. 2.2 and 2.3 of Chandernagore and Serampore, respectively.

Defining the Study Area

For the purpose of this research, a few factors were selected based upon the colonial influence and its long-term effect in a settlement's urban character, such as location of the river which was the main transport mode for mercantile purpose during colonial time, position of the current colonial core and its functional extent and street configuration which was laid by the colonial power for ease of access control. Based on the stated consideration, in both the cities, study area is defined as shown in Figs. 2.2 and 2.3. It is clearly visible from the figures that river being the most important spine of that time, the settlements flourished linearly along the course of the river. In the case of Chandernagore, the town has grown along the river and the Grand Trunk Road (G. T. Road) mostly. The area under noticeable transformation is also found in this segment largely (refer to Fig. 2.2). The proximity of the railway station is a bit more than that of Serampore (refer to Figs. 2.2 and 2.3) in this case which has triggered a linear development pattern along the transport spines in the rest of the town of Chandernagore. Whereas if we see the case of Serampore,

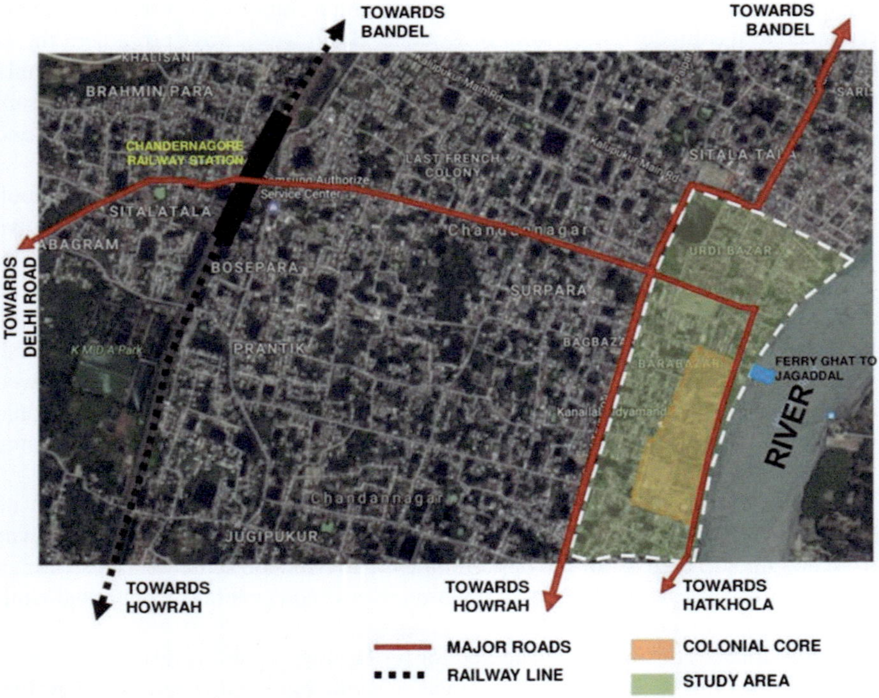

Fig. 2.2 Selected zone: Chandernagore

railway station and the river are very close by, so the development pattern here is sandwiched between the railway line and the river.

For appraisal purpose, Chandernagore has been subdivided into three subzones namely:

Zone I – Residential Area The main residential area surrounding the administrative hub is called as Padri Para. The word Padri means Priest, and Para is a neighbourhood; it can be collectively called a 'neighbourhood of Priests'. Residential building present here dates back to the colonial era, but a blend of local and colonial architecture is highly visible in the old residential buildings of this zone still now.

Zone II – Administrative Hub The administrative hub of Chandernagore falls under Burrabazar area. It also has the iconic Jora Ghat along the banks of Ganges which was the port during French rule. The Strand Road acts as the grand avenue of public place of the town, and the Burrabazar Main Road acted as the connector between the administrative centres to the other parts of the town where the bungalows of the French officials can be found.

Zone III – Mercantile Zone The French came in Chandernagore in 1696 and fortified it (calling it 'Fort d'Orleans') by 1701, which was later destroyed by the

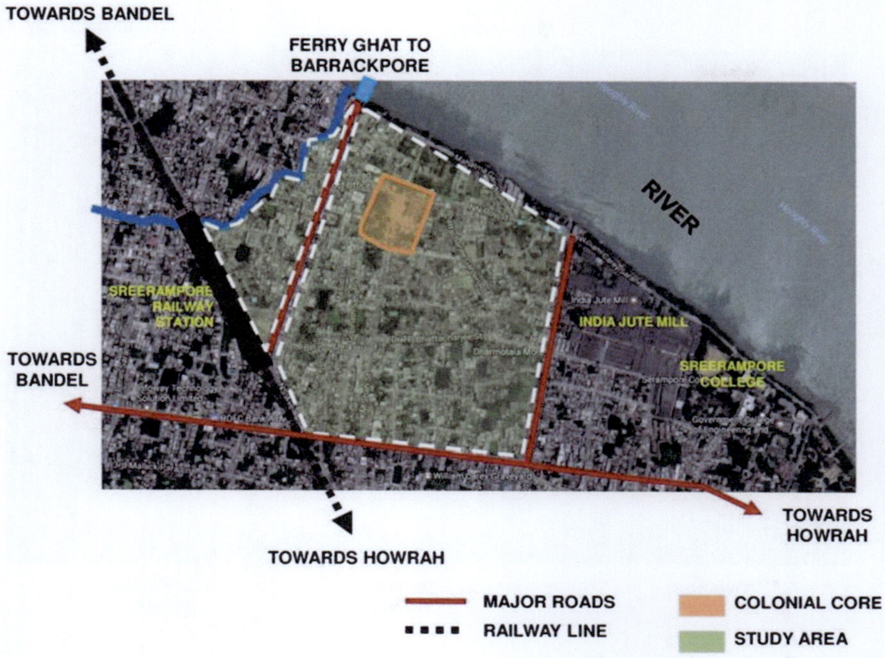

Fig. 2.3 Selected zone: Serampore

Britishers. The fort existed in the area which is now known as Urdi Bazar. Today, Urdi Bazar acts as major commercial hub (Fig. 2.4).

Similarly, Serampore has been subdivided into four subzones, namely:

Zone I – Riverside Area The two most oldest residential settlements of Serampore – Roy Ghat and Shill Para – are part of this area. The building typology and architectural expression in this area reflect the colonial heritage of the town.

Zone II – Administrative Area The most prominent colonial heritage area consisting old colonial institutional and religious structures is present here along with very few but architecturally rich residential buildings which make this area important for study.

Zone III – Sociocultural Zone This area is located western side of the Serampore Court. The major old and famous institutions of the town are present in this zone. Some of them are constructed in the mid-nineteenth century during Bengal's educational revolution and still functioning as anchors of knowledge. Residential buildings present in this area are also of equal important as heritage.

Zone IV – K.M. Shah Street Area Between K.M. Shah Street, G.T. Road and Rishi Bankim Sarani, the entire area is native residential area of the old town. Major

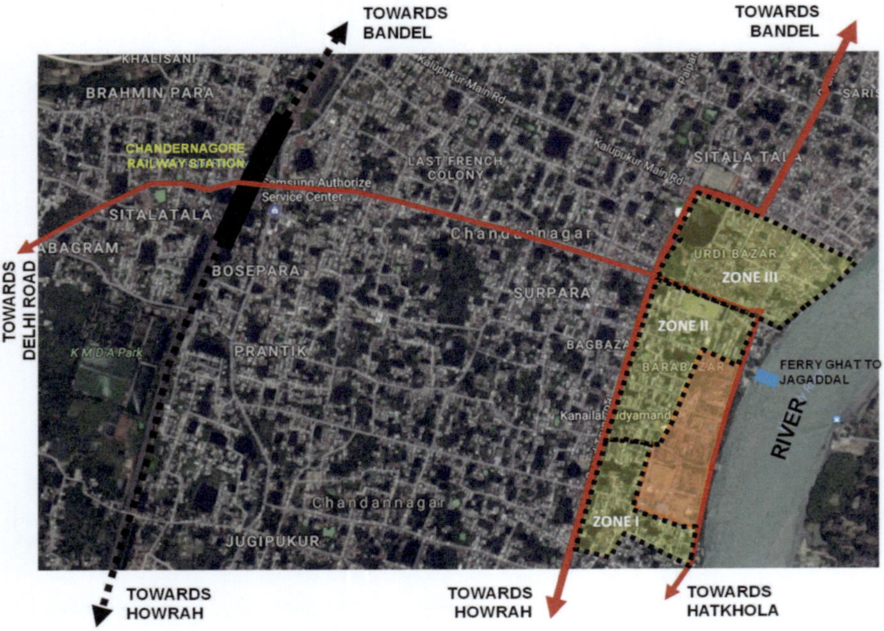

Fig. 2.4 Zonal subdivisions – Chandernagore

old Indo-Danish or Indo-British lavish residential precincts are present in this zone. From historical background, it could be found that this area was majorly developed by the traders of old days, and their villas were present in this area (Fig. 2.5).

Appraisal of the Current Scenario

Based on several literature explorations and reconnaissance surveys of the places, a threefold assessment method has been formed to evaluate the current conditions of the selected areas of the two towns, namely, accessibility analysis, character appraisal and legibility analysis. Accessibility analysis deals with identification of street network system, identification of mode of transport and availability of public transport system. It is done for both vehicular and pedestrian systems to understand the conflicts or successful nature of overlapping of both.

Character appraisal here is defined by deriving the built–open ratio, understanding the built form typology and identifying its architectural character. It is done majorly for residential buildings but in a residential cluster if any mixed-use or other functions are present that were also taken into consideration for an understanding of functional mix in dedicated residential areas.

Legibility analysis comprises of identification of the basic urban design way finding parameters, i.e. Lynch's five elements, namely, path, edge, node, district and

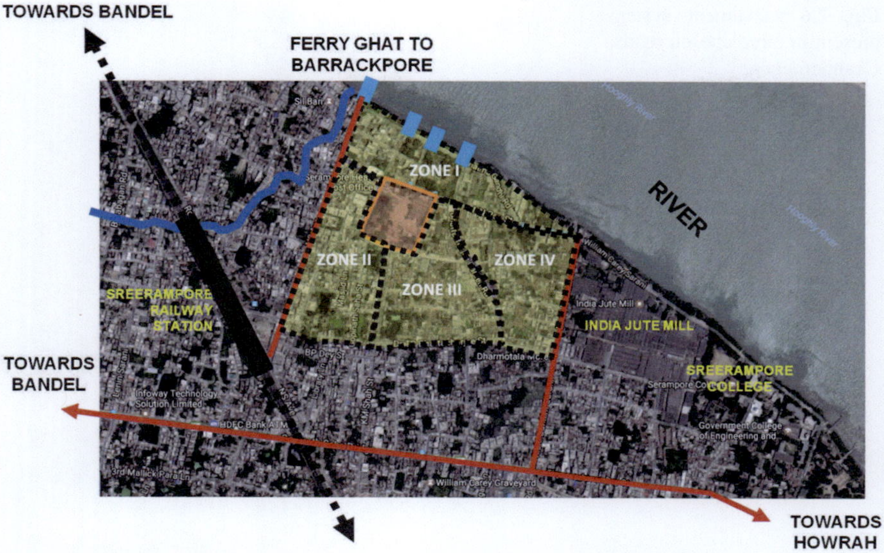

Fig. 2.5 Zonal subdivision: Serampore

landmark. Out of which path, edge and district were eliminated later as the selected zones were majorly residential districts subdivided into smaller parts depending upon their current functionality. Edge and path were eliminated as the major roads were considered in this pilot survey purpose.

Accessibility Analysis (Figs. 2.6 and 2.7 and Tables 2.1 and 2.2)

Being a small town, there is only one major spine that connects the neighbouring towns with Chandernagore, that is, the G.T. Road with four types of public transportation system: bus, auto, toto and rickshaw. In case of other two roads, Toto is the only mode of shared public transport available. Majorly people travel by their own vehicles like bicycle and motorcycle. Use of four wheeler is comparatively quite less. Pedestrian-vehicular conflict is extensively visible in G.T. Road and Burrabazar Main Road, whereas Strand Road has a wide walkway cum promenade which cuts down the conflict (Fig. 2.6 and Table 2.1).

A few private bus services from Serampore are available which connect the town with Kolkata. Frequency of bus service is more in this case. But this is also a fact that the buses are used for intercity connectivity. For intracity connectivity, this town is also dependent on shared Auto, Toto and cycle rickshaw as visible from the table above. Conflict of pedestrian and vehicular movement is extensively visible in all the roads. No proper pedestrian facility is available in the mentioned road though an enormous number of people travel as many places are within walkable radius (Fig. 2.7 and Table 2.2).

Fig. 2.6 Movement system present in three chosen roads, Chandernagore

Fig. 2.7 Movement system present in four chosen roads, Serampore

Table 2.1 Survey information, Chandernagore

Accessibility analysis	Chandernagore					
	Vehicular				Pedestrian	
	Motorized transport		Non-motorized transport			
	Volume	Direction	Volume	Direction	Volume	Direction
G.T. Road	Bus – 1/20 min	Towards Chinsurah or Bhadreswar	Rickshaw – 20/5 min	Towards Bagbazar or Jyotir more or convent road	10/min	Towards Strand Rd
	Auto – 25/5 min				8/min	Towards Bagbazar
	Toto – 30/5 min				10/min	Towards Jyotir more
Strand Road	Toto – 25/2 min	Towards convent Rd. junction or Hatkhola	Rickshaw – 25/2 min	Towards convent Rd. junction or Hatkhola	20/min	Towards Urdi Bazar
					15/min	Towards Hatkhola
Burrabazar Main Road	Toto – 10/5 min	Towards Vivekananda road or Church road	Rickshaw – 10–12/5 min	Towards GT Rd. or Strand Rd	6–7/min	Towards Church road
					10–12/min	Towards Bhudeb Mukherjee road

Table 2.2 Survey information, Serampore

Accessibility analysis	Serampore					
	Vehicular				Pedestrian	
	Motorized transport		Non-motorized transport			
	Volume	Direction	Volume	Direction	Volume	Direction
M. G. Road	Nil	Nil	Nil	Nil	5/min	Shill Para–Roy Ghat (both)
N. S. Avenue	3 bus – 1/30 min	Serampore station – Bag bazar (both)	Auto 6–7/min	Serampore Ferry Ghat–Serampore Station/super market (both)	50–60/min	Station/super market–T.C. Goswami Street (both)
			Toto 10–12/min			
			Rickshaw-5/min			
K. M. Shah Street	Auto – 2/min	Serampore Court-G.T. Road (both)	Rickshaw-5/min	Serampore Court–G.T. Road (both)	20–30/min	Serampore Ferry Ghat–T.C. GOSWAMI RD
	Toto – 3/min					
Rishi Bankim Sarani	3 bus – 1/10 min	Serampore Station–Bag bazar (both)	Toto 4–5/min	Serampore Ferry Ghat–G.T Rd./Walsh hospital	30–40/min	Serampore Ferry Ghat–G.T Rd/Walsh Hospital
			Rickshaw-5/min			

Character Appraisal (Figs. 2.8 and 2.9 and Tables 2.3 and 2.4)

Fig. 2.8 Character appraisal details in three chosen roads, Chandernagore

G.T. ROAD

STRAND ROAD

BURRABAZAR MAIN ROAD

Fig. 2.9 Character appraisal details in four chosen roads, Serampore

Table 2.3 Survey information, Chandernagore

	Chandernagore							
	Building details				Open space details			
Character appraisal	Number	Height	Ground coverage	Arch feature	Land value	Area	Pattern	Orientation
G.T. Road	**Number and height:**				**Land value:** 28–30 lakh per kottah			
	Bungalow – 39, old/new, active, 2–3 storey, aligned to the road				**Area**: Very less area is dedicated public open space			
	Apartment – 11, new, active, 3–5 storey aligned to the road				**Pattern:** Linear			
	Ground coverage: 60%				**Orientation:** North-South with respect to street alignment			
	Architectural features: Majority of the old buildings are demolished, and new apartment typology has come up with modern features							
	Existing old residences are villa type, build during the early twentieth century with front garden and balconies							
Strand Road	**Number and height:**				**Land value:** 35–40 lakh per kottah			
	Bungalow – 5, aligned to the road, 2–3 storey				**Area**: The entire street along with river front act as a dedicated public space			
	Apartment – nil				**Pattern:** Linear			
	Ground coverage: 60%				**Orientation:** East-West with respect to street alignment			
	Architectural features: Majority of the buildings are colonial structures with arches and colonnade							
Burrabazar Main Road	**Number and height:**				**Land value:** 25–30 lakh per kottah			
	Bungalow – 8, old, active, 1–2 storey, aligned to the road				**Area**: Street does not have dedicated open space			
	Apartment – 15, new, active, 4–6 storey, aligned to the road				**Pattern:** Linear			
	Ground coverage: 50–60%				**Orientation:** North-South with respect to street alignment			
	Architectural features: Old buildings are mansions with large gardens, carrying colonial architectural styles							
	New buildings are apartments with modern style							

Table 2.4 Survey information, Serampore

Character appraisal	Serampore							
	Building details				Open space details			
	Number	Height	Ground coverage	Arch feature	Land value	Area	Pattern	Orientation
M. G. Road	**Number and height:**				**Land value:** 10–12 lakh per kottah			
	Bungalow – 14–15, 2–3 floor, good condition				**Area**: Open space is available but unused			
	Apartment – 3, 4–6 floor, good condition				**Pattern:** Linear			
	Ground coverage: residential or public semi-public zone 60–80% and 20–40% coverage, respectively				**Orientation:** North-South with alignment			
	Architectural features: Indo-Danish architectural style is predominant in this stretch with remarkable buildings like Danish tavern, SDO residence, Roman Catholic Church, etc.							
N. S. Avenue	**Number and height:**				**Land value:** 15–20 lakh per kottah			
	Bungalow – 6–7, 2–3 floor, aligned to the road				**Area**: No open space available			
	Apartment – 4, 5 floor, aligned to the road				**Pattern:** Linear			
	Ground coverage:				**Orientation:** East-west with respect to street alignment			
	Architectural features: Old buildings are situated without front setback, whereas new buildings are having road facing balcony and set back as per rules							
K. M. Shah Street	**Number and height:**				**Land value:** 10–12 lakh per kottah			
	Bungalow – 10–12, 2–3 floor, good condition				**Area**: No open space available			
	Ground coverage:				**Pattern:** Linear			
	Architectural features: The main architectural feature of this stretch is the magnificent house of K M Sha with traditional Indian and European style of architecture				**Orientation:** East-west with respect to street alignment			
Rishi Bankim Sarani	**Number and height:**				**Land value:** 12–15 lakh per kottah			
	Bungalow – 20–22, 2–3 floor, good condition				**Area**: No open space available			
	Apartment – 1, 5 floor, good				**Pattern:** Linear			
					Orientation: North-south with respect to street alignment			

Table 2.5 Survey information, Chandernagore

	Chandernagore				
	Path	Node	Edge	Landmark	District
Legibility analysis	Type and directional quality	Focal point and generator	Type and extent	Volume and meaning	Continuity and homogeneity
G.T. Road	Main arterial road	**Node 1**	Residential towards the southern part mixed use on the middle part	All the following landmarks are different in character and expression	Being the main spine of the town the road acts as a linear continuous commercial district
	Public transportation facility available in a large scale	**Focal point:** Padri Para more	Commercial edge towards the northern edge	**HDFC Bank** – placed on a ground floor of building with signage	
	Clear directional quality	**Generator:** Transportation activity		**Sweets Corner** – placed on a ground floor of building, no predominant signage or character is present	

(continued)

Table 2.5 (continued)

	Chandernagore				
	Path	Node	Edge	Landmark	District
Legibility analysis	Type and directional quality	Focal point and generator	Type and extent	Volume and meaning	Continuity and homogeneity
		Node 2		**Vidyasagar Abasan** – a government housing with signage	
		Focal point:			
		Chobi ghar more			
		Generator:		**Petrol Pump –** almost an open space with signage	
		Transportation and commercial activity			
		Node 3			
		Focal point:			
		Bagbazar more			
		Generator:			
		Transportation and commercial activity			
		Node 4			
		Focal point Hospital more			
		Generator			
		Transport, school, hospital			
		Node 5			
		Focal point			
		Laxmigunj bazar Chowmatha			
		Generator			
		Commercial activity			

(continued)

Table 2.5 (continued)

	Chandernagore				
	Path	Node	Edge	Landmark	District
Legibility analysis	Type and directional quality	Focal point and generator	Type and extent	Volume and meaning	Continuity and homogeneity
Strand Road	Main arterial road	**Node 1**	Public semipublic facility on western side of the edge	Entire standard is full of French colonial structures like church, museum, court, schools, etc. along with new educational facilities like Chandernagore College, new complex, Rabindra Bhaban, etc. All of them act as city-level landmark	Complete public semipublic zone along the entire stretch with homogeneous uses but nonhomogeneous architectural expressions
	A pedestrian friendly spine with recreational and public, semipublic facility	**Focal point:**	River on the eastern side of the edge with public recreational facility		
	Clear directional quality	Jora Ghat			
		Generator:			
		Recreational facility			
		Node 2			
		Focal point:			
		Rani Ghat			
		Generator:			
		Public, semipublic and transportation activity			

(continued)

Table 2.5 (continued)

	Chandernagore				
	Path	Node	Edge	Landmark	District
Legibility analysis	Type and directional quality	Focal point and generator	Type and extent	Volume and meaning	Continuity and homogeneity
Burrabazar Main Road	Subarterial road with commercial and educational facility	**Node 1**	Mixed use on the northern part institutional edge towards the southern part	**Kanailal Vidya Mandir –** One of the major institutes of the town consisting colonial heritage structure	It is the most heterogeneous spine of the town with mix of different activities starting from religious to educational to commercial to administrative and residential
	Clear directional quality	**Focal point:**		**Old Post Office Building –** Its colonial-built expression makes it identifiable	
		Old post office junction		**Church –** colonial-built expression makes it identifiable	
		Generator:			
		Commercial activity			
		Focal point:			
		Church junction			
		Generator:			
		Transportation activity			

Legibility Analysis (Tables 2.5 and 2.6)

The three selected roads are acting as main spines of the town and are very well-defined. In each case almost all the five legibility elements are present. Being the main path of the entire town, these roads are acting as guiding factor for the whole town. Majority of the buildings present in Strand Road or G.T. Road act as city-level nodes or landmarks, whereas presence of city-level landmark is very less, and local landmarks, are also not found in abundance. But wayfinding is not a problem in all the three cases. For the selected study area, the river is acting as one edge and G.T. Road as another which are very distinct in nature. View and vista created by the street network allow the nodes and landmarks to be prominent from a distance also (Table 2.6).

Unplanned commercialization has led into congestion in some of the major spines like N.S. Road. But overall it can be concluded that legibility is moreover achieved with the presence of defined elements.

Table 2.6 Survey information, Serampore

	Serampore				
	Path	Node	Edge	Landmark	District
Legibility analysis	Type and directional quality	Focal point and generator	Type and extent	Volume and meaning	Continuity and homogeneity
M. G. Road	Subarterial road with unclear directional quality		**Edge:** North-eastern edge of the road is the river South eastern edge of the road is a combination of residential and public semipublic buildings	**SDO Bungalow –** colonial-built expression makes it identifiable	Non continuous character
			Extent River is a natural barrier, and it runs through the entire length of the road	**Danish Tavern –** colonial-built expression makes it identifiable	
N. S. Avenue	Main arterial road	**Node 1**	**Edge:** Commercial building and river	**SBI old building –** old colonial 2 storey building with associational value	Continuous commercial stretch with non homogenous character
	Public transportation facility available in a large scale	**Focal point:** Ferry Ghat	**Extent:** **Commercial -** throughout the street	**Satsang Society –** old building with religious value Serampore high school – Old colonial structure with institutional and associational value	
	Clear directional quality	**Generator:** Transportation activity	**River** At the extreme edge	**Golakdham –** old Indo-European style private villa with traditional architectural value	

(continued)

Table 2.6 (continued)

Legibility analysis	Serampore				
	Path	Node	Edge	Landmark	District
	Type and directional quality	Focal point and generator	Type and extent	Volume and meaning	Continuity and homogeneity
		Node 2			
		Focal point: Junction of Roy MC Lahiri street and NS Avenue			
		Generator:			
		Transportation and commercial activity			
K. M. Shah Street	Subarterial road	**Node 1**		**K. M. Sha's Residence** – the complex is a landmark, and its architectural value makes it unique	Continuous residential stretch with homogenous character
	Residential edge	**Focal point:**			
	Unclear directional quality	Junction of Panchu Gopal Bhaduri Sarani and K. M. Shah street			
		Generator:			
		Serampore DIB office, Police Station			
		Node 2			
		Focal point:			
		Junction of GT Rd. and K. M. Shah street			
		Generator:			
		Transportation and commercial activity			

(continued)

Table 2.6 (continued)

	Serampore				
	Path	Node	Edge	Landmark	District
Legibility analysis	Type and directional quality	Focal point and generator	Type and extent	Volume and meaning	Continuity and homogeneity
Rishi Bankim Sarani	Main arterial road	**Node 1**	**Edge:**	**India Jute Mill** – Colonial industrial structure, but the presence of the entire precinct makes the district different from the rest of the town for its activity and allied facility	Mixed stretch with several activities with heterogeneous character
	Public transportation facility available	**Focal point:**	Residential, mixed use and industrial	**Serampore Jail** – Colonial architectural structure with an high security expression	
	Clear directional quality	Bor Tala crossing	**Extent:**		
		Generator:	Northern part of the road is mostly industrial along with industrial residents quarter, and southern part of the stretch is mostly residential along with a few mixed use activity		
		Transportation activity			
		Node 2			
		Focal point:			
		Dharmatala crossing			
		Generator:			
		Transportation activity			

Ascertaining the Current Condition

Chandernagore was under French rule till the town got independence. Thus, the French colonial imprint is still visible to a large extent. Whereas Serampore was seized by the British from the Danes which has resulted into a mixed colonial expression for the town.

A brief discussion of the current scenario of the two towns based on the observation of the study has been done in subsequent sections.

Chandernagore: The French Sole

The entire colonial zone falls under the Burrabazar area comprising of the city administration, judicial sector, educational centres and cultural establishments. Thus, this area can be termed as the heart of the town. It can be alleged that the colonial town planning system of distinctive administrative core and residential area is still vividly present in this case which encompasses the essence of the colonial past.

The public areas along Strand Road, i.e. the promenade along with its commercial and recreational facilities, are still functioning in full swing and well maintained by the town administration which is a very prominent example of successful public realm. Unlike many colonial cities like Kolkata, New Delhi, etc., this city core doesn't become deserted after working hours but becomes a commercially successful public space.

The commercial values of areas close to G. T. Road are increasing day by day. Thus, the real estate sector is taking over the old buildings, and subsequently new developments are coming up, disregarding the inherent settlement character. Large commercial developments are taking place in this area, resulting in a homogenous urban environment not conducive to the celebrated image of the town. This is the generic and most obvious transformation process that majority of the Asian small towns are undergoing.

Serampore: The Danish-British Duo

The colonial core consisting of current judicial section of the town is still present in its old place keeping along with some residential and commercial area which makes the situation a little chaotic due to presence of conflicting activities together in an unintended approach.

Being a commercial centre of nearby places, large unplanned commercial development along major roads has squeezed the 'right of way', and therefore conflict between pedestrian and vehicular is increasing day by day in these roads. Moreover, uncontrolled introduction of battery-operated vehicles has degraded the condition in current days.

Serampore has a legacy of being an industrial town, and a few of the old industries are still functioning. The old areas are taken over by industrial workers in most of the cases. The living conditions in these areas have been blighted mainly due to huge illegal encroachment.

Conclusion

Serampore and Chandernagore are different from each other in the method of city development. Serampore has witnessed the rise and fall of feudal system, arrival and settlement of the Danes and a cultural renaissance brought about by the English by constructing the railway, leading to industrial developments. Chandernagore, on the other hand, experienced development under mostly one rule.

At present, both places have settled in a coherent development process that is driven by the market and has been spontaneous and chaotic. Both these towns are part of the larger development of the metropolis Kolkata and are part of the Kolkata Metropolitan Region (KMR). Their regional planning is now thus tied along with that of Kolkata.

The towns are now in a transitional mode either from planning type that is top down to something more organic or from one mode that is spontaneous and driven by market to a more stable one. New approaches should be made towards governance and development to strike a balance between the top-down and bottom-up schemes and the planned and unplanned urban components.

Robust guidelines to be developed by the concerned authorities to channelise the development in a desired direction and to enable multiple stakeholder participation. The guideline, that is to be developed, should consider the factors like – accessible street network, optimum population density, maintained public space, streetscape allowing social interaction, development of building typology to encourage mixed use functions – to maintain the identity and image of these towns.

As the existing environments in some areas are stable and not conducive for regeneration or new development, this research concludes that small-scale changes in Serampore will enhance the urban vitality. The aim is to bring some diversity to the existing environment and the lifestyle. In areas like Chandernagore or Pondicherry in Tamil Nadu, four- to five-level apartments' density is conducive to building up a sense of the urban. These places also serve to show that conventional street layouts consisting of open blocks and neighbourhood boundary help small businesses to prosper and develop; street life also flourishes under this compared to the tree-like system with isolated neighbourhoods.

Developmental authorities should from this realize the fact that the responsibility of building a sustainable and vital society shouldn't be the responsibility of the market. The duties of planning and administration should be promptly taken by the authorities. The small businesses that spring up and give birth to the street life can only partially develop a slice of urban space. A society of high quality also requires properly planned facilities, city centres and services and administration that benefit the public.

New towns that are coming up face both the challenge of development and testing if the planning is effective as well as approach of the government, to strike a balance between top-down and bottom-up strategies to develop new areas that are successful in the years to come.

Factors relating to social processes have been the strongest impact in the urban transformation period of the two towns. Illegal construction was a direct reflection of the fact that legal systems can't keep track of the social relations during the transition period. Urban form has been transformed during the transition period which led to the development of building as per private capital needs and construction land denationalised; and it is being evaluated as per market requirement, leading to change in the approach to urban space planning. The involvement of the owners and users of space with the process of the planning of urban space has become an important aspect of it.

Architectural design has a direct impact on urban space through the shape and the materialisation of which, depending on the degree of integration of the contextual conditions, considers the architectural creativity directly related to the degree of economic development of society. If the urban space has to be designed in a holistic method, one needs to take into consideration all factors in mind.

The present research work has focused on analysis of these two towns of the Little Europe only. The other adjoining similar towns also need to be considered and scrutinised upon, in order to get an overall dimension of these transformations. In the long run, it would be worthwhile to investigate further on this issue, so as to make these towns livable and meaningful in the near future.

Bibliography

Al Ani M (2012) Urban types and transformation of the City, ISBN: 978-3-9503110-3-7

Asian Development Bank (2014) Urban metabolism of Six Asian Cities

Brankica M (2012) Influential factors on the urban transformation of Cities in Bosnia and Herzegovina in the period of transition, 48th ISOCARP Congress 2012

Cetin M (2010) Transformation and perception of urban form in Arab City. Int J Civil Environ Eng IJCEE-IJENS 10(04):30–34

Clerici A, Mironowicz I (2009) Are landmarks essential to the city – its development? ISBN: 978–39502139-7-3

Danış D (2008) Background of the urban transformation

East – West Center (2008) The urban transformation in Asia, policy implications of decentralization

Gulersoy NZ, Gurler E (2011) Conceptual challenges on urban transformation. ITU A|Z 8(1):10–24. 2011-1

Home R (2000) Colonial cities: urban planning in the third world, Report of IDF meeting

Inceoglu A, Yurekli I (2011) Urban transformation in Istanbul: potentials for a better city, 23rd Enhr conference 2011–5-8 July, Toulouse

King AD (2009) Postcolonial Cities. Elsevier Inc, Binghamton

Li Y (2011) The complexity of urban transformation in China: new trends in current transitional era. ITU A|Z 8(1):155–168, 2011-1

Ministry of Urban Development, Government of India (2015) Atal Mission for Rejuvenation and Urban Transformation (AMRUT), Mission Statement and Guidelines

Mishra SA, Pandit RK (2013) Urban transformation and role of architecture towards social sustainability, e-ISSN: 2278-067X, p-ISSN: 2278-800X 5(7):16–20 (January 2013)

Montgomery MR (2008) The urban transformation of the developing world. Science 319:761–764

Njoh AJ (2008) Colonial philosophies, urban space, and racial Segregation in British and French colonial Africa. J Black Stud 38(4):579–599

Ozden K, Enwere C (2012) Urbanization and its political challenges in developing countries. Eurasian J Bus Econ 5(! 0):99–120, pp. 100–120

Sepe M (2014) Urban transformation, socio-economic regeneration and participation: two cases of creative urban regeneration. Int J Urban Sustain Dev 6(1):20–41

UN – Habitat, ESCAP (2015) The State of Asian and Pacific Cities 2015, Urban transformations shifting from quantity to quality, ISBN: (Volume) 978–92–1-132681-9

USAID Preparation of local area plans: pilot project for Delhi, India

van Roosmalen PKM (2003) Changing views on colonial heritage

van Roosmalen PK (2011) Designing colonial cities: the making of modern town planning in the Dutch East Indies and Indonesia 1905–1950. Newsl 57:7.9

World Bank Group (2015a) East Asia's changing urban landscape measuring a decade of spatial growth, urban development series, ISBN (paper): 978-1-4648-0363-5

World Bank Group (2015b) Ellis P, Roberts M, Leveraging urbanization in South Asia managing spatial transformation for prosperity and livability, ISBN (paper): 978-1-4648-0662-9

Chapter 3
Transformation of Commercial Centres and Urban Development Process in Global South

Sanghamitra Sarkar, Mainak Ghosh, and Sanjib Nag

Abstract Urban transformation of cities in response to sustainable economic development has been touted to be one of the defining features of development in the twenty-first century which is demonstrated in growth of commercial centres all around the world. Commercial centres have transformed in shape, form, scale and function over the years. In recent times, it is evident that commercial centres are undergoing drastic transformations. These transformations are creating immediate impact on the urban development process in a city. It is also apparent that this impact is mostly happening in cities of Global South. The aim of this paper is to understand the transformation of commercial centres with changing urban development policies with focus on countries in Global South. The objective is to study similar transformation of selected commercial centres and corresponding process of urban development in the cities of Africa and Asia.

Keywords Transformation · Commercial centres · Urban development process · Global South · Identity

Introduction

Transformation is essentially changes of space and time which generates adaptation. Adaptation creates domination, which with time gives rise to transformation (Mishra and Pandit 2013). Transformations are true both in macroscale, i.e. global, economic, political and social structures, and also at micro level, i.e. the way in which people fashion their everyday lives.

Urban transformation aims to improve the social, economic and spatial quality of cities. Changes in the urban structure are paving stones for growth of a city. Cities represent the most powerful economic engines in the world. Transitions in economic strategies and development have affected social and physical structures of

S. Sarkar (✉) · M. Ghosh · S. Nag
Department of Architecture, Jadavpur University, Kolkata, West Bengal, India
e-mail: sanghamitra.sarkar@jadavpuruniversity.in

© Springer Nature Switzerland AG 2020

M. Ghosh (ed.), *Perception, Design and Ecology of the Built Environment*, Springer Geography, https://doi.org/10.1007/978-3-030-25879-5_3

cities. Urban transformation of cities in response to sustainable economic development has been touted to be one of the defining features of development in the twenty-first century which is demonstrated in growth of commercial centres all around the world.

Commercial centres are centres where economic, social, cultural and administrative services of a city are concentrated. It is a complex of retail store and related facilities planned as an unified group which offer goods and services for profit and performance of various financial and office functions. The earliest commercial centres were marketplaces. Arthur B. Gallion states that marketplaces have always been the focal point of cities and centres for exchange of goods. Cities in ancient times developed along riverbanks and coastlines thriving on flourishing trading centres. Increased trading activities between new settlements and civilizations gave rise to economic development. From trade connections in ancient Mesopotamia to the *Agoras* of Greece to the *Souqs* of the Middle East and eventually to Canary Wharf, London, commercial centres have transformed in shape, form, scale and function. In recent times, commercial centres have become a symbol of image and identity of cities. Evidently, New York is defined by Times Square; Singapore is symbolised by the Merlion with a backdrop of the main commercial centre, etc. Commercial centres are thus the physical manifestations of advancement in economic and urban development (Figs. 3.1 and 3.2).

Urban development is the process of growth of an area situated in city. Historically urban development has come to symbolise physical, social and economic development reflected by emergence of particular types of human settlements, societal institutions and cultural forms. The spatial aspect/physical aspect can manifest itself

Fig. 3.1 Times square, New York. (Source: Author)

Fig. 3.2 Merlion, Singapore. (Source: Author)

either in the form of urban renewal on an existing urban area or a new development on a virgin urban area.

The process of urban development is the path which determines direction of future development of an urban area or a city. This process can be either planned or an unplanned. The planned path of development is known as urban planning. Various concepts of urban planning have dominated the process of urban development. One of the biggest drivers for urban development has been economic development which is manifested physically in growth of commercial centres. The impetus of such development has gradually shifted towards countries of Global South due to its potential as emerging markets and availability of cheap labour.

Global South comprises of Africa, Latin America and developing Asia including the Middle East. North–South divide is broadly based on a socio-economic and political divide. The North mostly covers the West and First World, along with much of Second World, while the South largely corresponds to Third World. The nations of Africa, Central and Latin America and most of Asia – collectively known as the Global South – face great challenges and offer real opportunities. Political, social and economic upheavals are prevalent in many of these nations. But the increasing populations of the Global South Nations and their emerging markets offer immense hopes for economic growth, investment and cultural contribution. Following the fall of the Eastern Bloc, which was commonly referred to as the Second World, many of its constituent countries were reclassified as developing, despite being geographically northern. At the same time, geographically southern nations previously considered "developing", such as the Four Asian Tigers, have joined the modern First World but are classified inconsistently in maps showing the North–South divide. In recent times, with the advent of globalisation and open markets, there has been a surge in the economic development of the countries in the Global South. This has affected the per capita income of the people in these countries eventually changing the urban fabric in terms of market typologies and building forms.

The aim of this paper is to understand the transformation of commercial centres with changing urban development policies with focus on countries in Global South. The objective is to study similar transformation of selected commercial centres and corresponding process of urban development in the cities of Africa and Asia and establish parallels between them.

Discussion

Transformation of Commercial Centres

Transformation of commercial centres refers to change or alteration in organisational, spatial, economic and social composition of commercial centres which is governed by modifications in trajectory of urban development. Currently, it has become one of the dominant physical processes in cities of Global South countries.

The current notion of traditional commercial centre roots from physical structure of medieval cities where markets formed one of the major components alongside the administrative centre, religious centre and surrounding settlement. Based on type of cities, the formation, structure and function of commercial centres change. Cities can be broadly classified as ancient, medieval, colonial, postcolonial and contemporary cities. Countries of the Global South have a predominance of medieval, colonial and postcolonial cities. Medieval marketplaces consisted of market streets which were organised based on different occupations of traders. Shops lined on both sides of the street made it thoroughly active in nature. Streets would predominantly follow either organic pattern. In contrast the colonial markets were organised under one roof into a marketplace. Shopping arcades were one of the defining features of colonial marketplaces. Postcolonial cities have adopted ideas of community and neighbourhood markets which are essentially the result of zonal planning concepts.

In contemporary times, commercial centres have evolved as image-makers for cities all over the world. There has been a major shift in the type of commercial activity which has resulted in change in the commercial pattern. This change of commercial pattern has transformed the built environment of commercial centres. Countries in Global South, especially in Asia like China and India and parts of Africa, have undergone this transformation quite visibly. In recent years, it has been seen that the emergence of large commercial centres have changed the identity of their respective cities. Cities like Shanghai, Hong Kong, Guangzhou, Delhi, Mumbai, Addis Ababa, etc. have come up as emerging commercial hubs of the world. It is also seen that there is not one but multiple centres of commercial exchanges indicating the decentralisation of such commercial centres. Transformation of commercial centres has thus not only influenced immediate built form and commercial character of an area of cities but has also the process of urban development and vice versa.

Commercial Centres with Respect to Urban Development

The morphogenetic school of thought propagated by Christopher Alexander and Nikos Salingaros states that urban development is a process similar to growth of cells in an organism where buildings, infrastructure, human beings, their activities and processes combine to form an urban superorganism. With evolution of technological know-how and knowledge economy, civilizations have advanced in different parts of the world.

The history of development of a city determines complexity of urban fabric demonstrated by a city. Economic trajectories like capitalist, socialist or modernist all have distinct influences on the immediate physical form of the city fabric. Another characteristic of urban development is the natural process of transformation that occurs due to sociocultural composition of a city. The existing physical structure and activities are a result of years of culture and traditions followed by its people.

The strategic placement and functions of commercial centres have determined their importance and gradually established them in city fabric. (Weinstein 1974) postulated three dimensions as being important contributors in a conceptual model for the social segregation of an Indian city (Fig. 3.3).

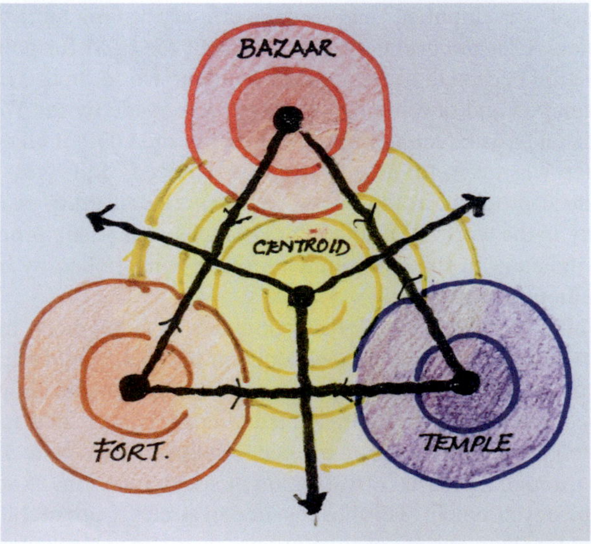

Fig 3.3 Traditional Islamic city model. (Source: Author)

These three dimensions were:

- Socio-economic dimension symbolised by the bazaar
- Political dimension represented by an administrative symbol
- Prestige dimension derived from the religious function of a temple

In contrast, traditional African city patterns were based on arrangement of residential areas for the clan where palace of the clan chief was of primary importance. Second most important component was main market. These two major anchors of the city would be bound together by a central plaza which was the major congregation space for the settlement. Shape of an African city was essentially circular in nature which had direct relationship to the circle of life.

Similarly different countries in Global South had their distinct sense of city building based on their sociocultural composition. But the economic centre of any city has always been a distinct feature. Economic development of these cities either was based on trade, agriculture, natural resources, etc. which were typical of the region. New urban and economic developments are transforming fundamental nature and structure of cities. The above-mentioned city types have undergone a sea change in their development processes and so have their commercial centres, bazaars and markets.

Urban Development in Global South

Countries in Global South are currently some of the fastest-growing economies in the world. Idea of "development" has been the primary driving force behind shaping of modern cities. After the events of World War II, the need for political and economic stabilisation became a necessity all around the world. In this regard, classifications of developed and developing countries were made by the World Bank and United Nations on basis of economic advancements and GDP of all countries of the world. This step was taken to identify countries in need of progress and countries which can induce progress in the above nations, thus ensuring peace and much-needed stability in the world. Till date this classification is vastly prominent and has been a major indicator to the development policies undertaken by both the developed and developing countries.

In recent times, the advents of globalisation and "open markets" have paved way for rapid information exchange and development of a penchant to acquire a global outlook. One of the most prominent policy approaches that encouraged this global venture was neoliberalism. This approach led to open, competitive and often unregulated market policy driven by action of social groups to drive state downsizing and public reform through increased privatisation of social functions to achieve optimal socio-economic development. This has resulted in strategic urban transformation of commercial centres by a variety of institutional restructuring to enhance their local economic growth capacities (Brenner and Theodore 2002).

All major urban centres of the world have strived to build a similar image, reflecting trend of becoming global villages which tends towards establishment of a single global tribe, rather than merely concerning ourselves with individual communities. The continents of Africa and Asia have been considered to understand changing trajectory of urban development processes and subsequent transformation of commercial centres.

Case of Africa

The idea of urban development and city making is comparatively new concepts for the continent of Africa. The ideas of urban development and city making are comparatively new concepts for the continent of Africa. According to historical records, Africa was composed of societies, states, kingdoms, decentralised societies and empires which were deep rooted in idea of community formation through kinship. The authoritarian position was taken up by head of family, and subsequent transfer of power would happen within the family. Compatible families had business ties with each other, and clashes between incompatible tribes were also common.

Between the 1870s and 1900s, Africa faced European imperialist aggression, diplomatic pressures, military invasions and eventual conquest and colonisation. At the same time, African societies put up various forms of resistance against the attempt to colonise their countries and impose foreign domination. By the early twentieth century, however, much of Africa, except Ethiopia and Liberia, had been colonised by European powers. The European imperialist push into Africa was motivated by three main factors, economic, political and social. It developed in the nineteenth century following the collapse of the profitability of the slave trade, its abolition and suppression, as well as the expansion of European capitalist Industrial Revolution.

Colonisation by imperialists introduced the concept of formal cities and city planning, but it came at a cost of territorial civil wars. Ideas of democracy and nation building were instilled after independence which in case of certain countries like Namibia happened during the 1990s. But Africa is a continent which is so deep rooted in their cultural values that the notion of democracy has always been a foreign concept. The trend of development aid has also hampered development process because the local ruling dictators profited more from the aid than common people. The trickle-down effect of development aid never came to fruition giving rise to a state of neo-colonialism.

Rapid urbanisation of African cities is the common trend. Cities like Addis Ababa, Nairobi and Dar-es Salaam have grown exponentially. The population of Dar-es Salaam has been recorded to grow almost six times since 1978. The major issues related to such high rates of urbanisation in Africa were increase in urban poverty, transportation and sanitation issues and rising inequality which primarily was derived from loss of identity of Africa as a continent. Imposition of westernised trend of development has resulted in unsustainable growth. Huge foreign investment and real estate have become the major driving force behind all development activities. This has led to significant change in the economic policies and development of commercial centres. The World Bank forecasts growth of 3.4% for the year 2018 which is more than 2.4% in 2017. The continent has also demonstrated rise in the per capita income rapidly in the last 18 years. As a result the consumption capacity of the population has increased. South Africa has ranked sixth in the world as the fastest in shopping centre development. Apart from the cities in South Africa, Nairobi is one of the fastest in development of malls in Africa. Nairobi has recorded the largest retail space across sub-Saharan Africa. In 18 months, the city has seen three shopping malls getting built which is a result of this fast growth. Various African scholars and urbanists have spoken out against such unplanned, unprecedented and rapid city development and stressed on the need for visualising its future of how to move ahead.

Case of Asia

Asia in contrast has a different approach to advancements in urban development. Similar to Africa, the Asian countries have also been colonised for almost 200 years. Historically there have been evidences of formal city formations in medieval times. The basis of formation of these cities was majorly to act as centres for trade routes and connect to the immediate hinterland which helped the colonists. The development of such cities and their intermediate infrastructure was initiated and promoted widely by the colonists across South Asia. The concept of urban was embedded in the sociocultural context of the settlements and had flourished into large cities before colonisation.

The advent of colonisation gave rise to colonial cities with distinct physical demonstrations of both authoritarian control and economic growth evident in the urban form of a city. After independence, the scenario was different in comparison to Africa. Many of the colonised nations did not immediately open their markets to the world which helped them to build their own intrinsic economic policies and markets to become self-sufficient. Even though this policy was good for acquiring stability, it failed to become an agent of growth. Meanwhile, the market-based capitalist economies in the West continued to flourish. In the world stage, these developing economies, due to their socialist-driven policies, fell behind and at times incurred huge debts which forced them to open up their markets. But as the markets opened, rapid urbanisation and city expansion become the preferred direction of growth. Formation of megacities and global cities has been the most popular development trend in this region.

The dawn of the twenty-first century has brought about a major shift in global urban development trends. From a majorly West-centric global power, the world scenario has shifted to a more East-centric powerhouse. China and India have come up as the fastest-growing world economies with China surpassing the USA in total world output in 2010. The recent projections on the growth of global urban population have indicated that China, India and Nigeria are going to account for 37% of the increase of nearly 2.5 billion urban population growth during 2014–2050. The megacities were once exceptional in South Asia but are now beginning to grow in the region. The megacities (exceeding 10 million inhabitants) or large cities (with 5–10 million inhabitants) are frequently distinguished for their size and concentration of economic activities. At the same time, many secondary cities have prospered which are continuing to fulfil the primary city functions within the context of the countries in which they are located. These secondary cities are performing a vital role in case of production, governance and logistical functions at the subnational or regional level in South Asian countries. This highlights the trend of formation of multiple decentralised urban centres as opposed to the earlier trend of polarised urban centres. One of the examples of such a development trend is development of smart cities in India. The Government of India declared formation of a 100 smart cities in an effort to jumpstart a major urban transformation process. This process aims at "comprehensive development" which occurs in areas by "integrating the physical, institutional, social and economic infrastructure" as mentioned in the smart city guidelines. As a result of this process, commercial centres are getting integrated to all the other city infrastructures to create a consolidated city function. This has spurred development of new city extensions like New Town, Rajarhat, near Kolkata, which is supposed to act as the future commercial centre of the city. Another such development is GIFT City in Gujarat. Not only have new centres of commerce cropped up; the old cities are undergoing major urban transformation. There has been a major drive for the urban regeneration of the old city regions of the chosen smart cities.

Transformation of Commercial Centres with Changing Urban Development Process in Global South

The similarities and differences between urban development and commercial growth between Africa and Asia are quite striking. The following section aims to establish this by illustrating some examples.

Case of Africa

As previously discussed, booming commercial activities in Africa were a result of need for colonists to transport goods, natural resources and labour back to their home countries. The primary difference was between "peasant" and "settler" economies (Amankwah-Ayeh 1996). Colonial cities were mostly made on trading outpost and were more often than not entirely new cities.

In recent age of rapid urbanisation, new cities are still being planned on virgin territory. One such example is Eko Atlantic City in Lagos. It is a project in Lagos, Nigeria, which has been proposed by Nigerian Federal Government. Eko Atlantic will include wide-scale residential, commercial, financial and tourist developments, easing the burden on exhausted infrastructure and overpopulation on adjacent

Victoria Island, which is currently the commercial centre of Lagos. Around 250,000 people are believed to be living in the city with 150,000 more commuting into work each day. The planning includes a business district with the boulevard inspired from Manhattan, New York.

Another example is that of the six new town projects in Nairobi, Kenya. The Nairobi Metro Vision 2030 is focussed on turning Nairobi into a globally competitive business, industrial and services region. The plan is for a peri-urban ring of 32,000 km^2 that is expected to swallow all 15 surrounding municipalities around Nairobi, doubling the city's population to 6.3 million. Konza City in Nairobi, Kigamboni in Dar es Salaam and Hope City in Accra are other latest projects of this kind, for which funding has been secured (Nairobi Metro 2030 Strategy 2008). They represent the exact opposite of compact, inclusive and functionally mixed urbanism which is intrinsic to the African culture.

At the neighbourhood scale, street markets and local shopping markets have also transformed. Keeping with the theme of city building and image building, a large number of big box malls have cropped up. A few examples of such commercial centres are Marina Mall in Accra, Mall of Arabia in Cairo, Addis Mercato in Addis Ababa, etc. The exclusive nature of mall typology is a far cry from the buzzing street markets that has been the tradition of African cities. Most of these commercial centres have cropped up in newly developed city extensions where car-centric planning dictates urban life. Life on the street defined by boundary walls and parking lots and is devoid of any commercial or interactive activity.

In all these examples, one has lost not only the original African market but also meaning of urbanism for Africa. Be it Cairo or Durban, Accra or Konzo, infiltration of globalised image of city building has never let the idea of African urbanism be conceived resulting in loss of identity for all her cities.

Case of Asia

The development trajectory of Asia shows that opening up of economies to effects of globalisation was taken much later. This resulted in formation of inclusive cities with functionally mixed urban character. But increase in population led to creation of new towns and satellite cities where commercial centres were segregated into central commercial centres, block markets and neighbourhood markets. For example, commercial areas of Chandigarh, India, were originally designed to have a planned hierarchical order, but in the actual process, they developed in a traditional way apart from the main commercial centre complex. "The shopping centres of the city have undergone internal and external transformations across the years, nonetheless the overall character remains the same" (Chandigarh Master plan 2031).

Similarly in China, traditional cities transformed into socialist cities and later to contemporary cities that are emerging in the current reform era. "One of the primary goals of urban planning in China was to achieve the transition from the 'consumer cities' of capitalism to the 'producer cities' of socialism. With the economic reforms of the late 1970s, however, planners began actively to promote expansion of the tertiary sector of urban economy" (Gaubatz 1999). They focussed on special economic zones for foreign investment that were relatively free of the bureaucratic regulations and interventions that hampered economic growth.

Prior to the early 1990s, the retail typology in townships was mainly composed of small, mainly informal retail businesses offering basic products and services to a relatively low-income consumer market. After 1994 there was a shift in consumption topology as the country has experienced a substantial economic uplift. There was an increase in per capita income resulting in increase in spending. Traditional retailing centres became saturated with the fast growing commercial activities and could not support the drive further. This development caused national retailers, especially supermarket chains, to increasingly focus on market expansion strategies in these emerging markets.

Currently cities like Shanghai, Guangzhou and Beijing are on the path to become global cities which revived themselves as new commercial centres in the East. The idea of economic advancement has seen commercial centres building up reflection of development based on the globalised concept of image-making. Shanghai financial district is a classic example of this globalised image of commercial centres. Modern built forms dominate the skyline. Segregated and exclusive activity patterns define the urban characteristic of that area. This is a distinct departure from the traditional Shanghai markets.

In comparison, traditional markets of Shanghai exhibits space making in the form of a sequence of streets accessible at various locations ensuring ease of access. It also caters to active and successful interactive spaces due to compact circulation pattern. The continuous shopping facade running all around the external edges of the market provides appropriate human scale for the approach and use. The spatial distribution is zoned for different activities which make the place simple and easily perceptible for the users and sellers alike (Figs. 3.4 and 3.5).

Fig 3.4 Traditional Shanghai markets. (Source: Author)

Fig 3.5 Shanghai financial district. (Source: Author)

Since 2014, China has recorded the largest development drive in the world exhibiting a total of 5.7 million sq.m of shopping centre space being constructed the same year. In those 12 months, 44% of global development completions were in China; however, this amount of new supply was down by 14% in comparison to 2013. Outside of China, three other Asian markets – Seoul, Kuala Lumpur and Manila – featured in the top 15 most active markets. The Lotte World Mall in Seoul – at 480,370 sq.m. – was the largest completion in Asia-Pacific region. New development in Asia-Pacific on the whole spreads over 32 million sq.m., distributed over 36 cities – over 80% of this is situated in China, Thailand and India.

This enormous development drive along this whole region is a rapidly growing trend which has brought forward various urban problems. In the macro scale, these problems range from increase in urban poverty, environmental degradation, mass scale housing shortage, polarisation of money and power, etc. In the micro level, there has been problems of rapid change in the urban landscape with large mega projects clouding the development activities. The consumption patterns of the people have changed and so has the commercial built form. Local markets have made way for largescale shopping centres, and the number of shopping centres has been on the rise.

From the above discussion, it has been seen that the economic development for both continents has been made possible by rapid urbanisation and decentralisation of market centres. Aided by the increase in per capita income and gradual change in consumption pattern of the rising population, urban form of the commercial centres has changed from local markets to shopping malls. Both the continents have shown evidences of drastic development drive of building largescale commercial centres in the past decade, making them a major building typology to define commercial centres. In an effort to redefine the cities into local development platforms, there is a definite loss of local identity for cities because the development patterns are guided by the Westernised model of development which has not been a model for sustainable development in these regions giving rise to stark inequalities, mass poverty and urban sprawl.

Conclusion

Thus, it can be postulated that presently development policies in Global South are all looking at advancement in economic development. Global image of urban development has been taken up by all the countries. In this day and age of polycentric and multi-cultural cities especially in the Global South, a singular notion of urban development will not be a sustainable solution. The image of "One World" that is being promoted by all economic policies and demonstrated by cities in their urban form is a generic solution to attain rapid development. This has resulted in massive transformation of urban character in cities especially of commercial centres.

In this regard, two aspects of development should be considered for advancement of cities. Firstly, cities should look at establishing ingenuity in their imagebuilding strategies. Commercial centres in Africa should promote and celebrate Africa not only in its physical form but also in the kind of spaces that it generates. The Indian bazaar, Arabian Souq and traditional Chinese marketplaces should not become obsolete as a result of urban economic development. Instead they should be redefined in various ways to make a smoother transition from postcolonial city to highly urbanised Global City. Secondly, the major characteristic of the original "Global South City" was inclusive development. Incorporation of inclusivity through urban design and architectural interventions in commercial centres should be carried out. This, in turn, will induce interaction within the resultant urban space while enriching the urban life of the city for one and all.

In this perspective, future directions of research should be made to explore transformation of commercial centres and process of urban development from an urban design viewpoint and thereby establishing a parametric relationship between them. Based on this further exploration should be carried out of similar transformation of commercial centres and corresponding process of urban development, applying the above relationship, thereby formulating relevant guidelines for the future and ensuring desired spatial/physical environment for cities in Global South.

Global South may have been underdogs in the race for development, but now the pace of development has skyrocketed towards producing the emerging superpower nations. The responsibility of redefining the idea of global urbanism should be taken up by the nations in order to pave the path for holistic development, restore identity and transform the image of commercial centres of Global South.

References

Amankwah-Ayeh K (1996) Traditional planning elements of pre-colonial African towns. New Contree J Hist Hum Sci South Afr 15(39):60–76

Alexander C (1977) Pattern language: towns, buildings, construction. Oxford University Press

Brenner N, Theodore N (2002) Preface: from the "New Localism" to the spaces of neoliberalism. Antipode 34(3):341–347

Chandigarh Master Plan (2031)

Gaubatz P (1999) China's urban transformation: patterns and processes of morphological change in Beijing, Shanghai and Guangzhou. Urban Stud 36:1495–1521

Mishra SA, Pandit RK (2013) Urban transformation and role of architecture towards social sustainability. Int J Eng Res Dev 5:16–20

Nairobi Metro 2030 Strategy (2008) Sustainable development knowledge platform, United Nations

Salingaros N (2002) Development of the Urban Superorganism, Katarxis N°3

Weinstein JA (1974) Madras: an analysis of urban ecological structure in India, Beverly Hills. Sage Research Papers, Social Ecology of the Community Series 90-015

Chapter 4
Issues and Challenges for Transit-Oriented Development in the Scenario of a Developing Country: The Case of Kolkata Metropolitan Area, India

Subrata Kr. Paul, Abhinanda Chatterjee, and Souvanic Roy

Abstract Transit-Oriented Development (TOD) is a well-acknowledged strategy to integrate land use and transportation for sustainable urban development. Discourse on TOD characteristically upholds its promise for creation of high-density, compact, and mixed-use developments, supported by public transits to minimise dependence on private vehicles and thus to reduce sprawl, congestion, and pollution. Conversely, several scholars cautioned that the absence of enabling circumstances would turn TODs into Transit Adjacent Developments (TADs), significantly swayed away from the goal of sustainable urban development.

Like many other developing nations, India is confronted with a host of urban transportation problems, such as growing use of private vehicles, traffic congestion, degrading urban environment, inequity in allocation of road space between vehicles and pedestrians, increasing road accidents, and such like. The Government of India relies on TOD for ameliorating the present urban transportation problems and, thus, emphasises on public transportation systems for proceeding towards the goal of 'growth of urban transport along low-carbon path'. This has, lately, resulted in emergence of Rapid Transit Systems (RTS) in several Indian cities, and competent authorities are proceeding towards framing appropriate TOD guidelines. While some of these cities seek to reap the TOD potential accorded by the recent investments in RTS, there are several debates regarding implementation TOD in India. Many of these debates focus on the strategies and potential of land development along transit corridors. Unfortunately, other issues and challenges related to functional connectivity of land use and transit, sustainability, equity, and implementations are less pondered upon.

S. K. Paul (✉) · S. Roy
Department of Architecture, Town and Regional Planning, Indian Institute of Engineering Science and Technology, Shibpur, West Bengal, India
e-mail: subrata@arch.iiests.ac.in

A. Chatterjee
Faculty of Architecture, Manipal Institute of Technology (A unit of MAHE), Manipal, Karnataka, India

© Springer Nature Switzerland AG 2020
M. Ghosh (ed.), *Perception, Design and Ecology of the Built Environment*, Springer Geography, https://doi.org/10.1007/978-3-030-25879-5_4

This article aims to identify the issues and challenges for TOD in the context of a developing scenario—in this case Kolkata Metropolitan Area (KMA) in India. The issues are identified, and the need for a multidisciplinary approach for arriving at context-specific and carrying-capacity-based TOD strategies—allowing development along the 'low-carbon path'—is suggested.

Keywords Transit-oriented development · Transit adjacent development · TOD · TAD · Smart development · Low-carbon development · Land use-transportation

Acronyms

BRTS	Bus Rapid Transit System
CSD	Complete Street Design
GDP	Gross Domestic Product
GNI	Gross National Income
HDI	Human Development Index
IRC	Indian Roads Congress
KMA	Kolkata Metropolitan Area
KMDA	Kolkata Metropolitan Development Area
LDC	Less developed country
MDC	More developed country
MRTS	Mass Rapid Transit System
NMT	Non-motorized Transport
NUTP	National Urban Transport Policy
PTS	Public transport system
ROW	Right of way
RTS	Rapid Transit System
TAD	Transit Adjacent Development
TOD	Transit-Oriented Development

Introduction

TOD refers to an urban planning strategy that encourages development of mixed-use, residential, and commercial areas close to—and well-served—by public transport systems. A TOD typically consists of a public transit station at its centre surrounded by moderate to high-density developments characterised by presence of civic spaces and environs, friendly to pedestrian and non-motorised vehicles, frequently bicycle.

Peter Calthorpe, a renowned architect and urbanist, codified the concept of TOD in the late 1980s (Carlton 2007) and defined TOD in his pioneering work 'The New American Metropolis' in 1993. Afterwards, concepts and definitions of TOD (Table 4.1) appeared in several discourses and policy documents. All of these defini-

Table 4.1 TOD definitions and concepts

Calthorpe (1993)	Mixed-use community within an average 2000-ft walking distance of a transit stop and a core commercial area that mixes residential, retail, office, open space, and public uses in a walkable environment, making it convenient for residents and employees to travel by transit, bicycle, foot, and car
Bernick and Cervero (1997)	A compact, mixed-use community, centred on a transit station by design, invites residents, workers, and shoppers to drive their cars less and ride mass transit more
Still (2002)	A mixed-use community that encourages people to live near transit services and to decrease their dependence on driving
Cervero et al.(2004)	TOD is a tool for promoting smart growth, leveraging economic development, and catering for shifting housing market demands and lifestyle preferences
Transit Oriented Development Institute (http://www.tod.org, viewed 10.03.2019)	TOD is Regional Planning, City Revitalisation, Suburban Renewal, and Walkable Neighbourhoods combined. The public has embraced the concept across the nation as most desirable place to live, work, and play. Transit-Oriented Development is also a major solution to the serious and growing problems of climate change and global energy security by creating dense, walkable communities that greatly the need for driving and energy consumption. This type of living arrangement can reduce driving by up to 85%
World Resources Institute (https://www.wri.org, viewed 10.03.2019)	TOD is a sustainable urban development solution that has been successful in creating mixed-use, dense, walkable communities with access to high-quality transport. TOD does not automatically translate into better livability and quality of life for citizens
US Department of Transportation (https://www.transportation.gov, viewed 10.03.2019)	TOD includes a mix of commercial, residential, office, and entertainment centred around or located near a transit station. Dense, walkable, mixed-use development near transit centre attracts people and leads to vibrant connected communities
Centre for Transit Oriented Development (http://ctod.org, viewed 10.03.2019)	TOD is commonly defined as high-density, mixed-use developments within walking distance (1½ mile) of a transit station. TOD provides a range of benefits including increased transit ridership, reduced regional congestion and pollution, and healthier, more walkable neighbourhoods. Neighbourhoods with a mix of both affordable and market rate housing can also provide many benefits, such as reducing income segregation and providing lower-income residents with opportunities to move up the occupational and social ladders
Maryland Department of Transportation (http://www.mdot.maryland.gov, viewed 10.03.2019)	A piece of relatively higher density that includes a mixture of residential, employment, shopping, and civic uses located within easy walking distance of a bus or rail transit centre

tions essentially rest within the concepts of new urbanism and are in response to the prevailing issues of road congestion, environmental pollution, climate change, increasing energy costs, rising transportation costs, increasing demand for urban living, shrinking household size, awareness and desire for sustainable and green living, etc. And the same are to be kept in sight while planning sustainable human settlements, in varied scales, ranging from a region to a neighbourhood (Austin et al. 2010).

TOD, as a planning concept, aims at creating livable environments, where people may access transit arteries to employment centres, services, and shelters with minimal use of personal automobiles. It, thus, promises sustainability through reduction of energy consumption for transportation. Three defining objectives of TOD are first, discouraging/reducing dependence on private vehicles and inducing increased use of public transport through design, policy measures and enforcement; second, providing easy access to public transport system within a walking distance to maximum number of commuters through appropriate land use distribution, densification, and enhanced connectivity; and third, creating an enhanced level of accessibility by increased use of non-motorized transport (NMT) modes to the areas beyond walking distance for connecting public transit stations and the actual points of trip origin or the final destinations.

While the discourse on TOD upholds its promise of sustainable development, several progressive cities and regions in the United States and Australia (MTC 2005, 2006; WMATA 2013; NSW Government 2010; QDIP 2005; State of Victoria 2008) have relied on TOD in policy terms; nevertheless, experience from the reported case studies are diverse, and the transition from planning to practice and implementation remains a key issue for discussion. A frequently referred, often anecdotal, discussion found in discourse on TOD success revolves around the question of whether a project or place is genuinely ‘transit oriented’ or only an incidental TAD, significantly swayed away from the goal of sustainable development on several counts (Hale 2014). A new distinction within ‘smart development’ literature is noted, which describes the difference between TAD and TOD (Belzer and Autler 2002; Cervero and Duncan 2002; Ditmar and Ohland 2004; Renne 2009). Like TOD, TAD also refers to an area within a 10-min walk, or half-mile radius around a major transit station, but unlike TOD, TAD fails to reap the benefits from its geographical advantage of being in close proximity to the station. Cervero and Duncan (2002, p.6) describes TAD as ‘Physically near transit [but] fails to capitalise upon this proximity … [It] lacks any functional connectivity to transit — whether in terms of land use composition, means of station access, or site design’.

The world has been—and continues to be—divided into developed and developing countries, sometimes also referred to as ‘haves’ and ‘have-nots’ or ‘more developed countries’ (MDCs) and ‘less developed countries’ (LDCs). Such designations are often condemned with the imperfection of definitional inadequacies, as the earth continues to be a dynamic place with many less developed countries continuously progressing on the path of modernisation and development. There is no absolute or universally accepted set of criteria to confirm where a country lies on the spectrum of development. However, the world community and development agencies have identified several variables, such as per capita Gross Domestic Product (GDP) (Hobbs 2009), per capita Gross National Income (GNI), Human Development Index (HDI) (UNDP 2008), life expectancy, infant mortality, level of energy use, and such like, which permit the distinction between the more developed and less developed portions of the world (Timothy and Nyaupane 2009, p. 3; Hobbs 2009).

Similarly, in the context of urban development, there are several common issues and challenges in the developing world. Solving transportation issues is one of the commonalities and often one of the major challenges confronting governments in developing countries.

Despite significantly increased investments in urban transport systems, transportation problems in developing nations continue to worsen due to poor planning, lack of awareness among people, corruption, poor governance, inadequate institutional capacity, and multiplicity of stakeholders with lack of shared understanding and coordination (Gwilliam 2003). Like many other developing countries, India is confronted with a host of urban transportation problems, such as rapidly growing use of private vehicles, increased travel cost, severe traffic congestion, deteriorating quality of urban environment owing to emission from transport sector, sharp inequity in allocation of road space between vehicles and pedestrians, and increasing rate of road accidents. As reflected in the current policies and the development programmes, the Government of India recognises the need for development of public transport systems adopting TOD as one of the approaches for mitigating the transportation problems, thus moving towards the goal of 'growth of urban transport along low-carbon path'.

With the above background, the focal intent of the current article is to identify the issues and challenges to be addressed in TOD planning in the case of KMA—the largest urban agglomeration in eastern India—and to avoid incidental TAD.

The rest of the article is organised in four subsequent sections: Section "TOD in the context of urban India" reviews TOD in the context of urban India; Section "The background of KMA" introduces the background of KMA within the selected study areas; and issues and challenges for TOD in the KMA are identified in Section "Issues and challenges", followed by the concluding remarks in Section "Concluding remarks".

TOD in the Context of Urban India

The state of urban public transportation systems in majority of the Indian cities have deteriorated over the last few decades (CUE 2017) in spite of significant investment in the sector. Rising population and inadequate public transportation has led to rapid growth in the number of personal vehicles, leading to traffic congestion and increased pollution levels. Scholars have reported that the transport sector in India is extremely energy intensive and needs methodical overhauling in mass transit systems to arrest the growing trend of private motorised mobility (Rizvi 2013; Yedla 2015). With the urban population projected to more than double in the next generation, the situation might go beyond control unless effective remedial measures are taken up.

The National Urban Transport Policy (MUD 2014) of India recognises that people occupy the centre stage in the city, and thus, all plans are to be for their common benefit and well-being. The policy also strongly emphasises the need for development of public transport systems at all levels of the government (EPC 2012). Several recent mission-based programmes of the Government of India—Jawaharlal Nehru Urban Renewal Mission (JNNURM), Atal Mission for Rejuvenation and Transformation (AMRUT), and Smart Cities (2015)—have emphasised on investment in public transport. This has resulted in the emergence of Mass Rapid Transit Systems (MRTS) and Bus Rapid Transit Systems (BRTS) in several Indian cities, such as Delhi, Mumbai, Kolkata, Chennai, Bengaluru, Hyderabad, Ahmedabad, Rajkot, Surat, Pune, Pimpri-Chinchwad, Hubli-Dharwad, Lucknow, Kochi, Jaipur, Bhopal, Indore, and others.

All these have led to the recent spotlight on TOD in India.

Recently, the Ministry of Urban Development has come out with the 'National Transit Oriented Development Policy' for the cities, which essentially identifies 12 guiding principles and 9 supportive principles (Table 4.2).

Historically, transit corridors have always been the driving factor of land use change across the world, including India. In the early major Indian cities—Calcutta, Bombay, Madras, and Delhi—developments were observed along the major transportation corridors, such as Chowringhee road in Kolkata, Mumbai-Thane railway corridor, and such like. However, these spontaneous developments, by and large, were more guided by obvious advantages of mobility owing to proximity of a transit system, speculations, and market forces rather than planned interventions. Consequently, these areas might have led to undesirable outcomes—overcrowding, traffic congestion, pedestrian-unfriendly narrow streets, and absence of public spaces—in terms of development.

The first ever modern TOD was observed for the designed city of Navi Mumbai. It started in 1992 to accommodate the growing population of Mumbai. The spread over was about 344 km^2 of land, including a 150-km-long creek line corridor. The need to link up workplaces to residential areas was the triggering factor behind the start of such development, and the concept of last-mile connectivity was reflected in the design of the city.

At present, TODs in India can be categorised into three types depending on the scale of development—first, station-level TOD, which involves only the area around the station, such as the TOD around Kochi; second, area-level TOD that involves policy for development in an area encircling a transit corridor, such as MRTS corridors in Mumbai, Delhi, and Ahmedabad; third, city-level TOD, such as Naya Raipur.

There are several debates regarding adoption and implementation of TOD in India. Petkar and Hamand (2013) argue that in western countries TOD was used for

Table 4.2 TOD guiding principles and supportive principles in India

TOD guiding principles	Multimodal integration, first and last mile connectivity, interconnected street network, complete streets, NMT network, traffic calming, mixed land use, optimised densities, street-oriented buildings, managed parking, informal sector integration, housing diversity
TOD supportive principles	Engagement of private sector, barrier-free environment, high-quality transit systems, land value capture, preservation and creation of open spaces, green buildings and infrastructures, right sizing of infrastructure, technology integration, safety and security

Source: TOD Policy Norms Guidelines, UTTIPEC-DDA

densifying certain areas, but India cities, particularly the inner cores and transit corridors, already have higher densities. Hence, TOD in Indian cities should be looked as a tool for improving quality of life and a means to provide infrastructure facilities. In India, several of the current mainstream TOD debates largely focus on the development potential of the land along transit corridors and the strategies for 'land value capturing' (Suzuki et al. 2015) along metro rail corridors.

But unfortunately, other issues and challenges related to functional connectivity of land use and transit and the themes sustainability, equity, and implementation are less focussed and rather late entrants in the discourse. Concrete attempts towards preparation of TOD plans through development plans, or standalone local area plans, are yet to be explored in most of the Indian cities.

The Background of KMA

KMA, with a population of about 14.112 million (Census of India 2011 A.D.), is a primary urban conurbation in the eastern part of India. It is also the second largest urban unit in India after Greater Mumbai.

The history of Kolkata, as an urban centre, can be traced back to the setting up of a trading-cum-port settlement by the British, on the east bank of the river Hooghly in the last decade of the seventeenth century (A.D.), when British coloniser Job Charnok first established a small trading post on the eastern bank of the river Hooghly near the three villages of Sootanuty, Kolikata, and Govindapur.

At present, spread over the two banks of River Hooghly, the KMA covers an area of 1785 km^2 (as recorded in 2001). It consists of the following administrative units: 3 municipal corporations; 38 municipalities, and 22 panchayat samities (administrative units in the rural fringe areas).

The physical setting and natural features have, to a large extent, determined the pattern of development of the KMA. It has grown in a linear form, along the two banks of approximately 95-km- long course of River Hooghly, extending from Kalyani to Baruipur on the east bank and from Bansberia to Uluberia on the west (Fig. 4.1). Over the centuries, the river has deposited large quantities of alluvial silt along its banks, forming natural levees of high land, suitable for human settlements, with a natural slope of land away from the river. Within a very short distance from either bank (about 3 km), the level falls quickly leading to a low-lying land. This low-lying land—being either perennial marsh and swamp or liable to annual inundation during monsoon—poses difficulties for large-scale urban development. The physical expansion of Kolkata on the east bank has been prevented by the immediate proximity to large areas of low-lying lands and marshes. On the west bank of the river (beyond Howrah and its neighbouring areas), the problems of periodic flooding, difficult drainage, and brackish water have restricted urban development to a narrow strip of high land along the riverbank.

The suburban rail and major road alignments (Fig. 4.1) developed parallel to the river course because of the above-mentioned topographical factors, which further

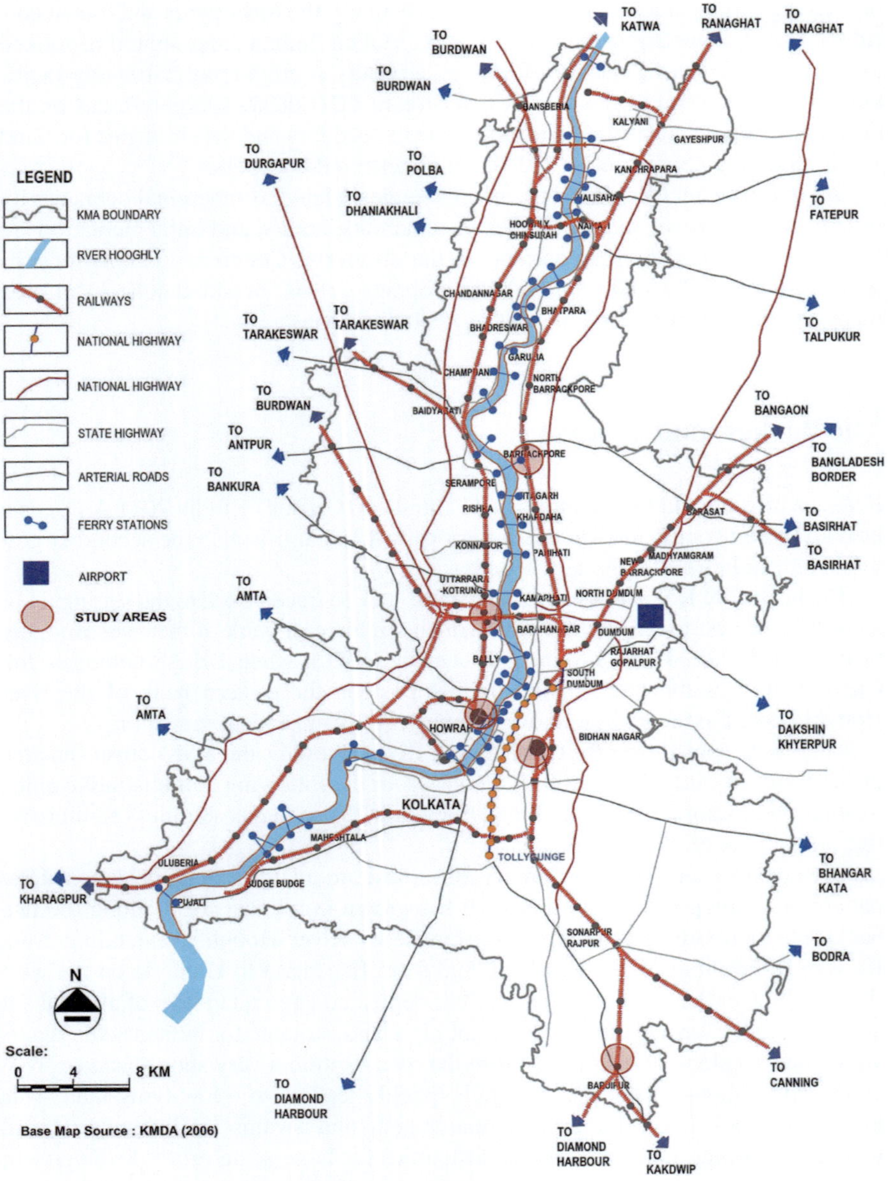

Fig. 4.1 Kolkata metropolitan area. (Data source: KMDA)

strengthened the linear pattern of the conurbation. There are 105 suburban railway passenger stations within the KMA placed at intervals of 2.5–5.5 km. The important railway passenger terminals (Sealdah, Howrah, Chitpur, and, under construction, Santragachi) are within Kolkata city, the metro core.

During the last four decades, several large-investment projects have been implemented as major inputs in the transport infrastructure of the KMA, such as construction of Circular Rail, Eastern Metropolitan Bypass, major bridges (viz. Iswar Gupta Setu, the Vidyasagar Setu, the Sister Nivedita Setu), several flyovers and grade separations, and large-scale parking facilities and introduction of organised ferry services across the river, several intra-city and intercity bus routes, and the Metro Rail alignments (probably, the most significant development) (Fig. 4.2).

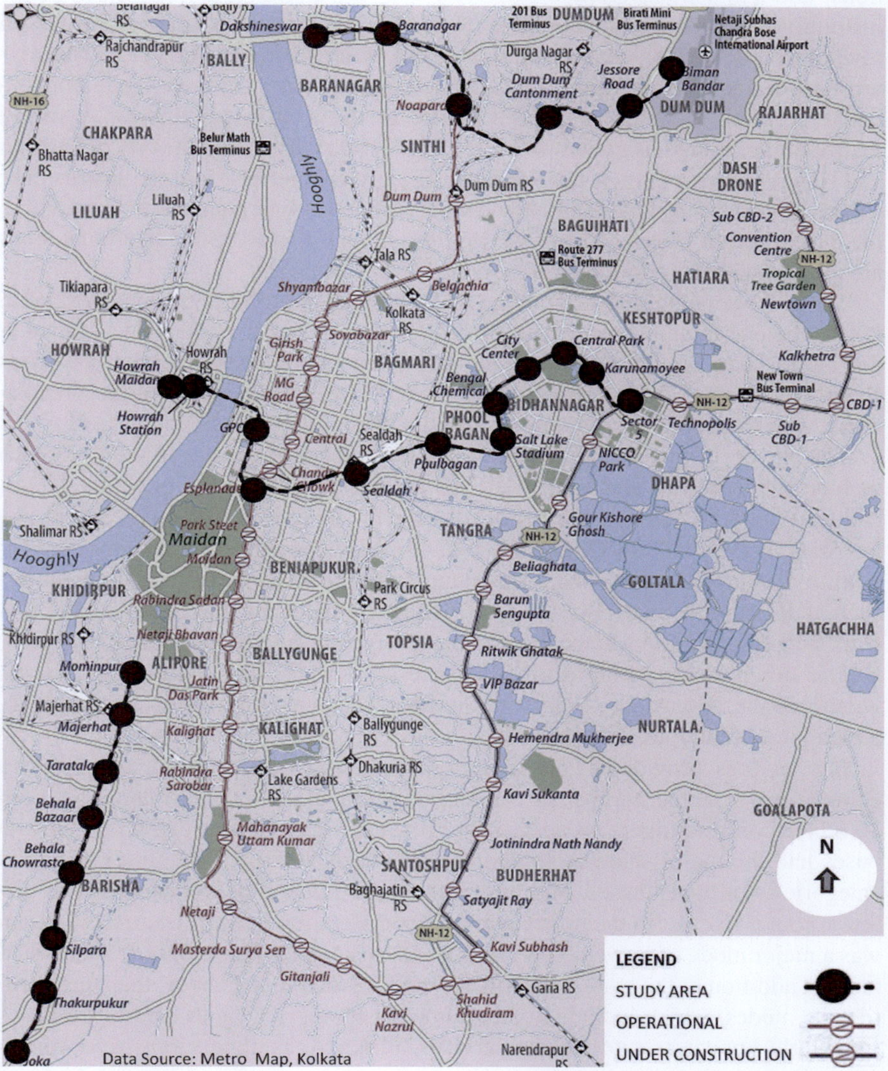

Fig. 4.2 Metro rail alignments in Kolkata

The generalised land use distribution in the KMA indicates that the residential use is predominant, comprising about 32% of the total land area. Other important uses like industrial, commercial, and institutional accounts for about 8%; organised open space in the form of park and recreational areas constitute about 1%; land use under transportation is about 7%; the rest of the land area (about 52%) is vacant land comprising cultivable land, forest and waste land, waterbodies, swamps, and marshes.

The *residential areas* follow mainly the suburban railway line (around the stations) and major transportation corridors. The residential use is often mixed with institutional, industrial, and commercial areas. The *major industries* are located in several well-defined patches near the river, rail, or the major arterial roads. The city-level *central commercial area* is located at the metro core and there are minor secondary centres at suburban areas to cater to the local requirements. A strong metropolitan centre has grown with a *central business district*, in the city of Kolkata and Howrah, which attracts huge number of daily commuters from the surrounding suburban areas. Salt Lake and Newtown-Rajarhat are two planned annexations to the city in recent times. Most of the *organised open spaces* and recreational areas are located within the city proper.

The gross population density in the KMA worked out to 8225 persons per km^2 in 2001. The maximum gross density of 24,718 persons per km^2 (2001) was found in the metro core, or the Kolkata Municipal Corporation Area. The density drops sharply as it moves from the metro core towards the rural fringes through suburban areas, with local peaks [high-density areas] at the secondary activity centres.

The passenger transportation in KMA is accomplished by several modes, which include suburban rail and metro rail, bus services, para-transits, private vehicles, and ferry services. A large number of passengers (about 3.39 million on an average weekday in 1997–1998 (KMDA 2004, p. 49) daily commute to the metro core for work by suburban rail service, originating from the nearby urban agglomerations such as Kalyani, Naihati, Barrackpore, Sodepur, Belgachia, and others on the east bank and Bandel, Chinsurah, Serampore, Bally, etc. on the west bank and other similar urban areas, within a time distance of 2.5–3 h, like Bardhaman, Katwa, Krishnanagar, Bongaon, Diamond Harbour, Canning, Kakdwip, Tarakeswar, Kharagpur, Medinipur, and such like.

Bus services, provided by public sector agencies as well as private owners, are a choice for a sizeable number of passenger trips in KMA. These catered to about 10.65 million passengers on an average weekday in 1997–1998, of which private buses had the major share at around 88% (KMDA 2004, p. 33). Buses in varied categories—ordinary buses, minibus, and air-conditioned deluxe bus—ply at different routes to cater to the travel demands of different groups of passengers. Tram was a major mode of movement in Kolkata in earlier days, but is almost abandoned now. In addition to the mass transport services, as mentioned above, there are para-transits modes, which include taxies, auto-rickshaws, rickshaws, and e-rickshaws (popularly known as toto). Ferry services, in KMA, facilitate trans-river movements in east–west direction. Presently there are thirty-six ferry routes, which are served by motorboats or launches (vessels). Private vehicles, which are increasing at an alarming rate since the last two decades in KMA, include cars and motorbikes.

A household survey—conducted by Kolkata Metropolitan Development Authority in 1997–1998 (KMDA 2004, p. 70)—reveals that on an average weekday about 22.6 million trips were generated within KMA. The work trips were found to be 53.70% and educational trips at 25.43% and the rest 20.87% were for other purposes like cultural trips, social trips, shopping trips, health trips, and such like. The trip length distribution of work trips indicated that about 95% of such trips were within a distance of 8 km; for educational trips, 84% were within 3 km of travel distance. Among the factors affecting the choice of mode, journey cost was the predominant one and about 76% of the total trips were guided by the out-of-pocket expenses associated with a trip.

For future development, Kolkata Metropolitan Development Authority (KMDA) envisages (KMDA 2006) a polycentric metropolitan structure for KMA, with the cities of Kolkata and Howrah as the nucleus and a number of hierarchical urban centres distributed in the area. Within the framework of the National Transportation Policy, the KMDA has formulated the Metropolitan Transportation Policy for KMA, which aims 'to develop an integrated transportation network within the KMA to cater to the movement of passengers and goods within the area and in the hinterland, taking into consideration the likely future scenarios up to 2025'. (KMDA 2006, p. 116).

Against this existing scenario of transportation network, trip characteristics, travel characteristics, and the envisaged future development of the KMA, the strategies of TOD—if implemented successfully—can take a significant role in guiding the future development of KMA in line with the NUTP 2014.

Issues and Challenges

The issues and challenges related to TOD in the KMA have been identified in the following sections based on the case studies taken up in three upcoming metro alignments (Fig. 4.2), namely, East-West Metro Corridor (Chatterjee 2018), Esplanade–Joka Metro Corridor (Sahoo 2016), and Dakshineswar–Barasat Metro Corridor (Paul 2012) and five major suburban railway stations (Fig. 4.1), namely, Howrah and Bally on the west bank and Sealdah, Barrackpore, and Baruipur on the east bank of the Hooghly river. Howrah and Sealdah are major suburban railway terminals in the KMA, while Bally, Barrackpore and Baruipur are important junction stations and transport nodes. The studies relied largely on secondary data; however limited primary surveys were undertaken whenever deemed necessary. Findings from the case studies are compared with cases in other cities based on secondary data.

Inadequacy in Public Transit Systems

Availability of a good quality public transit system (PTS) is a prerequisite for successful TOD; however, the supply of PTS in the country is grossly inadequate in terms of quantity as well as quality (Pucher et al. 2004). Very few Indian cities, currently, have organised, registered, and regulated PTS. The coverage of local commuter rail services is available only in the seven metropolitan cities in India of Mumbai, Delhi, Chennai, Kolkata, Bengaluru, Hyderabad, and Pune. Organised city bus services are available now in about 65 cities (NTDPC 2014)[1]. Leaving aside the recently introduced Metrorail/Light Rail Transits (LRTs) and a few BRTs, the overall condition of PTS in India is poor. It is plagued by absence/poor quality of infrastructure (shelter, stops, unclean vehicles, shared right of way), overcrowding leading to extremely uncomfortable and unsafe ride, low operational speed leading to delay, failure to connect key origins and destinations, lack of passenger information, failure to provide seamless travel, inconvenient locations of stops/stations, inadequate frequency, fear of crime in stations/buses, and other issues. The above issues are challenges for the KMA as well.

Further, PTS in the KMA is confronted with other serious issues at the end of the service provider, particularly the government organisations, such as lack of productivity, overstaffing, inefficiency, excessively high operating costs, and large subsidy needs.

While a growing attention and willingness in favour of PTS is noted in recent policies and mission programmes of the government (NUTP 2014), several challenges need to be dealt with both at planning and implementation stages for development of a good PTS in the KMA. Like other Indian cities, the major impediments in the KMA include typical factors such as lack of information and planning data, inadequate institutional and technical capacities, insufficient funding and financing, lack of political will at the local level, multiplicity of organisations[2] (Table 4.3) with inefficient coordination, geographical/physical limitations, and oppositions from key stakeholders, (operators, motorists), among other factors.

[1] At the Census 2011, there are 7935 towns in the country. Many of these towns are part of urban agglomerations (UAs) and the rest are independent towns. The total number of urban agglomerations/towns, which constitutes the urban frame, is 6166 in the country. Out of 468 UAs/towns belonging to Class I category, 53 UAs/towns each have a population of one million or above each (Source: Ministry of Home Affairs 2011)

[2] Responsibilities for planning, investment, operations, and management of PTS are divided into central, state, and local government organisations, which results in fragmented functional responsibilities, lack of local expertise, paucity of financial resources, and lack of private participation.

Table 4.3 Organisations involved in planning, operations, and management of PTS in India

Organisations	Functions	Relevant acts
Urban transport planning		
Ministry of Urban Development	Overall responsibility for urban transport policy and planning	
Land Development Authority, State Government	Land use allocation and planning	State Development Acts
Roads		
Transport Department, State Government	Licenses and controls all road vehicles, inspection of vehicles, fixing motor vehicles tax rates	Motor Vehicles Act 1988
Ministry of Surface Transport	Administers the Motor Vehicles Act and notifies the vehicle specifications as well as emission norms	Motor Vehicles Act 1988
State Transport Undertaking, State government	Operation of bus services	Road Transport Corporation Act 1950
Public Works Department, State Government	Construction and repair of state roads	VII schedule of the Indian Constitution (Article 246), List II (State List), Item 13
Urban local bodies	Construction and repair of roads, road signage, traffic lights, licensing and control of non-motorised vehicles, removal of encroachments, and land use planning	Constitution (Seventy-Fourth Amendment) Act 1992
Police	Enforcement of traffic laws and prosecuting violators	State Police Acts
Railways		
Ministry of Railways	Own and operate urban rail transit systems, wherever they exist	Railway Act 1989

Source: TERI (2015)

Declining Modal Share for PTS

Modal share [using public transport system] is recognised globally as an important indicator of successful TOD or a condition favourable for TOD (Hale 2014; Reene 2009). However, a sharp decline in the shares of public transport trip is noted in cities of all sizes in India (TERI 2015, p. 20), and so it is for KMA. While the National Commission on Urbanisation (1987) recommended that in cities with a population exceeding 5 million, the PTS share should be 80%, the scenario is not heartening when a study by RITES in 1994 is compared to a similar study by WSA in 2007 (Table 4.4).

The share of personalised trips, especially on two wheelers, have gone up in leaps and bounds, clocking 12% per annum increase in the past two decades, while public transport has generally dwindled (Fig. 4.3). There are reported instances of bus services, which have been pushed out of business in the last two decades.

Table 4.4 Trend of modal share of PTS in India

City category	City population range in million	Study by RITES, 1994	Study by WSA, 2007
1	Less than 0.5	14.9–22.7	0.0–15.6
2	0.5(+)–1.0	22.7–29.1	0.0–22.5
3	1.0 (+)–2.0	29.1–35.6	0.0–50.8
4	2.0 (+)–4.0	35.6–45.8	0.2–22.2
5	4.0 (+) – 8.0	45.8 – 59.7	11.2–32.1
6	More than 8.0	59.7–78.7	35.2–54.0

Note: The study by RITES included 21 cities and that by WSA included 30 cities. A number of cities selected In the latter study have no PTS and a high percentage of PT share (50.8%) is observed in Category 3, as Kochi, which falls in this category, is supplied with very good public transport.
Source: MOUD (2008), Traffic and Transportation Policies and Strategies in Urban Areas in India

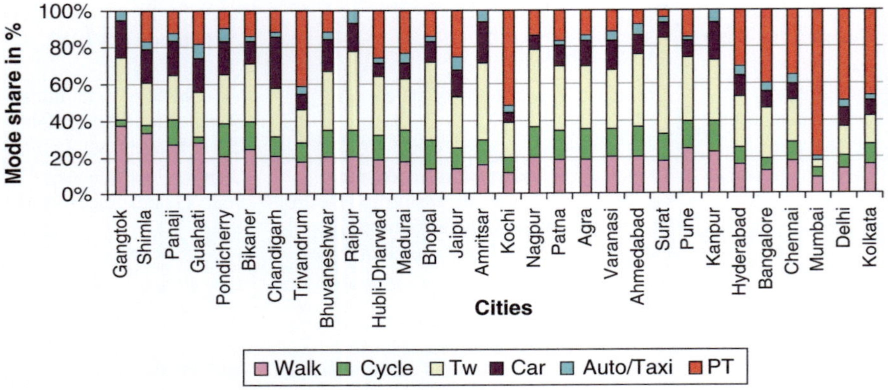

Fig. 4.3 Modal share in Indian cities. (Source: Study on Traffic and Transportation Policies and Strategies in Urban Areas in India—MOUD (2008))

This is further corroborated by the tremendous growth in the number of private vehicles (Fig. 4.4), particularly in the metro cities, while fleet sizes in nearly all public undertakings have declined rather than grown to meet the increasing travel demand.

The declining trend in the modal share of PTS may be attributed to the poor condition of the PTS on the supply side and several other reasons on the demand side, such as lack of awareness among people about the economic and environmental benefits of PTS at the individual as well as the societal level. Also, existing cultural and psychological values in favour of private vehicles (often seen as symbol of success and status), individual's pleasure in driving, increasing affordability of people, easy availability of finance for owning private vehicles, etc. (Steg 2003) have perpetuated the decline.

Therefore, preventing the shift of trips from public transport and non-motorised transport to individual motorised modes and reversing this present trend (that may happen only with development of good PTS) are two major challenges that need to be addressed for successful implementation of TOD in India.

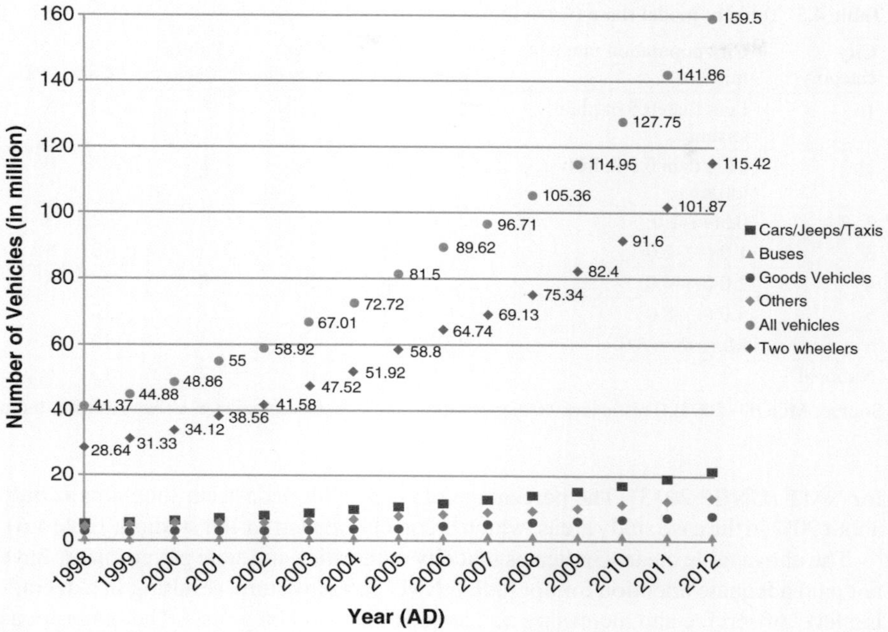

Fig. 4.4 Mode wise vehicular growth in India. (Data source: MOUD, Government of India)

Dwindling Non-motorised Transport and Poor Infrastructure

NMT modes, also known as 'Active Transportation', include walking, bicycling, and other variants, such as small-wheeled transport (push-scooters, skates, and hand carts) and wheelchair travel. Traditionally, in many Indian cities people who commute on cycles and/or simply walk outnumber (40–60% in 1980s, UNEP 2013) those who use private motorised transport. These modes are not dependent on fossil fuels and have minimal emissions. So, these are truly low-carbon modes.

Low-income households are dependent on these modes to access employment, education, and other essential services, even for commuting longer distance (Mohan and Tiwari 2000; UNEP 2013). Because of several environmental and health benefits, while the TOD principles advocates for use of NMTs for short-length trips and for connecting the 'last-mile', paradoxically the present trend indicates a drastic reduction (30% in 1994 to 11% in 2007) in NMT in urban areas, particularly the big ones with rising average income. The findings in KMA, too, clearly confirm the above trend. The modal shares for NMTs in 2007 in different categories of Indian cities are indicated in Table 4.5.

In the 1980s, use of non-motorised modes of transport in Indian cities was correlated with income levels. The use of NMT later (since early 1990s) declined with the increase in income levels (Replogle, 1992; UNEP 2013) and poor infrastructure

Table 4.5 Trend of modal share (%) in India—2007

City category	City population range in million	Walk	Cycle	Two wheeler	Public transport	Car	IPT
1a	Less than 0.5 in plain terrain	34	3	26	5	27	5
1b	Less than 0.5 in hilly terrain	57	1	6	8	28	0
2	0.5(+)–1.0	32	20	24	9	12	3
3	1.0 (+)–2.0	24	19	24	13	12	8
4	2.0 (+)–4.0	25	18	29	10	12	6
5	4.0 (+)–8.0	25	11	26	21	10	7
6	More than 8.0	22	8	9	44	10	7
National		28	11	16	27	13	6

Source: MOUD (2008), Traffic and Transportation Policies and Strategies in Urban Areas in India

for NMT (UNEP 2013). The percentage of roads with pedestrian footpaths is only about 20% in the case study areas, which is corroborated in earlier studies (Table 4.6).

The above table clearly indicates that city authorities and state governments have not paid adequate attention for upgrading NMT infrastructure, resulting in a degrading level of service and increasing risk to pedestrian and bicyclists. The major focus was on the provision of transport system to enable uninterrupted flow of motor vehicles. This approach has resulted in construction of foot-over bridges and subways, consequently increasing the distress of the vulnerable group—the pedestrian.

In view of the existing scenario, retaining the modal shares for walking and bicycling—encouraging them by choice—is a major issue in the successful implementation of TOD in the KMA, like in many other Indian cities.

Complete Street Design

Complete street are streets for everyone. These are designed to be safe for drivers, bicyclists, and other NMT users, transit vehicles and users, and pedestrians of all ages and abilities. The complete street concept focuses not only on individual roads but also on changing the decision-making and design process so that all users are routinely considered during the planning, designing, building, and operating of all roadways. It calls for change in policy and institution (Laplante and McCann, 2008). Complete Street Design (CSD) proffers an inclusive approach in terms of physical infrastructure and users and, thus, clearly articulates the concept of 'equitable allocation of road space'. It claims to create cultural vibrancy and social capital for neighbourhoods. Over the last three decades, the concept of CSD has been extensively adopted in smart growth initiatives across the globe, particularly in the developed countries and, thus, has become an essential and integrated element of TOD policy (Desai 2015).

Table 4.6 NMT infrastructure quality in Indian cities

City	Infrastructure quality		Source
	Pedestrians	Bicyclists	
1. Delhi, Vadodara, Jaipur, and Patna	Non-existent or poorly maintained footpaths. Pedestrians are therefore, directly conflicting with motorised traffic	Non-availability of specific infrastructure	Maunder and Fouracre (1989)
	Government authorities least concerned with low-income groups, who are highly dependent on NMT		
2.Pune		Increased use of MV resulting in increasing risk to bicyclists. Pune has been has been working to develop bicycle network	Replogle (1992)
3. Kanpur		City authorities experimented with yellow lane markings on certain main streets	Kuranami et al. (1995)
4. India	There is little provision of facilities for pedestrians and bicyclists. The existing transport infrastructure development programmes are designed for faster modes of transport claiming ROW by displacing NMT		World Bank (2002)
5. Chennai and Bangalore	Non-existent, broken down, and/or obstructed sidewalks; large height differences between sidewalks and frequent driveways/alleyways; danger at street crossings and distance between crosswalk locations; and flooding in monsoon seasons	Few exclusive-use lanes and on other roads bicyclists being pushed off from busy roads by motor vehicles	World Bank (2005)
6. Chandigarh	On some arterials, special pedestrian paths have been constructed	Constructed 160-km-wide cycle paths between 2001 and 2003	Pucher et al. (2004)
7. Bangalore	Most of the footpaths along the major arterial and sub-arterial roads need extensive repairs and upgradations		UNEP (2013)
	The footpaths lack continuity, with major portions, being in bad shape due to utility repairs and maintenance		

(continued)

Table 4.6 (continued)

City	Infrastructure quality		Source
	Pedestrians	Bicyclists	
8. Delhi	Presently, 40% of roads in the city do not have a pavement refuge, and those that do exist are often unusable, especially by the mobility impaired		Ghate and Sundar (2010)
	Zebra crossings are routinely ignored, and there is large spacing between light controlled crossings		
	More focus on providing hurdle-free mobility to MV by building foot-over bridges and subways		
9. Study areas in KMA	Presently, more than 40% of roads in the study areas do not any designated walkways. Existing ones are often unusable	Non-availability of specific infrastructure	Chatterjee (2018), Sahoo (2016) and Paul (2012)
	Constructions of sidewalks are under progress along some stretches, but the widths are often inadequate and issues of universal accessibility are frequently ignored. Encroachments of the footpaths by spilling commercial activities and street vendors are common		
	There is no comprehensive plan that would warrant complete and connected network for the pedestrians		

In India, CSD has been recognised as one of the key principles outlined in the NUTP as well as in the TOD policy. However, one of the key problems of majority of the streets in the KMA is that they are designed essentially for motorised vehicles, without taking into considerations the needs of all users. Frequently, the median is marked and a carriageway is constructed, and the rest of the ROW (right of way) is left undefined for other purposes. With inadequate provision on the one hand and weak enforcement of bylaws on the other, parking occupies a significant share of the road area. Consequently, pedestrians, trees, utilities, street vending, and social activities jostle for whatever space left. In many of the study areas, the left over space is too inadequate for accommodating the other essential elements of a complete street.

The traditional planning focus on improving mobility for private motor vehicles, by allocating more space, results in vanishing of the pedestrian walkway, which, in turn, forces people to walk on the carriageway. The same is true for cyclists, street vendors, and public transport. Thus, the resulting eventual street condition is inconvenient, uncomfortable, and unsafe for everyone.

Coordination with other utility services, such as water, electricity, drainage (storm water and sewage), optical fibre cables, gas lines, etc., which frequently depend on the ROW for their distribution (underground and/or overhead) is yet another challenge confronting CSD and its implementation in the KMA, particularly for existing roads. Undertaking a comprehensive street improvement/retrofitting work with CSD objective calls for taking into consideration not only vehicular, NMT, and pedestrian traffic but also multiple overlapping service providers, both public and private operators. This would demand extensive institutional coordination, which frequently is limited at present in the organisations responsible CSD planning and implementation. As a result, quite often, roads are intermittently and haphazardly dug up or filled in and remain under maintenance or repair and not suitable to be used by pedestrians at all. Another challenge is to find accurate documentation and historical data of previously laid utilities for the designers to work with, which is a significant impediment for mainstream CSD initiatives in the country.

Indian cities lack up-to-date and legally binding design specifications for urban streets. While available guidelines are fragmented and the central agency—the Indian Roads Congress (IRC)—recommends generic standards, city-specific design codes for CSD are absent. Other major challenges include lack of the stakeholders' awareness on the benefits of CSD and inadequate capacity for its implementation at the level of the local governments. In a scenario where the rate of NMT users are dwindling, it is natural that a major section of the population and people's representative do not recognise the benefits of CSD. Similarly, local technical hands and contactors in urban local bodies are frequently ignorant about the concept of CSD and operate with inadequate capacities.

Availability of Land and Land Acquisition

Being a relatively new policy design measure, TOD has not been a part of city development since its inception stage. The present development around the transit nodes are essentially guided by the prevalent speculations and real estate market forces, which, in absence of a strong enforcement, often flout the bylaws for profit maximisation. Thus, the present scenario, particularly, in the suburban towns in the KMA is often characterised by unplanned and haphazard growth with inadequate provisions for planned TOD projects and public spaces. The existing ROWs are too inadequate to accommodate a complete street and, thus, throw great challenges while planning connected pedestrian and NMT network with desired level of accessibility.

A strong will is needed for successful TOD projects and widening of existing ROWs—what's required is collaborative arrangements for land acquisition rather than the prevalent top-down acquisition and compensation formula, often resulting in delayed execution of public projects.

Parking Policy and Pricing

Inadequate land is always a constraint in a city like Kolkata. Thus, increasing use of private vehicles in the absence of a well-planned parking policy and pricing frequently results in skewed allocation of road space in favour of parking and/or encroachment of road space (Chatterjee 2018). Discourse reveals that parking policy is an important determinant of travel behaviours, regardless of proximity to transit (Hess 2001), and TOD parking supply and pricing policy seldom are structured to support transit ridership goals (Willson 2005). Where parking ratios for residents, shoppers, and commuters are positive and where private car use continues at former levels, it would be difficult to develop a sense of place and community essential for TOD to thrive. Researchers have called for minimum parking requirements and cashing out of parking subsidies (Shoup 2005; Willson 2000). However, in the Indian context, indiscriminate abandonment of the existing parking facilities and sharp hike in its pricing may not be appropriate until a good public transit system is developed. However, the challenge is to formulate short-term planning strategies for areas with high demands using smart technologies, such as multilevel and automated parking lots to ensure spaces for pedestrians and NMT users. At the same time, long-term policies should be formulated for parking and its pricing, which would discourage random use of private vehicles, such as Area Licensing Scheme (ALS) by Electronic Road Pricing (ERP) and Vehicle Quota Scheme (VQS). These and other measures have been implemented in Singapore, which demonstrates one of the successful models of smart development.

Gentrification

As reported in several discussions (Davis et al. 2012; Dutzik et al. 2014), gentrification would be a major challenge for TOD development in the KMA. It is clearly notable in the study areas that new developments along many transit nodes and corridors—shaped by speculation and the prevalent real estate market force—is resulting in emergence of 'gated communities', occupied by newly arrived rich class pushing out the poor, who are supposed to be the captive riders of public transportation. In addition, the newly arrived middle-income and high-income groups, who already own private motorised vehicle, would be reluctant to use public transits in the absence of adequate push factors in the form of high taxation and fuel prices. The issue of gentrification of this kind, if not addressed in the TOD policies, would force many habitants to overcrowded slums and squatters away from transit corridors completely defeating the low-carbon goals of TOD. This would ultimately lead to environmental, economic, social, and cultural unsustainability.

Amendment of Local Laws

The vision of TOD can only be materialised and translated into implementable actions if supported by town planning legislation of the state and the local building bylaws of the towns or cities. The absence of the required provisions—incorporating mixed use development in appropriate proportion, inclusion of standard walkways/bicycle paths, prohibition of on-street parking, green-space mandate, and enforcement of dilapidated building rules in the existing legislations and bylaws—is one of the primary challenges to implementing TOD at present. The problem is further aggravated by the fact that the amendment of these laws is a time-consuming affair, and their implementation is hindered by the lackadaisical attitude of the enforcers.

Carrying Capacity-Based TOD Policy

The present nature of development around the railways stations, transit corridors, and transportation nodes in KMA are notably heterogeneous (Fig. 4.5) on several parameters, such as population density, socio-economic profile, trip lengths, travel behaviour, vehicle ownership, availability of open spaces, ROWs, infrastructure provision, and such like. Therefore, a single set of standards and benchmarking is not appropriate for planning of TOD in the study area (Chatterjee 2018). The planning challenge is to estimate the right carrying capacity of the areas and, accordingly, formulate context-specific TOD goals.

Inadequate Institutional Capacity

Implementation of TOD would entail the involvement of various stakeholders and agencies for preparation of plans, reviewing the infrastructure and building regulations, making provisions for traffic management, adoption of creative financing model, and such like. Given the broad base of stakeholders, if the competing and conflicting interests are not suitably addressed, the success of TOD will be greatly undermined. Thus, there is a need for a robust regulatory framework with an autonomous body tasked with the responsibility of coordinating and implementing TOD.

Concluding Remarks

The authors feel that for the KMA TOD may prove to be successful in integrating land use and transport—triggering developments that would reduce the need for travel and where people would prioritise public transport and non-motorised

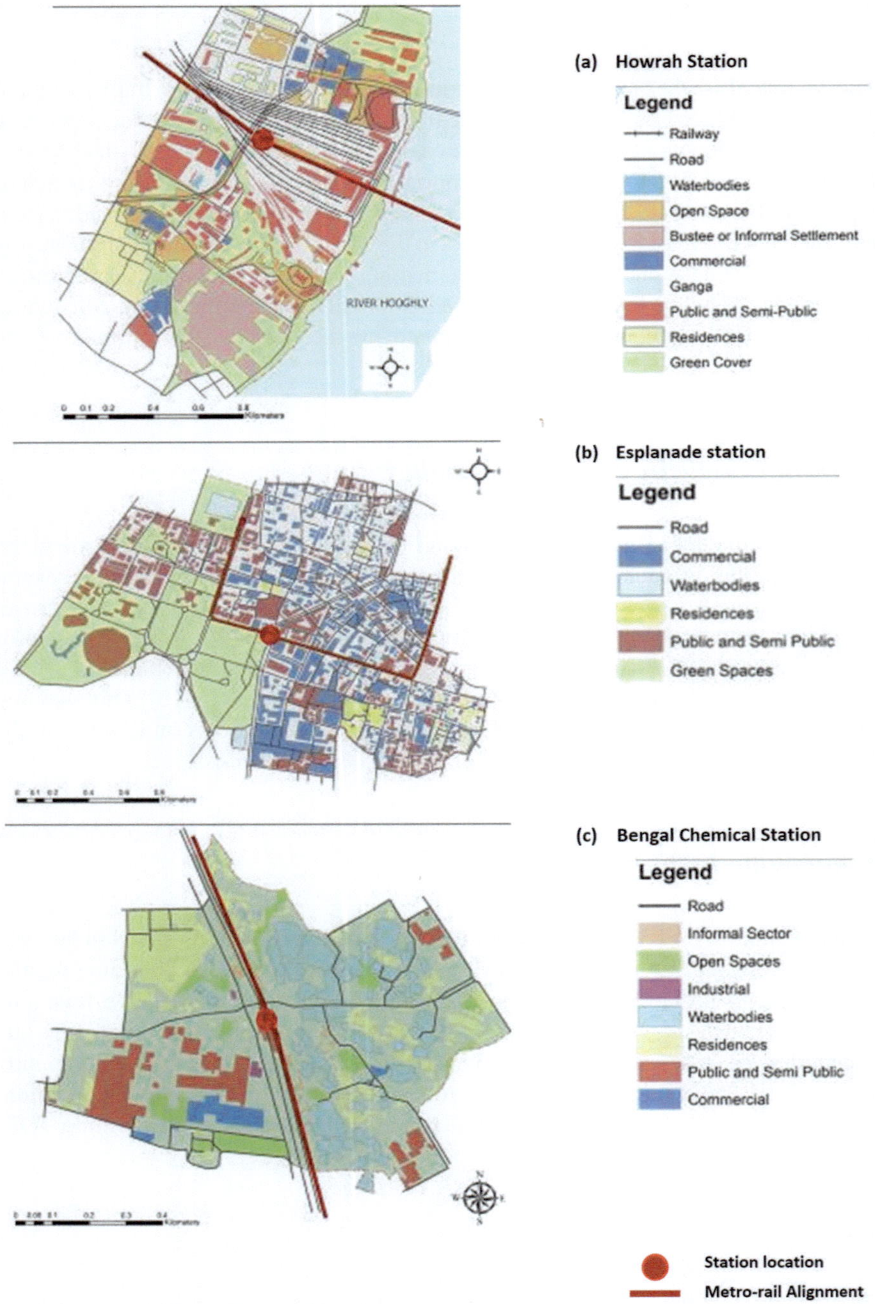

Fig. 4.5 Study of heterogeneity in development around (1-km radius) three metro rail stations along East-West Metro Corridor, Kolkata. (**a**) Howrah Station. (**b**) Esplanade Station. (**c**) Bengal Chemical Station. (Source: Abhinanda Chatterjee (Author) 2018)

transport when travel is necessary. However, it would call for resolving several issues and challenges, which are interrelated and complex in nature, as is the case for several urban areas in India.

The above discussions are meant to serve as a foundation for further work in addressing issues and challenges related to TOD in the Indian context in general and for the KMA in particular. It does not attempt to provide any solution, which needs to be responsive to the local needs and the context. The answer lies in systematic and concerted studies involving several disciplines, such as urban planning, transportation planning, sociology, economics, policy studies, urban administration, and management.

References

Austin M, Belzer D, Benedict A, Esling P, Hass P, Miknaitis G, Wampler E, Wood J, Young L, Zimbabwe S (2010) Performance-based transit-oriented development typology guidebook. Centre for Transit-Oriented Development, Berkeley

Belzer D, Autler G (2002) Transit oriented development: moving from rhetoric to reality, A discussion prepared for The Brookings Institution Center on Urban and Metropolitan Policy and the Great American Station Foundation, viewed 10 April 2019, https://www.brookings.edu/research/transit-oriented-development-moving-from-rhetoric-to-reality/

Bernick M, Cervero R (1997) Transit villages for the 21st century. McGraw-Hill, New York

Calthorpe P (1993) The next American metropolis: ecology, community, and the American dream, 3rd edn. Princeton Architectural Press, New York

Carlton I (2007) Histories of transit oriented development: perspectives on the development of the TOD concept, real estate and transit, urban and social movements, concept protagonist, working paper 2009-02. University of California, Berkeley

Centre for Urban Equity (CUE), CEPT University, Ahmedabad (2017) Transit-oriented development: lessons from Indian experiences, Working paper 36, viewed 10 April 2019, https://cept.ac.in/UserFiles/File/CUE/Working%20Papers/Revised%20New/36CUEWP-36_TOD%20Lessons%20from%20Indian%20Experiences.pdf

Cervero R, Duncan, M (2002) Benefits of proximity to rail on housing markets: experiences in Santa Clara County. J Public Transp 5(1):1–18. https://doi.org/10.5038/2375-0901.5.1.1

Cervero R, Murphy S, Ferrell C, Goguts N, Tsai Y-H, Arrington GB, Boroski J, Smith-Heimer J, Golem R, Peninger P, Nakajima E, Chui E, Dunphy R, Myers M, McKay S (2004) Transit-oriented development in the United States: experiences, challenges, and prospects, Transit Cooperative Research Program (TCRP) Report 102. Transportation Research Board, Washington

Chatterjee A (2018) Planning strategies for transit oriented development along MRTS corridor: case study east-west metro line, Kolkata, Masters thesis, Indian Institute of Engineering Science and Technology, Shibpur, Howrah, India

Davis B, Dutzik T, Baxandall P (2012) Transportation and the new generation: why young people are driving less and what it means for transportation policy, US PIRG Education Fund, viewed 10 April 2019, https://uspirg.org/sites/pirg/files/reports/Transportation%20%26%20the%20New%20Generation%20vUS_0.pdf

Desai M (2015) Reforming complete street: considering the street as place, Masters thesis, Graduate School of the University of Cincinnati, USA

Ditmar H, Ohland G (eds) (2004) The new transit town: best practices in transit-oriented development. Island Press, Washington, DC

Dutzik T, Inglis J, Baxandall P 2014. Millennials in motion: changing travel habits of young Americans and the implications for public policy, U.S. PIRG Education Fund Frontier Group, viewed 10 April, 2019, https://uspirg.org/sites/pirg/files/reports/Millennials%20in%20Motion%20USPIRG.pdf

Environmental Planning Collaborative (EPC) (2012) Station area planning: a guide to planning and implementing transit oriented development in Indian cities, viewed 10 April 2019, http://www.itdp.in/wp-content/uploads/2013/01/Station-area-panning.pdf

Ghate AT, Sundar S (2010) Section IV-Transport infrastructure: putting urban transport sector on a low energy and low carbon path: a focus on the passenger transport sector in million-plus cities, IDFC (ed), Oxford University Press, New Delhi, viewed 12 April, 2019 http://www.idfc.com/pdf/report/Chapter-16.pdf

Gwilliam K (2003) Urban transport in developing countries. Trans Rev 23(2):197–216

Hale C (2014) TOD versus TAD: the great debate resolved. Plan Pract Res 29(5):492–507. https://doi.org/10.1080/02697459.2012.749056

Hess D (2001) Effect of free parking on commuter mode choice: evidence from travel diary data, Working, paper 34, The Ralph & Goldy Lewis Center for Regional Policy Studies, viewed 8 March, 2018 http://city-council.cityofdavis.org/Media/Default/Documents/PDF/CDD/Downtown-Parking-Task-Force/Mtg7/Att-3%2D%2DHess%2D%2DEffect-of-Free-Parking-on-Commuter-Mode-Choice.pdf

Hobbs JJ (2009) World regional geography, 6th edn. Brooks/Cole, Belmont

Kolkata Metropolitan Development Authority (KMDA) (2004) Master plan for traffic and transportation in Kolkata Metropolitan Area: 2001–2025, Draft final report

Kolkata Metropolitan Development Authority (KMDA) (2006) Vision and perspective plan for Kolkata metropolitan area: 2025

Kuranami C, Winston BP, Bell DD (1995) Non-motorized vehicles in ten Asian cities: trends, issues, and policies, World Bank: Transportation, Water & Urban Development Department, Transport Division.

LaPlante J, McCann B (2008) Complete streets: we can get there from here. ITE J 78(5):24–28

Maunder DAC, Fouracre PR (1989) Non-motorized transport in third world countries. In Institute of British geographers Annual conference on managing our environment, 3–6 January 1989, Conventory Transport Research Laboratory

Metropolitan Transit Commission (MTC) (2005) Mobility for the next generation. MTC, Oakland, CA

Metropolitan Transit Commission (MTC) (2006) New places, new choices: transit-oriented development in the San Francisco Bay area. MTC, Oakland, CA

Ministry of Home Affairs (2011) http://censusindia.gov.in/2011census/censusdata2k11.aspx

Ministry of Urban Development (MOUD), Government of India (2014) National urban transport policy 2014, viewed 8 March, 2018, URL: www.iutindia.org

Mohan D, Tiwari G (2000) Mobility, environment and safety in megacities: dealing with a complex future. IATSS Res 24(1):33–46

Ministry of Urban Development (MOUD), Government of India (2008) Study on traffic and transportation policies and strategies in urban areas in India, viewed 8 March, 2018. http://mohua.gov.in/upload/uploadfiles/files/final_Report.pdf

National Transport Development Policy Committee (NTDPC), Government of India (2014) India transport report: moving India to 2032, vol. 3, part 2, pp. 382–471, viewed on 3 April 2019, http://planningcommission.nic.in/reports/genrep/NTDPC_Vol_03.pdf

NSW Government (2010) Metropolitan plan for Sydney 2036, viewed 5 April 2019, https://apo.org.au/system/files/93911/apo-nid93911-262151.pdf

Paul K (2012) Planning for land use and surface transport intervention along the proposed Dakhsineswar – Barasat Metro railway corridor, Masters thesis, Indian Institute of Engineering Science and Technology, Shibpur, Howrah India

Petkar AS, Hamand SS (2013) A study of transit oriented development in Indian context. In: Proceedings of international conference on advances in civil engineering. Elsevier, Pune

Pucher J, Korattyswaroopam N, Ittyerah N (2004) The crisis of public transport in India: overwhelming needs but limited resources. J Pub Trans 7(4):1–20
Queensland Department of Infrastructure and Planning (QDIP) (2005) South East Queensland Regional Plan and Program, Brisbane
Renne, J L 2009, 'From transit-adjacent to transit-oriented development', Local Environ, vol. 14, no. 1, pp. 1-15, DOI:https://doi.org/10.1080/13549830802522376
Replogle M (1992) Non-motorized vehicles in Asian cities, World Bank, Asia Technical Department Series (162)
Rizvi A (2013) Alternative approaches to economically sustainable mobility in India: comparing Ahmedabad bus rapid transit and Delhi metro systems. In: Global report on human settlements 2013. UN-Habitat, Nairobi
Sahoo C (2016) Strategies on land use and transportation around proposed Metro Station at Joka, Kolkata, Masters thesis, Indian Institute of Engineering Science and Technology, Shibpur, Howrah India
Shoup DC (2005) The high cost of free parking. American Planning Association, Chicago
State of Victoria (2008) The Victorian transport plan. State of Victoria, Melbourne
Steg L (2003) Can public transport compete with the private car? IATSS Res 27(2):27–35
Still T (2002, Winter) Transit-oriented development: reshaping America's metropolitan landscape, on common ground, 44–47, viewed 10 April 2019 https://www.nar.realtor/smart_growth.nsf/Pages/ocg_winter2002_transit_oriented_development
Suzuki H, Murakami J, Hong Y, Tamayose B (2015) Financing transit-oriented development with land values, adapting land value capture in developing countries, Word Bank Group, viewed 3 March 2019, http://openknowledge.worldbank.org/handle/10986/2174
The Energy and Resources Institute (TERI) (2015) Green growth and transport in India, New Delhi
Timothy DJ, Nyaupane GP (2009) Cultural heritage and tourism in the developing world. Routledge, New York
UNDP (2008) Human development report, viewed 9 October, 2012, http://hdr.undp.org/en/media/HDR_20072008_EN_Chapter2.pdf
UNEP (2013) NMT infrastructure in India: investment, policy and design. Magnum Custom Publishing, New Delhi
Washington Metropolitan Area Transit Authority (WMATA) (2013) WMATA joint development policies and guidelines. WMATA, Washington, DC. viewed 21 January, 2019 https://www.wmata.com/about/board/upload/Joint-Development-Policies-2013-07-25.pdf
Willson R (2000) Reading between the regulations: parking requirements, planners' perspective and transit. J Pub Trans 3(1):11–28
Willson R (2005) Parking policy for transit-oriented development: lessons for cities, transit agencies, and developers. J Pub Trans 8(5):79–94
World Bank (2002) India's transport sector: the challenges ahead, vol. 1: Main report
World Bank (2005) Towards a discussion of support to urban transport development in India, World Bank Energy & Infrastructure Unit South Asia Region
Yedla S (2015) Urban transport and the environment. Springer, New Delhi

Chapter 5
Transportation and Built Environment: Bus-Sense for Global South Based on a Case for Bringing Back Life in Bus Transport

Yogesh Dandekar

Abstract Cities in India are growing at unprecedented rate. The migration from the rural areas is creating pressure on the built environment and its systems. Cities are complex built environment which survive solely on many complex systems which keep it working. It's similar to our body, organs, and systems. One of these systems is the urban transport system. The Indian government is doing its best to build fly-overs, more roads, and metro rail systems, promoting smart infrastructure and creating policies. Urban transport is one of the highest capital intense investments. The developing world is visibly concerned of the financial rate of return and economic rate of return to ensure these infrastructure investments are not a burden on the country and the citizens. However, the current paradigm of measuring the livability index and quality of life is gaining more attention. Parameters for evaluating the quality of built environment are becoming center of discussions across forums.

The author brings attention to the impact on the quality of life for an individual who uses the good old, low tech yet agile bus transport for his daily commute while metro is being built in the city. Daily commute is the largest component of the trips in cities. Bus systems existed in many cities years and many decades before the focus shifted to building metro rail systems. These bus systems are quite flexible, and their dexterity allows it to perfectly respond to the organic nature of the urban development and is less capital intense. This article presents observations based on the daily commute of a person over a year. More than the quantitative data from the persons notes, the qualitative aspects are important. Reduced stress and opportunity to positively utilize this commute time is helping the author make his proposition. The author proposes that high capacity rail systems are inevitable for future, but the existing bus systems should not be neglected. In fact, to realize the vision of a multimodal integrated transport for the cities, attention and investments are required to establish good governance and robust systems, harness capacity of information technology, and design better user experience for these bus systems.

Y. Dandekar (✉)
sargdesignstudio, Pune, India
e-mail: yogesh@sarg.co.in

© Springer Nature Switzerland AG 2020

M. Ghosh (ed.), *Perception, Design and Ecology of the Built Environment*,
Springer Geography, https://doi.org/10.1007/978-3-030-25879-5_5

Keywords Bus transport · Urban transport in India · Commuting · PMPML · Pune · Livability standards of cities

Abbreviations

BRTS	Bus rapid transit system
MTR	Mass Transit Railway (MTR; Chinese: 港鐵; Cantonese Yale: Góngtit) is a **major** public transport network serving Hong Kong
NMT	Non-motorised Transport
PMPML	Pune Mahanagar Parivahan Mahamandal Limited
RATP	Régie Autonome des Transports Parisiens, is a state-owned public transport operator and maintainer headquartered in Paris, France
SMRT	Singapore Mass Rapid Transit
SPV	Special purpose vehicle
TOD	Transit oriented development

Built Environments and Its Lifelines

The human race evolved from foragers to an agricultural civilization and the thus sowed the seeds of urbanization. Homo sapiens have since come a long way. Frugally built huts, just for taking shelter and protection, have evolved into megacities of built spaces with complex structures and unimaginable network of systems. These systems of the built environments have now become essential to the survival of urban settlements. Increasingly getting smarter, faster, and adaptive, their performance drives the quality of life in these cities. Urban mobility is one of such crucial systems, the others being the electrical grid, communication, water supply, etc. Cities came first and the mobility evolved later. However, the future is turning into a reversal of the roles. Cities are being designed with mobility systems at the core.

Transit-oriented development (TOD) is now the latest buzz word across all the cities in India. Indian cities are growing at an alarming rate. Being one of the most populated countries in the world, Indians might find it difficult to handle this growth trajectory by the existing paradigms of planning, urban systems, and governance of the Indian cities. *In fact, the urban population in India is projected to reach 875 million by 2050 and comprises 58% of the total population of the country. Therefore, our cities need to be prepared to accommodate the large increase that is expected in their population, in order to survive, let alone thrive* (Institute of Urban Transport (India) 2014).

Urban transport in Indian cities needs a lot of work. The work on improving the mobility is happening across the Indian cities, and diverse approaches are seen with array of solutions to improve the living standards in a city. Government is playing a larger role and bringing in investment in infrastructure. The Indian government with an ambitious program is promoting the building of expensive rail-based mass transport systems (the Metros), and the Smart City Missions of all cities have sustainable mobility as one of the key focus areas. While the metro is definitely a good system considering the quantum of people to be moved efficiently, it is only a part of the solution. The existing bus system in many cities which have been serving the city much before this shift of focus happened is unfortunately losing out and being neglected. They are neither getting due attention for investments nor being looked at as a viable solution.

The bus systems currently work with an outdated management style and service methodology and not being able to keep up with the growing pressures of the expanding and congesting cities. They on their own are not able to initiate reforms. There are many like-minded people and organizations holding the torch and making themselves heard at multiple forums, advocating the building of sustainable cities through adapting NMT, making walkable cities and promoting the use of urban transport and revive the bus systems. It is a proven fact that the bus system would be very effective while using up only a fraction of the investment of a metro system. Several cities are seeing development of BRTS, but it is time to see these bus systems with new reformed structures and organized framework of service delivery.

Author of this article promotes designing and building these transport systems, keeping the experience of the customers/passengers at the core. It is important to know how people travel, but it is more important to know **who travels and why**. This leads to many more revelations to understand the users and design a system that responds to them. The real potential of the bus system in the city is going unnoticed. Pune is a classic case.

Pune is only one of the many Indian cities which are facing acute congestion, very low speed of moving vehicle, and unsafe roads. Like India, many other cities in Africa, Latin America, and the developing countries in Asia are facing similar situations. These countries are now being identified as Global South. Cities like Lima, Cairo, Accra, and Dhaka and cities in Pakistan are facing similar problems. In a presentation O'Neill (2010, p. 19) in his role as Lead Infrastructure Specialist at the World Bank has precisely pointed out the key shortcomings in resource at the institutions managing these transport systems. These are quite evident in the case of bus system and generally the urban transport in Pune. He quotes.

"Lack of essential technical skills in urban transport planning, management and operations. Lack of experience with market-driven rather than supply driven instruments/tools. Fragmented institutional arrangement without effective coordination. No or poor regulation of public transport services. Little money/funding, but drained quickly by expensive 'glamour' projects with no or little left for other more cost-effective improvements. There is no quick fix for urban transport problems, but sustainable development options are available. The urban transport issues cannot be just solved by Mega projects but need structural and institutional intervention across each touch point of urban mobility. Be it walkable streets, promote NMT, strengthen the bus system to support the High capacity rapid transit corridor and integrate across the 3 pillars: Information, Physical access and Fare collection."

Pune, Maharashtra, India

Pune is the second largest city in state of Maharashtra after Mumbai. It is the *ninth most populous city in India* according to the census of 2011 (Government of India, Census 2011). *Pune is ranked the number one city in India in the ease of living ranking index* (Times of India 2018).

This article is not just for those advocating better design, operations, and promotions of such bus systems but also for those who use buses for their daily commute. The impact created by each bus passenger gets ignored for the lack of collective representation of their conscious or causal usage.

Delhi Metro: The Torch Bearer

The Metros in India started with a paradigm shift to establish strong structure of planning, designing, building, and operating a metro system. This resulted in setting a well-defined process to deliver a complete system. The metro corporations set up as SPVs enjoy autonomy in their actions and having no predecessors in India were given a status of planning authorities. A good project management structure was laid down by Delhi Metro which has helped them to create their own island and reduce dependency on city infrastructure or administration. This has given them headway and attract investments. Delhi Metro paved the way and set the ball rolling. There are more than 15 cities in India which have been sanctioned with a metro rail project, 8–9 are already under construction. India is spending a lot of money. It may be effective but still begs the question should all the spending be done only on rail-based transport? It is an open secret that the Metros would perpetually seek subsidies from the government and have to work hard to barely meet their operational expenditures, leave alone the return on investment. *Taking note of substantial social, economic, and environmental gains of metro projects, the metro rail policy approved in 2017, stipulated a shift from the present "Financial Internal Rate of Return of 8%" to "Economic Internal Rate of Return of 14%" for approving metro projects, in line with global practices* (Ministry of Urban Development 2017). The larger question still remains, while we spend so much money, how efficient is it given the Metros' limited reach and inability to single-handedly solve the problem. The same policy is also stipulating the urban transport networks becoming more integrated and multimodal.

The Smart cities Mission of Government of India has arrived at a structure to measure livability standards of a city. This means that the focus is shifting to the quality of life than calculating the Economic Internal rate of Return. A document on livability standards in a city prescribes a total of 79 Indicators (57 Core Indicators and 22 Supporting Indicators). While the Core Indicators are considered an essential measure of liveability of cities, the Supporting Indicators supplement the Core

Indicators by adding value to them. Weights will be assigned to Category Indexes during the calculation of the City Liveability Index, depending upon the pillar of comprehensive development. "Physical" pillar has been given the highest weightage, followed by "institutional" and "social" pillar. The weights also recognize the extent to which City Governments can actively make improvements in the indicators. It is a clear indicator that "economic" pillar cannot be influenced by the actions of City Governments alone and has been assigned the lowest weight. No wonder that this document gives a weightage of 12 for the transport and mobility which is 30% more than the next category of power supply (Ministry of Urban Development, Government of India 2017).

This article presents Pune's bus system as a case study and wishes to underscore that it is high time PMPML (*Pune Mahanagar Parivahan Mahamandal Limited*) gets the required attention from the government. PMPML if revamped at the institutional level integrated seamlessly as a part of multimodal network can work wonders for Pune. This is also true for bus systems across other Indian cities. Strengthening the existing bus transport makes a good proposition at a lower investment, and many Indian cities can benefit.

This is the author's attempt to demonstrate the definitive impacts on quality of life by using a Bus for your daily commute.

A City with More Vehicles Than People

How sensible is using the Bus for daily commute? In the context of Pune, this article highlights an unpleasant truth. Pune is a city where the vehicles have outnumbered the population. While the population stands at approximately 3.5 million (35 lakh), the number of registered vehicles for the area is now 3.62 million (36.2 lakh) under the Regional Traffic Office MH-12 (Business-standard 2018).

We don't need an expert to decode the future of congestion and the desperate need for modal shift from personal transport to public transport.

The author is recounting the experience of a regular citizen and helping to corroborate it with indicative data (Table 5.1). All this data is based on the information collected from the person whose journey has been described. The data and figures have not been decisively authenticated with any sources as these are from the traveller's personal notes and it does not negate the possibility of having some discrepancies. The aim of this article is not to present a quantitative analysis but a more qualitative one to demonstrate the benefits of using a public transport. It is an attempt to bring back the attention of the government on the bus systems so that they get their due credit, receive investment for development, and positively impact the livability index.

Table 5.1 Comparative chart

Days travelled	**127**						
Distance travelled	**5588** km						
	Commuting time (everyday)	Total spend	Amount spend per person	Diesel/petrol consumed per person (Lit)	Cost of diesel/ petrol per person	Stress level	Quality of travel time
Bus (shared by 25 people)	**135 min**	Rs. 8370 (ticket purchase of 1 person)	1.5 Rs/km	74	4440	Negligible barring uncertainity which is due to service inefficiencies	11.8 days made productive
Car (shared by 4 people)	100 min	Rs. 46,566 (fuel + car ownership)	2.08 Rs/km	77.5	4650	Very high for driver. He gets company, but there are many preconditions to arrive at effective pool rides	8.82 days lost by driver. Co-passenger gets good comfort, but this condition is heavily dependent on many preconditions
Ride sharing (shared by 3 people)	120 min	Rs. 19,050 (ride fare for 1 person)	3.4 Rs/km	103.5	6210	Very less but uncertainty of time due to unplanned pickups and drops	10.58 days comfortable ride. The key value proposition for Uber and Ola
Two wheeler	80 min	Rs. 6916 (fuel + bike ownership)	1.23 Rs/km	62.08	4656	Very high and very high risk	7.05 days spent riding on a small seat stressful ergonomic posture and risk

References
1. Average cost of fuels in duration of making these observations petrol Rs. 75 per lit. Diesel Rs.60 per lit
https://www.iocl.com/Product_PreviousPrice/DieselPreviousPrice.aspx
https://www.iocl.com/Product_PreviousPrice/PetrolPreviousPrice.aspx
2. Bus's fuel efficiency is of 3 km/lit and 25 people share bus at any given point of time. The bus ownership cost is ignored as it is a capital expenditure shared across entire city
3. Cost of car 8.0 L, average life 10 years. Cost of ownership per day Rs. 220, fuel efficiency 18 km/lit
4. Cost of bike 60 K, average life 10 years. Cost of ownership per day Rs. 17.8, fuel efficiency 90 km/lit
5. Lowest offer rates for ride share by Ola and Uber for 22 km on my route – Rs. 75 one way. This is bare minimum

The figures given here are based on the personal experiences and methods of data collection and observations made during the time of study by the traveller. There may be some discrepancies/assumptions in the data the author is open to discuss on any queries. The important aspect is the qualitative use of time during commute

135 Minutes for 127 Days

Story of 127 Days of Commuting for Work by Bus in the City of Pune, Maharashtra, India

Our traveller is a very well-placed senior manager in a multinational Indian company with a stable job. He is a well-travelled, educated, and respected professional in his domain of work. A family man in his mid-forties, he distributes his time between his official commitments and family responsibilities. He owns an apartment, a car, and a two-wheeler but is an avid cyclist. He is a conscious citizen who strives to lead a lifestyle to maintain a low carbon footprint. Such a lifestyle is a mental tussle between the societal pressure of acquiring the next gadget on the market and the latest car versus slowing down the pace of life in a more meaningful manner. This story is about his experience of using the bus for his daily commute as a conscious well-thought-out decision by keeping his car at home. He does not drive to work, instead relies on the city buses managed and operated by PMPML (*Pune Mahanagar Parivahan Mahamandal Limited*).

This is his experience and a unique travelogue of commuting by bus for 127 days out of 483 days between 14 February 2016 and 12 June 2017. Other days include holidays, business tours out of Pune, and some days when he commutes with his car. On some days, while he used the bus but did not buy a daily pass, he instead purchased tickets on shorter rides for going to meetings at other places than office. These commutes have not been documented by him as they were minimal. The cover image of the article shows his archival of the 127 daily bus passes (Fig. 5.1).

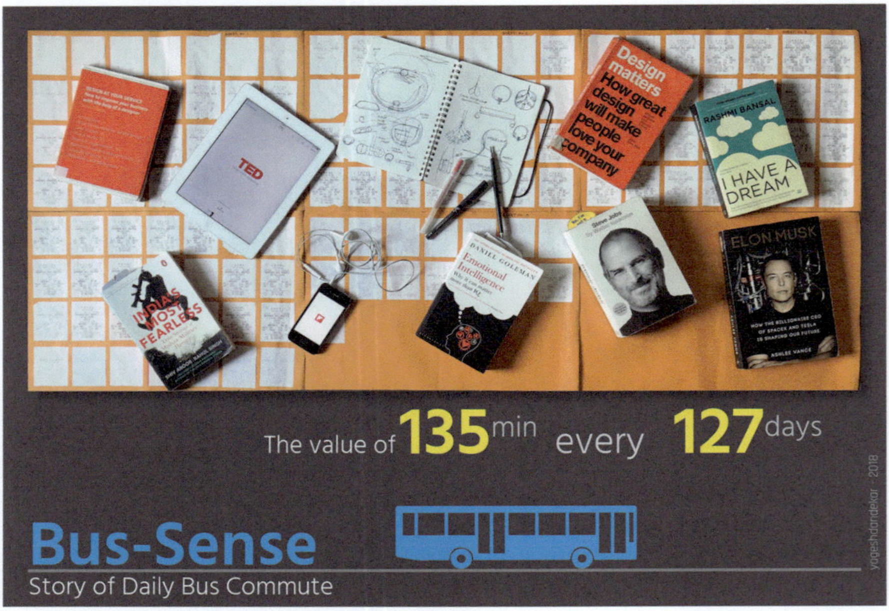

Fig. 5.1 Story-of-bus commute

The Beginning of the Trip

The Decision to Use a Bus for Commute

At 8:05 am, he leaves home for his office which is 22 km away and starts his day at 9:00 am at office. If he starts at 8:05 am by car, it takes about 35–45 min in the city traffic. Using a bus takes an hour. He feels that it is ridiculous to use a two-wheeler for commuting this distance daily, even though it does make perfect economic sense and may help him complete many chores and access many places on the way and back from office. He thinks the stress of riding a two-wheeler on busy roads is not worth the benefits of running errands on daily commute. Pune is known for extensive use of two-wheelers which are proven to be a high-risk mode of transport (Institute of Urban Transport (India) 2014, p. 12). Using a car for a single person is no doubt expensive, stressful, and impractical, even though it feels safe and his company may have helped him purchase a car and may even have compensated for the fuel expenses. He chooses not to ride a two-wheeler or drive for other reasons too. In fact, he doesn't use a two-wheeler at all. All small errands within a radius of 4–5 km from home are completed on cycle. Weekend travel with the family is either mostly by car or at times by bus.

The bus stop nearest to his home is less than a minute's walk. It was also one of the crucial reasons for selecting his apartment – staying close to a bus route. It makes perfect sense to stay within a kilometer of a bus stop or a mass rapid transit access point. We are talking of a transport oriented life style (TOLS) here. Many people don't realize that they are within this range from a mass transit corridor. They have never even noticed the nearest bus stop or thought of taking the bus for their daily commute. People in Pune are almost habituated to use a two-wheeler in order to avoid walking. The use of two-wheelers is so deep rooted in the social psyche that, for every family, it becomes an obligation to buy a two-wheeler for every teenager in the house as he starts going to junior college. Parents being unaware of the advantages of the urban transport and its accessibility are turning their blind eye towards it. The unpopular unattractive image of bus system is also adding to the reduced usage and neglect by the younger generation. There are many social and perceptional barriers for using urban bus transport in Pune, and they are brushed away under the excuse of inconvenience, uncertainty, inefficiency, and unpunctuality.

Agreed, that using the PMPML service could be an inconvenient and unpleasant experience as a user. The systems' management is equally responsible for its neglect in not taking decisive steps in making it user-friendly. But shouldn't using the public transport be a priority for every citizen if it is going to create a better city? This topic if discussed further will delve into people's values behind making choices. Across all age groups and genders, and in fact across countries, there is an increased awareness of staying healthy and concentrate on one's fitness. Getting up early in the morning for exercise needs you to change your lifestyle and undergo inconvenience or discomfort. Since exercise is important for a healthy lifestyle, people are ready to face these discomforts. But how many do it without a warning?

Increasing use of urban transport seems like a sensible decision for a better city life – reducing carbon emission, congestion, and pollution. But, how many are ready to face the inconvenience and discomfort for the larger good of the city and think beyond self? The more the users, the better business sense it will make to improve the service. The conscious decision by our friend is commendable considering that he is putting city's interest before his own convenience. A deeper understanding of such a behavior is required to help design and plan the urban transport inspiring a modal shift from one's personal vehicle to the urban transport. It is not an unidimensional problem with a single solution. It needs multiple solutions from diverse directions to culminate into a single goal. How can people in a city be thinking of shared urban transport whenever there is a need for travelling within the city?

The average waiting time in the morning for the bus is about 10–15 min. It is that time of the day when the buses on his route change their frequency from 20 min to 10 min for the peak hour. A bus network needs to be understood well. The planning and routes definitely have a structure and designed to respond to the way city behaves all through the day.

For him, bus travel works best as he buys a daily all-route pass. It was Rs. 50/– ($ 0.72) during the promotional period but was later increased to Rs. 70/– ($ 1.0) The advantage is that one has to pay only once for all the bus rides within the city limits for the entire day. His regular one-way ticket is Rs. 40/– ($ 0.58), and, thus, the daily pass makes perfect sense, saving him Rs. 10/– ($ 0.15) daily. The daily pass brings in another great advantage. It makes it possible to hop on and hop off buses on the way at interchanges, thus cutting down waiting times. It helps him to move faster in smaller steps to his destination. To him, cost advantage is less important than the ability to move faster. Carrying an all-route pass, it is possible to get things done across other places by taking detours. This also works efficiently and seamlessly. He is waiting for the time when we would be able to carry a wallet card and pay electronically for tickets like many cities across world. Till that happens in Pune, the good old thermal paper daily passes are good enough to get the work done. He has stored these daily bus passes of the 127 journeys to prepare this documentation. These 127 passes which otherwise would have met their fate in a trash bin are helping him to substantiate his findings and help reveal the softer and deeper advantages of the bus system. Although he travels only 44 km, the pass has the potential to help one travel anywhere in the city. If one travels a minimum of 30–35 km daily, pass cost is recovered.

The Interchange

Direct Bus Is a Myth in Todays Connected and Sprawled Urban Cities

The bus services started as point-to-point service in the old days. They connected pockets of development with prominent areas in the cities like market place, railway stations, bus stands, and government offices. En route they connected many places, but the starting and ending points of the buses remained prominent points to remember the buses. People visualized the buses as a point-to-point connection rather than lines along the route. If we ask people for their reason to avoid using the bus, the most common answer was – "We don't have a direct bus." The communication and representation of the bus system is not able to change this image and understanding of the bus system.

There is no direct bus to his office, and hence, his story fascinates many in Pune who consider his venturing into a bus, daily, as a great achievement.

A "direct bus" is a myth leading to reduced usage of bus transport. People will use the bus only if there is a direct service and with a decent frequency. They are not to be blamed, because a change in bus means delay, increased uncertainty, and increase in expense. Compounded by this perception are the absolutely cluttered and disorganized interchanges. Neither have people understood the concept of interchange nor is it considered important by bus operators and development authorities. Interchanges can in fact be converted into big revenue drivers, and this aspect has never been considered in Pune. In Pune, providing infrastructure for bus transport is the responsibility of the city administration – the municipal corporation. Their actions demonstrate the lack of understanding about how a bus service should be operated. They believe that marking some space on the footpaths and installing a contraption of steel with loud advertising represent the bus stop. Mostly, the design is left to the fabricators and agencies who are more interested in the advertisement revenue rather than the convenience of the bus passengers. Ideally, bus stops are the entry points into a highly effective and versatile mobility system for everyone across the city. They have the power to connect the city which no other shared system does. But, instead of welcoming the passengers, they discourage them. They are more for selling shampoos, biscuits, undergarments, cell phones, real estate, loans, jewelry, and even cars and two-wheelers, rather than representing the bus system. A transport network as is evident in its name is a network of lines and routes connecting points across the city and intersecting on the way. These intersections are the points which work as interchanges for people to switch routes and change directions. Metros do give considerable emphasis on these interchanges because they have very few and probably have an advantage of quantum of people passing through.

Our friend has to change his bus at the municipal corporation bus stand which is one of the biggest in Pune city. It is completely chaotic for a first-time visitor or those who wish to find a new connecting bus other than their regular daily commute. The corporation bus stand probably has many times the footfall of Pune Airport but receives the least attention for making it a better and more efficient place.

The plight of most bus stops in a city is more or less the same: an absolutely unsafe pedestrian environment, no segregation of spaces for buses and the regular traffic, undeniably illegible signs with no information on network, and no touch point for guidance on the next bus. The reason this place works is due to all the hard work of the PMPML staff. Drivers avoid the crisscrossing pedestrians and traffic to keep things moving, bus conductors herd passengers to the right bus, and controllers keep time, tracking the bus trips and doubling up as personified timetables to help passengers know the correct stop, bus route number, and its time. They are the real backbone of the bus service, and their efforts are truly admirable.

The Last Mile

The Journey Ends But the Day Starts Fresh

The bus stop near the office is on the BRT line that makes the last 4.5 km of his total journey. The travel time for this stretch, particularly in the evening, reduced from 40 min to 20 min when buses started using the dedicated lane. Whenever he drives to office, he envies those travelling by buses that zoom past him even as he waits many extra minutes for the traffic signal to turn green, only to be stuck again at the next signal. When the tables are turned and he is on the bus, he is thrilled and feels privileged to be able to breeze through the traffic.

Although he completely understands why the general road users oppose the BRT, it is actually a boon for this part of Pune. It would have been more effective if the bus network effectively connected the areas beyond 1 km on both sides of the BRT route. The feeder buses should ideally work as ring routes, picking up passengers right from the BRT stops. Currently, there are no feeder services. His office is a 600-m walk from the BRTS stop. Getting off at the BRTS stop and using the underpass is a very safe walk, to and from the office. This walk is largely a good experience barring the littered and sometimes dark subway and the hurdles of numerous two-wheelers and *chaiwalas* (street tea sellers) with a throng of customers and littering on the footpath. It is truly painful to see many people jumping the BRT barricades and endangering their lives on road just to avoid the subway. They should take a moment to consider why the government is spending money on subways. The integrated seamless transport and developments favoring the urban transport also have to make more walkable streets. This is another key component in the promotion of urban transport. The first barrier in reaching the nearest mass transit entry point is the obstacle-laden footpath, which discourages a person from walking on the streets and kills the idea at the very beginning. The unpaved footpath, street vendors, parked vehicles, and encroachments make it difficult to walk. Something substantial has to be done for reclaiming footpath from encroachments and designing them better.

He definitely vouches for the fact that – while commuting by bus is tiresome considering how crowded they are and how he has to walk more and it is hard at time to get a seat, it definitely is more physical stress to start ones day. But the mind is not stressed.

While the journey in the morning takes about 60 min, the return journey takes anywhere between 75 and 90 min. The stress of driving through the traffic snarls combined with the tiresome day at work adds to the road rage and chaos. But in the bus, it's very comfortable. Although returning from work does take 15–30 min more, one is disconnected and shielded from the road rage and stress of driving.

Additionally, one can probably finish a few pages of a book or read about the latest technology and catch up with the latest news of the day.

A total of 135 min on an average are spent everyday in commuting to office by the bus. The qualitative merits of this time spent in the travel add a positive dimension to the story of his commute.

While one may argue that he is probably spending less time with his family as he prefers not to compromise on productive time in the office hours, he starts earlier and reaches later than others who commute in their cars or bikes. These are the 135 min of his day that he would have otherwise spent holding the steering wheel or balancing his bike breathlessly working brakes, accelerator, and clutch with a dollops of honking. It can be argued that prima facie, bike riders would spend less money and time, but it is a very high-risk proposition. They should probably pay more for their insurance premium. The advantages of spending less money on bus commutes might well look like a side effect, if the real qualitative value of the time spent in the two modes is compared.

While his daily commute is not very unique, many do so everyday. He probably represents lakhs of people who have chosen bus transport as their primary mode for daily commute over a personal mode of transport. Some would have opted out of compulsion, and others are waiting to own a personal vehicle. But there are a few who use it as a conscious decision and must be commended for their choice. Whatever the reason may be, they all would have similar stories to tell. It would make sense to document such stories in support of the bus system and in the larger good of Pune city to create an efficient network as an economically viable solution.

Table 5.1 in this article gives a comparative analysis on data gathered during his commute to work. The data compares the use of bus, carpooling (rather rare due to practical considerations), shared mobility, and two-wheelers.

Insights and Observations

Observations and insights around commuting by bus can be a great starting point for a bus system to promote its usage and gain public confidence. Peoples' support is critical in making a stronger case for government to take notice and act decisively to improve the public bus service.

As one can see, although there is definitely a cost advantage in using the bus, it isn't a great one if we compare quantitative data (Table 5.1). The figures that one gets in the bus travel are most steady and impact all commuters alike. In fact, as the number of users grows, the figures improve, whereas, in other modes, to increase the numbers, vehicles have to be added, leading to congestion and pollution. While bus travel is not remarkably cheap or very cost-effective, fuel consumption per person is less even when we consider only 25 passengers sharing the ride. In fact, this number is higher during peak hours. The more crowded the bus, the better the efficiency of the service and its environmental benefits.

Quality of Commute Time

The Me Time Which Everyone Craves for in Today's Fast-Paced World

The most important point to be considered here is the remarkable qualitative impact travelling on a bus as against a personal vehicle can have on an individual's time. The point where bus usage beats the other travel modes hands down is in the stress-free travel and opportunities it opens up for an individual to utilize this time of commute. In fact, it has a lasting effect on a person's health as it opens up possibilities to stimulate the brain while keeping it free of stress, not to mention the added advantage of keeping one away from inhaling toxic fumes at traffic signals if one uses a two-wheeler.

Barring the problems of uncertainty which are created due to the inefficient design of network and management of the bus system, there is a bright side for each passenger, and it has the ability to become brighter if the efficiency of the system improves.

Buses are a great way to observe and know the city. The advertising can thus be distributed along the roads instead of cluttering the junctions. The vehicle drivers barely see the advertisements; bus commuters definitely do. One can connect with new people from diverse walks of life everyday and exchange views on news, sports, business, and culture over simple conversations. Those who do not relish this can take to their own worlds of books and music, read online articles, listen to podcasts, stream videos, and play games. While in a bus, one can sketch, draw, and meet many interesting people from all walks of life from carpenters and construction workers to college students and teachers and coders in tech companies to restaurant owners.

One can also write a blog or publish articles while riding a bus. This article too has taken shape during one such commute. During the commute time, one gets time to prepare for the day ahead with a stress-free mind. One can also take business calls and practically start office work while commuting. This is the best time to catch up on social media enabling one to avoid getting distracted by it later through the office hours. Yes, 45 min of family time may be lost, but this "me time" on the bus frees up your entire attention for the family. The daily dose of TED Talks, listening to various podcasts, reading and documenting important information on Flipboard, catching up on the news, or even chatting with friends on WhatsApp is indeed putting time on the bus to good use. At the end of the day, there is greater satisfaction on being able to contribute by reducing congestion on the roads, using less fuel, meeting new people, reading something new, growing knowledge, and most importantly staying stress-free.

It is also evident from the data in Table 5.1 why shared mobility is popular and is attracting investments. It has been evolving and getting packaged as a great experience. But, it still is expensive and affordable to only a few. While it may give you quality time, more and more shared vehicles of smaller sizes will lead to congestion and make travelling more miserable.

It is a great opportunity for urban transport to learn from the service model of these ride-sharing services. The blueprint of the service design framework employed by these ride-sharing services is a great starting point for bus operators to enhance their systems to respond better to people's needs and requirements.

This article has been written with a dual objective. It has many useful inputs for team operating bus services to promote their services and also utilize the opportunity to improve the service conditions. It is also aimed at making citizens more aware on the advantages of using public transport for daily commute. Of course, there is a lot needed to build and operate an efficient bus service. If one is able to convince passenger on how bus travel helps them well beyond simply taking them from point A to B, they will begin to trust the bus service as they would be able to see themselves in it. Most marketing cutting across all brands does this effectively. Every consumer is finding a self-nourishing value in the products and services that they choose to buy and use. They connect them with their own identities, thereby owning them and becoming their promoters. Urban transport systems like the Underground in London, RATP in Paris, SMRT in Singapore, and MTR in Hong Kong have played a crucial role in creating a unique identity for their respective cities. These cities are identified by their mobility systems.

Every bus ticket purchased should help a person to get carbon credits. If the bus systems improve bus efficiency and the system is designed to increase ridership, they are bound to make greater impact in improving the quality of life in the cities and pave the way for a sustainable future. The bus system in Pune should be well integrated with the upcoming metro and support the transit-oriented development planned for the future.

Authorities need to take a stand on making the mobility system a critical and integral part of the city's identity. Citizens should make a wiser choice. People opting for urban transport will make the authorities take notice. This problem has to be solved from both ends simultaneously.

Like various global examples, Pune can definitely achieve a great integrated seamless urban transport. Urban transport as the most favored choice would no longer surprise people; instead they will take to proactively advocating it. Residents of Pune will invite guests to their city and proudly hand over a common mobility card and an integrated map to experience a clean, open, walkable, safe, enjoyable, and fun-filled city.

The author is working closely with the PMPML team to make the bus services commuter-friendly. PMPML is also modernizing its fleet of buses with intelligent fleet management system and electric buses. Figure 5.2 shows one of the electric buses procured in 2019.

Fig. 5.2 PMPML electric bus

Bibliography

Business-standard (2018, April 5). www.business-standard.com. Retrieved from https://www.business-standard.com/article/news-ians/vehicles-figure-in-pune-overtakes-human-population-118040500811_1.html

Government of India, Census (2011) Retrieved from https://www.census2011.co.in/city.php

Institute of Urban Transport (India) (2014) Review of urban transport in India. Institute of Urban Transport (India) and Center for Study of Science, Technology & Policy, Delhi

Ministry of Urban Development (2017) Metro rail policy 2017. Ministry of Urban Development, New Delhi

Ministry of Urban Development, Government of India (2017) Liveability standards in cities. Smart cities Mission

O'Neill P (2010) Urban transport in developing cities challenges, strategies and examples. Lead infrastructure specialist. World Bank, Washington, DC

Times of India (2018, August 14) timesofindia.indiatimes.com. Retrieved from https://timesofindia.indiatimes.com/city/pune/pune-ranked-no-1-city-in-country-in-ease-of-living-rankings/articleshow/65394696.cms

Chapter 6
From Grey to Green: Rethinking Setback and MGC Rules as a Sustainable Growth Strategy of Residential Areas – A Case Study of Anannya Residential Area of Chittagong, Bangladesh

Rezuana Islam, Kanu Kumar Das, and Samira Binte Bashar

Abstract At the moment urban agglomeration sees how cities grow and expand within a shorter period by overlooking the existence of natural ecosystem and green spaces. Natural and built components of the urban environment are the main focal point for sustainable regeneration strategies of a city, but unfortunately the small-scale green areas which are closed to people in the cities are given poor attention and as a result various environmental problems are created. This phenomenon raises globally growing concern for sustainable urban development. To ensure sustainable development and healthy living environment, every country sets some obligatory codes of development. In Bangladesh to control urban development and boost living environment, Bangladesh National Building Code (BNBC), a comprehensive document, is enacted as officially obligatory codes of development in urban areas, in which setback and maximum ground coverage (MGC) are the two major and foremost important rules to maintain for better area planning. Chittagong is the second largest city and commercial capital of Bangladesh and is enriched with mountains, forest, rivers, canals, coastal belt, ancient shrines, and historical structures. In Chittagong city, urban areas become a junk of concrete with its grey pockets lacking green spaces resulting from insensible planning and irresponsible behavior of its citizen. The Anannya Residential Area was developed in 2004 by Chittagong Development Authority (CDA) in Chittagong City which is not fully developed yet. The main concern of this chapter is to rethink residential setback and MGC rules of BNBC from a sustainable regeneration strategy perspective and to suggest several ideas of creating green spaces that can be taken into consideration to recover and enrich the sustainable environmental quality of the urban areas.

The original version of this chapter was revised: The original version of this chapter was published with some citations missing which has been included now. The correction to this chapter is available at https://doi.org/10.1007/978-3-030-25879-5_27

R. Islam · K. K. Das (✉) · S. B. Bashar
Chittagong University of Engineering & Technology, Chittagong, Bangladesh

© Springer Nature Switzerland AG 2020

M. Ghosh (ed.), *Perception, Design and Ecology of the Built Environment*,
Springer Geography, https://doi.org/10.1007/978-3-030-25879-5_6

Keywords Building setback · Grey pocket · Green spaces · Sustainable regeneration strategy

Introduction

Provision of environmental and ecological services as well as social and psychological development of human societies is dependent on urban nature which enhance human life with meanings and emotions. But rapid and uncontrolled urbanization is associated with change in urban ecosystem as the development phenomena always overlook the existence of green spaces though availability of natural environment is one of the major conditions toward healthy and quality urban living (Majumder et al. 2007; Karim 2006; Urban Design Compendium 2007).

Unplanned urban growth is experiencing various sorts of problems related to urban development and planning issues and management. One of the major problems is the lack of strong implementation of development control rules. Due to population increase, recent urban expansion in Bangladesh experiences the growth of buildings without much approval of the authority. However, if there is any, the proper rules are not maintained after approval. Setback, one of the major and foremost important rules which control the spacing between buildings, is included in building code thinking its environmental benefit.

Chittagong city is located toward southeast of the capital city of Dhaka which is around 280 Km. from the capital. Chittagong city is situated on the bank of Karnaphuli River, and the city is surrounded by rich natural resources like the green Hilly Terrain and the Bay of Bengal on the west. With a population of 4.1 million people in 168.1 square kilometers of land, the density of Chittagong is 24,390 persons per sq. km, which is very high. The city is the commercial capital and largest port of Bangladesh situated 16 km upstream from the Bay of Bengal and faces rapid urbanization, which has resulted in a loss of the city's existing open spaces in the last few decades very rapidly (Jafrin and Beza 2018).

In the first regional plan in Chittagong in 1961, this problem was noted and tried to be addressed for the first time as the city's lack of public parks, playgrounds, and small children's play areas. The document provided several strategies at planning level to mitigate such as proposed 26 new parks and other open spaces to meet the needs of the city's population, consisting 2935 acres of land scattered around the city in small parcels (Jafrin and Beza 2018). From 1989 to 2001, open space was reduced by 76% as a result of urbanization process in Chittagong which was further reduced by 72% between the year 2001 and 2013 (Chisty 2014).

Chittagong being the 2nd largest city of Bangladesh faced with tremendous pressure of overpopulation. To accommodate the large population, the local government organization Chittagong Development Authority (CDA) has developed and has been developing planned residential areas in Chittagong. But like other developing countries, these residential areas lack green spaces resulting in a junk of concrete because of insensitive planning and negligent behavior of its inhabitant.

WHO suggests that the availability, accessibility, quality, and security of public green spaces may vary in metropolitan regions, and this so-called green environment is one of the indicators of healthy cities (WHO 2012). WHO suggests a standard of 9 square meter per capita, and the cumulative existing open space in Chittagong is 0.18 square meter per person (Jafrin and Beza 2018).

Though all of these strategies to mitigate lack of open space especially green areas are majorly implemented at planning level, small steps with building by-laws or urban design guidelines can often create a cumulative effect for a positive outcome. From this perspective, regeneration of green space by rethinking the residential setback rules of building bylaws may be useful. For individual plots this area is insignificant, but in a combined view in a neighborhood, this change can cause noteworthy differences in the appearance and environment of the neighborhood.

The level of urbanization in Bangladesh is low, but the rate of growth of urban population is very high, and due to large total population, the absolute number of population is also large. The impact of large influx of population in the city has been quite strong on the overall environmental situation in the city. The rather poor environmental situation in the city is evident from secondary data, field observation, and from opinions of the residents (Majumder et al. 2007).

In Anannya Residential Area, Chittagong, unlike other residential areas of Bangladesh, the planned residential plots are arranged in grid-iron pattern having two rows of plots between two roads. That means that every individual plot has a front road, and each backside is actually the backside of another one. Literature review shows that for protection of the property from the encroachment of the adjacent one, the common tendency among people is to raise a boundary wall around the plot. Therefore when one keeps the coded setback according to BNBC, it becomes a shallow useless space occupied by extension of parking or accommodating with poor greeneries or simply a space for dumping waste making grey pockets of residential areas. So that the coded setbacks are not well utilized and projected setbacks are not entirely contributed with its neighboring context as the people don't feel the inherent sense of the rule and hardly show any interest to obey this setback rule properly, though if implemented properly, the adjacent green setback of two plots can be served as small community green spaces reviving the natural ecology of the area as well as enhancing the living quality of the people of that area. The main aim of this chapter is to find out feasible planning of residential block considering all the officially obligatory codes regarding setback and MGC of BNBC which will help enhance green spaces within the residential area from a convenient point of view.

Floor Area Ratio (FAR) and Maximum Ground Coverage (MGC)

The general definition followed around the globe for FAR and MCG are as follows:

Floor area ratio (FAR) means the quotient obtained by dividing the total covered area in all floors, excluding the areas as exempted by rules, by the plot area, i.e.,

$$\text{FAR} = \frac{\text{Total Covered areas in all floors}}{\text{Plot Area}}$$

Maximum ground coverage (MGC) means the maximum area of the building footprint at ground level, considering all horizontal projections in all floors, excluding the cornices, chajjas, and architectural features up to certain level of projection as exempted by rules.

$$\text{MCG} = \frac{\text{Area of the plot covered by building when vertically projected on ground level} \times 100}{\text{Total area of the plot}}$$

BNBC and Building Construction Rules of CDA

BNBC was prepared to regulate the technical details of building construction and to maintain the building construction standard. CDA follows the "Imarat Nirman Bidhimala, 2008" and all kinds of BNBC rules for residential construction and implementation. In BNBC the minimum side, rear, and front open space requirements and MGC of a plot for buildings of various occupancy are specified in Appendix 1 and 2, respectively. Tables 6.1 and 6.2 show only the rules of setback and MGC which are relevant to this study. Besides these according to BNBC, one can construct boundary wall around the plot not higher than 3 m of which maximum 1.75 m can be solid and rest of the portion will be grill or net, and the height will be measured from the top point of the adjacent road.

Table 6.1 Setback requirement of a plot

Building height (all occupancy type)	Plot size (sq.m.)	Front setback (m)	Rear setback (m)	Side setback (m)
Up to 10 stories or 33 m	Over 134 to up to 201	1.5	1	1
	Over 201 to up to 268	1.5	1.5	1
	Over 268 to up to 1206	1.5	2	1.25

Table 6.2 Road width, MGC, and FAR for residential building

Plot size (sq.m.)	Plot size (katha)	Road width (m)	FAR	MGC (%)
Over 201 up to 268	Over 3 up to 4	6	3.5	62.5
Over 268 up to 335	Over 4 up to 5	6	3.5	60
Over 335 up to 402	Over 5 up to 6	6	3.75	60

Selection of Study Area

Aiming to enhance usable green spaces within a residential area, Anannya Residential Area has been selected as case study for this chapter. It's an ongoing residential plot project beside Oxygen-Kulgaon road (Fig. 6.1) with a total of 174 acres of land area. The minimum plot size of this project is 201 sq.m (3 katha) and maximum 402 sq.m (6 katha). In this project a total of 57.75% land is used for residential plot. The land use details are given in Table 6.3. The main reason for selecting this project as case study is that it is an ongoing project with a huge number of residential plots, and this project has less green open space where still there is scope to enhance the usable green space through initialization of different policies.

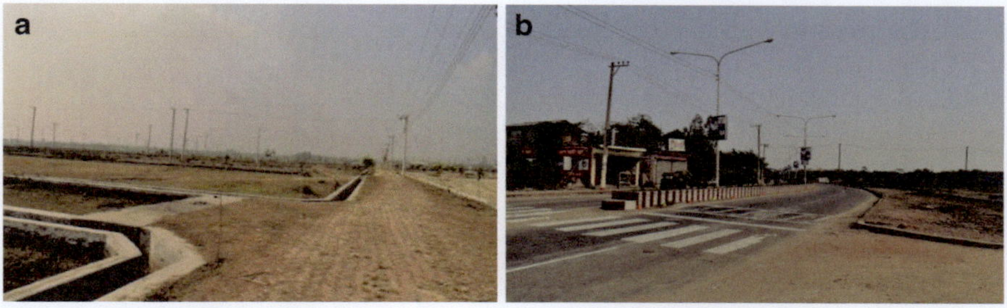

Fig. 6.1 (**a**) Internal road and plot of the site, (**b**) main road of the site

Table 6.3 Land use details of Anannya Residential Area, Chittagong

Sl no	Description	Total number	Total area, acre	Percentage area
A	**Residential**			
1	Plot area, 3.00–3.99 Katha (normal)	482	1522.77	57.37
2	Plot area, 3.00–3.99 Katha (corner)	102	325.47	
4	Plot area, 4.00–4.99 Katha (normal)	605	2448.55	
6	Plot area, 4.00–4.99 Katha (corner)	174	713.65	
7	Plot area, 5.00–5.99 Katha (normal)	129	648.39	
8	Plot area, 5.00–5.99 Katha (corner)	47	242.16	
B	**Apartment saleable**			
	16.00–25.00 Katha	07	155.16	1.50
C	**Commercial and community**			
	Shop, specialized hospital, commercial plot, taxi parking, kitchen market	58	636.44	6.19
D	**Special utility plots: saleable on special price**			
	Pump house, substation, P.O., P.S., etc.	06	41.16	0.04
E	**Plot schedule (utility): non-saleable**			
1	Utility services	14	718.14	34.54
2	Road and canal		2879.00	
	Total			100

Source: Chittagong Development Authority

Comparison with Other Similar Cities Bylaws Regarding FAR and MGC

1. **Kolkata**

Kolkata is the capital of West Bengal district of India formerly a river port city and capital of British India. The Kolkata Municipal Corporation has a population of 44,96,694 (Census 2011) in an area of 205 sq.km. Thus the density of the city is approx 22,000 persons per sqkm, which is similar to Chittagong. Kolkata is also has a similar climate, being a part of the same river delta.

FAR rules as per Kolkata Municipal (Building) Rules, 2007, as amended in 2016 are as follows:

Sl no.	Width of means of access (m)	FAR for residential buildings
1	Up to 2.4	0.00
2	2.4–3.5	1.25
3	3.5–7.0	1.75
4	7.0–9.0	2.00
5	9.0–14.5	2.25
6	14.5–20.0	2.50
7	20.0–24.0	2.75
8	Above 24	3.00

Setback rules as per Kolkata Municipal (Building) Rules, 2007, as amended in 2016 are as follows:

Height of building (m)	Front open space (m)	Open space side 1 (m)	Open space side 2 (m)	Rear open space (m)
7.0	1.2	1.2	1.2	2.0
7.0–10.0	1.2	1.2	1.2	3.0
10.0–12.5	1.2	1.2	1.5	3.0
12.5–15.5	2.0	1.5	2.5	4.0
15.5–20.0	3.5	4.0	4.0	5.0
20.0–25.5	5.0	5.0	5.0	6.5
25.5–40.0	6.0	6.5	6.5	8.5
40.0–60.0	8.0	8.0	8.0	10.0
60.0–80.0	10	15% of the height of building or 11.0 m whichever is less	15% of the height of building or 11.0 m whichever is less	12.0
Above 80.0	12.0	15% of the height of building or 14.0 m whichever is less	15% of the height of building or 14.0 m whichever is less	14.0

Comparing the setbacks of two cities, it is evident that the setbacks for a ten-storied building (33 m approx) and the setbacks required for Kolkata are much higher in spite of having similar densities. Therefore the surrounding areas can be much effectively planned after leaving circulations for car parking and other utilities. The 2 m rear setback for the same class of building in Chittagong is actually very inadequate for vehicular circulation as well as natural ventilation and lighting. The FAR in Kolkata is also much less than that of Chittagong for similar class of building. For example, for a road width of 6 m irrespective of plot size, FAR in Kolkata is only 2.00, but the same for Chittagong is 3.5–3.75 depending on the plot size. So not only the incidental open space will be high, but the height of building will also be much higher in order to consume maximum FAR. The incidental open space may in this case look and feel like narrow and very high corridor.

Therefore one general suggestion may be to revise the setback rules as per environmental conditions.

Findings

In the Anannya Residential Area, as a common strategy to apply for arrangement of plot planning, the total master plan follows "grid-iron" pattern (Fig. 6.2). There are footpath, open pucca drain, and service line by the side of the road, and road width includes the drain width and footpath width. The north-south main axis road is 22 m and the main internal road is 12 m wide and maximum other internal road is 9 m and few roads are 7.5 m wide. Due to such arrangement of plots, all plots, being separated with adjacent plots creating a barrier with surrounding neighborhoods, have a front road and a backside with a back of another plot. In the total master amount out of 174 acres, only 6–7 acres of land are proposed as play area. Not only that total master plan lacks the necessary green spaces. In this condition with a boundary wall of 1.75 m solid portion, if an owner obeys all the respected codes of BNBC and CDA regarding setback and MGC, still it will not be possible to ensure usable green within the residential area.

Discussions

Nowadays global effort is toward promoting more and more green for sustainable living environment. Just after the Local Agenda 21 (UNSD 1992) consultations, some cities have been developing their own sustainability indicators focusing to boost the quality of living environment. Aspects such as "amount of public green spaces per inhabitant," "public parks," and "recreation areas" are often given important priority to make the city inhabitable, pleasing, and attractive for its people. Enhancement of the abiotic and biotic aspects alone can't stand for sustainable city development but also about the social aspects of city life and about people's

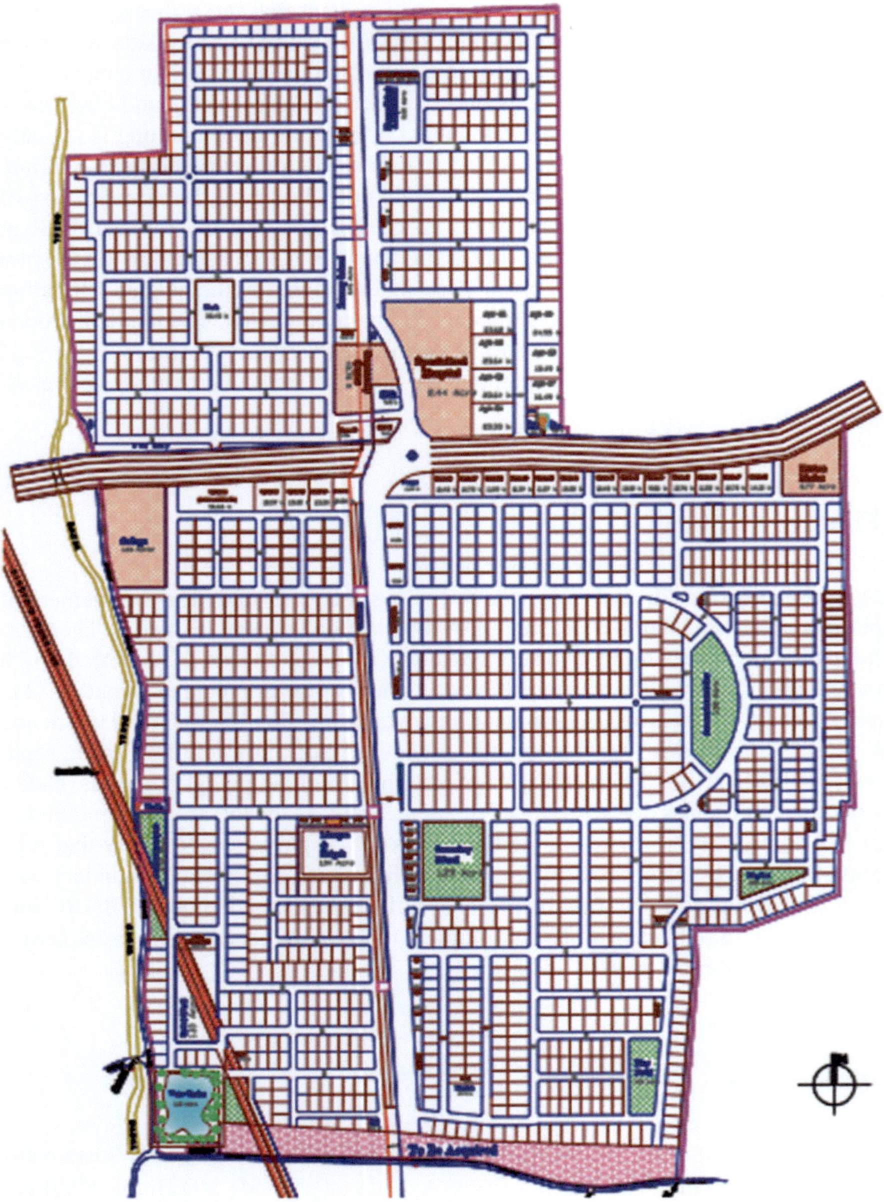

Fig. 6.2 Anannya Residential Area master plan (Source: CDA)

satisfaction, experiences, and perceptions of the quality of their everyday environments (Chiesura 2004; Beer 1993).

As a step forward toward sustainable living city development, the BNBC, setback, and MGC rules can be rethink from enhancement of social aspect perspective. In the study area based on plot size, minimum and maximum FAR is 3.5 and 3.75 with a MGC of 62.5 and 60%, respectively. From that perspective it can be said that buildings will not be higher than eight stories (ground for parking which is FAR excluded + maximum seven floors for residential purpose which is FAR included). For discussion this study considers the situation of maximum MGC with minimum setback condition as the given guidelines will be applicable for the other plots of the study area. Plot size 201–268 sq.m. falls in this category. This study assumes a mean plot with a size 234.5 sq.m.

Figure 6.3 considers the current trend of residential plot development. In the figure it is seen that with a boundary wall in each 234.5 sq.m. plot, possible green area ranges from 37.5 to 40.0 sq.m. but as per code.

Amount of permeable or soak able green that is needed to be kept in each individual site is 47.25 sq.m. It is also clear from the figure that because of the column footing the total green is segmented resulting a useless green space. That's why in practical situation, it is really hard to maintain proper permeable green space within the site area. And most of the time, these areas actually turned into a residential grey pocket and space for dumping or illegally occupied hard surface areas.

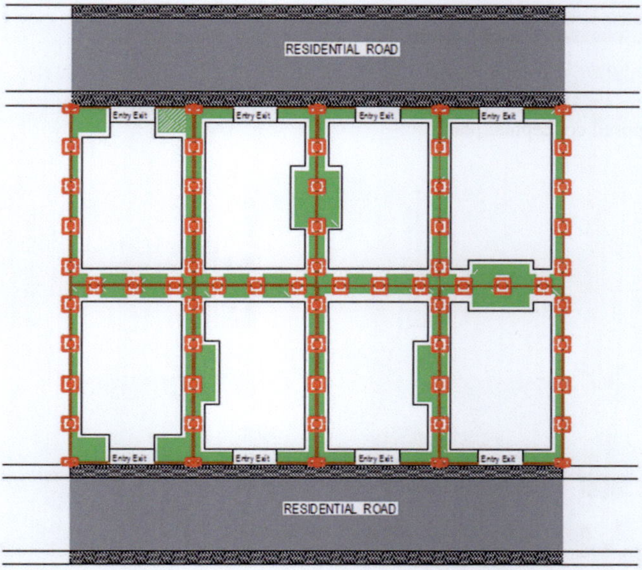

Fig. 6.3 Conceptual plan of current trend of individual plot development

Recommendations for Transforming Grey to Green

Figure 6.4a shows a possible layout option for plot development. In this layout the conventional idea of boundary is excluded to create a continuous green space connecting the side and rear open space of different plots. If the soak able green of four adjacent plots are connected together at the rear side of the building with the previous one, a large chunk of green space can be provided which will contribute as (Fig. 6.5) neighborhood communication space, children play area, pedestrian walkway, and most importantly as a deliberate space for community greening space. In this option green area (48.5 sq.m) results in 36.25% increase in each individual plot.

As another possible option (Fig. 6.4b), soak able green of the two individual plots can be joined at the front side of the plot to create a chunk of usable green. At the same time, if the side and rear open spaces are connected together, it will result in continues residential green corridor. The road side green chunk can be used as

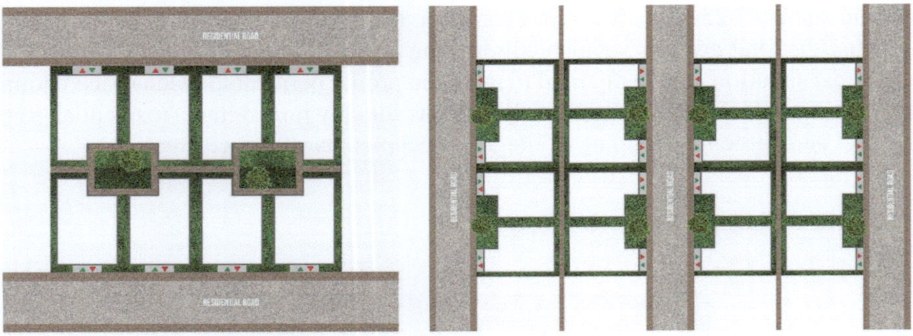

(a) *OPTION 01* (b) *OPTION 02*

Fig. 6.4 Proposed conceptual plan

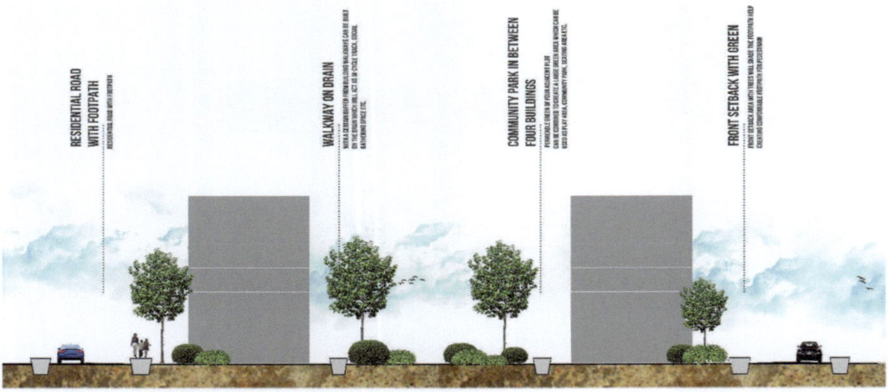

Fig. 6.5 Section (only setback area) of proposed building of plot without boundary wall

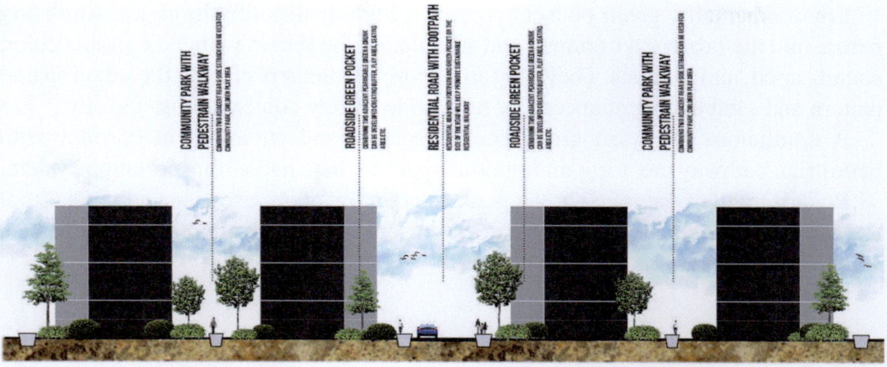

Fig. 6.6 Section (mandatory open space) of proposed building of plot without boundary wall

community interaction area, small roadside park, or space for trees. This option results in an increase of 58.42% green space (53.77 sq.m.) at each individual plot (Fig. 6.6).

Benefits of Green Pockets in Urban Sustainability: Environmental and Social Impacts

Transforming residential grey pockets into green one has twofold benefits. Firstly, the small park located in between buildings can be used either as a community gathering area or as community adhesive and will accommodate other social activities in it as a park is not only a green area of the city but also a place where people do activities. Urban park space contributes to fulfill various social functions and psychological needs of citizens. Children use it as play area, while adults and aged people of the community use it as their social interaction space. On the other hand secondly, the transformation of the grey spaces to green spaces will have a direct positive impact on the city environment. If the small grey spaces in between buildings can be turned into green one, it will help introduce biodiversity, reduce water clogging, and also reduce ambient temperature (through evapotranspiration process of trees).

When green pockets produce a green network, it helps to reconnect people with nature, so they provide immediate benefits for people and extend the effectiveness of these benefits to communities. Ecologically, green pockets develop small scattered habitats and ecosystems, they prevent soil erosion, and they absorb rainwater, thereby improving drainage and protecting against the urban heat island.

Socially, green pockets may have recreational uses like a “parklet” which is a very common concept in many developed countries: a place to play, meditate, gather, or rest. They give people the sense of place, of identity, and of belonging and enhance feelings of family kinship and solidarity and increase the sense of community.

Environmentally, green pockets provide a high-quality life through assimilating nature into the urban environment and stimulating the senses with their simple color, sound, smell, and motions. They play an important role in reshaping the urban spatial pattern and establishing connectivity for a wide variety context across the city.

A continuous backyard green pocket as a natural environment corridor with pedestrian pathway can form an important part of sustainable transportation system.

Regarding these benefits, green pockets are a sensitive response and crucial tool toward urban sustainability, which are at present unutilized, but once used can be a source of rejuvenation.

Conclusion

Compared to the vast range of core residential lands, grey pockets are often negligible, but combined effect can't be overlooked as it ultimately become a huge environmental concern for the city. The ignorance toward these developments comes not only from the city authority but also from the private owners. With proper implementation of urban planning initiatives as well as detailed pioneering ideas mentioned above, a huge enhancement can be made ensuring the environmental sustainability of these spaces. The setback and MGC rules and codes of BNBC can be reviewed to incorporate these neglected spaces to the sustainable residential development. To better shape the society, responsiveness from all sectors and consciousness from very early stage are necessary. Being a prototype, the design ideas can serve as a successful example that can be repeated in the prospect. Moving toward a better sustainable social and environmental living can be possible by providing proper care to this space.

Acknowledgment Dr. Qazi Azizul Mowla, Professor Department of Architecture, BUET
Mr Amit Imtiaz, Lecturer, Department of Architecture, CUET (for Image Render)

Appendices

Appendix 1

Building height (all occupancy type)	Plot size (sq.m.)	Front setback	Rear setback	Side setback
Up to ten stories or up to 33 m	134 or less	1.5	1	0.8
	Over 134 up to 201	1.5	1	1
	Over 201 up to 268	1.5	1.5	1
	Over 268 up to 1340	1.5	2	1.25
	Over 1340	1.5	2	1.5
Over ten stories or above 33 m	Any size	1.5	3	3

Appendix 2

Plot size		Building type (residential)		
Sqm	Katha	Road width (m)	FAR	MGC %
134 or less	2 or less	6.0	3.15	67.5
Over 134 up to 201	Over 2 up to 3	6.0	3.35	65.0
Over 201 up to 268	Over 3 up to 4	6.0	3.50	62.5
Over 268 up to 335	Over 4 up to 5	6.0	3.50	62.5
Over 335 up to 402	Over 5 up to 6	6.0	3.75	60.0
Over 402 up to 469	Over 6 up to 7	6.0	3.75	60.0
Over 469 up to 536	Over 7 up to 8	6.0	4.00	60.0
Over 536 up to 603	Over 8 up to 9	6.0	4.00	60.0
Over 603 up to 670	Over 9 up to 10	6.0	4.25	57.5
Over 670 up to 804	Over 10 up to 12	9.0	4.50	57.5
Over 804 up to 938	Over 12 up to 14	9.0	4.75	55.0
Over 938 up to 1072	Over 14 up to 16	9.0	5.00	52.5
Over 1072 up to 1206	Over 16 up to 18	9.0	5.25	52.5
Over 1206 up to 1340	Over 18 up to 20	9.0	5.25	50.0
Over 1340	Over 20	12.0	5.50	50.0
Any size	Any size	18.0	6.00	50.0
Any size	Any size	24.0	6.50	50.0

Bibliography

Anon, United Nations Conference on Environment & Development (UNCED)", Rio de Janerio. *UNSD-United Nation Sustainable Development*, "Agenda 21', 1992: Brazil

Anon (2007) Urban Design Compendium (UDC), "Urban design compendium-1,". London: Llewelyn Davies: English partnerships and the housing corporation.

Beer AR (1993) Urban greenspace and sustainability. In: van der Vegt et al (eds) Proceedings of a PRO/ECE-workshop on sustainable urban development: research and experiments, Dordrecht, Netherlands,

Chiesura A (2004) The role of urban parks for the sustainable city. Elsevier 68:129–138

Chisty K (2014) Landslide in Chittagong City: a perspective on hill cutting. J Bangladesh Inst Plann 7: 1–17

Jafrin M, Beza BB (2018) Developing an open space standard in a densely populated city: a case study of Chittagong City. Infrastructures 3(3): 40

Karim AKMR (2006) Best practice: a perspective of "clean and green" Chittagong. In: First 2006 workshop population and environmental protection in urban planning. Asian Urban Information Centre of Kobe, Kobe

Majumder AK, Hossain ME, Islam MN, Sarwar MI (2007) Urban environmental quality mapping: a perception study on Chittagong metropolitan city. Kathmandu Univ J Sci Eng Technol 1(iv):35–48

World Health Organization (WHO) (2012) World Health Organization (WHO). Health indicators of sustainable cities. Rio de Janeiro, Brazil, s.n

Chapter 7
Coastal Climate Readiness and Preparedness: Comparative Review of the State of Florida and Cuba

Haris Alibašić and John D. Morgan

Abstract Property owners living near the coast benefit from geographically specific amenities. However, residents of coastal zones are often faced with the risk of natural hazards, such as hurricanes and the increased risks stemming from the threats of climate change including rising seas. To effectively manage coastal zones, and better equip them for climate resilience and preparedness, decision-makers depend on information with which they balance risk, benefit, and use.

The work described in this book chapter focuses on one highly developed coastal areas in Florida, more specifically regional government. Informed by regional evacuation studies and broader compliance with Florida Law, hurricane evacuation zones have changed several times since historically active 2004–2005 hurricane seasons. These zones should communicate risks to the property market, and by assuming efficient markets, they should be incorporated into home prices. We evaluate the trends of policies in place in India by comparing a difference in approaches to the risks and policies in place to prevent large-scale disasters, costs, and mitigation of damage.

This chapter uses the comparative analysis of varying approaches to issues geographical position and the balance between ecosystem service amenity (e.g., distance-to-shore and view), sub-national policies, and natural hazard risk. We provide an overview of current policies in place, a review of the amenity-based analysis of the risk averseness, explicit incentives, and governance structures to make the housing market account for low-probability catastrophic risk scenarios. We look how effective markets compensate for incorporated natural hazard risks.

The book chapter draws on comparison to the emergent practices in regional settings in Florida and compare the emergent practices in Cuba and/or Carribean region and provides for potential applicability of models. These are all places with similar risk exposure and morphology but vary on national and sub-national institutions.

H. Alibašić (✉)
Public Administration Program, University of West Florida, Pensacola, FL, USA
e-mail: halibasic@uwf.edu

J. D. Morgan
Earth and Environmental Science, University of West Florida, Pensacola, FL, USA

© Springer Nature Switzerland AG 2020

M. Ghosh (ed.), *Perception, Design and Ecology of the Built Environment*, Springer Geography, https://doi.org/10.1007/978-3-030-25879-5_7

Keywords Climate change · Climate resilience · Sustainability · Climate preparedness · Climate readiness

Overview

The proximity to the coast has its benefits and disadvantages. Studies have found that there is a "healthy coast" effect, where those living within proximity to coast have relatively higher health and well-being, compared to interior residents (Wheeler et al. 2012). However, residents and businesses in coastal zones often face the risk of natural hazards, such as hurricanes and the increased threats stemming from the climatic changes including rising seas, heat waves, and beach erosion. To effectively plan for climate-resilient coastal zones, and better equip planners, administrators and first responders plan for climate resilience and preparedness. Decision-makers depend on information with which they balance risk, benefit, and land use in planning. The book chapter compares the risks and practices in the regional settings of Florida to the practices in Cuba with the mention of the Caribbean with the potential generalization of strategies and policies. While the regions differ significantly in political, population, economic, and even geographical arrangements and structure, the coastal areas of both the State of Florida and Cuba face similar threats.

The review provides an analysis of enduring risks and practical approaches to climate readiness with the consideration of plans to address climate risk assessments in these distinctly geographically, politically, and economically dissimilar regions. The two regions are exposed to similar risk but vary in responses at a national and sub-national level of governments. The specific interest is in the existing practices, and policies in place are included in the review with specific recommendations for improvement to processes and policies. The authors of the book chapter examine the current state of climate preparedness in the specific coastal areas of Florida and the Caribbean with a particular focus on Cuba. The similarity of threats provides for a unique understanding of the responses and preparedness strategies deployed in these areas under consideration.

Defining Climate Readiness and Preparedness

The study of various processes and tools available to coastal areas and regions affected by hurricanes and frequent natural hazards and also influenced by the climate change offers insight into collaborative nature of emergency preparedness and disaster mitigation. In recent years, the increased strength and frequency of devastating impacts of hurricanes renewed the debate and research into climate preparedness and readiness. Beyond storms, far-reaching effects of extreme weather events prompt the necessary responses by administrators to confront the impacts with the specific economic, environmental, governing, and social characteristics. The four effects may best be labeled through the Quadruple Bottom Line lens, described as the organizational and community aptitude to entrench well-defined practices and programs to "address economic, social, environmental, and governance aspects of sustainability, whereas governance is defined through fiscal responsibility and resilience, community engagement for efficient service delivery, transparency, and accountability" (Alibašić 2017, p. 41).

Due to their geography and location, Florida, Cuba, and the remainder of the Caribbean are prone to prolonged impacts of natural hazards, further increased due to climate risks. A deployment of preparedness- and readiness-related measures to address the community's emergency preparedness and disaster recovery is of paramount importance. Cities and regions in Cuba, Florida, and the rest of the Caribbean are on the front lines of natural hazards, and they must plan for the natural disasters. Resilience may best be described as the organizational and systematic capability "to respond and adapt quickly to societal demands, transformational, internal and external shocks, and governing, economic, environmental, cultural, and social changes" (Alibašić 2018d, p. 2). Moreover, resilient organizations and communities not only withstand shocks and disasters but continue to thrive in the post-disaster environment (Alibašić 2018d).

Within the parameters above, the readiness and preparedness plans and strategies must and should include all aspects and levels of infrastructure and institutional knowledge and support to incorporate available climate and resilience information. Being climate resilient and prepared supports the longevity of a community and enables them to rebound back from disasters. The social conditions and connectedness within the regions affect regional readiness and preparation in the face of natural disasters. Beyond practical implications, local, state, and regional planners and administrators have an ethical role and responsibility to prepare for and address the effects of climate change and enable communities to thrive after the disasters. The ethics of resilience is embedded into the core governance structure as administrators bear responsibilities and ownership of the inherent nature of planning and preparing for disasters (Alibašić 2018e). On a global level, there is a direction by national and sub-national level governments to seek resilience in planning to harden infrastructure and to improve governance of the preparedness and readiness. Building resilient infrastructure, promoting inclusive and sustainable industrialization and fostering innovation, is one of the goals in the disaster risk reduction and resilience in the United Nations 2030 Agenda for Sustainable Development (UNORR 2015).

Climate Change Risks and Vulnerabilities and Responses

The impact of natural hazards and extreme weather on most vulnerable population has been noted for severity and prolonged impacts. Among others, Alibašić (2018a, b, c, d), Blaikie et al. (1994, 2014), Bullock et al. (2014), López-Marrero and Tschakert (2011), Luber et al. (2014), Nordgren et al. (2016), Smit and Pilifosova

(2003), The Task Force (2014), Wisner et al. (2004), Wamsler (2014), and Wisner et al. (2012) offered insights into the impact of natural hazards and extreme weather effects on the population in disaster-prone areas and the roles and responsibilities of local, state, and national governments in assessing and preparing to protect vulnerable population and infrastructure. The climate change exacerbated the effects and consequences of the flooding, sea level rise, wildfires, heat waves, and more intense and frequent hurricanes.

The case of the catastrophic impact of the hurricane Maria on Puerto Rico provided ample examples of the importance of planning for natural hazards and incorporating long-term preparedness and resilience strategies. The long-lasting impact on lives and infrastructure from the recent Hurricane Maria on the US island of Puerto Rico are significant while still being documented and analyzed (Zorrilla 2017). A further investigation into the precise social, economic, and environmental effects of the Hurricane Maria is warranted. Lessons from disasters such as Maria can serve to further climate readiness and preparedness plans in the future. For example, Guion et al. (2007) offer a social marketing approach to disaster preparedness and response lessons from Hurricane Katrina.

The threats and consequences from climate change impact are well recognized and documented (Fletcher 2013; IPCC 2014a, b; Karl et al. 2009; Mach et al. 2016; Malcolm et al. 2006; Pecl et al. 2017; Segan et al. 2015; Stott et al. 2016; Silva et al. 2013; Urban 2015; Visser et al. 2014). The risks of increased frequency and length of hurricanes are also being acknowledged (Mann et al. 2017). Moreover, Wuebbles et al. (2017) in the Fourth National Climate Assessment report for the US Global Change Research Program offered the conclusive evidence of climate change impact in the USA and the effect it has on people in all corners of the nation.

Hsiang et al. (2017) attempted to answer the fundamental question of the cost of climate change to the USA economy, and concluded "the bulk of the economic damage from climate change will be borne outside of the United States, and impacts outside the United States will have indirect effects on the United States" economy (p. 1369). In that sense, it is useful to compare and contrast various disaster threats and emergency preparedness responses between various regions and countries. Climate change will increase the pressure on economies globally (O'Brien and Leichenko 2000). NCEI's 2018 report shows the extent of the economic impact from various natural disasters from 1980 to 2018 as of July 9, 2018, across the USA from events "with losses exceeding $1 billion (CPI-Adjusted) each across the United States" (Fig. 7.1).

Finally, in addition to the damage from natural disasters such as hurricanes and flood, measuring in billions of dollars annually and human casualties, climate change will increase such predicaments. The annual damage indicated in the Hsiang et al. (2017), Fig. 7.2, is showing county-level median scenario for the predicted climate damage in the period 2080–2098; the economic impact of climate change will be significant in the state of Florida.

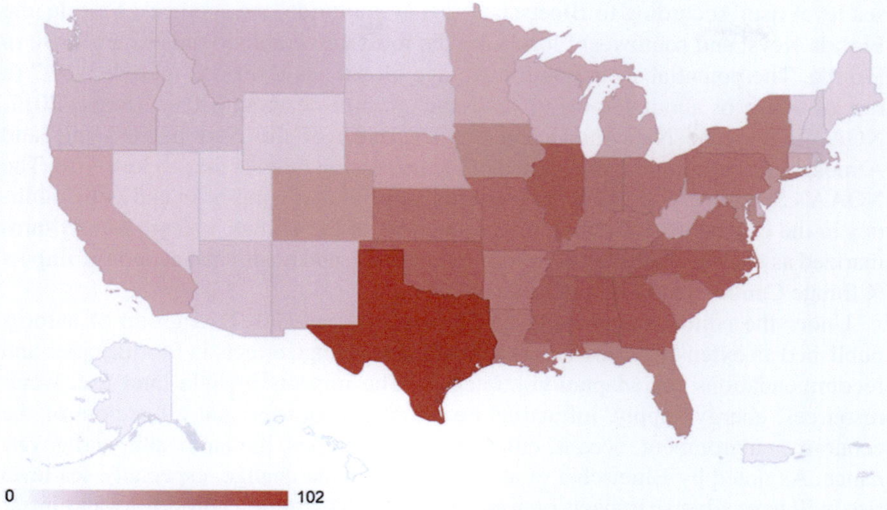

Fig. 7.1 NOAA National Centers for Environmental Information (NCEI) US Billion-Dollar weather and climate disasters state-by-state analysis for the 1980–2018 period as of July

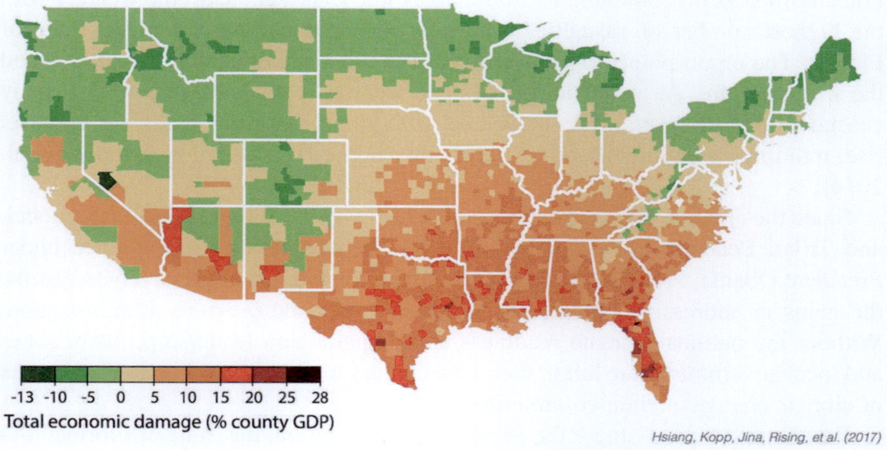

Fig. 7.2 County-level projected scenarios of climate change data for 2080–2098

Climate Readiness and Preparedness in Florida

The global risks with severe consequences of climate change are real and well documented. Reese (2017) provided an overview of the National Oceanic and Atmospheric Administration assessment of the risk to coastal communities such as sea level rise. According to Bloetscher (2012) in Florida, the southeast Florida, the Florida Keys, and southwest Florida are the most vulnerable to the sea level rise in Florida. The potential scenarios for average global sea level rise include "low" (a rise of 0.3 m or about 1 foot) to "extreme" (2.5 m or about 8 feet) (Reese 2017, NOAA n.d.). The National Ocean Service (n.d.) of the National Oceanic and Atmospheric Administration warned of the increasing rates of the sea level rise. The NOAA's Sea Level Rise Viewer is advantageous in assessing risks and vulnerabilities in the coastal area. Common consequences of the climate change can be summarized as coastal and beach erosion, sea level rise, and health and economic impact (Climate Central and ICF International 2015).

Under the editorial leadership of Chassignet et al. (2017), a group of authors published an extensive assessment of the climate change effects in Florida cities and recommendations for adaptation strategies. The impacts include land use, water resources, energy supply, infrastructure, human resources, and all sectors of the economy, environment, oceans, cities, and communities, human health, and governance. As noted by Bloetscher et al. (2017), "Climate change, especially sea level rise, will have adverse impacts on water, sewer, transportation and stormwater infrastructure, putting properties and the economic opportunities at "risk of failure of these systems" (p. 311).

Furthermore, the Bureau of Epidemiology's report from 2015 pointed out the significant impacts on communities from deadly hurricanes and extreme weather with the highest number of casualties occurring from such events are in the State of Florida. The report points to prolonged risks to the most vulnerable population and the infrastructure. As noted in the Climate Central report and projections, many coastal cities face several risks associated with global warming, including sea level rise, multiplying extreme flood of uncertainty in the coastal regions (Strauss et al. 2014).

Since the creation and recommendations made by the White House State, Local, and Tribal Leaders Task Force on Climate Preparedness and Resilience under President Obama, President Trump's administration worked aggressively to reverse the gains in addressing climate change made under the previous administration. Without the national climate readiness and preparedness leadership, many states and local governments are left to their own devices to plan for the inevitable impacts of climate change on their communities.

For the past 8 years under the previous administration, the State of Florida government has abdicated its responsibility and efforts to confront the climate change. The Florida Center for Investigative Reporting documented the decision by the previous Governor's administration not to allow the state of Florida employees to use climate change and global warming in reports and communication (FCIR 2015). Korten (2015) noted that Florida Department of Environmental Protection officials "…have been ordered not to use the term 'climate change' or 'global warming' in any official communications, emails, or reports, according to former DEP employees, consultants, volunteers and records obtained by the Florida Center for Investigative Reporting." The effect of such blatant disregard for scientific facts is that the state of Florida has not made the necessary investments to prepare the coastal regions for the impacts of climate change (Dennis and Fears 2017; Caputo

2017). Most climate readiness and resilience efforts are fragmented, and the majority of the burden of facing the impacts of climate change is passed onto the local and county governments. Some examples of successful regional climate preparedness planning include the efforts underway in the counties of Broward, Miami-Dade, Monroe, and Palm Beach Counties (Southeast Florida Regional Climate Change Compact Counties 2012; Broward County 2015, 2017). The fragmentation of planning leads to less desirable consequences of exhausting local funds for plans and investments beyond the necessary service provisions of public safety and infrastructure as perceived by the local government constituents. In particular, Torres et al. (2018) offered the insight into perspectives on climate change adaptation capacity from Broward County, Florida. Similarly, Patterson et al. (2017) compared the Broward County's adaptation activities to the ones in the UK and Brazil. Other exemplary planning activities are taking place in places from Miami-Dade County to Satellite Beach to Sarasota.

Finally, in areas most prone to the devastating impacts of the hurricane and other natural disasters, and impending sea level rise such as Pensacola and other parts of North West Florida, there is no regional or local plans to prepare and ready the community for climate change. This lack of planning is a direct consequence of the lack of leadership. Only recently the city council of the City of Pensacola appointed the Task Force on Climate Adaptation and Mitigation to make recommendations to the council on climate preparedness and readiness. However, the work of the Task Force was not endorsed or supported by the Mayor of the City. As noted by Alibašić (2018b), "Chief features of climate change actions at the local governments' level are the cities' ability to prepare their respective communities to be more agile and adaptive to extreme weather events and disasters, with 'resilience efforts [that] are flexible' and strong leadership commitment" (p. 4).

Cuba and Caribbean Regional Disaster Assessment

The natural hazards and vulnerabilities to extreme weather in coastal Cuba are examined in depth. With their research findings, Alonso and Clark (2015), Carter et al. (2014), Fernández and Pérez (2009), Pichler and Striesnig (2013), Kelman (2017), Lizarralde et al. (2015), López-Marrero and Wisner (2012), and Sims and Vogelmann (2002) deliver an improved understanding of the practical and policy implications surrounding climate preparedness and responsiveness to the climate threats in the coastal communities in Cuba and the broader context of Caribbean region. Figure 7.3 from the Centre for Research on the Epidemiology of Disasters (2018) represents a broad spectrum of natural disasters and the number of population affected by those. Figure 7.4 shows the total damage from storms from 1990 to 2018.

The threats and damage from natural disasters will further intensify in intensity due to climate change. Alonso and Clark (2015) pointed out that "for a small island nation such as Cuba, it is particularly important to identify as precisely as possible the most probable impacts of climate change, and thus determine scientifically-

CUB_Natural_1900_2018_total_affected

sep=			
Disaster No	Type	Date	Total affected
2017-0381	Storm	08-09-2017	10000000
2001-0612	Storm	04-11-2001	5900012
2005-0351	Storm	08-07-2005	2500000
1998-9210	Drought	00-01-1998	820000
1993-0083	Flood	24-11-1993	532000
1985-0134	Storm	18-11-1985	498891
2008-0352	Storm	29-08-2008	450019
1996-0231	Storm	17-10-1996	336676
2002-0636	Storm	01-10-2002	281470
1999-0400	Storm	14-10-1999	254990
Created on: July 18, 2018			
Source: EM-DAT: The Emergency Events Database - Universite catholique de Louvain (UCL) - CRED, D. Guha-Sapir - www.emdat.be, Brussels, Belgium			

Fig. 7.3 Total affected population by natural disasters in Cuba from 1900 to 2018

CUB_Natural_1900_2018_total_dam

sep=			
Disaster No	Type	Date	Total damage ('000 US$)
2017-0381	Storm	08-09-2017	13200000
2016-0355	Storm	28-09-2016	2600000
2008-0352	Storm	29-08-2008	2072000
1998-0381	Storm	22-09-1998	2000000
2008-0384	Storm	08-09-2008	1500000
2005-0351	Storm	08-07-2005	1400000
1993-0012	Storm	13-03-1993	1000000
2004-0415	Storm	14-08-2004	1000000
2005-0585	Storm	19-10-2005	700000
1963-0030	Storm	04-10-1963	500000
Created on: July 18, 2018			
Source: EM-DAT: The Emergency Events Database - Universite catholique de Louvain (UCL) - CRED, D. Guha-Sapir - www.emdat.be, Brussels, Belgium			

Fig. 7.4 Total damage from natural disasters from 1900 to 2018

based actions to prevent or reduce their adverse consequences and formulate the necessary adaptive measures."

Unlike the US political system, Cuba is defined and controlled by a single-party system and is one of the very few remaining communist countries in the world. The planning, including economy, emergency preparedness, and disaster response, is centralized. All climate readiness and preparedness actions are initiated and conducted with tacit approval from the party in power.

The United Nations Development Programme (UNDP) suggests that Cuba serves as the model for readiness and preparedness in a resilient Caribbean region. While the emphasis of this chapter is on Cuba, it is noteworthy a body of research focusing on other parts of the Caribbean region. The studies produced by Gamble (2014), Lam et al. (2014), Pichs-Madruga (2017), Taylor et al. (2012), and Tompkins (2005) offer a comprehensive examination of the state of the climate preparedness and readiness in the Caribbean islands beyond Cuba.

Conclusion

Regional and local governments have a role and responsibility to the residents and business, to confront the challenges and the threats of the changing climate and natural hazard. As a result of several direct and indirect factors, including but not limited to the economic and social factors, in Florida local governments and regional planning agencies deploy programs and policies to mitigate and adapt to the climate change. The more centralized planning role in Cuba indicates less flexibility on the part of planning agencies to adopt certain types of climate adaptation or mitigation measures. The range of climate resilience and readiness planning is evolving in nature, reflecting the changes in circumstances. Beyond the ethical responsibilities in alleviating the risks of climate change and reducing the negative environmental impacts, administrators and planners utilize climate preparedness and readiness to predict and void potential costly infrastructure and casualties (Alibašić 2018f, g).

The politics of climate change are on full display in the USA. Counterintuitively, on one hand, there is a democratically elected governor in the state of Florida who for political and dialogical reasons forbids the use of phrase climate change by state employees and hinders attempts to prepare and plan for climate change. On the other hand, there is an repressive, one-party system which openly promotes and supports climate readiness efforts, or at least it is not deliberately sabotaging them for ideological and political purposes.

Furthermore, in comparing the fragmented and localized approaches to climate preparedness and readiness in Florida to those of more centralized in Cuba respectively, the authors were able to discern the established patterns of confronting climate change:

- Recognizing the threats are tangible, and why and how these specific pressures need to be addressed.
- Examining, analyzing, and determining the localized data related to climate change threats and impacts.
- Building adaptive and mitigating frameworks for addressing climate change at all levels of government.
- Engaging, relevant stakeholders at local, regional, state, and national levels in both the public and the private sector.

The imperative for climate preparedness and resilience stems from devastating effects the climate change has on coastal cities and regions. The lack of adequate and urgent readiness and climate preparedness planning will lead to devastating consequences for communities around the world. Despite the difference in approaches, communities and local governments on the ground are the most effective when there is strong and unequivocal support from all levels of government including state and national administrators and elected officials.

Bibliography

Alibašić H (2017) Measuring sustainability impact in local governments using the quadruple bottom line approach. Int J Sustain Policy Pract 13(3):37–45. https://doi.org/10.18848/2325-1166/CGP/v13i03/37-45

Alibašić H (2018a) Sustainability and resilience planning for local governments: the quadruple bottom line strategy. Springer, Cham. http://www.springer.com/us/book/9783319725673

Alibašić H (2018b) Leading climate change at the local government level. In: Farazmand A (ed) Global encyclopedia of public administration, public policy, and governance. Springer International Publishing AG, Cham. https://doi.org/10.1007/978-3-319-31816-5_3428-1

Alibašić H (2018c) Paris climate accord sans the USA: the role of local governments in addressing the climate change and increasing resilience. Glob Stud J 11(2):33–39. https://doi.org/10.18848/1835-4432/CGP/v11i02/33-39

Alibašić H (2018d) The Nexus of sustainability and climate resilience planning: embedding climate resilience policies in local governments. Int J Clim Change ImpactsRespon 10(2):21–33. https://doi.org/10.18848/1835-7156/CGP/v10i02/21-33

Alibašić H (2018e) Redesigning organizations for maximum resiliency in dynamic environments. In: Farazmand A (ed) Global encyclopedia of public administration, public policy, and governance. Springer International Publishing AG, Cham. https://doi.org/10.1007/978-3-319-31816-5_3431-2

Alibašić H (2018f) Ethics and sustainability in local government. In: Farazmand A (ed) Global Encyclopedia of public administration, public policy, and governance. Springer International AG, Cham. https://doi.org/10.1007/978-3-319-31816-5_3427-1

Alibašić H (2018g) Ethics of resiliency in crisis management. In: Farazmand A (ed) Global Encyclopedia of public administration, public policy, and governance. Springer International Publishing AG, Cham. https://doi.org/10.1007/978-3-319-31816-5

Alonso G, Clark I (2015) Cuba confronts climate change. Int J Cuban Health Medicine 17(2):39–43

Blaikie P, Cannon T, Davis I, Wisner B (1994, 2014) At risk: natural hazards, people's vulnerability and disasters. Routledge, New York.

Bloetscher F (2012) Protecting people, infrastructure, economies, and ecosystem assets: water management and adaptation in the face of climate change. J Water Water 4:367–388. https://doi.org/10.3390/w40x000x

Bloetscher F, Hoermann S, Berry L (2017) Adaptation of Florida's urban infrastructure to climate change. In: Chassignet EP, Jones JW, Misra V, Obeysekera J (eds) Florida's climate: changes, variations, & impacts. Florida Climate Institute, Gainesville, pp 311–338. https://doi.org/10.17125/fci2017.ch11

Bullock JA, Haddow GD, Haddow KS, Coppola DP (2014) Living with climate change: how communities are surviving and thriving in a changing climate. CRC Press, Chicago

Bureau of Epidemiology and US Center for Disease Control (2012) Health effects of tropical storms and hurricanes in Florida http://www.floridahealth.gov/environmental-health/climate-and-health/_documents/tc-profile.pdf

Caputo M (2017) Florida governor remains unsure about climate change after Hurricane Irma. https://www.politico.com/states/florida/story/2017/09/14/florida-governor-remains-unsure-about-climate-change-after-hurricane-irma-114498

Carter LM, Jones JW, Berry L, Burkett V, Murley JF, Obeysekera J, Schramm PJ, Wear D (2014) Ch. 17: Southeast and the Caribbean. Climate change impacts in the United States: the third national climate assessment, J. M. Melillo, Terese (T.C.) Richmond, and G. W. Yohe eds, U.S. Global Change Research Program, pp 396–417. doi:https://doi.org/10.7930/J0NP22CB

Centre for Research on the Epidemiology of Disasters (2018) EM-DAT: the emergency events database – Université catholique de Louvain (UCL) – CRED; www.emdat.be, Brussels, Belgium

Chassignet EP, Jones JW, Misra V, Obeysekera J (eds) (2017) Florida's climate: changes, variations, & impacts. Florida Climate Institute, Gainesville. https://doi.org/10.17125/fci2017

Climate Central and ICF International (2015) States at risk: America's preparedness report card: Florida report. http://assets.statesatrisk.org/summaries/Florida_report.pdf

County B (2015) Climate action plan: local strategy to address climate change. Retrieved from http://www.broward.org/NaturalResources/ClimateChange/Documents/BrowardCAPReport2015.pdf

County B (2017) Environmental planning and community resilience division. Organizational Chart. Retrieved from http://www.broward.org/NaturalResources/Documents/orgchartepcrd.pdf

Dennis B, Fears D (2017) Florida governor has ignored climate change risks, critics say. Retrieved from https://www.washingtonpost.com/national/health-science/florida-governor-has-ignored-climate-change-risks-critics-say/2017/09/08/04a8c60a-94a0-11e7-aace-04b862b2b3f3_story.html?utm_term=.144a8fe5607f

Fernández MA, Pérez R (eds) GEO Cuba (2009) Evaluación del medio ambiente cubano. Havana: Ministry of Science, technology and the environment (CU). Chapter 3 Impacto sobre el medio ambiente. Spanish

Fletcher C (2013) Climate change: what the science tells us. Wiley, Heboken

Gamble D (2014) The neglected climatic hazards of the Caribbean: overview and prospects in a warmer climate. Geogr Compass 8(4):221–234

Guion DT, Scammon DL, Borders AL (2007) Weathering the storm: a social marketing perspective on disaster preparedness and response with lessons from hurricane Katrina. Am Mark Assoc 26(1):20–32

Hsiang S et al (2017) Estimating economic damage from climate change in the United States. Science 356:1362–1369

Intergovernmental Panel on Climate Change (IPCC) (2014a) Summary for policymakers. In: Field CB, Barros VR, Dokken DJ, Mach KJ, Mastrandrea MD, Bilir TE, Chatterjee M, Ebi KL, Estrada YO, Genova RC, Girma B, Kissel ES, Levy AN, MacCracken S, Mastrandrea

Intergovernmental Panel on Climate Change (IPCC) (2014b) Summary for policymakers. In: Field CB, Barros VR, Dokken DJ, Mach KJ, Mastrandrea MD, Bilir TE, Chatterjee M et al Climate change 2014: impacts, adaptation, and vulnerability—part a: global and Sectoral aspects, edited by, 1–32. Contribution of working group II to the fifth assessment report of the intergovernmental panel on climate change. Cambridge University Press, Cambridge

Karl TR, Melillo JM, Peterson TC (eds) (2009) Global climate change impacts in the United States. US global change research program. Cambridge University Press, Cambridge. Accessed 25 Oct 2017. http://www.iooc.us/wp-content/uploads/2010/09/Global-Climate-Change-Impacts-in-the-United-States.pdf

Kelman I (2017) How can island communities deal with environmental hazards and hazard drivers, including climate change? Environ Conserv:1–10

Korten T (2015) In Florida, officials ban term 'climate change'. Florida Center for Investigative Reporting (FLIR). Retrieved from https://fcir.org/2015/03/08/in-florida-officials-ban-term-climate-change/

Lam NSN, Arenas H, Brito PL, Liu KB (2014) Assessment of vulnerability and adaptive capacity to coastal hazards in the Caribbean region. J Coast Res 70(sp1):473–478

Lizarralde G, Valladares A, Olivera A, Bornstein L, Gould K, Barenstein JD (2015) A systems approach to resilience in the built environment: the case of Cuba. Disasters 39:s76–s95. https://doi.org/10.1111/disa.12109

López-Marrero T, Tschakert P (2011) From theory to practice: building more resilient communities in flood-prone areas. Environ Urban 23(1):229–249

López-Marrero T, Wisner B (2012) Not in the same boat: disasters and differential vulnerabilities in the insular Caribbean. Caribb Stud 40(2):129–168

Luber, G., K. Knowlton, J. Balbus, H. Frumkin, M. Hayden, J. Hess, M. McGeehin, N. Sheats, L. Backer, C. B. Beard, K. L. Ebi, E. Maibach, R. S. Ostfeld, C. Wiedinmyer, E. Zielinski-Gutiérrez, and L. Ziska (2014). Ch. 9: human health. Climate change impacts in the United States: the third national climate assessment, J. M. Melillo, Terese (T.C.) Richmond, and G.W. Yohe, Eds., U.S. Global Change Research Program, pp 220–256. https://doi.org/10.7930/J0PN93H5.

Mach KJ, Mastrandrea MD, Bilir E, Field CB (2016) Understanding and responding to danger from climate change: the role of key risks in the IPCC AR5. Clim Chang 136(3–4):427–444. https://doi.org/10.1007/s10584-016-1645-x

Malcolm JR, Liu C, Neilson RP, Hansen L, Hannah L (2006) Global warming and extinctions of endemic species from biodiversity hotspots. Conserv Biol 20(2):538–548

Mann ME, Peterson TC, Hassol SJ (2017) What we know about the climate change-Hurricane connection. Scientific American. Accessed 2 July 2018. https://blogs.scientificamerican.com/observations/what-we-know-about-the-climate-change-hurricane-connection/

National Ocean Service (n.d.) Is sea level rising? Retrieved from https://oceanservice.noaa.gov/facts/sealevel.html

National Oceanic and Atmospheric Administration (NOAA) (n.d.) Sea level rise viewer. Retrieves from: https://coast.noaa.gov/slr/

NOAA National Centers for Environmental Information (NCEI) U.S. Billion-Dollar Weather and Climate Disasters (2018). https://www.ncdc.noaa.gov/billions/

Nordgren J, Stults M, Meerow S (2016) Supporting local climate adaptation: where we are and where we need to go. Environ Sci Pol. https://doi.org/10.1016/j.envsci.2016.05.006

O'Brien KL, Leichenko RM (2000) Double exposure: assessing the impacts of climate change within the context of economic globalization. Glob Environ Chang 10(2000):221–232

Patterson SK et al (2017) Size does matter: city scale and the asymmetries of climate change adaptation in three coastal towns. Geoforum 81:109–119. https://doi.org/10.1016/j.geoforum.2017.02.014

Pecl G, Araújo M, Bell J, Blanchard J, Bonebrake T, Chen I-C, Clark TD et al (2017) Biodiversity redistribution under climate change: impacts on ecosystems and human well-being. Science 355(6332):eaai9214. https://doi.org/10.1126/science.aai9214

Pichler A, Striesnig E (2013) Differential vulnerability to hurricanes in Cuba, Haiti, and the Dominican Republic: the contribution of education. Ecol Soc 18(3):31. https://doi.org/10.5751/ES-05774-180331

Pichs-Madruga R (2017) Climate resilience and sustainable development: challenges and options for small islands developing states. In: Ninan KN, Inoue M (eds) Building a climate resilient economy and society: challenges and opportunities. Edward Elgar Publishing, Cheltenham

Reese, A (2017) Sea-level rise poses a rising risk to U.S. Shores. Scientific American. Retrieved from: https://www.scientificamerican.com/article/sea-level-rise-poses-a-rising-risk-to-u-s-shores/

Rist L, Campbell BM, Frost P (2012) Adaptive management: where are we now. Environ Conserv 40(1):5–18

Segan DB, Hole DG, Donatti CI, Zganjar C, Martin S, Buthart SHM, Watson JEM (2015) Considering the impact of climate change on human communities significantly alters the outcome of species and site-based vulnerability assessments. Divers Distrib 21(9):1101–1111

Silva RA, West J, Zhang Y, Anenberg SC, Lamarque J-F, Shindell DT, Collins WJ et al (2013) Global premature mortality due to anthropogenic outdoor air pollution and the contribution of past climate change. Environ Res Lett 8(034005):1–11. https://doi.org/10.1088/1748-9326/8/3/034005

Sims H, Vogelmann K (2002) Popular mobilization and disaster management in Cuba. Public Adm Dev 22(5):389–400

Smit B, Pilifosova O (2003) From adaptation to adaptive capacity and vulnerability reduction. In: Smith JB, Klein RJT, Huq S (eds) Climate change, adaptive capacity, and development. Imperial College Press, London, pp 9–28

Smit B, Wandel J (2006) Adaptation, adaptive capacity and vulnerability. Glob Environ Chang 16:282–292

Southeast Florida Regional Climate Change Compact Counties (2012) Regional climate action plan. A region responds to a changing climate. Retrieved from https://southeastfloridaclimate-compact.files.wordpress.com/2014/05/regional-climate-action-plan-final-ada-compliant.pdf

State, Local, and Tribal Leaders Task Force on Climate Preparedness and Resilience (The Task Force) (2014) Recommendations to the President. Accessed 1 July 20178 https://obamawhitehouse.archives.gov/sites/default/files/docs/task_force_report_0.pdf

Stephenson K, Taylor MA, Stephenson TAC, Bezanilla A, Charlery J (2018) The regional climate science initiative. Caribbean Quarterly 64(1):11–25. https://doi.org/10.1080/00086495.2018.1435314

Stott PA, Christidis N, Otto FEL, Sun Y, Vanderlinden J-P, van Oldenborgh GJ, Vautard R et al (2016) Attribution of extreme weather and climate-related events. Wiley Interdiscip Rev Clim Chang 7(1):23–41. https://doi.org/10.1002/wcc.380

Strauss B, Tebaldi C, Kulp S, Cutter S, Emrich C, Rizza D, Yawitz D (2014) Florida and the Surging Sea: a vulnerability assessment with projections for sea level rise and coastal flood risk. Clim Cent Res Rep:1–58

Taylor MA, Stephenson TS, Chen AA, Stephenson KA (2012) Climate change and the Caribbean: review and response. Caribb Stud 40(2):169–200

The United Nations Office for Risk Reduction (UNORR) (2015) Disaster risk reduction and resilience in the 2030 agenda for sustainable development. http://www.unisdr.org/files/46052_disasterriskreductioninthe2030agend.pdf

The World Bank (2012) The Sendai report: managing disaster risks for a resilient future. Retrieved from https://openknowledge.worldbank.org/bitstream/handle/10986/23745/80608.pdf?sequence=2&isAllowed=y

Tompkins EL (2005) Planning for climate change in small islands: insights from national hurricane preparedness in the Cayman Islands. Glob Environ Chang 15(2):139–149

Torres H, Alsharif K (2017) What it means to become "more resilient": an analysis of local resilience-building approaches in three Florida communities. Weather Climate Soc 9:405–419. https://doi.org/10.1175/WCAS-D-16-0101.1

Torres H, Alsharif K, Tobin GA (2018) Perspectives on adaptive capacity to climate change in hazardous environments: insights from Broward County, Florida. Weather Climate Soc 10:361–372. https://doi.org/10.1175/WCAS-D-17-0094.1

United Nations Development Programme (UNDP) (2015a) A Cuban model for a resilient Caribbean. Retrieved from: http://www.undp.org/content/undp/en/home/presscenter/articles/2015/02/25/a-cuban-model-for-a-resilient-caribbean.html

United Nations Development Programme (UNDP) (2015b) Climate change adaptation – Cuba. Retrieved from http://adaptation-undp.org/explore/caribbean/cuba

Urban MC (2015) Accelerating extinction risk from climate change. Science 348(6234):571–573

Visser H, Petersen AC, Ligtvoet W (2014) On the relation between weather-related disaster impacts, vulnerability and climate change. Clim Chang 125(3–4):461–477

Wamsler C (2014) Cities, disaster risk, and adaptation. Routledge, New York

Wheeler BW, White M, Stahl-Timmins W, Depledge MH (2012) Does living by the coast improve health and wellbeing? Health Place 18(5):1198–1201

Wisner B et al (2004) At risk: natural hazards, people's vulnerability and disasters, 2nd edn. Routledge, London

Wisner B, Gaillard JC, Kelman I (2012) Framing disasters: theories and stories seeking to understand hazards, vulnerability and risk. In: Wisner B, Gaillard JC, Kelman I (eds) The Routledge handbook of hazards and disaster risk reduction. Routledge, London, pp 18–33

Wuebbles DJ, Fahey DW, Hibbard KA, DeAngelo B, Doherty S, Hayhoe K, Horton R, Kossin JP, Taylor PC, Waple AM, Weaver CP (2017) Executive summary. In: Wuebbles DJ, Fahey DW, Hibbard KA, Dokken DJ, Stewart BC, Maycock TK (eds) Climate science special report: fourth national climate assessment, volume I. U.S. Global Change Research Program, Washington, DC, pp 12–34. https://doi.org/10.7930/J0DJ5CTG

Zorrilla CD (2017) The view from Puerto Rico — Hurricane Maria and its Aftermath. N Engl J Med 377:1801–1803. https://doi.org/10.1056/NEJMp1713196

Chapter 8
Post-tsunami Reconstruction and Panchayats: Political Economy Barriers to Effective Implementation. Independent Consultant and Urban Environmental Specialist

Savitha Ram Mohan

Abstract This research chapter presents barriers to post-tsunami housing reconstruction in a developing country context specifically focused on the issues of peri-urban areas and villages located in coastal areas. This chapter can be used as a model to understand post-tsunami reconstruction issues in the Global South. It also presents institutional management and governance challenges that can significantly delay successful implementation of post-tsunami housing projects. India has been hit by a spate of natural disasters to include the Chennai floods of 2015 and the recent floods of Kerala of 2018 and therefore is extremely relevant to understand post-disaster reconstruction challenges.

The tsunami of 2004 affected 3,415,000 people in India alone, killing 2,75,000 persons in Southeast Asia. Although various levels of government and nongovernmental organizations participated in reconstruction, the maximum impact of the tsunami was on villages and therefore the traditional Panchayats (local governance institution). They were the lead organizations that had to make key decisions in post-disaster reconstruction. However, Panchayats faced with limited institutional capacity and experience were overwhelmed. This chapter aims to understand a. the specific role of Panchayats in post-tsunami reconstruction and assess their performance b. to identify the sociopolitical economy factors that affect Panchayats's effectiveness in reconstruction program implementation c. to identify remedial measures to address the identified barriers and recommend performance improvement measures.

Keywords Post-tsunami reconstruction · Inequity · Global South · Governance · Political economy · Panchayats · India

---

S. R. Mohan (✉)
Independent Researcher and Urban Environmental Specialist, Bangalore, Karnataka, India

© Springer Nature Switzerland AG 2020

M. Ghosh (ed.), *Perception, Design and Ecology of the Built Environment*, Springer Geography, https://doi.org/10.1007/978-3-030-25879-5_8

The specific methodology used includes a. secondary research through literature survey, b. case study analysis through study of the performance in the state of Tamil Nadu, and c. application of system, structure, and policy analysis which hypothesizes that for any policy program to be effective, the socioeconomic system in which it operates and the organizations that administer the policy in the form of structural elements need to be facilitative for the success of the program.

In rural India, there are numerous political economy barriers that prevent Panchayats from being effective administrative agencies. State-level agencies are not keen to give greater autonomy to local governance agencies to protect their own interests. Private entities in the form of land developers and real estate agencies affect reconstruction program design in order to promote their land development priorities in prime coastal land, while nonprofits are often driven by the need to attract international funding, improve their public image than truly assist affected communities. Local Panchayats too are often headed by power-hungry leaders who often are willing to sacrifice community reconstruction priorities for monetary gains. Given this situation, the chapter explores ways to improve the performance of Panchayats to deliver reconstruction programs that meet community needs.

Panchayats are greatly impaired in their ability to deliver programs due to lack of sufficient training to handle post-disaster reconstruction programs, lack of clear devolution of function and authority due to multiple organizations that duplicate efforts, and lack of financial and technical capacity to administer reconstruction programs in addition to several other identified critical factors. Given these barriers the Panchayats' ability to become powerful organizations at the grass roots is greatly restricted. Steps need to be taken to address these barriers as recommended.

Panchayats have to be given their due recognition as local institutions of self-governance. They have to be consulted in decision-making especially in the context of disaster management. Panchayats are capable of handling problems at the local level but need technical and financial capacity building. Internal administrative issues and relation with state and central governments have to be addressed for improvement. If these changes are made, the Panchayats can successfully deliver planned reconstruction programs in disaster-affected areas.

Background

The tsunami of 2004 was devastating affecting up to 3,415,000 people in India alone, killing 275,000 persons in Southeast Asia. The waves of fury were generated by an underwater earthquake measuring 9.1 on the Richter scale, the second largest to hit the planet. The government of India reported a loss of 10,881 people and injury to 6913 persons with 5792 persons missing.

The tsunami generated a tidal wave measuring between 3 and 10 m affecting 2260 kilometers of the mainland Indian coastline excluding Andaman and Nicobar Islands. The four southern states of Andhra Pradesh, Tamil Nadu, Kerala, and Pondicherry were badly hit with Tamil Nadu registering the maximum number of deaths of 7983 persons out of total deaths of 10,881. The maximum impact of tsunami was borne by poor fishing communities. Rural areas were the worst hit by the tsunami.

A total of 861 villages were affected by the tsunami. The total number of dwellings lost amounted to 110,826, and the number of cattle lost was 9116. Fishing harbors, fishing boats, and equipment were badly hit. There was considerable damage to physical infrastructure. The Asian Development Bank, World Bank, and United Nations assessed the overall tsunami rehabilitation and relief needs at US$1.2 billion in the needs and damage assessment report of 2005.

The preliminary damage assessment report estimated total damages at $421.7 million. Maximum damage was caused to livelihoods, which came to $567.8 million, followed by housing damage costing $228.5 million. The least damage was in the areas of health and education with cost of restoration as low as $23.6 million, although it is a considerable amount. In terms of states, Tamil Nadu suffered the maximum damage with total damages of the order of $437.8 million and losses of the order of $377.2 million and total of damages and loss being $815 million.

In Tamil Nadu, the total numbers of districts affected were 13 with an agricultural area of 8460.34 ha covered by sand. In addition an area of 5211.79 ha under crops was damaged, while 669.82 ha of land under horticulture were damaged. The total lives lost were 7995, the highest in the country.

Different levels of government agencies worked immediately to restore normalcy. Institutions at the center, state, and local level were involved. Although various levels of organizations participated, the maximum impact of the tsunami was on villages. Traditional Panchayats were the lead organizations that had to effect immediate evacuation and administer relief and reconstruction to affected persons.

Despite the large volume of funds received for reconstruction and rehabilitation, a number of issues came to surface in the relief and reconstruction phase. Although the process seemed perfect in paper in terms of funds being more than adequate for managing the scale of disaster, the implementation of reconstruction program suffered from various deficiencies. Based on primary and secondary research conducted, it emerges that the implementation process suffered largely from lack of public participation and skewed decision-making. Some persons did not receive relief, others did not get housing, and some others were totally left out of receiving any form of assistance.

Panchayats were the institutions that could guide development at the local level and play an important role. However in several cases, they became the key perpetrators of injustice. In some villages they served as exemplary models of local self-governance, while in many others they formed coalitions with developers and nonprofits to promote housing development that did not cater to people's needs. Women and children were not considered equitably in the process and were neglected. An analysis of the causes for these reveals that although policies for handling disasters are in place, and government agencies exist to implement the same, the structures or organizations that implement the same are inefficient.

Stakeholders in Disaster Management

The tsunami of 2004 in India saw numerous stakeholders come into play. This included government agencies, business corporations, nongovernmental organizations, and international development organizations.

At the center, the Ministry of Home Affairs (MHA) was the nodal agency for handling immediate relief. The MHA opened a control room with public lines for information. A national crisis management committee drew up plans for handling the crisis. Individual ministries also took measures and supported relief through financial contributions. The National Calamity Contingent fund contributed a total of $627.81 million to all affected states.

The state governments provided relief through revenue departments under the direction of relief commissioner. District collectors were directed to oversee immediate rescue and relief. State police, fire, and medical services were pressed into action. The army, navy, air force, and coast guard also helped in disposal of dead bodies and evacuation of rescued persons. Immediate measures were taken to prevent epidemics and provide medical attention. Relief camps were organized in Kerala, Tamil Nadu, Andhra Pradesh, and Pondicherry. In addition to the central government assistance, state governments and the Union territory administration also contributed financially.

A number of international, national, and state-level nongovernmental organizations (NGOs) participated. Some of the leading international NGOs included World Vision India, CARE (India), Catholic Relief Service (India), Project Concern International, Echo, and Oxfam. NGOs were involved in a variety of activities.

Private sector contributed in no small measure toward restoration. A total amount of $9.2 million was donated to the prime minister's relief fund. They contributed to prime minister's relief fund as well as to some NGOs. From the international community, the UNICEF (United Nations Institute) was appointed as the lead agency. The UNDP assisted in coordinating the efforts of other UN agencies. While the governments do not approach UN, any assistance given suo moto is accepted.

The Panchayat had a role at the village level but limited to functioning under the umbrella of state and national governments. All of the stakeholders are guided by guidelines set by the National Committee on Disaster Management (NCDM).

National Disaster Management System

The government of India had no provisions for managing disasters until the 2001 Gujarat earthquake.[1] In 2005, the National Committee on Disaster Management (NCDM) was set up. This committee has the prime minister as its chair. In the same year, the disaster management act was promulgated. A national executive

[1]Gujarat earthquake occurred on January 26, 2001, with its epicenter in Kutch, Gujarat, India, leading to loss of 13,800 lives as per estimates of the Bhuj Development Authority, though initial toll was predicted to be as high as 30,000. 167,000 people suffered injury and 1 million houses were damaged. http://www.bhujada.com/

committee was formed to assist the NCDM. This committee prepares the national disaster management plan, which is reviewed and modified every year. The national disaster management plan suggests ways by which disasters can be prevented and sets in place a set of guidelines for minimum relief measures in camps in terms of food, water, sanitation, and compensation for widows.

NCDM clearly delineates functions of state and central government. States manage relief and reconstruction, while finance and logistics are handled by the central government. The extent of government intervention depends on the intensity of the disaster, scale of relief operations required.

State governments are required by NCDM to establish a state disaster management authority. The chief minister heads this agency and has the power to constitute an advisory committee in case of emergency. The state authority frames the state disaster management policy. State executive committees have the power to control traffic, entry and exit of a person. The committee should also provide relief and rehabilitation as per the standards laid down by the national authority and state authority. The chief secretary is responsible for administration. In many states, secretary of revenue administration oversees disaster management.

The disaster management committee also functions at the district level with the district magistrate acting as the exofficio chairman. The district collector is the officer in charge of implementing state-level directives. All the agencies are responsible for coordinating government's policies, response to disasters, and post-disaster relief and rehabilitation.

Central government is authorized to assist by deploying army, navy, and air force and coordinate with international agencies, governments of foreign countries, and institutes of training and research. The act empowers the central government to set up an institute for disaster management. This institute would be responsible for training, research, and documentation of national information on policies, management, and mitigation.

The national disaster response fund was constituted by the central government for emergency response. The fund is available to the national executive committee. The act provides for punishments to noncompliance. The local, district, and state government are authorized to direct audiovisual media to make announcements relating to disasters to inform them. At the state level, a calamity relief fund was set up under the eleventh finance commission. The government of India contributes 75% of the funds, and 25% is contributed by the state. Relief assistance to affected persons is distributed from the calamity relief fund. From the above description, it becomes clear that at the Panchayat level, there is no formal disaster management agency. Neither is there any significant allocation of funds for Panchayats.

The disaster management act although in place has numerous areas for improvement in terms of implementation. First of all most of the plan addresses post-disaster activity than pre-disaster preparedness. Secondly there may be delays in coordination and communication between state and central governments. Such a delay can prove detrimental during a disaster. There is no component of community training in the plans. Another major drawback is lack of field data to handle disasters. Most of the data available is at the district level. Most importantly all the infrastructure in place is directed toward other goals. It is necessary to create a special infrastructure that serves disaster management exclusively (Source: World Bank Institute of Distance Learning).

Traditional Role of Panchayats

The 73rd amendment to the Indian constitution establishes a system of local democracy through Panchayats. India has a total of 600 district Panchayats, 6000 block Panchayats, and 250,000 gram Panchayats. There are about 3 million elected representatives at all levels of Panchayat in India today. These village- and district-level institutions died out in the 1960s after the first two 5-year plans. They were revived in 1992 by an act of parliament to empower local communities and bring social justice. The amendment requires transfer of power and finance to local institutions to enable grassroots participation. The Panchayat is elected every 5 years with one-third of the members being women. The amendment created a three-tier system of village level (gram Panchayat), block level (Panchayat samiti), and district level (Zillah Panchayat) (Source: http://www.thp.org/sai00/india/Panchayat.htm).

There are 29 areas of responsibility. These functions include civic, development, land management, and land reforms and cover the areas of agriculture, natural resource management, infrastructure, industry, and social welfare management and development. Of these women have been given primary responsibility in some areas.[2] As is obvious, disaster management is not included in the traditional functions of the Panchayats.

Panchayats have fixed sources of revenues. These include tax on trade, profession, property tax, license fees, fines, and watch and ward tax. Panchayats may receive grants for a portion of land revenue from the state government. In the second case, Panchayats can collect revenue for land and thirdly can make income from common lands, ponds, etc. In state-sponsored programs, Panchayats have to bring in their own funds to support some portion of the activities. The total revenue that Panchayats have is minimal and insufficient to handle disaster management programs.[3]

Role of Panchayats in Tsunami

During the tsunami of 2004, the Panchayats were overwhelmed by the magnitude of the disaster and were in no position to assist the people. They had no previous training and technical and financial capacity. They saw an onslaught of relief and a rush

[2]Women have been given the right to agriculture development, irrigation, fishery, social forestry, minor forest produce, drinking water, fuel and fodder, poverty alleviation program, education, adult education, health and sanitation, family welfare, women and child development, social welfare, and maintenance of community assets.

[3]Source: http://education.nic.in/cd50years/15/8P/8A/8P8A0802.htm

of NGOs offering relief. Although they were effective in organizing sites for construction of temporary shelters, resuming drinking water supply, collating information on damage to life and property, and distributing relief materials, there was a lot more they could do. Panchayats are the local governance institutions that are faced with disasters much before other levels of governments get to know. Some of the main issues that arose in managing disasters are described below:

Key Issues and Mitigation Measures

(a) ***Absence of Pre-disaster Preparedness***

India does not have a disaster prevention strategy in place. Several countries worldwide have established such plans.[4] India following the tsunami has felt the need to establish an early warning system (EWS). The director of the UN International Strategy for Disaster Reduction has emphasized the need for developing an early warning system[5] (EWS). While there is one for the Pacific area, there was none in place for the Indian Ocean region. UNESCO has launched efforts to create an early warning system for the Indian Ocean region as well. In addition to EWS, it is essential to conduct awareness programs among villagers. Evacuation drills are a necessary activity to train people for emergencies. Panchayats have the ability to develop evacuation programs for villagers.

During the tsunami of 2004, a large number of people surveyed reported that they were taken by surprise and it took more than a few minutes to realize the need to run. Absence of EWS reduces chances of escape. Panchayat leaders have a critical role to play. In some villages, the leaders were the first to leave, while in others they were the last to leave. Panchayats have the responsibility to ensure safety of the residents. They also have the power to guide residents to safety. However if they have no tools, they cannot be blamed.

Mitigation Measures

In order to equip Panchayats to address pre-disaster preparedness, early warning systems have to be installed. Unless villages have information about a potential disaster, there is no way they can plan for evacuation. Panchayats need technological facilities to access information from EWS.

They should also be linked through Internet and telephone service to the nearest district officer and the state government. Such connections would enable Panchayats to seek help ahead of time and move people before the damage is done. Thus Panchayats' physical and technical capacities have to be increased.

---

[4]Japan, a country exposed to earthquakes and tsunami frequently, has a prevention and mitigation plan in place.

[5]A tsunami early warning system comprises of instruments that can detect tsunamis and works through a network to communicate the information to local, state, national, and international agencies to enable early evacuation.

Infrastructure facilities in terms of communication networks, roads, and transport are essential to implement an evacuation plan.

(b) ***Multiple Panchayats: Katta Panchayats Versus Gram Sabha***

In a number of villages, duplication of Panchayats created problems. While Gram Sabha is the government-instituted Panchayat at the village level, the fishing communities have their own Panchayat called the Katta Panchayat. These Panchayats are different from a fishermen's cooperative. Katta Panchayats operate through taxes and are called Varikaarar Panchayat as well where "Vari" means tax.

Three main criteria are employed in selecting members. Members should be fishermen, male, and local residents. All members have to pay taxes to maintain their membership. Women are not considered members. During the recent tsunami, women were left out of the relief packages and housing schemes as the Katta Panchayat does not recognize women members. Katta Panchayat plays an extremely critical role in defining the role of women, and their rights are highly restricted. Villagers felt that Panchayats misused powers as they included and excluded people receiving relief based on their personal connections. Katta Panchayats and their discretionary rules have to be addressed.

Mitigation Measures
The duplication of institutions at the village level has to be addressed. Katta Panchayats have to be integrated with Panchayati Raj institutions. Multiple institutions result in mismanagement of issues. Katta Panchayats have rules that exclude women. These rules have to be modified. As per the 73rd amendment, one-third of the members of the Panchayat have to be women. Katta Panchayats do not adopt election system. In several cases it is more of father to son power transfer. Such a system has to be disabled and democratic process instituted. The district and state governments should intervene to modify or integrate Katta Panchayats with mainstream institutions.

(c) ***Women Are Neglected in Relief and Rehabilitation***

Women as a rule were denied relief and rehabilitation. Since most women are not recognized as members of Panchayat if they did not have a male member in the family, plight of widows was the worst. Although these women have a source of income, they are not allowed to contribute to taxes. Women are not even allowed to voice their opinions in the presence of Panchayat leaders. Due to male-dominated Panchayats, a large number of women did not get housing following the tsunami.

Mitigation Measures
The local, state, and central governments have to intervene to redress the problems faced by women. Gross violation of provisions of the 73rd amendment should not be tolerated. It is appalling to see the condition of widows whose losses are more than widowed men but fail to receive any compensation. A separate committee to monitor the rights of women in villages should be set in place to ensure equal rights for women.

(d) ***Distribution of Relief Following Tsunami***

Panchayats are the agencies that coordinated relief activities. A number of issues arose in relief management. First of all there was unequal distribution of relief. Double counting of individuals led to great variation. Some individuals were left out, and some got more than their share. Second relief did not reach women and orphans. Third some villages did not reveal that they had received relief already. In some cases responsible Panchayats directed surplus relief to neighboring villages, which had been totally left out.

Mitigation Measure

Most villages lack inaccurate database on population. As a result, there is no way to identify the beneficiaries. Panchayats have the ability to collect information on population. Since they are directly in contact with the people, they should be given the responsibility of collecting baseline data on villages with assistance from the census bureau. Data related to persons per household; number of men, women, and children; and deaths and births should be accurately recorded and computerized. Individuals should be given identity cards to regulate distribution of relief.

(e) ***Land Distribution for Reconstruction Efforts***

Following the tsunami land was required for reconstruction activity. Reconstruction in many villages was in situ. In such cases the Panchayat had to acquire land from villagers. Land acquisition was an issue as some landlords owned large parcels of land, while others held very small parcels. The land from the rich landlords had to be acquired and redistributed into uniform parcels of land. The landlords felt the land was forcibly seized, and they were reduced to stay in houses as small as their poor neighbor. Although compensation for land acquired was promised, some landlords complained of not receiving any. Panchayats can handle such issues as they have the authority to do so.

Mitigation Measure

Land is the main source of income for cultivators. In many fishing villages, fishermen and farmers work through barter system. They exchange grains for fish. In one of the villages that was sampled, the construction of housing was the main industry. Fishermen traded fishes for construction of houses. Landowners were reluctant to part with land for two reasons. Loss of productive land would lead to decline in income. Secondly land signifies power within community. While Panchayat leaders were unwilling to part with land they owned in a village, they expected other people to do so.

In order to address these problems, Panchayats have to develop a land redistribution strategy along with compensation packages with the help of state government. Secondly in situ construction should be avoided as far as possible. A large number of undamaged houses are also lost due to such measures. Thirdly people should be given a chance to choose between rebuilding and retaining their old structures. Panchayat leaders cannot impose decisions. Participatory community-based approach to reconstruction should be adopted.

(f) ***Inadequate Consultation in Decision-Making by District and State Governments***

A number of Panchayats complained of being left out of the decision-making process. The district revenue officers were the key decision-makers. Although NGOs interested in adopting villages had to meet with the district officers and the Panchayat before starting relief activities, a large number of Panchayats felt they were bypassed.

Mitigation Measures
Panchayats have to be given their due regard as the local institutions of self-governance. For long they have been sidestepped in decision-making. They are mere puppets in the hands of state and district organizations. Increased devolution of power can only take place through a legislative act or bill introduced by the central/state government. Although all rules are in place, implementation is weak. Implementation mechanisms must be set in place.

(g) ***Financial Situation of Panchayats***

Most Panchayats do not have sufficient revenues to improve infrastructure of their villages. Unless basic infrastructure is in place, evacuation during disaster situations cannot be achieved.

Mitigation Measures
Panchayats must be given access to increased financial resources. The present revenues from use of land are totally inadequate. Panchayat leaders should be paid for their contributions. They should also be treated as government servants with due compensation. Policy and legal measures have to be taken.

(h) ***Lack of Technical Capacity***

Panchayats need technical capacity to administer housing construction by NGOs. This may include training sessions on housing options, materials, technology, etc. A number of NGOs constructed houses with reinforced cement concrete and adopted a minimalist design of spaces. Fishing villages have evolved housing typologies that are sensitive to their way of life over centuries. For example, most houses have kitchens outside. Such open kitchens allow smoke to dissipate. Without conversion to liquefied petroleum gas, villages cannot shift to houses with integrated kitchens provided by NGOs. Houses constructed by NGOs are modular and repetitive. All houses have the same number of rooms. Fishing families have varying number of family members ranging between 2 and 10. For large families NGO-supplied houses are inadequate. Panchayats are unable to raise these issues since they don't have bargaining powers. Panchayats can be at a loss when faced with critical decisions related to programs and policies recommended by NGOs. They do not know what is best for them.

Mitigation Measures
Panchayats should be trained to handle administrative issues and use of information technology. Capacity building programs should be initiated. Local labor, materials,

and technology should be employed to the maximum in order to ensure fruits of labor reach villagers. Environment friendly traditional materials like coconut ropes, casuarinas poles, thatch roofs, and Mangalore tiles should be advocated to NGOs. Damage from these materials during earthquakes is minimal. RCC (reinforced cement concrete) is known to cause maximum damage to people due to its density and lack of flexibility. Nonpartisan technical experts should provide advisory services to Panchayats.

Conclusion

Panchayats have to be given their due recognition as local institutions of self-governance. They have to be consulted in decision-making especially in the context of disaster management. Disaster management is possible through effective public participation only. Panchayats are capable of handling problems at the local level but need technical and financial capacity building. Internal administrative issues and relation with state and central governments have to be addressed for improvement. Basic infrastructure facilities including access to roads, water, and sanitation have to be provided before Panchayats can develop effective village-level disaster management plans. Women, orphans, and minority community roles and rights have to be addressed. Traditional hierarchical structures of Panchayats have to be broken down. With these changes, the Panchayats functioning in tsunami-affected areas will certainly improve and become more effective.

However the barriers to implementation of these suggestions are high. The forces that prevent effective operation of Panchayats may not allow change in power and functions. If Panchayats become truly strong agents of change at grass roots, they may leave agents at other levels redundant. Agents that believe in exploitation of rural people to further their own interests of increasing their unlimited individual financial accumulation may place as many obstacles in the way to ensure weaker sections continue to be exploited and power holders benefit. Political will to eliminate these inefficiencies is missing. Forces of capitalism would not favor loss of cheap labor and decreased access to natural resources. Capitalism would not fail to exploit the opportunity for real estate forces to take over cheap rural land and trade it with rich developers to make it a profitable venture. Panchayat leaders would not give up the chance to become rich men overnight by striking deals with developers to make village land available in exchange for the huge commissions they are promised. NGOs would benefit by advertising their work, building their public image and getting more funding from international donors. The villagers too contribute to the process as they benefit from free houses, new roads, and fresh water, which have long been denied to them. The only people who suffer in the process and clearly don't benefit are lower-caste persons, women and orphans. They are considered as unproductive and rejected by all. They do not serve to benefit the power holders and relegated to the bottom so that power holders can grow by climbing on their shoulders while pushing them down below and convincing them of their need to stay on at the bottom. This section serves the contented class and remains in that wrung of economic ladder as determined by the rich and powerful.

Inferences

Social disparities are not a result of sudden changes or random events. Instead it is something that is built into a society through the set of policies that are framed, the structure that exists, and the systems that support it. Policies are the most evident ones. We use it on a day-to-day basis and interact with it all the time. In the case of disaster management, there are a range of policies in place, which give directions on what needs to be done, when, and how. Structure refers to the dominant social groups that are responsible for implementation and framing of policies. In most societies such groups get differentiated on the basis of economics. High-income groups exert greater control on societal decisions and policies than low-income groups. Systems are the underlying core values of a society that drives everyone. This can be capitalism, feudalism, or socialism. Each of these systems strongly dictates which section of society is powerful and which is not and is absolutely responsible for the policies that are framed.

In order to understand the disparities in post-tsunami reconstruction and rehabilitation, we need to look at the structure, systems, and policies of the country. Panchayats are the agents of local self-governance in India. They serve as the interface between national- and international-level organizations, nonprofits, and other stakeholders translating funds into decisions that help people that form the grass roots of any society. Panchayats have been in existence prior to India's independence. They went through a series of troughs and crests in terms of their powers and functions until they were strongly reinstituted back as strong agencies of local self-governance under the 73rd and 74th amendment to the constitution. In order to understand the Panchayats and their functions in relation to disaster management, it is important to understand the institutions that govern them, the policies that operate, the revenues they receive, and specific problems that surfaced due to lack of sufficient capabilities of Panchayats.

Bibliography

Barenstein JD, Housing Reconstruction in Tamilnadu one year after Tsunami. Scoula Universatria Professionale della Suizzera Italiana, 2005, Switzerland

Chandran P (2004) Role of national disaster management system in the context of South Asia tsunami. World Bank Institute, Washington, DC

Gaonkar S (2004) Critical assessment of India's disaster management system in response to the tsunami disaster of 2004. National Institute Disaster Management and World Bank Institute, Washington, DC

Gautham S, Katta Panchayats' denying relief to women, 2007, India Together. http://indiatogether.org/2007/apr/soc.kattarule.htm#continue

Gomathy NB, Pattinavar Panchayats-II, Post-Tsunami Tri-Net Workshop, 2006

Government of Tamilnadu, *India Emergency Tsunami reconstruction project*, Environmental and Social Assessment framework, Tamilnadu

Meduri A, *Disaster Management Act*, ICFAI University Press, ICFAI Journal of Environmental Economics, Vol. V, No.2, 2006

National Consultation on Disaster Management, Chennai, 2005

Raghunandan TR, Strengthening Grassroots Level Planning- policy and Institutional issues

South Asia Regional Knowledge Program *Role of Panchayati Raj Institutions in Post-tsunami Reconstruction and Rehabilitation*, 2005

South Asia Regional Office, *Project Appraisal Document on Karnataka Panchayats Strengthening Agriculture and Rural development*, 2006, World Bank

World Bank, Asian Development Bank, *India Post Tsunami Recovery Program, Preliminary Damage and Needs Assessment*, Asian Development Bank, World Bank, 2005, New Delhi, India

Chapter 9
Design for Resilience: Traditional Knowledge in Disaster Resilience in the Built Environment

K. Mitra

Abstract Design is a loaded term that encompasses diverse viewpoints. Loon (Inter-organizational design: a new approach to team design in architecture and urban planning. In Proceedings of the 5th Design & Decision Support Systems Conference in Architecture and urban Planning. Nijkerk, Netherlands. August, 2000) interprets the term 'designer' to include anyone who has an impact on design, irrespective of the individual's professional background. It follows then that optimum design is the consensual design solution that is considered optimum for the largest number of people. People will have diverse responses to what constitutes optimum. These responses are likely to be dependent on a host of factors including gender, profession, occupation, health, race, religion, age, environmental experience and attitudes, to name just a few. Thus, 'optimum' will not necessarily be the 'best looking design' or the 'most economic design' or even the 'most functional design'; it will be the solution that best balances issues considered important to the largest section of people. Such a solution should ideally ensure maximum comfort and sense of well-being for all participants. This chapter looks at design within the domain of traditional knowledge systems and shows how communities residing in some of the most disaster-prone areas in the world, such as the Himalayas, have "designed" resilient environments that have withstood the ravages of hazardous events, for example, earthquakes. Unfortunately, these traditional design skills which were handed down through generations are no longer evident in their places of origin. The easy availability and economy afforded by reinforced concrete in even the most remote parts of the country, along with the associations of permanence (of the home) and prosperity (of the family) with this material, have resulted in the hybridization of traditional masonry constructions in different seismic zones of India.

Experiences from several past earthquake.s have shown that in many cases, traditional structures have performed remarkably well, while newer, "engineered" structures have not. Traditional construction, in this discussion, does not refer to

K. Mitra (✉)
Department of Architecture, Town and Regional Planning, Indian Institute of Engineering Science and Technology, Shibpur, Howrah, West Bengal, India
e-mail: keyamitra@arch.iiests.ac.in

© Springer Nature Switzerland AG 2020
M. Ghosh (ed.), *Perception, Design and Ecology of the Built Environment*, Springer Geography, https://doi.org/10.1007/978-3-030-25879-5_9

historic structures—though there are many examples of good earthquake performance in this category of buildings—but rather encompasses the vernacular residential constructions made with locally available materials and using indigenous knowledge. A number of such traditional earthquake-resistant practices exist in the Himalayan region, one of the most tectonically active in the world. Some of the most effective of these are *Dhajji-diwari* and *Taq*, around the Srinagar area in Kashmir, *Ikra* construction in Assam, and *Shee-Khim*, in Sikkim. This chapter describes some of these traditional construction techniques and shows how these are effective as earthquake-resilient systems.

Keywords Earthquake resilience · Traditional knowledge · Design typologies

Introduction

Anyone who has an impact on design, irrespective of the individual's professional background, can be termed a 'designer' (Loon 2000). It may be reasonably argued then that optimum design is the consensual design solution that is considered optimum for the largest number of people. These responses are likely to be dependent on a host of factors including gender, profession, occupation, health, race, religion, age, environmental experience and attitudes, to name just a few. Thus, 'optimum' will not necessarily be the 'best looking design' or the 'most economic design' or even the 'most functional design'; it will be the solution that best balances issues considered important to the largest section of people. Such a solution should ideally ensure maximum comfort and sense of well-being for all participants. This chapter looks at design within the domain of traditional knowledge systems and shows how communities residing in some of the most disaster-prone areas in the world have 'designed' resilient environments that have withstood the ravages of hazardous events, for example, earthquakes.

Built Environment and Resilience

Earthquake-resistant construction practices are not new to India or, for that matter, to human civilization. The city of Knossos (the Minoan capital) had in-built disaster resilience mechanisms such as locating buildings away from the reach of tsunamis, avoiding valleys for construction purposes on account of their vulnerabilities to floods and tsunamis and using of timber beams and joints for improving resilience against earthquakes (Main and Williams 1994). Inca settlements in the Andes addressed issues of seismic safety through restricting size of settlements, ensuring that buildings were located well apart from each other to avoid damage due to pounding, eliminating low walls, interlocking stone blocks for better structural bonding and other measures (Main and Williams 1994, p.17). In India, as far back as 1931, S.L. Kumar, a young railway engineer, successfully built several bungalows with earthquake-resistant features. These structures performed well during the

1935 earthquake in Quetta, Balochistan, that caused widespread devastation in the built environment (Jain 2005).

Traditional Knowledge Systems

Traditional knowledge systems (TKS) have been a part of the mainstream narrative in the fields of medicine, ecology and social sciences. The research literature acknowledges the importance of the conservation of this knowledge (eg., Gadgil et al. 1993; Folke 2004). It has been established that, on the one hand, traditional knowledge and related institutions increase capacity to cope with change, while on the other, traditional knowledge and beliefs tend to erode with adoption of modern technology. Drawing a parallel with the built environment, traditional knowledge systems in the design and construction of the built environment have received some attention globally, as it has been proven time and again that these are often the most optimum for the societies where they have evolved. For example, traditional practices in seismic areas, which have evolved over time, using locally available materials have offered increased seismic resistance along with good climate control. Some international examples of earthquake-resistant architecture include the *Himis style* of construction in Turkey (Gülkan and Langenbach 2004; Güçhan 2007), *Bahareque* construction in El Salvador (Bommer et al. 2002; López et al. 2004), timber houses in Nepal (Dixit 2004; Shakyaa et al. 2012), adobe houses in Yugoslavia and other parts of Eastern Europe (Dutu et al. 2012; Hrasnica 2009), confined masonry construction in Latin America and Central Europe (Brzev 2007; Langenbach 2007; Wood et al. 1987; Audefroy 2011) and *Dhajji-diwari*, *Taq*, *Shee-Khim* and *Ikra* in different parts of the earthquake-prone Himalayan Belt in India (Jigyasu 2002; Alkazi 2014).

Traditional Earthquake-Resistant Construction in the Himalayan Belt

The Himalayan Belt represents the boundary between two major tectonic plates (Indo-Australian Plate and Eurasian Plate), with the Main Boundary Thrust (MBT) and Main Central Thrust (MCT) coinciding with the Himalayan arc that forms the northern border of the Indian subcontinent spanning a distance of approximately 3200 km, stretching from Kashmir in the north-west to Arunachal Pradesh in the north-eastern tip of India. This tectonic plate boundary is a convergent boundary where the Indian Plate (at the north-western tip of the Indo-Australian Plate) is actively subducting into the Tibetan Plate (part of the Eurasian Plate). This belt has witnessed some of the largest earthquakes throughout geological time, 1897 Assam (M8.7), 1905 Kangra (M8.0), 1934 Bihar-Nepal (M8.3) and 1950 Assam-Tibet

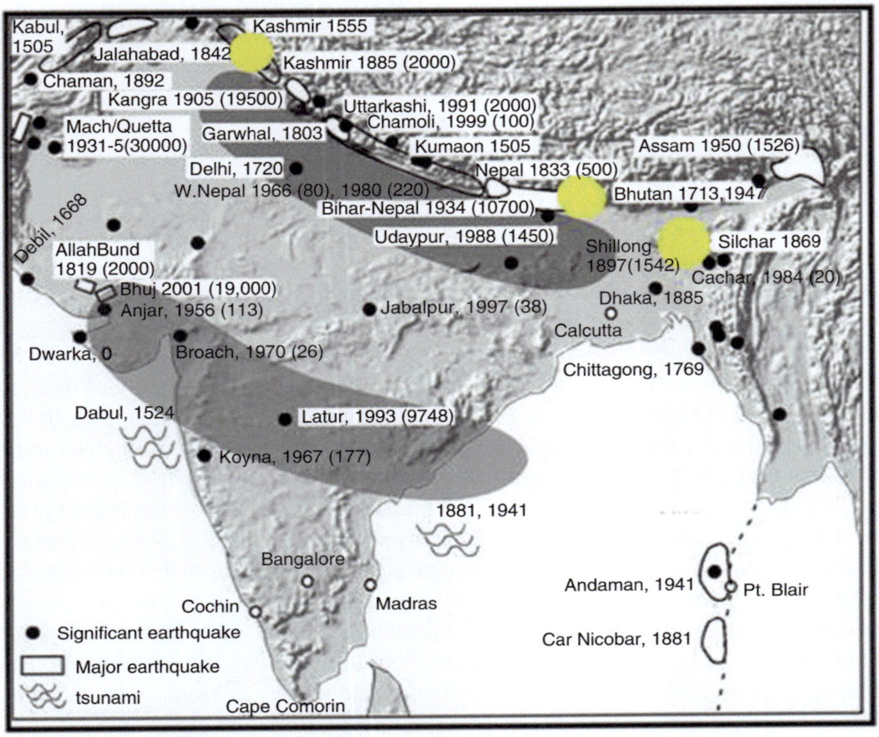

Fig. 9.1 Historical record of earthquakes in the Indian subcontinent showing the locations of three traditional earthquake-resistant systems. (Adapted from Bilham 2004)

(M8.6), and more recently, 1988 Bihar-Nepal (M6.6), 1991 Uttarkashi (M 6.4), 1999 Chamoli (M6.6), 2005 Kashmir (M 7.6), 2011 Sikkim (M6.9) and 2015 Gorkha Earthquake (M) and its aftershocks. The communities inhabiting these regions have, over time, developed resilient systems against earthquakes and other natural hazards (Fig. 9.1).

Dhajji-Diwari *and* Taq

The *Dhajji-diwari* style of construction is typically found in the Srinagar area of Jammu and Kashmir, in seismic zone V of the seismic zone map of India (IS1893:Part 2 2016), representing exposure to the most severe seismic hazard. The building typology uses three locally available materials, namely, stone, timber and clay bricks, in different combinations along the height of the building. Thus, stonemasonry is used in the plinth level, and also sometimes in the lower storeys, while a combination of brick masonry confined with timber members is used in the upper storeys.

Construction at the foundation and sometimes up to the plinth and even ground floor levels is with local stones, usually rounded in profile, as they are sourced from the river beds. The brick masonry in the upper storeys uses a lime-based mortar (locally known as *lime surki*), which uses lime and *surki*, which is an aggregate made of crushed bricks. These brick masonry panels are confined with timber members, placed vertically, horizontally and diagonally. *Taq* construction techniques are also seen wherein the vertical and diagonal wooden framing members are not used and the walls are made of brick with horizontal wooden bands placed at regular intervals along the entire height of the structure. The sloped roofs are constructed by putting panels of corrugated galvanized iron (CGI) sheets on a timber framework of trusses that rest on the confined brick masonry walls. In rural areas, the timber framework is often topped with thatch, finished with an application of a wet clay paste. In houses with more than one storey, the intermediate floors are made of timber planks above a timber floor grid that rests on the walls. Seismic safety can be achieved in a building through proper conformance to some well-known architectural and structural concepts that have proved to be beneficial in improving the seismic performance of structures.

The wood frame houses in the Srinagar region are typically two to four storeys tall. Plan configurations are compact and centralized. Floor plans are typically rectangular. The main entrance doorway leads into a small, square space which has a narrow dog-legged staircase opposite to the entrance. In the upper floor, the staircase leads to the lobby space with two rooms on either side of it like in the ground floor.

The foundation is strip foundation with stonemasonry. Often, the plinth masonry is 900 mm thick and made with locally available large-sized stones, coursed or uncoursed irrespective of the material used for the walls in the upper storeys (Fig. 9.2). Sometimes the ground floor walls are also of random rubble stonemasonry with a thickness of 600 mm, but usually brick masonry is used above plinth level where the brickwork is framed within a set of wooden members in the traditional *Dhajji-diwari* style.

The plinth masonry is laid in a shallow trench marginally wider than the wall width and about 600–750 mm deep. The plinth stops at about 600 mm from the natural ground level. A plinth beam made of timber is placed on the plinth masonry at plinth level. The plinth beams in two orthogonal directions are secured by nails or metal plates for better connections (Fig. 9.3).

Walls in the ground floor including plinth masonry are often made of stone-masonry, both, random rubble and dressed and coursed/uncoursed. In the upper floors, thin bricks are laid in horizontal courses with horizontal, vertical and diagonal ties that reduce the area of unreinforced masonry panels and help to confine the infills (Fig. 9.4) and prevent out of plane collapse during earthquakes.

The thicker walls in the ground storey often do not use diagonal members, while these are quite common in the thinner walls in the upper storeys, as the drift due to earthquake shaking increases along the height of a structure (Fig. 9.5). Buildings have CGI roofs on wooden trusses though in earlier times roofs were made of timber planks. Many variations in geometry are evident for roof design including (a) gable,

Fig. 9.2 Use of large stones and continuous wooden plinth bands. (Photos: Author)

Fig. 9.3 A metal plate connecting the timber plinth beam in two orthogonal directions. (Photos: Author)

Fig. 9.4 Two-storeyed house, rectangular in plan in the ground floor with diagonally placed room projections on the upper floor. (Photo: C.V.R Murty)

Fig. 9.5 Vertical and horizontal wooden frame members on the upper floor with diagonal framing at the corners of both faces of the building. (Photo: Author)

Fig. 9.6 Horizontal, vertical and diagonal bracing members in the upper section of the gable wall for improving out of plane performance

(b) split roof and (c) roof with dormer windows (Fig. 9.6). The roof structure is a truss with closed triangles that are desirable for best performance. Roof bands are used, and a variant of this is the use of the crossbeam with the roof band.

The suitability of the *Dhajji-diwari* system for earthquake resistance lies in the structural integrity that is achieved through the use of timber horizontal, vertical and diagonal bracing members that can deform without losing their strength and that do not allow the infill panels to fail out of plane. The strength of the system therefore lies in the adequacy of the bracing members and their appropriate placement in the most vulnerable sections of a structure, namely, corners, overhangs, gable ends, around openings, etc. However, not all *Dhajji-diwari* buildings have performed well in past earthquakes. Some common deficiencies of poorly constructed *Dhajji-diwari* structures include use of low-quality materials, inadequacy of wooden bracing members in all vulnerable locations, use of incomplete trusses on the roof and lack of connections between the structural elements such as foundations, walls and roofs. Need for adequate insulation necessitates the use of double walls, which, in *Dhajji-diwari*, are typically brick walls, where baked bricks are used for the external *wythe*, while the interior panels are made of unbaked bricks. Lack of connections between the two *wythes* contributes to the vulnerability of the system as the walls behave as independent panels with low out of plane strength and are liable to collapse or get severely damaged during strong earthquake shaking.

Ikra and the Assam-Type Construction

The typical housing in the hills of the Guwahati Region in northeast India is built both on flat lands and hill slopes. These single-or, at most, two-storeyed structures are built on raised plinth as a safeguard against surface runoff.

'Ikra' structures are built mainly for the residential purposes of the common people. The so-called Assam-type structures owe their origin to *Ikra* construction. *Ikra* typology makes use of a range of materials used for structures ranging from non-permanent (*kachcha*) to semi-permanent to permanent (*pucca*) structures. Basic features of the *Ikra* house are thatched roof, bamboo walls plastered with a mixture of mud and cow dung and bamboo splints woven together and fitted inside the wooden frame plastered with mud mortar. The bamboo adds stiffness to the mud, and being a flexible material, it also brings ductility to the system.

Ikra houses are usually low-rise structures, not more than two storeys in height (Figs. 9.7 and 9.8). In two-storeyed structures, the ground storey is made of conventional load-bearing construction, while the upper storey has lighter construction using wooden members. Simple rectangular plans are used for smaller structures, while L-and C-shaped structures are used for multifamily houses or larger structures. The sloped roofs with tall gable walls are needed to allow quick runoff during heavy rainfall. Roofs are usually sloped with a high gable to drain off the heavy rainfall.

Fig. 9.7 Single-storeyed *Ikra* structure. (Photo: Author)

Fig. 9.8 Double-storeyed *Ikra* house. (Photo: Hemant Kaushik)

Ikra structures are made largely using wood-based materials. A weed, called 'Ikra', grows wildly in river plains and adjoining lakes across the state of Assam, and this material is extensively used in the walls and roof (Fig. 9.9). The wall panels are made of bamboo with infill panels made of vertically oriented, mud-plastered shoots of the *Ikra* reed (Fig. 9.10). The covering on the roof truss is a thick stack of *Ikra* reed or, for the more affluent, metal sheets.

The wooden framework for the *Ikra* panels is made of either bamboo or wood. The wooden frames are plastered on both the sides with mud mortar. Three layers of plaster are applied one after another, waiting for a coat to dry before application of the next coat. After all the layers of plasters are fixed and firm, a final finishing is given with a coating of a liquid mixture of mud and cow dung (Fig. 9.11).

These structures have no formal foundations, as such. The main wooden verticals of the superstructure continue below the ground to depths of about 600–900 mm. In more formal constructions (so-called Assam-type structures), the main wooden posts of the house are supported on masonry or plain concrete pillars constructed over the ground up to plinth or sill level (Fig. 9.12). The connections between wooden posts and the pillars are achieved using steel bolts and U-clamps. Later foundations were made of plain concrete mats (generally in plain cement concrete (PCC) of grade 1:3:6) over which pedestals of same grade were raised up to plinth level of buildings. Wooden posts were fixed to these pedestals with the help of iron clamps (Fig. 9.13).

Fig. 9.9 Bundles of 'ikra'

Fig. 9.10 Schematic sketch of *Ikra* panel

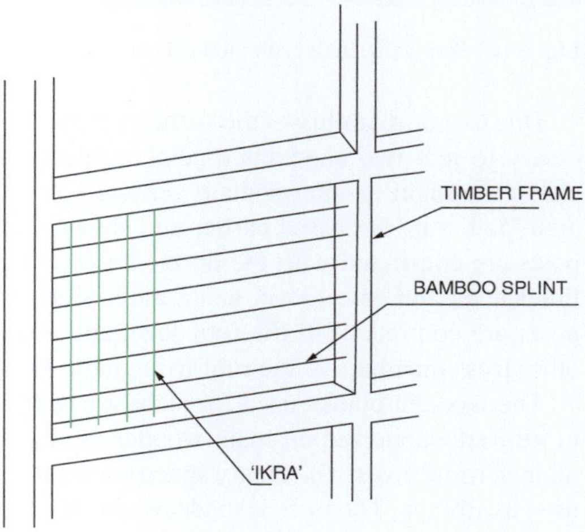

An important aspect of this typology is the joinery between the various elements – the posts, wall panels, roof truss and roofing elements. In Assam-type structures, connections were achieved with nails and bolts. In the informal construction, coir ropes are used to connect the various elements. The latter raises concerns on durability of the connection materials and thereby on the safety of the house. One of the most important connections is at the plinth level between the vertical main posts and the supporting pedestal. The connection is achieved by U-clamps and bolts.

Fig. 9.11 *Ikra* wall panels with mud plaster

Due to unavailability of the sufficient length of vertical posts, it is sometimes necessary to join two elements together (splicing), using bolts (Fig. 9.14a). In some cases, the main or intermediate vertical posts are embedded inside the plain concrete pedestals discussed earlier and shown in Fig. 9.11. The vertical intermediate posts are connected with the horizontal wooden members at floor level, sill level, lintel level and eaves level using nails, steel clamps and bolts. The main vertical posts are continued till the roof level and connected to the horizontal rafters and other truss members of the roof using nails, bolts and steel clamps (Fig. 9.14b–g).

The wooden planks used for slabs are supported on intermediate rafters, which in turn are supported on main wooden beams at ends that transfer the load to the main vertical posts. The empty space between the slab and the roof truss is generally used as storage. The truss is made of wooden members that support the tin or asbestos roofing.

Ikra structures are known to have a number of strengths that influence earthquake safety of the house. These include:

(a) Architectural aspects: good plan shape, small openings, appropriate location of openings, e.g. away from corners, and small projections and overhangs
(b) Structural features: light mass of walls and roofs, good wall-to-wall connection (in case of formal construction), good quality and strength of materials used
(c) Flexible connections (bolting, nails, grooves, etc.) between various wooden elements at different levels

Fig. 9.12 Brick masonry pedestals supporting the vertical timber posts

Moreover, in *Ikra,* bamboo is used as the main structural element. Bamboo imparts ductility in the system leading to good earthquake performance. The lightweight material, owing to lower seismic weight, helps to reduce the earthquake-induced inertia forces in the structure leading to better seismic performance.

The system does, however, have a number of shortcomings. The choice of wood as the basic construction material and thatch (in rural areas) as roofing material of the house draws high maintenance and is vulnerable to fire. To a large extent, the fire hazard to the house is mitigated, when the kitchen is separated from the main house but placed within the courtyard of the house. But the use of electricity in such houses leaves possibilities of fire due to short-circuit during earthquake shaking. In urban areas, the roof has long been converted to metal roofing; hence this is not an area of concern.

The mud-dung plaster on walls requires a lot of maintenance and frequent application. During summers, it becomes brittle and then falls off easily during the rainy season. In rural areas, the thatch on the roof is vulnerable to suction under strong winds.

When the wooden vertical posts are directly plugged into the ground without any foundations, structures have sunk up to 300 mm. Sometimes, differential sinking of

Fig. 9.13 Connection details between vertical wooden posts and plain concrete plinths

the vertical posts can lead to lateral sway of the house and tearing. The problem is aggravated in sites with high water table but can be mitigated by providing the vertical posts with stone piers or plain cement concrete as a foundation.

Use of *Ikra*-type construction in hill slopes has some inherent problems. On hill slopes, the unequal lengths of the vertical posts lead to unsymmetrical shaking.

Shee-Khim Construction of Sikkim

Shee-Khim is the traditional style of construction, practiced in Sikkim, and most prevalent in Upper Sikkim. *Shee-Khim* houses are of single-storeyed wooden plank construction. These are made of wooden frames and planks, supported on wooden posts. Random rubble masonry is used in foundation. The floors and double-pitched roofs were of timber construction, using single post beam system.

Traditional structure in Sikkim can be classified based on the type of material used. The two predominant categories are (1) wood houses, e.g. *Ikra* and *Shee-Khim* (Fig. 9.15), and (2) masonry houses.

Shee-Khim structures have four types of plinths according to the slope profile. These are (1) random rubble masonry (RRM) with and without mud mortar; (2) dry

Fig. 9.14 (**a**) Splicing; (**b**) connection between vertical posts and horizontal rafters at verandah; (**c**) connection between vertical posts and horizontal rafters at eaves level; (**d**) connection between vertical post and wooden slab; (**e**) connection between vertical post, horizontal rafters and inclined roof member at eaves level from inside; (**f**) connection between vertical post, horizontal rafters and inclined roof member at eaves level from outside; and (**g**) connection between vertical post, horizontal rafters and asbestos sheet used for roofing

Fig. 9.15 *Shee-Khim* house. (Photo: Sutapa Joti)

dressed stonemasonry; (3) dressed stonemasonry; and (4) dressed stonemasonry with pointing. In the hilly slopes, tapered stone plinth is preferred, while uniform masonry made of stone and mud mortar is used in flat ground. Both mud and lime-based mortar are used for bonding. Mud with fine river sand plaster is used in interiors, while the exterior has exposed stone finish without plaster.

Traditional *Shee-Khim* structures performed excellently in both the earthquakes of 2006 and 2011 (Kaushik and Dasgupta 2012) due to a number of factors. Symmetrical and simple geometric configuration in plan is excellent for earthquake resistance. The horizontal bands help to increase lateral strength capacity, while the closely packed wooden frames prevent the spread of diagonal crack, while the closely spaced vertical members also restrict the diagonal shear and out of plane collapse. The reduction of mass in the upper storey results in lower earthquake-induced inertia forces at roof level.

Concluding Remarks

The traditional earthquake-resistant systems discussed in the preceding sections fall within a rich sample of such typologies that have evolved over centuries in some of the most earthquake-prone regions of the world. These can all be said to belong within the broad umbrella of confined masonry systems with excellent connections, flexibility and ductility, stability, strength and structural integrity. Earthquake resistance is achieved through damping and shock absorption, horizontal tying actions, reduction of span between supports, enhancing out of plane stability and consequent containment of masonry. Well-constructed samples of these traditional systems are amongst the best examples of vernacular earthquake-resistant construction and may be taken up as models for new vernacular constructions in earthquake-prone areas. Rural housing programmes such as the Pradhan Mantri Awas Yojana (PMAY) initiative, for example, add huge numbers to the housing stock, involving enormous fiscal allocations. Aspiring house builders in this owner-driven programme should be encouraged to construct their houses in these earthquake-resistant technologies, using locally available materials through the construction of model PMAY houses in these technologies in district headquarters and by conducting masons' training programmes. Traditional housing construction techniques should be brought back in a significant way. Appropriate research is required to be undertaken to develop better understanding on critical aspects of the said traditional housing, especially the quantitative understanding of the earthquake resistance of such housing. These practices need to be given wider publicity amongst the various stakeholders – homeowners, NGOs, governments, artisans, financial institutions and contractors.

Acknowledgements The author gratefully acknowledges C.V.R Murty, Hemant B Kaushik and Sutapa Joti for some of the photographs and the support from BMTPC for conducting surveys in Srinagar and Guwahati.

Bibliography

Alkazi F (2014) Srinagar: an architectural legacy. Roli Books Private Limited

Audefroy JF (2011) Haiti: post-earthquake lessons learned from traditional construction. Environ Urban *23*(2):447–462

Bilham R (2004) Earthquakes in India and the Himalaya: tectonics, geodesy and history. Ann Geophys 47(2-3)

Bommer JJ, Benito MB, Lemoine A, López-Menjívar MA, Madariaga R, Mankelow J, de Hasbun PM, Murphy W, Nieto-Lovo M, Rodriguez-Pineda CE, Rosa H (2002) The El Salvador earthquakes of January and February 2001: context, characteristics and implications for seismic risk. Soil Dyn Earthq Eng 22(5):389–418

Brzev S (2007) Earthquake-resistant confined masonry construction. NICEE, National Information Center of Earthquake Engineering, Indian Institute of Technology Kanpur

Dixit AM (2004) Promoting safer building construction in Nepal. In: 13th World Conference on Earthquake Engineering.

Dutu A, Ferreira JG, Guerreiro L, Branco F, Goncalves AM (2012) Timbered masonry for earthquake resistance in Europe. Mater Constr *62*(308):615–628

Folke C (2004) Traditional knowledge in social–ecological systems. Ecol Soc *9*:3

Gadgil M, Berkes F, Folke C (1993) Indigenous knowledge for biodiversity conservation. Ambio:151–156

Güçhan NŞ (2007) Observations on earthquake resistance of traditional timber-framed houses in Turkey. Build Environ *42*(2):840–851

Gülkan P, Langenbach R (2004) The earthquake resistance of traditional timber and masonry dwellings in Turkey. In: 13th World Conference on Earthquake Engineering, pp 1–6

Hrasnica M (2009) Damage Assessment for Masonry and Historic Buildings in Bosnia and Herzegovina. In: Damage assessment and reconstruction after war or natural disaster. Springer, Dordrecht, pp 333–356

IS 1893(Part 1) (2016) Indian Standard Criteria for Earthquake Resistant Design of Structures. Bureau of Indian Standards, New Delhi

Jain SK (2005) The Indian earthquake problem. Curr Sci 89(9):1464–1466

Jigyasu R (2002) Reducing disaster vulnerability through local knowledge and capacity, The case of Earthquake Prone rural communities in India and Nepal.PhD thesis, Faculty of Architecture and Fine Art Department of Town and Regional Planning.Norwegian University of Science and Technology

Kaushik HB, Dasgupta K (2012) Observations on performance of structures during 18 September 2011 Sikkim (India) Earthquake.15 WCEE, LISBOA

Langenbach R (2007) From "Opus Craticium" to the "Chicago Frame": earthquake-resistant traditional construction. Int J Archit Herit 1(1):29–59

Loon PP (2000) Inter-organizational design: a new approach to team design in architecture and urban planning. In: Proceedings of the 5th Design & Decision Support Systems Conference in Architecture and urban Planning. Nijkerk, Netherlands. August 2000

López M, Bommer J, Méndez P (2004, August). The seismic performance of bahareque dwellings in El Salvador. In: Proceedings of the Thirteenth World Conference on Earthquake Engineering, Vancouver, paper (No. 2646).

Main H, Williams SW (eds) (1994) Environment and housing in third world cities. Wiley, Chichester

Shakyaa M, Varuma H, Vicentea R, Costaa A (2012) Structural vulnerability of Nepalese Pagoda Temples, 15WCEE LISBOA 2012

Wood SL, Wight JK, Moehle JP (1987) The 1985 Chile earthquake: observations on earthquake-resistant construction in Viña del Mar. University of Illinois Engineering Experiment Station. College of Engineering. University of Illinois at Urbana-Champaign

Chapter 10
Sustainable Planning Interventions in Tropical Climate for Urban Heat Island Mitigation – Case Study of Kolkata

Santanu Bajani and Debashish Das

Abstract Urbanization plays an important role in the development of urban heat islands (UHI) in urban areas. UHI is a phenomenon where the urban built landscape shows a higher air temperature conditions than the surrounding hinterland and rural landscapes. Heat island also shows a lesser fall in temperatures than the adjacent outskirts and rural areas. The atmospheric concentration of GHGs has led to global climate change and because of the frequency of heat waves led to heat stress-related mortality and morbidity in mega high-density and compact cities located in the tropical and subtropical region of Southeast Asia. It is also observed that the city is divided into various microclimatic zones, and these zones are always in the process of continuous change due to unplanned and haphazard growth in urban built form.

In this study, the openness of the urban built form which can be measured by the sky view factor ratio and calculated by various techniques to find the distribution pattern in these various microclimatic zones is calculated and analyzed. SVF is the ratio of the radiation received (or emitted) by a planar surface to the radiation emitted (or received) by the entire hemispheric environment called the sky view factor or SVF (Watson ID, Johnson GT: Int J Climatol 7:193–197, 1987). The scale of SVF is ranged between 0 and 1, where SVF is 0 which means sky is fully obstructed and SVF is 1 which means there are no obstructions. Urban environments appear as uneven artificial terrains with building materials partly different from those of natural surfaces. Generally, its strongest development occurs at night when the heat, stored in the daytime, is released (Landsberg HE: The urban climate. Elsevier, New York/London, 1981; Oke TR: Boundary layer climates. Methuen Publishers, Lagos, 1987; Wienert U, Kuttler W: Meteorol Z 14(5):677–686, 2005). In Kolkata, narrow streets and high buildings create deep canyons. This 3D geometrical configuration plays an important role in regulating long-wave radiation heat loss. Due to the fact that only a smaller part of the sky is seen from the surface (because of the horizontal and vertical unevenness of the surface elements), the outgoing long-wave radiation loss here is more restricted than in rural areas.

S. Bajani · D. Das (✉)
Department of Architecture, Jadavpur University, Kolkata, India

© Springer Nature Switzerland AG 2020

M. Ghosh (ed.), *Perception, Design and Ecology of the Built Environment*,
Springer Geography, https://doi.org/10.1007/978-3-030-25879-5_10

This study outlines the urban built form typology in various microclimatic zones in Kolkata and its impact on land surface temperature and urban heat island variations. The chapter analyzes the relationship between the aspects of urban built form and sustainable planning interventions which are proposed to ameliorate the impact of urban heat island in Kolkata for improving the quality and usability of outdoor spaces in terms of human health and sustainable future.

Keywords Sky view factor · Local climatic zones · Land surface temperature · Urban heat island · Sustainable planning

Introduction

Urban local climate has been dramatically changed by human interactions. Intensive urbanization has caused impervious surfaces to sprawl (Chatterjee et al. 2019). Thermally massive structures installed in the cities absorb, store, and reradiate the extra solar energy per unit area as compared to the green (vegetated) surfaces that dominate their rural surroundings (Wang et al. 2016). The proportion of latent heat transfer through evaporation reduces due to the increased surface runoff from the impervious surfaces of roads, pavements, and buildings (Sodoudi et al. 2015). Modified urban fabric directly affects the urban climate, unbalanced surface energy budget, observance of the urban heat island (UHI) effects, and consequently the effects on microclimate of the city. Tropical cities are reported to be more sensitive to global warming phenomena (Cutter et al. 2014). In respect of UHI characteristics, the tropical cities are different from the temperate cities which remain warmer during a larger part of the day throughout the year (Van Hove et al. 2015). It is also important to note that, in general, the UHI effect is less intense in the tropics compared to that in the midlatitudes and that the cooling effects in a tropical climate are modulated largely by wetness (Charabi and Bakhit 2011; Ojeh et al. 2016).

The intensity of the UHI depends strongly on the land use and urban parameters (e.g., built-up ratio, green surface ratio, sky view factor (SVF), etc.) characterizing the immediate environment obscured by the surroundings of the site of the measurement (Golany 1996; Oke 1987; Unger 2004). Sky view factor (SVF) is an important component to understand the UHI effects of any urbanized region. SVF can be simply defined as “a measure of the degree to which the sky is obscured by the surroundings for a given point” (Grimmond et al. 2001). SVF values are dimensionless and range between 0 and 1, representing totally obstructed and open spaces, respectively (Oke 1988). The air temperature in urban canyons is not dependent solely on urban and street geometry but is also governed by complex and regional factors (Bärring et al. 1985).

Local climate zones (LCZ) are defined as regions of uniform surface-air temperature distribution at horizontal scales of 10^2–10^4 m (Stewart and Oke 2009). The interaction between the urban surfaces and the atmosphere is governed by the surface heat flux (Dousset and Gourmelon 2003; Stewart et al. 2012). Based on the characteristic of the urban surface, we used the LCZ classification system proposed by Stewart and Oke (2012).

Remote sensing, although challenged by the spatial and spectral heterogeneity of urban environments (Jensen and Cowen 1999; Herold et al. 2005), seems to be an appropriate source of urban data to support such studies (Donnay et al. 2001). Reduced sky view factor, replacement of soil cover by concrete/asphalt surfaces, and emission of a large amount of waste heat from transportation, commercial, residential, and industrial sectors of all form conditions favorable conditions to the rise in temperatures over the urban areas as compared to surrounding of rural areas (Pandey et al. 2012).

The Kolkata Municipal Corporation (KMC) exhibits fast and haphazard urban expansion in the last decade, in relation with certain thermal behavior owing to some characteristic features that distinguish it from the other cities. Kolkata belongs to the tropical savanna climate ("Aw" type as per Köppen's scheme). This chapter aims to analyze and evaluate the relationship between various urban surface and urban built form characteristics by developing various local climate zones in Kolkata and understanding the urban heat island pattern with respect to sky view factor and land surface temperatures.

Methodology

The research proposes the study of sky view factor (SVF) and land surface temperature (LST) variations in different microclimate zones of Kolkata, an organically evolving, mostly unplanned and densely populated Indian metropolitan city. To investigate all the selected parameters, a number of methods and techniques have been used. Existing literature on the said components is studied to identify and calculate the zones or values. Remote sensing and GIS techniques were used during the analysis stage.

Small subsets from the selected built-up areas have been opted in order to maintain the analysis and manage the computational capacity. The spatial dimension of all the selected areas has the width and length of 100 m. Google Earth images were also used for creating and editing files. The analysis was done on a typical cool day of 11th November 2017. A large number of urban parameters with their complex interrelationships such as urban typology, urban materials, and urban vegetation that influence the microclimate of urban built environments need to be examined in order to formulate plans for making climate-resilient cities (Perini and Magliocco, 2014; O'Malley et al. 2015). These factors are mutually and reciprocally affected by each other in the spatiotemporal context (Ozkeresteci et al. 2003). The parameters have been assessed and evaluated. A comparison has been made among different built forms.

Methods to Identify LCZ

This study employs a microclimatic analysis using WUDAPT (World Urban Database and Access Portal Tools) methods for Kolkata Municipal Corporation, India, estimated using Google Earth image, Landsat image, and SAGA GIS to identify the local climate zones (Figs. 10.1 and 10.2).

The boundary of LCZ has been identified and estimated using Google Earth and field verification regarding surface structure and surface cover. The specific type has been assigned to all the LCZ. The climate zones have been identified and categorized as per LCZ classification developed by Stewart and Oke in 2012 (Table 10.1).

Methods to Estimate SVF

SVF consists of a dimensionless value between 0 and 1 obtained as a ratio between the radiation received by a planar surface and that from the entire hemispheric radiating environment (Watson and Johnson 1987). SVF is calculated on selected locations by using SVF mapping tool V1.1 developed by Urban Climate Research Group, University of Szeged, Hungary, and sky view factor calculator software developed by Fredrik Lindberg and Björn Holmer of the University of Gothenburg, Sweden. The equipment used for collecting SVF photographs is the Nikon DSLR (D610) camera with the Nikkor fish-eye lens attachment.

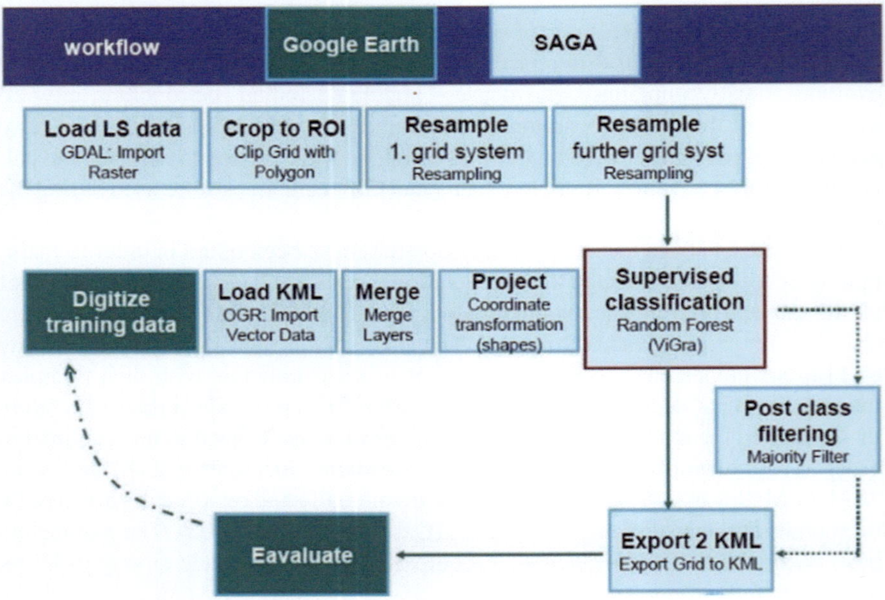

Fig. 10.1 Workflow of the LCZ classification in SAGA

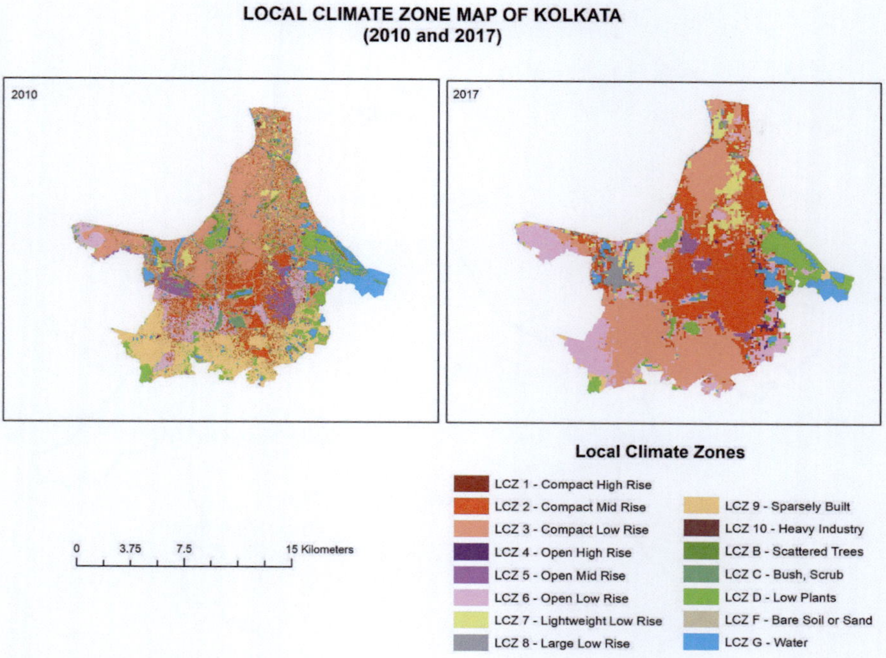

Fig. 10.2 Local climate zone map in the year 2010 and 2017

The **fish-eye photographs** were processed by sky view factor calculator software developed by Fredrik Lindberg and Björn Holmer of the University of Gothenburg. Five sequential steps were needed to calculate SVF value from a photo graph (Fig. 10.4).

Besides fish-eye photographs methods, **GIS-based techniques** were also used to calculate SVF, for the same region. For this calculation "SVF mapping tool V1.1" developed by Urban Climate Research Group, University of Szeged, is used. For the analysis, we need building polygon in shape (.shp) file format with height field and point layer for the location from where SVF value is calculated. The output is in point shape format (Fig. 10.5 and Table 10.2).

Methods to Estimate Land Surface Temperature

The earth surface composed of different materials which arise some complication for land surface temperature estimation (Qin and Karnieli 1999). LST is one of the most important parameters in the physical processes of surface energy and water balance at local through global scales (Brunsell and Gillies 2003; Kustas and Anderson 2009), and UHI is a consequence of LST.

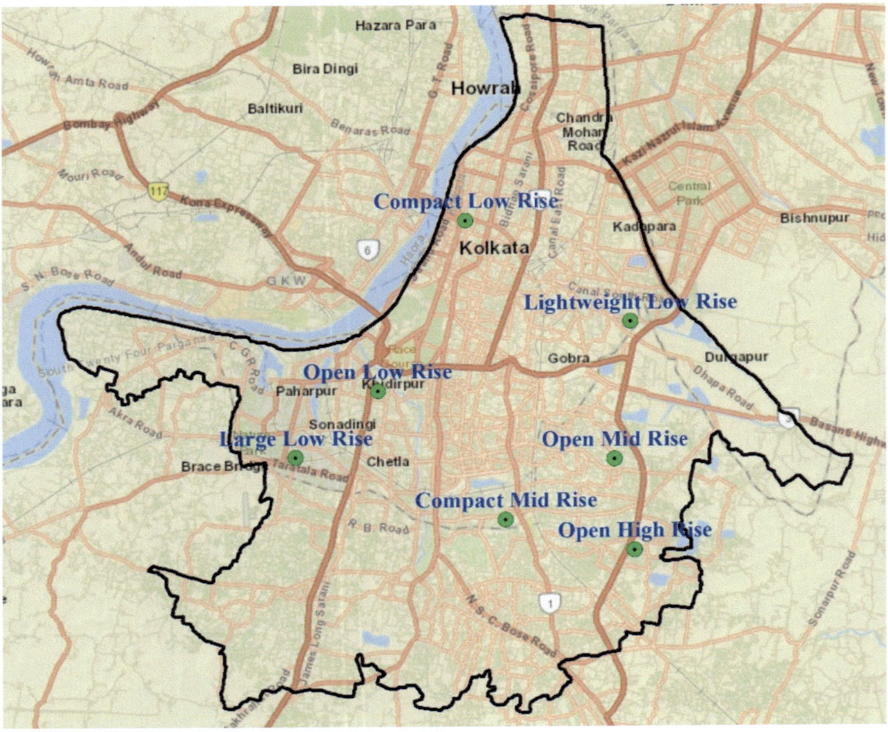

Fig. 10.3 Locations of the sample case study areas

Table 10.1 LCZ class and locations of sample case study area

LCZ class		Location
LCZ2	Compact mid-rise	Jadavpur
LCZ3	Compact low rise	Ahiritola
LCZ4	Open high rise	EM Bypass
LCZ5	Open mid-rise	Ruby Park
LCZ6	Open low rise	Outskirt Alipur
LCZ7	Lightweight low rise	Chowbaga
LCZ8	Large low rise	Garden reach

LST was calculated by applying a structured mathematical algorithm, viz., split-window algorithm. It uses brightness temperature of two bands of TIR (thermal infrared) and mean and difference in land surface emissivity for estimating LST of an area.

Accurate values of LST can be only obtained if surface emissivity is well characterized (Jimenez-Munoz et al. 2012); as a matter of fact, a variation of 0.01 units in emissivity implies an error of 0.5 K in the temperature. There are alternative option to estimate emissivity from images, i.e., the TES (Gillespie et al. 1998) and normalized

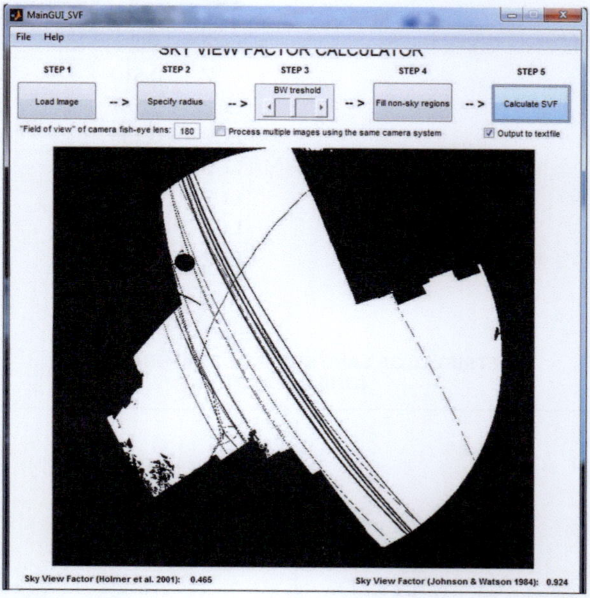

Fig. 10.4 Layout of the sky view factor calculator

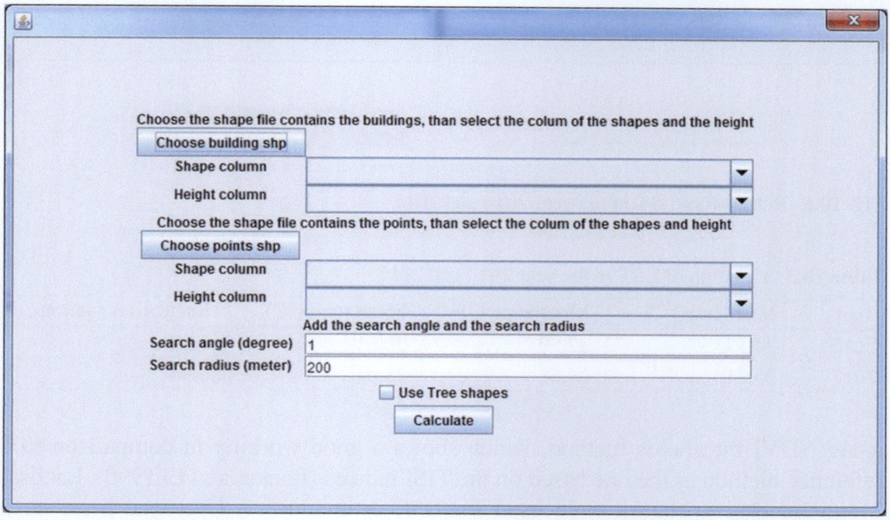

Fig. 10.5 Layout of the SVF mapping tool V1.1

difference vegetation index (NDVI) threshold methods (Sobrino et al. 2004); nevertheless, the best approach to mapping emissivity over urban areas is classification-based (Jimenez-Munoz et al. 2012). In these study, emissivity map was elaborated by

Table 10.2 SVF values in the selected study area

LCZ's	SVF value in fish-eye lens	SVF value in SVF calculator
LCZ 2	0.510	0.795
LCZ 3	0.587	0.765
LCZ 4	0.736	0.856
LCZ 5	0.725	0.815
LCZ 6	0.621	0.818
LCZ 7	0.546	0.639
LCZ 8	0.570	0.598

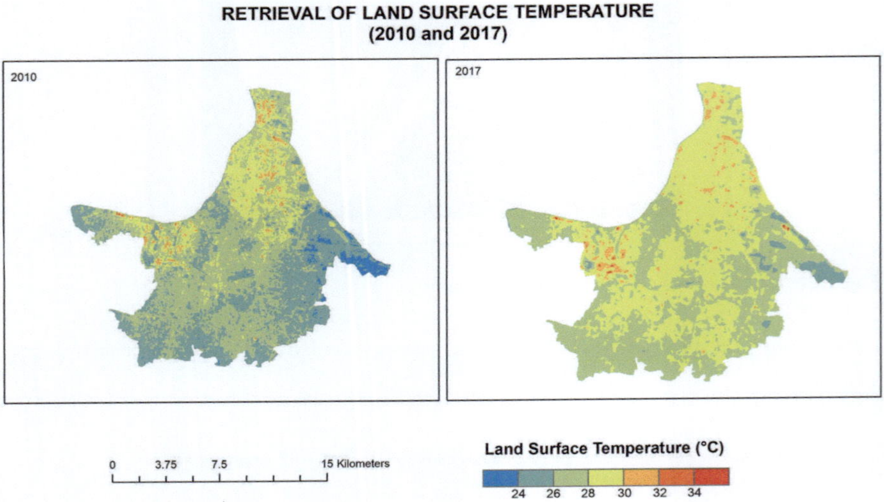

Fig. 10.6 Retrieval of LST in the year 2010 and 2017

Table 10.3 Variation of LST in the year 2010 and 2017

Year	Max temp(°C)	Min temp(°C)	Mean temp(°C)	Standard deviation
2010	34.13	22.59	26.61	1.50
2017	33.81	24.68	28.04	0.96

using NDVI thresholds method, which shows a good working in comparison to a reference method as the one based on the TISI indices (Becker and Li 1990). Landsat 7 and Landsat 8 images were used for LSE estimation on Universal Transverse Mercator projection system (datum WGS84, UTM zone N45) and resampled using the nearest-neighbor algorithm with a pixel size of 30 m (Fig. 10.6, Tables 10.3 and 10.4).

Table 10.4 Mean LST and LCZ variations within the selected study sample area in the year 2010 and 2017

2010		2017	
LCZs	Mean LST (°C)	Mean LST (°C)	LCZs
LCZ2	26.2	28.01	LCZ2
LCZ3	28.23	29.48	LCZ3
LCZ4	24.44	27.58	LCZE
LCZ5	25.07	27.48	LCZ6
LCZ6	26.11	27.43	LCZ6
LCZ7	28.49	28.89	LCZ7
LCZ8	29.36	31.46	LCZ8

Results and Discussion

Evaluation and Validation of Study

Analyzing the related literature and analysis of the data, the evaluation methods for the relationship between the screen level (air) temperature or UHI intensity (ΔT) and SVF were established. The KMC area is mainly characterized by compact urban built form with concrete road pavements and a high percentage of compact mid-rise and compact low rise. The microclimate zone was examined using WUDAPT methods and a variety of meteorological parameters. To validate LCZ, numerous sampling points are compared with the corresponding point on Google Earth images in the same period. Besides this LST was validated by using local meteorological data of the same period and the SVF calculated using the two algorithms, one is by using fish-eye lens and another is GIS-based SVF calculator software.

In the following, the results of some important investigations are summarized. Estimation of SVF for an entire urban environment in the real world is never an easy task, for instance, seven sites are selected as the sample area. From these points SVF values were calculated with both methods resulting in two SVF values for each point; UHI intensity was calculated and also identified LCZs.

The Relationship Between SVF and LST

Using known surface geometry metrics, a rapid and straight forward method was used for calculating the SVF in Kolkata. The relationship between LST and SVF is investigated through statistical correlation analysis. In Fig. 10.7, the scatter plots represent this relationship between SVF and LST (i.e., x- and y-axis, respectively). The qualitative analysis of the results reveals a positive relationship between LST and SVF. SVF values differ because of the varying height of buildings and their degree of compactness. Compact mid-rise building areas had the minimum SVF due to their lower base compactness and a low urban canopy layer. From the selected

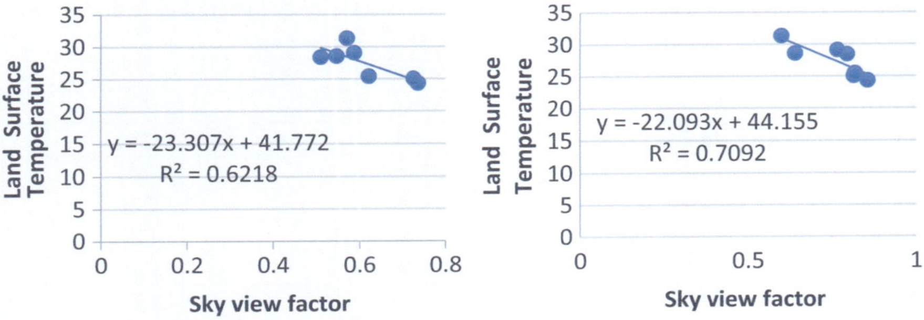

Fig. 10.7 Correlation coefficient between LST and SVF in photographic method and GIS method (2017)

seven-sample area, the SVF value is retrieved, and the ranges are within 0.510–0.736 (fish-eye photo), which corresponds to the old dense urban core comprising more values, and the function trend decreases significantly. Within the sample sites, the maximum SVF value of 0.736 and 0.856 is found at Open high rise as per photographic calculation method and GIS calculation method, respectively. There are some variations on SVF values in photographic and GIS calculation method because of the presence of vegetation cover and overhead urban facilities (i.e., electric posts, cables, hoardings, etc.). In GIS calculation methods, these facilities are ignored.

Figure 10.7 shows the relationship between SVF and LST of 2017 in both photographic and GIS methods. The correlation value is slightly stronger in GIS methods ($R^2 = 0.7092$) than the SVF value of the photographic method ($R^2 = 0.6218$). In contrast with this, the compact mid-rise and compact low-rise built-up area consists of less SVF value. The deciduous leaf cover is most likely responsible for the unusually low mean SVF value. This relationship reveals that open urban building environment received a greater amount of solar energy and shading effect through the trees could reduce the solar exposure, which would eventually result in a cooler environment at a diurnal scale. The urban thermal heterogeneity in the large low-rise urban environment consists of lower SVF (0.598) value but absorbs more temperature (31.34 °C) due to mainly lower green cover and metal roof. Hence, as per analysis, the open building environment was cooler than that in compact building areas.

The Relationship Between LST and LCZ

The distinctive LST patterns are associated with the thermal characteristics of the land cover types (Lu and Weng 2007 and Lo 1997). The analysis herein is based on the assumption that individual LCZs should demonstrate certain features typical of a given LST regime. LST fields were overlaid with LCZs, and typical LSTs were calculated for each zone. Differences between mean LCZ temperatures were evaluated by one-way analysis of variance.

Kolkata is highly urbanized with large areas classified as built-up LCZs according to the LCZ maps. Downtown areas of Kolkata Municipality, especially north and center part, are extremely dense and compact and are mainly classified into LCZs 2, 3, 5, 6, 7, and 8 which show the potential high UHI intensity. There is no obvious gap between the downtown area and the suburbs, representing the high urbanization and a large area of potential UHI distribution. However, the samples in some classes such as LCZs 1, 4, 5, and 7 and B and E only have a few pixels. The LST map identifies hot areas in the downtown areas in the southwest and north of Kolkata. It is also noted that large areas of high temperature occurred in the east Kolkata, over Dhapa. From Table 10.5, it can be observed that there is a large LST variation across LCZ classes. In particular, LCZ 8 (large low rise) has the highest LST among the built-up LCZs. From Fig. 10.8 it's observed that LCZ E (rock or paved cover) has the highest LST among the land covering LCZs. There is a declining trend of LST from LCZ 3 to LCZ 6; this may be due to the complex and diverse urban morphology of these LCZs in Kolkata, resulting in great temperature differences within the LCZ classes. Some of the LCZ E in Kolkata represents the port and areas built up with concrete and steel, so they may have similar thermal characteristics with the built-up LCZs. LCZ G (Water) has the lowest LST among all the LCZs (Table 10.6).

The Relationship Between SVF and LCZ

Kolkata has a high building density with a moderate green cover and many low-rise buildings. Around 60% area is covered by compact mid-rise (LCZ 2) and compact low rise (LCZ 3). LST is strongly related to the solar access of a surface. The surfaces of the densest areas warm more slowly, given the presence of buildings and obstacle shadows. Despite the differences among the results, one of the advantages of SVF analysis is the possibility to obtain a general view of an urban area in opposition to an actual analysis. Regarding the results of this research, by analyzing the existing morphologies in some selected place of Kolkata, the average sky view factors were shown in Table 10.1.

Compact mid-rise and compact low rise have the lowest SVF values, indicating the denser urban morphology and lower height-width ratio with respect to open low rise and open mid-rise. Compact high rise (LCZ 1) is observed in the eastern part of the city with some scattered patches. Kolkata is characterized by mainly moderate SVF values. The increased building height greatly increases the surface roughness over the cities, thereby enhancing the frictional drag against the flow of wind and affecting UHI.

Conclusion and Recommendations

Climate modifications at the local level are the most familiar influences caused by land use/land cover changes due to urbanization. Very often, temperatures over the central parts of a city are found to be significantly higher than those in the

Table 10.5 Temperature map as per selected LCZ (2017)

LCZ	Actual view from Google Earth	Temperature range in different LCZs	Temperature scale
LCZ 2 Compact mid-rise			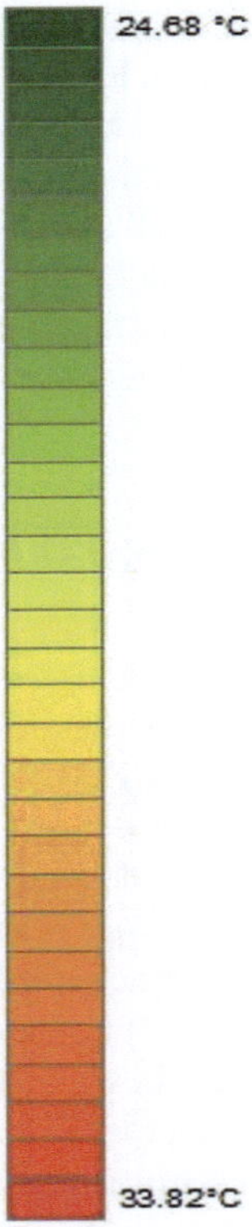
LCZ 3 Compact low rise		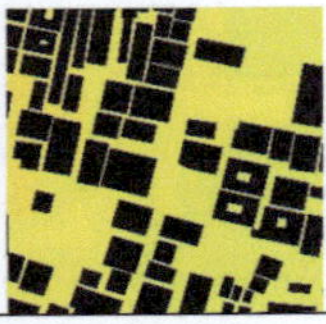	
LCZ 4 Open high rise		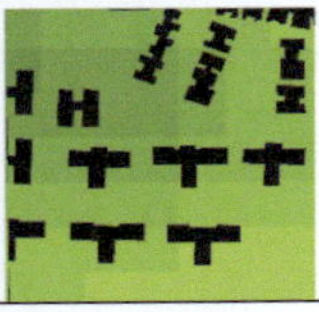	
LCZ 5 Open mid-rise			
LCZ 6 Open low rise		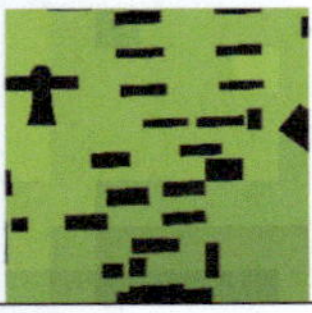	
LCZ 7 Lightweight low rise		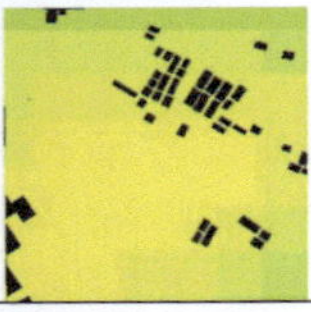	
LCZ 8 Large low rise			

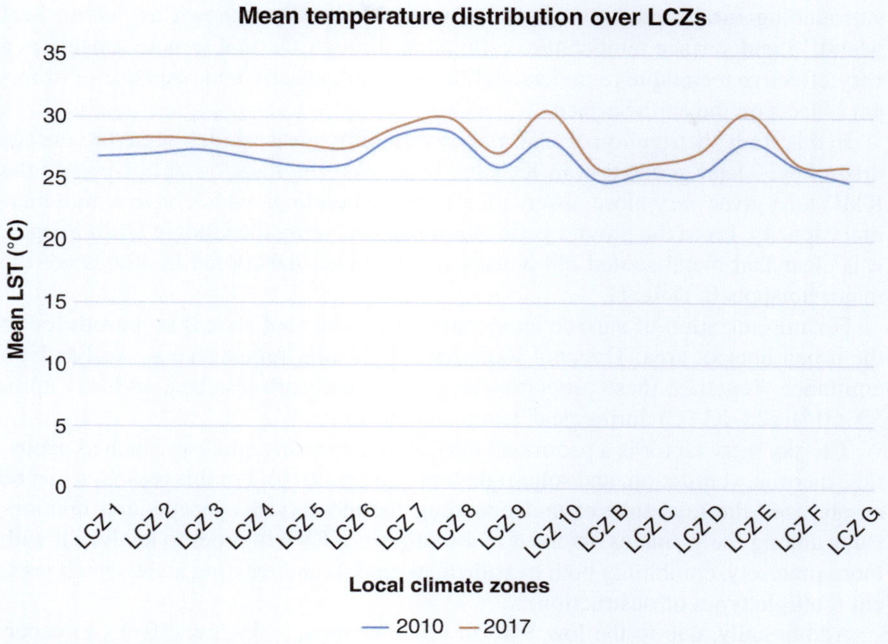

Fig. 10.8 Mean temperature distribution over LCZs (2017)

Table 10.6 Mean LST variations as per LCZ category in the year 2010 and 2017

LCZs	Mean LST (°C)2010	Mean LST (°C)2017
LCZ 1	26.87	28.15
LCZ 2	27.06	28.40
LCZ 3	27.39	28.28
LCZ 4	26.76	27.79
LCZ 5	26.10	27.82
LCZ 6	26.32	27.56
LCZ 7	28.64	29.29
LCZ 8	28.92	30.03
LCZ 9	26.02	27.15
LCZ 10	28.25	30.56
LCZ A	–	–
LCZ B	24.93	25.88
LCZ C	25.56	26.26
LCZ D	25.43	27.15
LCZ E	28.05	30.36
LCZ F	25.97	26.25
LCZ G	24.84	26.04

surrounding rural areas, giving rise to what is popularly known as "urban heat island." Land surface temperature estimation through thermal remote sensing is a very effective technique to understand the thermal capacity and response of different objects on the earth surface.

In this study distribution of LST and its variability gives an idea about the surface urban heat island generation in Kolkata. Moreover, the location of hotspots in the KMC area gives very close observation of those buildings which have a high thermal capacity. From the ground verification and analyzing the Google Earth images, it is clear that metal-roofed old houses and extended flat-roofed buildings are the major hotspots in Kolkata.

For minimization of surface temperature, the cool roof should be introduced in the urban hotspot area. The cool roofs have high solar reflectance as well as high emittance. Together, these properties help roofs to absorb less heat and stay up to 50–60 °F (28–33 °C) during peak summer weather.

The sky view factor is a parameter that influences many analyses, such as acoustics, thermic, ventilation, and solar radiation (Cheng 2010). For this reason, it can be emphasized the relevance of understanding the theoretical concepts and relationships among the variables involved in the subject of SVF in order to analyze it with more precisely, combining both hypothetical models and existing areas which present multiple types of obstruction.

Additionally, due to the low SVF in urban centers, a shadow effect can occur, which actually results in urban cooling especially during periods of low sun angle in the early morning. Due to these discrepancies, great care must be taken to not associate low SVF values directly with high levels of urban heating, even though some studies have found a correlation between SVF and urban air temperatures (Svensson 2004). Instead, a comprehensive approach which analyzes a range of additional factors such as albedo, heat capacity, surface emissivity, and surface roughness in tandem with SVF values would be more appropriate for accurately evaluating the magnitude of the UHI effect. The study considers the impact of building geometry on the urban microclimate. Other determinants, such as vegetation, topography, anthropogenic heat release, and transportation-related heat fluxes, are expected to reveal more comprehensive understanding in future works. From the comparative study of 2010 and 2017 of building and the urban environments, microclimate study of urban built form is one of the most important approaches toward planning for climate-resilient cities alongside experimenting with mitigation strategies.

References

Bärring L, Mattsson JO, Lindqvist S (1985) Canyon geometry, street temperatures and urban heat island in malmö, sweden. J Climatol 5(4):433–444

Becker F, Li ZL (1990) Temperature-independent spectral indices in thermal infrared bands. Remote Sens Environ 32(1):17–33

Brunsell NA, Gillies RR (2003) Length scale analysis of surface energy fluxes derived from remote sensing. J Hydrometeorol 4:1212–1219
Charabi Y, Bakhit A (2011) Assessment of the canopy urban heat island of a coastal arid tropical city: The case of Muscat, Oman. Atmos Res 101(1–2):215–227
Chatterjee S, Khan A, Dinda A, Mithun S, Khatun R, Akbari H, Kusaka H, Mitra C, Bhatti SS, Van DQ, Wang Y (2019) Simulating micro-scale thermal interactions in different building environments for mitigating urban heat islands. Sci Total Environ 663:610–631
Chen L, Ng E (2011) Quantitative urban climate mapping based on a geographical database: a simulation approach using Hong Kong as a case study. Int J Appl Earth Obs Geoinf 13(4):586–594
Cutter SL, Ash KD, Emrich CT (2014) The geographies of community disaster resilience. Glob Environ Change 29:65–77
Donnay JP, Barnsley MJ, Longley PA (2001) Remote sensing and urban analysis, 1st edn. Taylor & Francis, London/New York
Dousset B, Gourmelon F (2003) Satellite multi-sensor data analysis of urban surface temperatures and landcover. ISPRS J Photogramm Remote Sens 58(1–2):43–54
Gillespie A, Rokugawa S, Matsunaga T, Cothern JS, Hook S, Kahle AB (1998) A temperature and emissivity separation algorithm for Advanced Spaceborne Thermal Emission and Reflection Radiometer (ASTER) images. IEEE Trans Geosci Remote Sens 36(4):1113–1126
Golany GS (1996) Urban design morphology and thermal performance. Atmos Environ 30(3):455–465
Grimmond CSB, Potter SK, Zutter HN, Souch C (2001) Rapid methods to estimate sky-view factors applied to urban areas. Int J Climatol 21(7):903–913
Herold M, Hemphill J, Dietzel C, Clarke KC (2005) Remote sensing derived mapping to support urban growth theory. In: 3rd international symposium remote sensing and data fusion over urban areas (URBAN 2005) and 5th international symposium remote sensing of urban areas (URS 2005)
Jensen J, Cowen D (1999) Remote sensing of urban/suburban infrastructure and socio-economic attributes. Photogram Eng Remote Sens Photogrammetric Eng Remote Sens 65(5):611–622
Jimenez-Munoz JC, Sobrino JA, Gillespie AR (2012) Surface emissivity retrieval from airborne hyperspectral scanner data: insights on atmospheric correction and noise removal. IEEE Geosci Remote Sens Lett 9:180–184
Kustas W, Anderson M (2009) Advances in thermal infrared remote sensing for land surface modeling. Agric For Meteorol 149(12):2071–2081
Landsberg H (1981) The Urban Climate 28, 275
Lo C (1997) Integration of landsat thematic mapper and census data for quality of life assessment. Remote Sens Environ 62(2):143–157
Lu D, Weng Q (2007) A survey of image classification methods and techniques for improving classification performance. Int J Remote Sens 28:823–870
Ojeh V, Balogun A, Okhimamhe A (2016) Urban-rural temperature differences in Lagos. Climate 4(2):29
Oke TR (1987) Boundary layer climates, 2nd edn. Methuen Publishers, Lagos
Oke TR (1988) Street design and urban canopy layer climate. Energ Build 11(1–3):103–113
O'Malley C, Piroozfar P, Farr ERP, Francesco Pomponi F (2015) Urban Heat Island (UHI) mitigating strategies: a case-based comparative analysis. Sustain Cities Soc 19:222–235
Ozkeresteci I, Crewe K, Brazel AJ, Bruse M (2003) Use and evaluation of the Envi-Met model for environmental design and planning: an experiment on linear parks. In: Proceedings of the 21st international cartographic conference (ICC). Durban, South Africa, pp 10–16
Pandey P, Kumar D, Prakash A, Masih J, Singh M, Kumar S, Jain VK, Kumar K (2012) A study of urban heat island and its association with particulate matter during winter months over Delhi. Sci Total Environ 414:494–507
Perini K, Magliocco A (2014) Effects of vegetation, urban density, building height, and atmospheric conditions on local temperatures and thermal comfort. Urban For Urban Green 13(3):495–506

Qin Z, Karnieli A (1999) Progress in the remote sensing of land surface temperature and ground emissivity using NOAA-AVHRR data. Int J Remote Sens 20:2367–2393. https://doi.org/10.1080/014311699212074

Sobrino J, Jiménez-Muñoz JC, Paolini L (2004) Land surface temperature retrieval from LANDSAT TM 5. Remote Sens Environ 90:434–440

Sodoudi S, Fallah B, Walter BS, Schubert S (2015) The ability of mesoscale climate model COSMO-CLM with the double canyon urban canopy scheme to simulate the urban heat island in Berlin. Toulouse, France, s.n

Stewart ID, Oke TR (2009) Newly developed "thermal climate zones" for defining and measuring urban heat island "magnitude" in the canopy layer. In: Eighth symposium on urban environment. J8.2A

Stewart ID, Oke TR (2012) Local climate zones for urban temperature studies. Bull Am Meteorol Soc 93(12):1879–1900

Svensson MK (2004) Sky view factor analysis – implications for urban air temperature differences. Meteorol Appl 11(3):201–211

Unger J (2004) Intra-urban relationship between surface geometry and urban heat island: review and new approach. Clim Res 27:253–264

Van Hove LWA, Jacobs CMJ, Heusinkveld BG, Elbers JA, Van Driel BL, Holtslag AAM (2015) Temporal and spatial variability of urban heat island and thermal comfort within the Rotterdam agglomeration. Build Environ 83:91–103

Wang Z, Xing W, Huang Y, Xie T (2016) Studying the urban heat Island using a local climate zone scheme. Polish J Environ Stud 25:2609–2616

Watson ID, Johnson GT (1987) Graphical estimation of sky view-factors in urban environments. Int J Climatol 7(2):193–197

Wienert U, Kuttler W (2005) The dependence of the urban heat island intensity on latitude A statistical approach. Meteorol Z 14(5):677–686

Chapter 11
Factors Leading to Disposal of Toxic and Hazardous Sacred Waste and Its Effect on Urban River Contamination: Case of Adi Ganga, Kolkata, India

Srijita Chakrabarty

Abstract Urban rivers are very crucial for any city and often have to bear the burden of pollution and various other discharges from its catchment area. There are many such waste streams which flow into urban rivers, out of which one is that of sacred waste. It is the waste generated out of worship or puja offerings, mostly from Hindu ritual practices. In Hinduism, rivers are treated as gods and nature is given high respect. In fact, that is the reason that the largest and the national river of India, Ganga, is referred to as mother, but strangely enough it is one of the five most polluted rivers in the world. It is this conflict that forms the basis of the research. For this research, Adi Ganga is chosen which is considered as the original flow of the Ganga to the Bay of Bengal as it flows past Kolkata. However, with time, this channel became more of a gutter, and the main flow of the Ganga is now the Hooghly by Kolkata. In spite of this, Adi Ganga is still considered extremely holy, and all the rituals associated with Ganga is still being performed here. Now, the sacred waste has few toxic and hazardous components, and this research tries to explain the factors leading to disposal of toxic and hazardous sacred waste disposal.

To study this, case study approach was decided upon with three sub-cases. The sites were chosen such that basic characteristics remain same but the scale varies. From the literature two main factors came up – individual or internal and situational or external factors. While there are other factors in play, the most prevalent individual factor for Site A is noticed to be values and personal norms, while for Site B and Site C, the most prevalent individual factor is (lack of) knowledge. The most prevalent situational factor for all the sites was found to be subjective norm. As an outcome of the research, the recommendations are to consider sacred waste as a separate waste stream for policy framework, subsidise eco-friendly paints, provide alternative options before banning plastic, and raise awareness regarding environment, amongst others.

Keywords Urban river · Sacred · Waste · Rituals · Idols

S. Chakrabarty (✉)
Institute for Housing and Urban Development Studies, Erasmus University Rotterdam, Rotterdam, The Netherlands

© Springer Nature Switzerland AG 2020

M. Ghosh (ed.), *Perception, Design and Ecology of the Built Environment*, Springer Geography, https://doi.org/10.1007/978-3-030-25879-5_11

Abbreviations

KMC	Kolkata Municipal Corporation
WBPCB	West Bengal Pollution Control Board
CPCB	Central Pollution Control Board

General Phrases and Their Description

Ganga/Ganges	River Ganga
Gangajal	Water of River Ganga
Hooghly/Hugli/Baro Ganga	River Hooghly; as River Ganga is known in West Bengal from Farakka down south to the Bay of Bengal
Nullah	Canal
Tolly's Nullah	Often synonymously referred to Adi Ganga
Calcutta	Old name for the city of Kolkata
Ghat	Steps leading to river or any waterbody which forms the interface between land and water and people can directly access the waterbody at such places
Puja	Worship; usually refers to the Hindu practices of worship
Durga puja	The biggest Hindu festival of Bengal, usually held in the month of September or October, encompasses the worship of Goddess Durga along with her two sons and two daughters. The Goddess is envisioned as a woman coming back to her maiden home with her children for the annual vacation, and this sentiment gives this festival the human touch and is celebrated by people from all religions and communities, though the actual worship by means of mantras and shlokas is carried out only by the Hindus. The main duration of the festival is 5 days though the festivities often last longer. This happens at community levels and also in individual homes, but the scale of the former is larger
Lakshmi puja	Worship of Goddess Lakshmi, goddess of wealth and prosperity and daughter of Durga, usually held in September or October after the Durga puja and on all Thursdays in individual homes. This happens in individual homes and also at community levels, but the number of such pujas happening at the community level is relatively lower
Saraswati puja	Worship of Goddess Saraswati, goddess of knowledge and daughter of Durga, usually held in February in individual homes and educational institutions and at community levels
Chhat puja	Worship happens by a waterbody, mostly Ganga where available. Women from the Indian state of Bihar or ancestry mostly conduct this
Sradhha	Ritual performed usually by the eldest son to free the soul of the dead and help it attain salvation or moksha; happens around 10–15 days after the death of the individual based on the caste and community
Mandir	Temple

Introduction

Rivers play a key role in defining a place while maintaining its relationship with nature at the same time. Sadly, urban rivers have often been abused and undervalued, often overlooking their importance in urban health systems, aesthetics and the community assets these rivers are (Beatley 2013). Rivers are considered as life-givers and given the stature of gods and goddesses and are not seen merely as water-bodies in the Indian culture (Sadhguru 2017).

Sixty three percent of the urban sewage flowing into rivers (some 62 billion litres a day) is untreated according to reports from India's Central Pollution Control Board. Moreover, riverbanks, floodplains and wetlands have been claimed over time by slums, infrastructure, offices and housing developments, which have narrowed natural river channels and distorted their flow, thereby reducing the ability of India's rivers to buffer flooding to a great extent. This has in turn taken a toll on biodiversity. The cost of this abuse has mounted over years. A recent study linked increasing cases of typhoid, diarrhoea and hepatitis in New Delhi to severe pollution in the urban river flowing past the city, River Yamuna, which is the main source of drinking water for New Delhi. Large stretches of the Yamuna, as well as other urban rivers of Indian metro cities like Mumbai's Mithi and Ulhas rivers and Chennai's Cooum, are considered dead zones, with too low oxygen levels to support most fish life. (Chandrashekhar 2018) India's rivers and mostly urban rivers are facing a severe crisis, and something really needs to be done before things go completely out of hand.

Being the longest and the second greatest river by water discharge in the world and the national river of India, the Ganga was ranked amongst the five most polluted rivers of the world in 2007 (Rai 2013). River Ganges, also known as the Ganga, has been an inseparable part of India and her people since time immemorial with this river basin sustaining more than 40% of the country's population. The relationship of the people with the river, seen as one of the most important facets of Hindu rites and rituals like religious bathing, offerings, cremations, etc., is witnessed mostly in the ghats or the stepped landings at the land-water interface. With increasing urbanisation and rise in population, the Ganga has witnessed an alarming increase in its levels of pollution. To address the same, the Ganga Action Plan (GAP) was launched in 1985 by the Government of India with the cleaning of the river as its main objective. This was not very successful and had negligible public participation. GAP was later followed by the National Ganga River Basin Project with the same objective, launched in 2011 and funded by the World Bank. (Das and Tamminga 2012)

According to Hindu mythology, Ganga is worshipped as a goddess, and with this river, there are many myths and beliefs associated which make the river a means of attaining salvation. The article mentions that many other religions and cultures like the ancient Greeks, medieval Christians and a Jewish scholar named Josephus also believed the Ganga to be the first river of Eden, stating its universal importance. Ganga has been equally respected in the Muslim community as well. The Mughal King Akbar is said to have high values towards the Ganga water for its power, being considered to bring immortality, and used only this water for drinking. (Sampat 2009)

Apparently some studies show that there may be some scientific facts which account for the religious basis that this river has purifying powers. All rivers inherently have a cleansing mechanism which depend a lot on factors like dissolved oxygen (DO), biological oxygen demand (BOD), etc., but it has been revealed in a study that the decomposition rate of organic waste is 15–25 times faster in the Ganges than other rivers and the monsoon rains also helps in flushing away the waste. However, with growing population and industrial growth, the pressure on the Ganga is growing, and the river carries nearly 27 tonnes of sedimentation per hectare per year, while the amount is 8 tonnes per hectare for the Nile and 13 for the Amazon. This huge load often leads to change in the course of the river, largest sedimentation happening towards the mouth of the river in the Indian state of West Bengal and also floods during monsoons. (Sampat 2009)

The Adi Ganga, also known as Tolly's Nullah in some parts, is an important tidal creek draining into the river Hoogly from the left in the vicinity of the city of Kolkata. One important factor which affects the drainage potential is the effect of tides, which creates a fluvial phenomenon, "tidal bore," consisting of the head wave of the advancing tide at the junction of the estuary and the river, often exceeding 2.1 m in height. The difference between the highest point of high water in the rainy season to the lowest point of low water in the dry season is reported to be more than 6 m. (Government of West Bengal 2016) This leads to the "sarasarir baan" in the canal during the month of Bhadro (monsoon) leading to local flooding every year. Navigable till the first half of the twentieth century, Adi Ganga has been reduced to a mere drain heading to fast extinction (Deb Sarkar 2017).

According to historians and river experts, Adi Ganga was the main channel of the Ganga flowing into the sea till the Hooghly became the main one much later. William Tolly in the 1770s supervised the dredging of Adi Ganga making it a vital link between the Ganga, that is, Hooghly to the west and Vidyadhari to the east, that eventually flows through the Sundarbans into the Bay of Bengal. Encroachments and multivarious human activities along the banks of Adi Ganga have made the river to disappear in certain stretches. Central agency has been appointed to trace the original course of Adi Ganga. (Times News Network 2017)

In the last decade, the city has lost 53% of its wetlands and Adi Ganga, which is the main river flowing through the city in a severe state with untreated sewage and solid waste. The rejuvenation of the link between Adi Ganga and the wetlands has thus been highlighted with utmost importance to make Kolkata resilient to flood (South Asian Forum for Environment 2017). Even though issues are being addressed locally and also with the help of action plans, a more holistic approach needs to be taken at the policy level which would make the city resilient and sustainable in the long run. The existing canals and drainage channels are addressed to be improved, interlinked along with improvement of internal drainage network as per Vision 2025 (Kolkata Metropolitan Development Authority 2005). A more long-term integrated plan is very important keeping in mind the global climate change. However, in India much work is yet to be done and implemented coherently. This brings forth the importance of Adi Ganga as an important urban river in the heart of Kolkata.

Banks of rivers or waterbodies hold great significance for rituals as most of these rituals are conducted there, and the authors note that religious rituals form one of the causes of degradation due to increase in pollution of such waterbodies along with various other causes like dumping of domestic, hospital and industrial waste and along with domestic activities like bathing and washing (Billore and Dandawate 2015; Giripunje et al. 2014; Bhattacharya et al. 2014; Das et al. 2012; Bhatnagar et al. 2016; Bhatnagar and Sangwan 2009; VishnuRadhan et al. 2017; Shrivastava et al. 2011; Watkar and Barbate 2014; Haseena et al. 2017; Dana Gopal et al. 2014). The once main channel of Bhagirathi, as the Ganga is called in West Bengal, is now reduced to a black gutter stretching 75 Km., dumping polluted water into the Hoogly – the main course of the Ganga that now flows by Kolkata. The garbage-choked old river has been the source of malaria, dengue and cholera. Dredging and other measures have been taken but with no proper impact. Adi Ganga restoration has been included under the National Ganga River Basin Project under the central government. (Mitra 2017) The National Green Tribunal in October 2015 had directed the state authorities to remove the unauthorised structures from the banks of the canal within 2 weeks. Factories and godowns along the banks also release the effluents directly into the water. (Government of India 2015) There are other reports which could not prove the miraculous power of the river but that has not stopped the people from taking a dip in the holy sites (Rai 2013).

The conflict between the concept of purity that Ganga personifies and the actual pollution of the river is amusing as people do not tend to perceive the pollution and continue considering the river as pure and hence go on with their ritual rites and offering to the gods and to the river goddess herself. In the scientific world, purity of a water is judged by its water quality and pollution levels, whereas to the believers, the degraded state of the river does not stop them from continuing their rituals (Alley 1994), which is harming the river (Billore and Dandawate 2015; Giripunje et al. 2014; Bhattacharya et al. 2014; Das et al. 2012; Bhatnagar et al. 2016; VishnuRadhan et al. 2017; Shrivastava et al. 2011; Watkar and Barbate 2014) and is also harmful to the people as the river is highly degraded. Reports mention that during the cholera plague in Bengal, the disease was imported to Kolkata by pilgrims and bathing in the impure watercourse of Tolly's Nullah, associated with the puja rituals of Kalighat temple, aggravated that (The British Medical Journal 1905). In the older days, desecration of a river was perhaps one of the greatest crimes, and now despite knowing the consequences of pollution, most of the people are hardly taking any heed about the ecological system, and with this situation, the policy-makers have to address the concern tactfully as the spiritual role of this river is impossible to overlook (Rai 2013).

The main objective of the research for this chapter was to understand and explain the factors that lead to disposal of toxic and hazardous sacred waste in urban rivers, considering the case of Adi Ganga, mainly in the stretch around Kalighat, Kolkata. Why people contribute to the river pollution, knowingly or unknowingly, has been the primary interest here. The research also tries to understand what are the toxic components in sacred waste and what influences the usage of such components and whether it is mandated in the scriptures, led by belief systems, made commercialised by the sellers and the market or influenced by the users themselves and their behaviours. It is also to be noted that while literature mentions about various waste streams affecting urban rivers, like industrial or domestic waste along with religious rituals (Billore and Dandawate 2015; Giripunje et al. 2014; Bhattacharya et al. 2014; Das et al. 2012; Bhatnagar et al. 2016; Bhatnagar and Sangwan 2009; VishnuRadhan et al. 2017; Shrivastava et al. 2011; Watkar and Barbate 2014; Haseena et al. 2017; Dana Gopal et al. 2014), this research focuses only on the religious activities and the sacred waste they generate as the author could not come across any concrete policy framework aimed only to cater this kind of waste stream.

The literature mentions two main types of factors that work behind whether a person will do a pro-environmental action or not, and these are broadly categorised as individual or internal factors and situational or external factors (Gatersleben et al. 2014; Gaspar 2013; Joshi and Rahman 2015; Krajhanzl 2010; Steg et al. 2014). This research attempts to understand those in detail as per the case in concern.

It also needs to be noted here that the urban river contamination basically refers to the types of contaminants produced by sacred waste rather than the water quality in this context because in a flowing waterbody subject to tidal effects, it is very difficult to practically segregate out the changes in water quality only due to a certain type of waste stream. However, water quality data can surely add to the knowledge at a higher level to note the changes over time for observing an overall improvement.

Significance of the Study

The research is significant from the viewpoint of both – the professional and the academia. The concept of sacredness is from religious and cultural side, whereas the contamination and pollution are a burning issue based on scientific measurements and waste is studied as both a sociological issue and also from the scientific aspect, but very little has been done on all the aspects together. Thus, this research surely has a unique viewpoint having a holistic explanation of how cultural and religious beliefs relate to the environment and hopes to have a different angle at the overall situation. This will enrich the academic side, and hopefully, the professionals can get a takeaway and look for better workable solutions as well.

This is applicable to all rivers in India; the added sentiment in this case is the holiness of the river and hence the specific rituals meant for Ganga. As per Hinduism, water of the river Ganga is considered holy and is used for ceremonial rites and dips and also for disposing off the wastes from the rites along with immersion of idols. Heritage structures along the canal, like the bathing ghat in Kalighat stretch (Kolkata Municipal Corporation 2009), along with informal settlement with open sewerage of nearly 3000 people on both banks, make the situation more critical to address.

Thus, the ecology of the river is directly influenced by the society, which in turn affects the society back (Sutton and Anderson 2010).

Currently, a fully dedicated restoration project aimed at the Adi Ganga is not known and available in the public domain apart from overview on a few dredging and beautification projects under the national Clean Ganga Mission which includes Ganga and all its tributaries and channels. This research can be considered as a pilot study before generalising the research to the entire Ganga river basin and can also be extended to all waterbodies in India in general, as many forums talk about the Ganga pollution and management practices associated with it.

From the academic perspective, this research will help in addressing the policy gaps and the measures to be undertaken in similar contexts. Policies are often implemented in a top-down manner and often lack grass-roots involvement. Identification of the source of the toxic components as mentioned later ahead in the chapter, from other studies (Giripunje et al. 2014; Bhattacharya et al. 2014; Das et al. 2012; VishnuRadhan et al. 2017; Shrivastava et al. 2011; Haseena et al. 2017; Dana Gopal et al. 2014), will help in identifying gaps in the procedure and could help policy-makers address the cause at the root, thus bringing out more workable solutions by bottom-up approach for better management of urban blue infrastructure, which is immensely crucial considering the evergrowing urban population. Understanding why people do harm to the environment consciously or unconsciously based on the literature (Gatersleben et al. 2014; Gaspar 2013; Joshi and Rahman 2015; Krajhanzl 2010; Steg et al. 2014) that this research is based on will help address the policy gaps.

Moreover, in 2018, 55% of the world's population resides in urban areas, which is projected to go up to 68% by 2050. India along with China and Nigeria is slated to account for 35% of this growth in world population between 2018 and 2050. (United Nations 2018) Currently, India being the second most populous country in the world, this particular kind of waste stream cannot be overlooked with regard to urban rivers. The issue becomes more significant as 79.8% of India's population consists of practising Hindus as per the last census (Government of India 2011).

Being an urban river, Adi Ganga has multi-faceted links to the overall ecology in the city region and beyond. Due to time limitation, the focus of this research was on the Kalighat stretch, with Kalighat as the central site, one upstream and another downstream of Kalighat, as this entire stretch is right in the heart of the city, is a multi-functional and multi-cultural area, with tourism potential and also houses one of the most powerful Hindu temples. Sites for the research have been chosen based on varying density of temples along this river as described in detail later; the entire stretch could not be studied thoroughly due to time constraints. Effects of urbanisation, economy and direct politics were not taken into account to limit the scope for this study. Even though there may be many causes and dimensions to this, the focus of this study was kept on the sacred or holy waste because sewerage, informal settlements and many of the other factors probably need more of direct policy-level interventions to begin with, whereas understanding people's behaviour may probably help bring a positive change from the grass-roots levels and at the source.

Understanding the Urban River

This section discusses the concepts, theories and perspectives that portray the relationship between sacred waste and urban river contamination and attempts to link them. This eventually laid the basis for the final data collection, analysis, conclusions and recommendations of this research. Several perspectives had been studied which provided the concepts and the basis of this research. The conceptual framework for this research is presented at the end of this section.

Sacred Waste

Defining Sacred Waste

The term sacred is a concept that comes with its own set of peculiar logic, understanding and belief system. It is a fundamental structure seen in human culture in some form or the other irrespective of a specific place and time and can take various forms like pilgrimage, fasting, meditation, etc. Such forms of religious behaviour are culturally based on the idea to keep one's physical self separate from the daily mundane routine social life. The logic behind sacred characteristics of traditional practices and activities cannot often be explained by conceptual framework of religious traditions themselves but need a different perspective altogether. (Anttonen 2000) There are multiple schools of thought towards explanation of sacred, and as mentioned by Anttonen (2000, p. 272), "the sacred is comprehended as numen, a dynamic force that manifests itself in feelings of religious awe, in inexplicable sentiments of horror and dread, on the one hand, of overwhelming ecstasy and fascination, on the other" as followed by the phenomenologists.

Sacred is often seen as an element as experienced by a religious person and can be defined as a subjective mind's state of feeling determined by its distinctive behavioural rules and is also open to public observation. By explaining sacred in this methodological way, the reason as to why people behave in a certain way with regard to the environment can also be understood. However, the concept of treating sacredness as an emotion as put forwards by phenomenologists is often criticised on anthropological and philosophical grounds. Sacred is also seen as a form of value based on religious, social or ethnic groups which creates cognitive boundaries distinguishing it as unique. Symbolic representations and rituals undertaken by the members of the community enhancing their sense of integration that often go beyond individual consciousness may also be termed as sacred. The author also notes that sacred and ritual are often treated as analogous and become part of cultural practices. Thus, sacred can be seen as an individual quality holding collective meaning, often having culture-specific references to the concept of God or to higher principles of life like equality, freedom, love, etc. as in the case of nontheological concepts.

Cultural myths, forms of rituals and collective life values hold much importance in understanding what sacred is. (Anttonen 2000) With regard to this, ritual refers to a standardised set of behaviour or custom, usually involving religion, and may be irrational as well, though not always. Rituals or acts for doing sacred things can have two positive characteristics – the respect and the attitude with which the rituals are carried out form the basis of sacredness and also that the actions are symbolic and not intrinsic. However, it is to be noted that ritual is a religious action in its primary sense, whereas similar actions and expressions of feeling not restricted to religious settings are usually termed as ceremonial. (Goody 1961)

Discarded or excess matter is referred to as waste in simple terms. However, there are several emotions that waste can provoke supported by more complicated meanings. While rubbish, litter, etc. usually refers to everyday life by-products, waste can also mean medical waste, nuclear waste, etc. with a different paradigm altogether. The author mentions that people do not really like seeing waste in the open and like to maintain a distinction of public and private with regard to waste, making most of the waste management practices functioning outside of normal sight. Waste education and practices also has meaning and can impart a sense of sacred or virtue when making it into a habit or order. Another aspect is the absence of guilt when people do not abide by the waste management norms laid down by the municipality and they keep doing that. Thus, waste is very much linked to habit and sensibilities and the sense of being by means of rituals. By proper waste education campaigns and awareness, cultivating a waste ethos can help in taking the responsibilities and of dealing with waste beyond disgust or moral duty. (Hawkins 2006)

Also scholars have pointed out that a thing sacred by itself might not hold a special meaning but, when attached to the other ideologies and context, embodies the idea of sacred (Anttonen 2000). For example, the rubble of Ground Zero hold special meaning to the families of the bereaved even though it is waste otherwise. Similarly, the heaps of flowers offered as tributes to Princess Diana at royal sites, spirit houses of Thailand, tattered items retrieved after the tsunami in Japan, holy statues from abandoned churches of Belgium can all be depicted as sacred waste. There are many such cases which have religious, emotional or moral references while being leftovers without apparent significant value at the same time which makes the subject little difficult to perceive for others and often a reason for conflicts. Thus creating a general idea about sacred waste can bring out the significance of such wastes and help bring out the true meaning out of the same. Probably due to this lack of ambiguity in defining and lack of meaning of waste, the production of waste has not received much attention in religious studies. A better understanding of the meaning of the waste generated in such cases of rituals will help address the outcomes and also the various other dimensions linked with it. Thus special treatment is needed for dealing with sacred waste as it is linked with attachment value, and irrespective of materials or origin, it may be preserved or be ritually neutralised. The author also points out the example of sacred waste being regularly produced during Eucharist ritual performances and mentions that a certain protocol needs to be followed in such cases. She also mentions that proper attention needs to be given for dealing with such waste not only for the sentimental reason but also because inappropriate treatment may lead to severe consequences due to the materials used and their effects. (Stengs 2014)

Religious practitioners consider that the waste generated out of religious practices is of a special category as it contains spiritual energy and the disposal of the same is also part of the ritual, leading to a mixed opinion on public health, environmental consciousness and spiritual contamination as this type of waste is usually not considered waste but rather as offering which is purifying by itself (Wirtz 2009). While other scholars have stated that sacred is derived out of consciousness and termed more or less pure, the author also asserts that even though an object is manifested as sacred, it still continues to be what it is, like a sacred stone is still a stone though it can hold a sacred meaning to the believers and there lies the paradox between sacred and profane (Rennie 2007).

The above research materials in turn lead to importance of culture and its influence of prescribing what is considered sacred and what is not and in what cultural views. The word culture was established in English by Tylor in 1871 though it did not find a proper place in a dictionary right at that point of time (Kroeber and Kluckhohn 1952). Culture, however, can be put forwards as a manifestation of all the combined social habits of a community, the way individual habits affect the community and outcome of such activities. Culture, essentially, is an amalgamation of all such ideas and behaviours of the people constituting the society which have been acquired by imitating or through instructions. It is this complex system of traditional behaviour, developed by the human beings, that is being learnt by the successive generations and is often characterised by the society or groups of it, time period and also geographical area. (Brumann 1999) Kroeber and Kluckhohn (1952, p. 181) gave one of the earliest definitions of culture: "culture consists of patterns, explicit and implicit, of and for behaviour acquired and transmitted by symbols, constituting the distinctive achievements of human groups, including their embodiments in artifacts; the essential core of culture consists of traditional (i.e., historically derived and selected) ideas and especially their attached values". This complex system of ideas and its attached values have been compared to an onion and its layers; deeper values being more stable and being passed down the generations (Enserink et al. 2007). According to Keesing (1974, p. 75), "cultures are systems (of socially transmitted behaviour patterns) that serve to relate human communities to their ecological settings. These ways-of-life-of communities include technologies and modes of economic organization, settlement patterns, modes of social grouping and political organization, religious beliefs and practices, and so on". It is also noted that culture is often brought down to and linked with the customs and way of life of the people (Keesing 1974). To summarise, "the 'cultural' aspect refers to shared ways of thinking and acting (ideas, values, beliefs, behaviours), which differ from one culture to another and even within cultures" (Cliff et al. 2009, p. 3).

Sacred Geography

Perception is influenced by knowledge, which in this respect reflect to the traditional and cultural knowledge. In this research, the cultural perception is tied to the human-nature relationship, which is gathered into a form of local knowledge. The cultural perception, which this research is aiming to draw direct link with, is how human treat the environment using their indigenous knowledge or perhaps lack of the same. These perceptions are based on earlier historical experiences, which later gets formulated into the mentioned forms of knowledge. This knowledge is not scientific but indigenous, related to context of the local condition, which at the same time concerns about the ecology. The forms of this knowledge are categorised into resources management practices, internalisation (rituals and myth), worldviews and values, artefacts, five senses wisdoms (Berkes et al. 2000).

Land has often been linked with traditions, culture and spiritual significance and norms which have been passed down from generations to generations. For example, the Australian aboriginals worshipped rocks, and those in turn made the place significant and sacred. (Berkes 2012) Similar associations happen in Hinduism as well.

Locative strand is one of the oldest strands of Hinduism where the ritual traditions are associated with geographical place, like top of hills, confluences of rivers, etc. In such cases, the place itself becomes significant and sacred. In the wider perspective of Hindu tradition, such places of traditional significance, mostly the ones associated with water, are often called tirthas, pilgrimage to a tirtha is one of the most sacred and prominent aspects of Hindu religious life. The flow of life from birth to death is compared to the river, while reaching the other end of the river is analogous to the goal of the spiritual traveller. Likewise, the Vedas gives two images of crossing the river. The universe is considered to be three-storied with heaven being the topmost level, atmosphere in the middle and the earth in the lowest level. One transcends from the earth to the heaven or from heaven to earth in the case of the Gods. The atmosphere in between is often visualised as a vast expanse of river connecting the heaven and the earth. For the second image, Hinduism considers that rivers originate in the heaven, flow past the atmosphere and finally come down on the earth. Rivers thus being the connection between the heaven and the earth are considered sacred. The word tirtha means a bathing place or even a passage or path. The word in present context primarily means such holy places of pilgrimage which connect with the gods and goddesses and is a living embodiment of the sacred geography of India attracting countless number of pilgrims. In one sense, each temple can be considered as a tirtha. The main seven rivers of India believed to be released from the heaven by the God Indra are identified with the heavenly drink of Soma and is considered to be nourishing and purifying, Ganga being the foremost of all. (Eck 1981) There are many such tirthas, and deserving special mention are 10 jyotirlingas for God Shiva and 51 Shakti Peethas for the goddess (Natarajan 2013); Kalighat temple or Kalitirtha being of the 51 Shakti Peethas is one of the most sacred pilgrimages in Hinduism (McDermott 2011). The Ganga flowing by this area makes this geographical setting all the more sacred.

Such physical aspects of nature, which can be certain mountains, rivers, water-bodies, etc. that hold special sacred meaning based on spiritual or cultural context, represent sacred geography, and sacred waste is mostly generated in such locations as offerings are made here (Berkes 2012). So essentially, sacred geography is where sacred waste is most likely to be found.

As put forwards by Sutton and Anderson (2010, p. 3), "Ecology is the study of the interaction between living things and their environment. Human ecology is the study of the relationships and interactions amongst humans, their biology, their cultures, and their physical environments". The author also notes that religion often being linked to economics directly or indirectly can be treated as part of environmental adaptation. How things are perceived and cognition of the same form an integral part of this adaptation. Ritual management can be considered as a passive method of environmental control where these rituals and customs passed down for generations contribute to resource management, even when it was not the original intent. Ownership in such cases may work as a conservation technique. The tragedy of the commons as coined by Hardin (Hardin 1968) is of utmost importance here. This happens when a common resource faces overexploitation unintentionally looking at the short-term gain overlooking the long-term return leading to unregulated use. This way, a valuable resource often gets misused, and none is left to enjoy the resource anymore. Religion, which used to be the means of environmental morality in earlier days, still has some virtues, and those can help environmentalists reach out to social morality and persuade people in understanding the balance of human needs and the environment's capacity to withstand the same. (Sutton and Anderson 2010)

Sacred sites are believed to have special powers, like purification, etc., and it is considered that the more the number of visitors, the more sacred a place is and sacred sites often have buildings like mazhar, etc. that add more to the association of sacredness. Why people visit such places and the kind of rituals they practise hold special significance in this aspect, and sacred/spiritual commons is considered as a subpart of cultural commons that withholds the importance of such sacred sites. The author mentions that a sacred site is no longer considered special in Kyrgyzstan if it is polluted or mistreated. (Samakov and Berkes 2017)

Different environments may result in variations amongst cultures, and thus, culture can define how people perceive river at a particular place and time. Thus, public attitude over time plays a very crucial role in understanding the essence of river management across cultures and can also help for future actions. Thus a specific universal approach cannot address the issues of river management in a realistic manner. Rather, an approach that reflects the influence of culture will be more practical while giving out the recommendations. Hence, increase in awareness regarding the past cultures and their characteristics is very much crucial. (Gregory 2006) It is seen that waste management is often more of an implementation problem than a technological issue. As observed in the case of Purwokerto City, the city faced issues with societies' involvement when the municipality initiated a program on waste collection. However, when it adopted the local wisdom of *kerigan* pattern for community participation to manage household waste, it showed immense success, and the city got the cleanest city award in Indonesia in 1995 and 1998. Thus it shows the importance of traditional knowledge and cultural wisdom in dealing with human behaviour, attitude and their perception. (Suyanto et al. 2014)

Flowing for about 2550 Km. eastwards from its source in the Himalayas, draining an area over a million sq. Km. with a population of over 451 million, Ganga has been given the title of national river of India in 2008 by the then Prime Minister (Bhadula and Joshi 2014). In 2017, the Uttarkhand High Court had given the Ganga and her tributaries and all the natural water streams and channels in continuous or intermittent flow with the main river the status of a legal person with its own set of duties, rights and liabilities. The purpose of this was to ease the communication between the multiple stakeholders and help in the protection of the river, keeping in mind the importance of the natural resource and also the faith of the society. The Supreme Court of India's jurisprudence recognises Hindu deities as legal persons managed by trustees who are entrusted with the possession of the deities' properties. There have been debates on this case, and the Supreme Court had stalled the order of the local court though this thought surely gives a new insight of realising the significance of religion, river and culture and how these can be used for a greater good of protection of rivers. Similar incident of giving legal status can also be seen with the Whanganui River in New Zealand under Te Awa Tupua Act of 2017. (Pecharroman 2018) China has seen an increase in use of modern knowledge and technology in accessing funds, and cultural traditions have simultaneously been given the right support by providing which suitable trees can be planted along the river that in turn are being properly taken care of and also prevent erosion along the banks (Sayers 2002).

Activities Related to Sacred Waste

The river Ganges, locally known as Ganga, is beyond just a natural resource and has immense spiritual, cultural, economic and ecological influence in its river basin. Being considered sacred mainly by the Hindus and respected by people of other religions as well, Ganga unanimously is considered as the mother, the faith being as old as the Indian culture itself. The Ganga water or Gangajal is supposed to have miraculous healing and regenerative powers and have been noted to be used and given significant importance by the Mughal King Akbar, the Maharaja of Jaipur and many other influential people. According to Hindu mythology, this river signifies purification, and taking a dip in the river or sprinkling some is believed to freed oneself of all sins. Gangajal is also used for all auspicious Hindu rituals, and all Hindu homes have at least a bottle of water of the same. Considered as a path to salvation, dying on the banks of Ganga is believed to help attain moksha. Thus, Hardwar and Varanasi are popular funeral sites of the Hindus. It can also be noted that crematoriums are still located by the Ganga or its parts, and the ashes of the deceased are floated in the river. Observed to remain fresh with a capacity of self-purification, this river water is miraculous indeed. Observations noted that it stayed fresh for 3 months on the way to England by sea, and also in 1896, it was stated that the cholera-producing bacteria could be killed within 3 h when treated with

Gangajal. However, much scientific study needs to be done to prove the self-cleansing and pollution-removing powers of the river and anxiety that such research may counter the beliefs and sentiments of the people often have stalled the process. (Kumar 2017) Even though religion and geography might seem to have little correlation, it is to be noted that religious traditions like bathing in the Ganges by the Hindus leave a mark on the geography of the area (Park 2004). Holy dips mainly on auspicious days like "amavasya" or the new moon day and on solar eclipses are believed to help one absolve all sins (Bhatnagar et al. 2016; Bhatnagar and Sangwan 2009).

Religion, ritual, sacred and holy have often been used unanimously and mostly refer to imaginary beings demanding respect by means of symbolic representations (Stausberg 2017). Puja is a principal ritual of Hinduism which basically refers to the worship or the prayer being done to the god or the goddess, as the case maybe (Ellis 2015). Hindu culture is based on karma or actions. These actions or acts or the rituals become habits which in turn shape the lives of the practitioners to a great extent. Food offered to deities is also quite common (Frazier 2012). Arati or fire worship is an integral part of Hindu worship. There are also specific family deities or Kuladevata worshipped based on a certain lineage or caste and option for choosing to pray to a particular deity or ishtadevta, for each member of the family, and these are mostly for the domestic worships. (Frøystad 2012) In Hinduism, the puja or worship exudes a sense of attachment and protectiveness forming an intimate bond, and the rituals are manifested in various physical forms like adorning the deity, offering food, etc. It also needs to be noted here that the deities need to be evoked first before the real worship can begin (Ellis 2015).

The author (Dwivedi 1993) here points out that religion can invoke a sense of awareness though it might be different from the scientific explanation while referring to the ancient moral teachings in Hinduism and Buddhism where even the kings showed their patronage and took steps personally to create environmental awareness. The idea of religion that life goes much beyond materialism can help in environmental policy planning, education, etc. and make people more aware of the moral values of the environment. He also mentions that as per Hinduism, maintaining proper sanitation is a part of the code of conduct, and the Hindu society do not consider throwing dirt on a public pathway. Water is considered sacred and thus sacred water is used during rituals. It is because of this idea of water in any form being sacred, prayer is offered to the deity of water, Ganga. According to an ancient Hindu scripture Padmapurana, any person engaging in killing creatures, polluting waterbodies or destroying nature goes to hell. However, the effectiveness of this concern towards the environment eventually depends on the faith of the believers. Hinduism talks about dharma or the way of life and is about ingraining such concepts into everyday life. Finally, the author mentions that in present day, people are so preoccupied with economic needs that nature often takes a backseat. (Dwivedi 1993)

In summary, the activities related to this research are floral offerings, bathing in the holy river water, collecting Gangajal, idol immersion, etc. (Bhatnagar et al. 2016; Bhatnagar and Sangwan 2009; Watkar and Barbate 2014).

It would be apt to note here that even though the river Ganga is respected as a goddess, it is a reality present in its physical state in the form of a river, thus demanding special treatment and concern, not only from the emotional perspective of the people but also from the environmental aspect. This is also the sad reality of almost all rivers in India.

Factors Behind Toxic and Hazardous Sacred Waste

The highly social, self-aware, cooperative and technological characteristics of human beings make them unique, and it is these distinct characteristics that make their interactions with the environment all the more interesting and complex at the same time. Humans affect the environment, which in turn affects human beings themselves. Thus, the relationship between humans and the environment is very much interdependable. (Sutton and Anderson 2010)

For centuries, the environment that has protected humans and other organisms is now bearing the negative consequences, direct and indirect, of many of man's technological advancement and survival strategies, many of such problems being the outcome of man's individual actions and is resulting in the degradation of the environment. It is high time that the root causes, connections and patterns of the degrading environment is studied and addressed to, as wellbeing of the environment is an intrinsic part of the human life and the world economy. In order to better understand this, the role of behavioural models and theories plays a major role. These theories and concepts can also explain why people take part in different behavioural measures that influence the environment. However, a single theory cannot give a clear picture of the interactions amongst the variables that influence human behaviour with regard to environmental preservation. The following theories and models can help in a better understanding. Many studies have been done on possible relationship between cultural perception and urban river conditions, like "Primitive models (Behavioural change model, Environmentally Responsible Behaviour model, Reasoned/Responsible Action theory), Planned behaviour theory, Environmental Citizenship model, Model of Human Interaction with the Environment, The Value-Belief-Norm Theory of Environmentalism, Model of Diffusion of innovation and Health Belief Theory" (Akintunde 2017, p. 122), focusing on the human behavioural change towards the environment. The models of this relationship evolved through time and differ on perspectives and behaviours of human as actors and the course of cycle or stages the human relates to nature. However, these models do not reflect and pin point what the author is trying to convey with the aim of this research.

Literature (Gatersleben et al. 2014; Gaspar 2013; Joshi and Rahman 2015; Krajhanzl 2010; Steg et al. 2014; Hardin 1968) mentions that any behaviour can be called environmental behaviour, just that the influence can be positive, negative, greater or smaller. Often it can also happen that an individual's behaviour may be unintentional where the person is not even aware that his action can harm the environment in any way.

Realisation and concern towards the harm caused to the environment have paved way towards sustainable development and emergence or revival of green products which seek the incorporation of eco-innovation throughout the entire supply chain of the products. Such green consumption is often related to environmentally responsible behaviour of mostly the consumers as studies say that 40% of the environmental damage is the effect of consumer household purchases. However, research also shows that even though people are becoming more and more conscious about the environment, it often does not reflect on the increase in sales of such eco products which state that a minor role is being played by environmental considerations with multiple other factors that determine the consumer decisions of such products which try to explain the green attitude-behaviour gap or the gap between the actual purchases and the consumers' thinking of the same. For example, a study showed that while 67% wanted to buy organic food products, only 4% eventually made the purchase. To explain this gap, the authors (Gatersleben et al. 2014; Gaspar 2013; Joshi and Rahman 2015; Krajhanzl 2010; Steg et al. 2014; Hardin 1968) came up with two broad categories of the determining factors pertaining to pro-environmental behaviour, which again have their subcategories, as shown below.

Individual/Internal Factors

This relates specifically to the individual decision-maker or the consumer.

- Emotions
  Consumer emotions on environmental concern in particular, be it a positive way or out of guilt, are shown to have a direct impact on purchase intention and behaviour of green products.
- Habits
  Habit has mostly been observed as an obstacle though the details need to be studied.
- Perceived consumer effectiveness
  This is the consumer's evaluation or perception of the product of the extent it could bring a difference to the overall scenario, and this is seen to have a positive impact on the purchase behaviour.
- Perceived behavioural control
  An individual's capacity to actually perform or contribute to a given behaviour also plays a crucial role though further details need to be studied for its contribution towards green purchase behaviour.
- Values and personal norms
  Social, ethical and moral values also play a key role.
- Trust
  The expectation or the belief that the product in concern will actually do good to the environment is one of the deciding factors.

- Knowledge
  Knowledge and general awareness regarding environmental benefits also matter a lot in bringing about a positive change.
- Other individual factors
  Other factors like response efficacy, looking for variation, etc., may also influence the decisions.

Situational/External Factors

This mostly pertains to the situation or the other forces outside the individual which influences the decision.

- Sociocultural factors/subjective norm/social norm and reference groups
  Social and reference groups, peer group, etc. strongly influence an individual's behaviour towards a certain action.
- Infrastructure
  Presence or absence of suitable infrastructure can also have an effect on such behaviour.
- Price
  Price is seen to play one of the biggest key factors as higher price in most cases outweighs the ethical considerations and people go to the affordable options even if they are not good for the environment which further widens the attitude-behaviour gap.
- Product availability
  Limited availability of an eco-friendly product or lack of alternatives hampered the purchase decisions negatively as people prefer to buy easily accessible products.
- Product attributes and quality
  The product quality, look and feel along with the functional environment-friendly aspects also mattered according to the study.
- Store-related attributes
  The retail atmosphere and where and how the products are sold also aided to the purchase decisions.
- Brand image
  Overall impression along with the sustainability concerns also plays a role.
- Eco-label and certification
  Certifications informing the green characteristics of the product with proper labelling and manufacturing procedures make the people to make more informed purchases by gaining customer trust.
- Other situational factors
  Other factors like media, relevant structures and services, regulations, laws and local involvement of a consumer in pro-environmental work also aid to making such conscious purchases.

Attitudes play a very important role in determining pro-environmental behaviour with values and identities being crucial indicators for explaining individual behaviours, values here referring to cultural values. The authors (Gatersleben et al. 2014; Gaspar 2013; Joshi and Rahman 2015; Krajhanzl 2010; Steg et al. 2014; Hardin 1968) note that attitudes can be measured with a target object or event and can change easily as attitude towards a particular behaviour in a certain domain may not necessarily be linked to another behaviour in a separate domain. This can be observed in the following example: a person exhibiting positive attitude towards recycling, which is surely a pro-environmental behaviour, needs not necessarily be cycling to work even though it still is a pro-environmental behaviour, but the domains are different. This thus relates to the concept of self-identity or how much environmentally friendly one perceives himself or herself to be. This is basically a subjective assessment of individual behaviour by the person himself or herself. On the other hand, values are basically the beliefs or concepts of certain behaviours based on specific situations and can be guided by relative importance. Values make one aware of the consequences and help them act accordingly. Materialism or ownership can also play a pivotal role in pro-environmental behaviour. The authors (Gatersleben et al. 2014, p. 377) note three main factors that predict the intention to behave in a certain way: "attitudes (is it a good or bad thing to do?), subjective norms (what do others think I should do?) and perceived behavioural control (can I do it?)".

Literature (Gaspar 2013) also mentions a third factor apart from internal and external factors that basically is an amalgamation of these two factors with one's influence on the other depending on the case or context. Thus many of these come together to decide why people do certain actions that harm the environment that can also be by buying harmful materials knowingly or unknowingly. It is often noticed that the eco-friendly version of the same product costs multiple times higher than the one which is not environmentally friendly in all respects. Thus, a holistic idea is required to judge the overall situation.

Essentially the development of right attitudes, intentions, information, knowledge, etc. come together to make people take a pro-ecological or pro-environmental behavioural path where the individual and situational factors all play crucial combined roles. "Sooner or later, consciously or subconsciously, he senses that he has received two communications and that they are contradictory" (Hardin 1968, p. 1246) and this is where the individual factors take over from the situational factors in making a conscious behavioural choice, and as rightly put forwards by (Gaspar 2013, p. 2961), "it is considered here that process approaches focus on the mental activities associated with cognitive, social and emotional dimensions of human functioning, and explain the (1) conditions for them to take place and interact in a conscious and unconscious way; and (2) their role in mediating/moderating the influence of 2.1) characteristics of the physical and social environment (situational) and 2.2) individual characteristics (dispositional), over people's behavioural goals/intentions and behaviours". The authors state that all these factors eventually add up to either facilitation or inhibition effect and the final manifestation depends on magnitudes of behavioural constraints which lower the strength of a pro-environmental behaviour and behavioural barriers which inhibit the activation of such behavioural goals. For example, "unrealistic optimism about environmental degradation or uncertainty about resources level of availability and of how many people are cooperating in a pro-ecological way can be considered behavioural constraints." (Gaspar 2013, p. 2962)

These earlier studies (Gatersleben et al. 2014; Gaspar 2013; Joshi and Rahman 2015; Krajhanzl 2010; Steg et al. 2014; Hardin 1968) are used to list down the individual/internal and situational/external factors pertaining to this research.

To summarise, cultural norms and values have been seen to strongly affect human behaviour. Certainly such norms are internalised, and this capacity to internalise norms has likely evolved when man tried to address certain social issues amongst other challenges. A behaviour that the society expects from a person in a certain situation is known as a social norm, which is often learnt from parents, friends, acquaintances, religious and educational practices, books and media. Multiple factors like informed social instincts, material cost-benefit analysis, cultural norms, values and institutions and personality often play a significant role in deciding human social behaviour which are usually followed as expected by the society, but there may be violations by individuals under certain conditions as well. (Gavriletsa and Richerson 2017)

The authors mention that culture is a crucial factor in understanding human behaviour and the values and beliefs that people acquire play a major role in the way they react. (Richerson and Boyd 2005) "Local beliefs played a major role in institutionalised responses to environmental change" (Gómez-Baggethun et al. 2012, p. 3). However, it has also been noted that a combined body of knowledge has often been transferred through generations and some of such local ecological knowledge were adopted in resource management for social regulations. Such traditional knowledge had similarities with adaptive management and included the unpredictable and uncertain factors of ecosystems. (Berkes et al. 2000) Transient population in urban areas also play a crucial role in making the urban areas prone to high damage potential, more than necessary, as they have low awareness of potential hazards. Assimilation of literature and theories will be helpful, but adaptation and modifications based on the site context are very crucial. Thus, level of human tolerance is certainly required along with acceptable environmental performance standards. (Brown et al. 1976)

Urban River Contamination Related to Toxic and Hazardous Sacred Waste

Urban River Water Contamination

Rivers play a major role in shaping civilisations throughout the world by supporting a diverse ecosystem, while waste disposal and human activities have often lead to their degradation (VishnuRadhan et al. 2017). Findlay and Taylor (2006, p. 133) define an urban stream as "a stream where a significant part of the contributing catchment consists of development where the combined area of roofs, roads and paved surfaces results in an impervious surface area characterising greater than 10% of the catchment". The management of urban river systems through integrated river basin management often misses out on the cultural perceptions during the implementation phase. A holistic approach will hopefully help understand and address the graveness of the situation in a better and wiser manner. (VishnuRadhan et al. 2017)

Urban river and its natural landscape are one of the most significant elements in urban development pertaining to environmental and social sustainability, cultural aspects, aesthetics and public psychological health amongst multiple other facets. Its impact on people and urban design and visual qualities cannot surely be underestimated. Natural urban landscape greatly affects people's lives, and this surely gives further impetus to urban renewal projects. Urbanisation comes with a baggage of severe damage to the natural resources and environmental quality often leading to degraded natural landscapes which in turn makes the city suffer and also lose its inherent identity and lack of public spaces. (Shafaghat et al. 2017)

Urban heritage needs to be valued and preserved for the future generations, and the cities should thus recover their sense of identity and community. Urban heritage not only means historical elements but goes beyond only the tangible buildings and finds meaning in traditional cultures, characteristics and representations of public memory and surely public spaces, river and landscapes that give meaning to the place. Rivers are definitely one of the most significant elements of natural urban landscapes as they are vital for human survival and often hold special meanings and mental image for people and affect their perception as well. Due to rapid development, it has been noted that people's connection to river is rapidly diminishing. Standardised approach to urban development is making the city lose its identity and links with its local history, traditions and culture affecting the sense of the place in a negative way. To attend to this cause, nations are coming up with urban riverscape development projects and its preservation though its implementation in heritage cities and areas is a bit controversial. Of late, government policies worldwide are giving immense weightage on river's natural environment and its resource evaluation, redesigning the urban riverscape and its management keeping in mind the crucial role rivers have on human perception. (Shafaghat et al. 2017)

Educating and empowering people is considered as an important step as environmental stewardship mostly depends on the sense of ownership (Frey and Berkes 2014). Fiscal incentives have often been noted to change the behaviour of the people overnight, be it the levy on plastic or driving to the city square, but that is not a permanent solution as behaviour and attitude are different things affecting one another, and it is more likely that changes in attitude will eventually lead to behavioural change. Therefore, when people leave out their self-interest and commits to the common good, there will be real change to environmental behaviour, and citizenship model gives the essence of that by giving the people the rights and the responsibilities that come along with it. (Dobson 2007)

Thus, it is important for people to realise that rules enforced and governance alone cannot be the final changemaker; it has to be the people as well. People need to realise that even though a river is considered as a goddess, it is a river in its physical state. That Hinduism believes in respecting the nature, as seen in the literature (Sampat 2009; Eck

1981; Kumar 2017), still needs to be shown in practice because it is after all these human interactions with the riverscape under the broad umbrella of socioecological system that is empowering and acknowledges the duty of society towards the ecosystem, in turn creating a better and sustainable world. Customs, rituals, festivities, etc. also need to be addressed while chalking out the frameworks (Laborda-Pemán 2015).

More details on sacred waste contributing to river contamination is explained later.

Standards of Water Quality

The result of the study done by the authors (VishnuRadhan et al. 2017) showed that the side channel was more polluted and had lower measures of water quality when compared to the mean river channel. The inference was that flow retention in the side channel led to less dissolved oxygen due to lack of aeration and particulates settling down which eventually prolonged the presence of inorganic and organic constituents of the waterbody. For self-purification and good health of a river, ambient water quality plays a crucial role which largely depends on waste assimilation capacity (WAC) of the water. Increasing levels on human-derived contaminants change the WAC beyond the standard capacity of the water leading to deterioration of the waterbody. Therefore, each variable of water quality helps to get a detailed overview of the overall environmental conditions. As it is practically very difficult to measure all the parameters of water quality, the influential ones are chosen and measured, and based on a system, a single value for the water quality of the entire sample is derived. Water Quality Index (WQI) is a commonly used tool used for determining water quality status. The authors used WQI and the Malaysian standards based on beneficial uses of water – National Water Quality Standards (NWQS). The NWQS has five defined classes based on descending order of water quality for classification of rivers or river segments, Class V being the worst quality and Class I being the best. WQI summarises large amount of water quality data into simpler values for a specific river and takes the following six parameters of chemical oxygen demand (COD), biochemical oxygen demand (BOD), dissolved oxygen (DO), total suspended solids (TSS), pH and ammoniacal nitrogen (NH_3-N). (VishnuRadhan et al. 2017)

For successful functioning of any city, water, sewerage, drainage along with flood defence infrastructure form the major key aspects which provide valuable information on how cities and their natural environments relate to each other. Water provides such a major tangible infrastructure in everyday life that links both the urban infrastructure and the hydrological landscape extending beyond. Thus, it is not only about flood risk and rainfall but the relationship of the people to the landscape they are part of, which is equally important for the future of the city. However, with increase in impermeable surfaces in the cityscape and higher frequency of intense storm events arising out of climate change, combined sewer overflows have also increased. The significance of green infrastructure and urban drainage is of utmost importance in managing pluvial flood risk. (Bell 2013)

Urban rivers and ecosystems face graver disturbances from human activities than rivers in non-urban areas and it is not completely possible to restore it back to the pristine state. The idea, thus, is to have a rational trade-off keeping in mind the socio-economic activities, the intensity of utilisation and exploitation, optimisation and reasonable regulation. (Zhao et al. 2007) It has been observed that 10% of catchment imperviousness lead to degradation of the streams. Waterways which was a valuable resource to human life for water supply, wastewater disposal and also flood mitigation degraded a lot over time leading to degradation of urban water systems, and this case of heavily degraded urban river systems is a familiar situation worldwide now. Social, political and economic issues have strong roles in the urban river rehabilitations. Environmental factors are equally important as well. (Findlay and Taylor 2006)

Indicators Determining Water Quality

Since the contamination affects the food chain and the ecosystem, the authors conducted the experiment in three lakes in Nagpur, India, where tilapia fish were analysed by coupled plasma optical emission spectrometry for heavy metals. As DO is one of the prime indicators for the health of an aquatic ecosystem, the levels of it were measured, and the results showed low levels of DO, that is, 0.89–2.34 mg/L, after idol immersion. Concentrations of calcium and magnesium salt levels have also been noted to increase after the immersion ritual even though the value of these salts was found to be within the limits of the Bureau of Indian Standards (BIS). Scholars have mentioned that magnesium is not poisonous, but it increases the hardness of water and oxygen solubility decreases exponentially with increase in salt content in waterbodies. The authors also found an increase in hardness levels in their study with values going up from 198.10 to 275.70 mg/L and 78.1 to 154.35 mg/L before and after idol immersion. However, the recommended limit of hardness for drinking water set by the Bureau of Indian Standards or BIS was reported to be 300 mg/L, and the results were found to satisfy that criterion. Plaster of paris and sulphates were also found to increase by about threefold even though the final values (144.5, 160.1 and 187.3 mg/L) were below the BIS limit. Heavy metal concentrations were found to increase. (Giripunje et al. 2014)

Bhadula and Joshi (2014, p. 310) in their attempt to understand the impact of religio-touristic activities on the water quality of Ganga have taken the following indicators, namely, "temperature, pH, transparency, turbidity, Total Dissolved Solids (TDS), Dissolved Oxygen (DO), Bio-chemical Oxygen Demand (BOD), Chlorides, Sodium and Potassium". Conductivity and salinity are other parameters for the water quality, and guidelines formulated by the Central Pollution Control Board need to be followed for pollution control. Turbidity is found to increase as the water column is disturbed during the immersion. It is also to be noted that the hardness is not a parameter for pollution even though it indicates water quality, and while the BIS limit is 300 mg/L, the World Health Organisation (WHO) has the limit as 100 mg/L. Presence of organic sources of pollution is determined by the average values of DO and BOD. (Bhattacharya et al. 2014) The pH of water affects the solubility of nutrients and is hence important for the study on waterbodies, while rise in temperature reduces the solubility of gases in the water by affecting the chemical reactions and biological activity. (Billore and Dandawate 2015) Surface water quality has been studied by many scholars, but odour is hardly given much importance, except for the case of drinking water, but at the same time, it is to be noted that even though odour is not life-threatening, it indeed is unpleasant and annoying and causes aversion. The authors found that unpleasant odour is associated with hypoxia and anaerobic decomposition of organic substances which in turn is linked with DO, BOD and COD levels. Higher DO concentration meant that nitrogen compounds gets degraded to ammonia and then to nitrate during aerobic decomposition, and with DO concentration less than 2 mg/L, it is called hypoxic condition. Even though odour is very much linked with individual perception which is pretty subjective, it does affect the image of the waterbody in concern and also the tourism and other related activities that might be associated with it. (Sado-Inamura and Fukushi 2018)

Toxic and Hazardous Sacred Waste

Literature (Giripunje et al. 2014; Bhattacharya et al. 2014; Das et al. 2012; VishnuRadhan et al. 2017; Shrivastava et al. 2011; Watkar and Barbate 2014) mentions the different toxic and hazardous sacred waste that stands relevant to this research. Idols or statues or images of God used in Hindu rituals are mostly made out of clay, plaster of paris, cement, plastic, thermocol, jute, paper wood, cloth, small iron rods, bamboo, plastic, cement, synthetic paints, varnishes, adhesive materials, cosmetics, ornaments, etc., and out of all these materials, thermocol, cement, plaster of paris, ornaments are non-biodegradable, while heavy metals like mercury, lead, arsenic, zinc, chromium and cadmium are often found in the paints used leading to increase in heavy metal pollution after the immersion which affects the food chain along with heavy metal pollution in the waterbodies. A traditional red-coloured powder called 'sindoor', worn by Hindu women as a sign of marriage and also used in religious rituals, contains lead and chromium, also gets mixed in the water during the various rituals. Incense sticks, camphor, ash, etc. are also released in the water during idol immersion. The authors point out that it is not possible to stop or change such religious activities, being such a sensitive issue. Hence they suggest that raising awareness within the society and use of eco-friendly biodegradable materials can at least help reduce the pollution.

As seen in the Betul city in Madhya Pradesh, 50 tonnes or 500 Durga idols and 120 tonnes or 2500 Ganesha idols were immersed in the Machna river, an important source of potable water, in the year 2009, and it has been noted that immersion practices lead to contamination and sedimentation in rivers which attribute to the degradation of the waterbody and its water quality (Shrivastava et al. 2011).

Polythene bags and plastic have also been noted as one of the major sources of water pollution (Haseena et al. 2017). Being cheap, it is overused, and careless

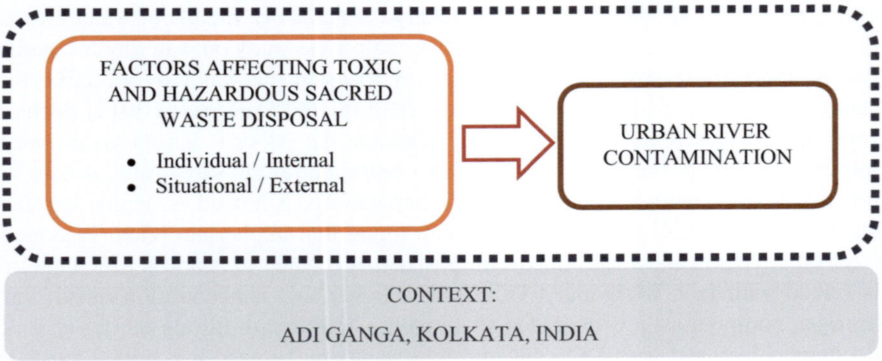

Fig. 11.1 Conceptual framework. (Source: Author 2018)

disposal of plastic bags clogs waterbodies like canals and rivers mostly in urban areas (Dana Gopal et al. 2014) which adds to the contamination and further pollution.

Conceptual Framework

The theoretical framework puts forwards the structure that captures the theory in a research study (Akintunde 2017). Figure 11.1 shows the conceptual framework for this research.

Research Design and Methods

This part of the chapter describes the operationalisation and research design that are used to answer the research question and also discusses them. It starts with the research questions and then includes further information on the research strategy, site selection, data collection methods, instrument, the sample size of the study, the way they have been approached and how the data has been analysed. Validity and reliability of collected data and conducted research along with the limitations of the study are also mentioned in this section.

The main and sub-research questions are in Box 11.1.

The concepts from the conceptual framework were operationalised here by translating the theoretical concepts into realistic measurements to aid the data collection and analysis, as shown in Table 11.1. Also, based on the findings, indicators from the literature have been updated, and only those which found their place in the data collected are being tabulated here for the final report.

Box 11.1: Research Question

Main research question:

Which factors explain the disposal of toxic and hazardous sacred waste which adds to the urban river contamination in Adi Ganga, Kolkata, India?

Sub-research questions:

Which individual factors explain the disposal of toxic and hazardous sacred waste in this river?

Which situational factors explain the disposal of toxic and hazardous sacred waste in this river?

What kind of contamination happened related to toxic and hazardous sacred waste in this river?

Furthermore, water quality data has been taken from WBPCB (Government of West Bengal 2018), and they conduct the tests at certain fixed station points only on quarterly basis. Site B for this research, described later, is the only station point of WBPCB that matches with the research sites. As the main purpose of the third sub-research question is to find out the types of contaminants, only the total suspended solids are considered in this case as those can be distinguished from other waste streams by means of observation, while other indicators of water quality, like BOD, COD, DO, etc., are combined effect of all waste streams falling onto this river and not sacred waste alone.

The main research question is explanatory in nature, and the aim of the research here was to explain the relationship between the independent and dependant variables in the mentioned context. Qualitative research strategy was used for this study. The research strategy gives the logical procedure that the research will follow, that is, the overall design of the research (van Thiel 2014). The idea was to analyse the different factors that contribute to the disposal of toxic and hazardous sacred waste and how what leads to urban river water contamination gets revealed in the process. For this, the research at this level required more depth than breadth in order to get a better insight. The number of units needed to be selected based on the body of existing knowledge. Considering the above concerns, case study has been the selected research strategy. (van Thiel 2014) Literature mentions that case study strategy can be used for explanatory researches even though there is a notion that this strategy is used for exploratory approaches (Yin 1981). Multiple case design was taken up to get a better understanding of the factors figured out from the literature and to analyse whether they reflect any considerable change based on different site characteristics. This was helpful to create the base for a pilot project on this case of Adi Ganga, learnings from which can then be applied for other urban rivers. Primary data was collected as mentioned in later sections. Alongside this, a number of reports from the State and Central Pollution Control Boards and concerned municipal corporation for this river in concern and court cases under National Green Tribunal, etc. were also studied which helped in the validity of the primary data collected.

Table 11.1 Operalisation table

Variables	Sub-variables	Indicators	Description	Source of data	Analysis
Factors affecting disposal of toxic and hazardous sacred waste (Gatersleben et al. 2014; Gaspar 2013; Joshi and Rahman 2015; Krajhanzl 2010; Steg et al. 2014)	Individual	Emotions	Feelings that make one take a decision	Interview, report and observation (photographs)	Atlas Ti
		Habits	A continuous process which has become a part of life		
		Perceived consumer effectiveness	Perception regarding impact of a certain action or product		
		Perceived behavioural control	Perception of the individual's behaviour by the person himself in advance in regard to a context		
		Values and personal norms	Social, ethical and moral values and ideals of an individual which aid in taking an action		
		Knowledge	Previous learning and general awareness about a certain action or product		
	Situational	Sociocultural factors/ subjective norm	The social system forming a certain image or idea about a certain action or product		
		Infrastructure	Presence or absence of infrastructure that helps in decision-making and execution		
		Price	Cost of an item; relates to the affordability of a product		
		Product availability	Relates to availability of alternate options		
		Product attributes	The components of a certain product		
		Eco certification	Marked under a certain category as environment-friendly by relevant authorities		

Urban river contamination (Giripunje et al. 2014; Bhattacharya et al. 2014; Das et al. 2012; VishnuRadhan et al. 2017; Shrivastava et al. 2011; Watkar and Barbate 2014; Haseena et al. 2017; Dana Gopal et al. 2014)	Toxic and hazardous waste	Types of toxic and hazardous waste	Names of the different types of toxic and hazardous waste as per the research, like idols, flowers, leaves, etc. derived out of literature and was studied in detail from the data collected	Interview, report and observation (photographs)	
	Water quality	TSS	Measure of suspended solids	Report and observation	
	Estimation of volume of sacred waste	Weight per year	Estimated weight of sacred waste dumped in the river in the said sites per year	Interview, report and observation (photographs)	

As already mentioned earlier, Ganga forms an integral part of India and also Hindu rituals. Thus, if this river can be studied at least in a part within the limitation of this research, the learning from that can be extended to other rivers as well. To fit in the constraints of this research, Adi Ganga, the original flow of the Ganga to the Bay of Bengal, an urban river flowing through the city of Kolkata, (Times News Network 2017) which is presently downgraded to almost a drain (Deb Sarkar 2017), was selected. Furthermore, influence of tide in Adi Ganga (Government of West Bengal 2016) and its link to the wetlands in the eastern fringe (South Asian Forum for Environment 2017) in the city makes Adi Ganga all the more significant.

Figure 11.2 shows the position of the Adi Ganga in brown, whereas to the west of this is Hooghly River which is the main flow of Ganga to the Bay of Bengal in its current state. The red star within the figure shows Kalighat, which is a very significant Hindu temple and Site B for this research, details of which are provided later.

This place has a unique character being in the heart of the city of Kolkata, well-connected by metro and multiple modes of transport that caters to the residential, commercial and the multitude of characters that this place withholds. The Kalighat

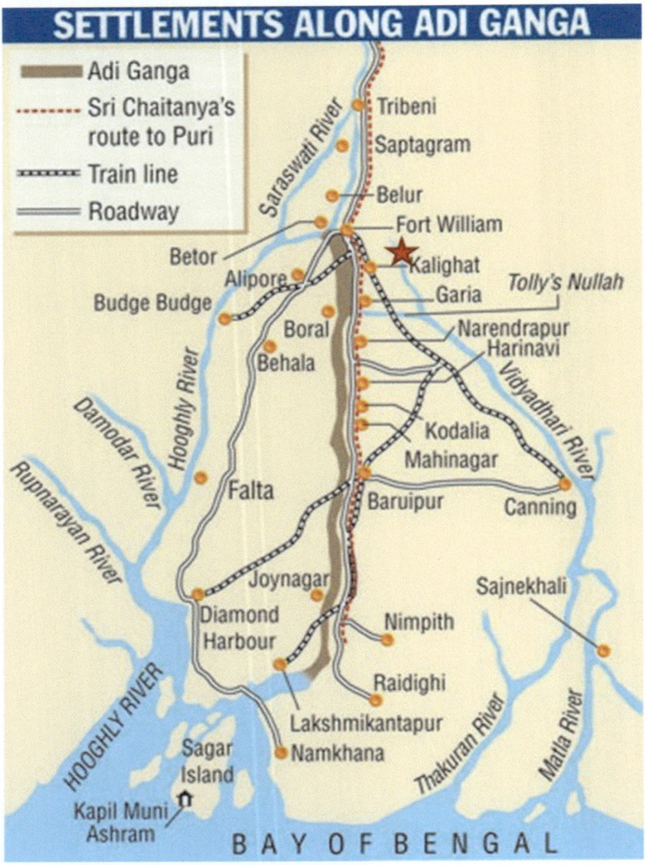

Fig. 11.2 Map showing the course of Adi Ganga. (Source: Das 2009)

Kali Temple is considered one of the most holy Hindu temples in the country, and right next to it is Nirmal Hriday – Mother Teresa's home. Little to the east is the Greek Orthodox Church. Thus, this place is a true amalgamation of faith and culture, though the Kali Temple often is considered as the prime focus. This draws people from various parts of the country and world making this place a major tourist attraction of the city, and the state of the river and the cultural perceptions of the people reflect on the image of the city and thus need much attention.

Three sites have been chosen for this research – Kalighat, one downstream of Kalighat and another upstream of it. This will help in studying the differences, if any, in the upstream and downstream areas. Also, the sites were chosen such that basic characteristics of temples, shops selling puja offerings and residential character remain almost similar. However, variation in density of the temples is a prime factor in the site selection, which helped in the research. In the upstream, the site chosen was Site A, Balaram Bose Ghat, followed by Site B, Kalighat, and then downstream of Kalighat is Site C, Prasannamoyeeghat. Also, it is to be noted that ghats, or steps, form the physical connection between the land and the water (Das and Tamminga 2012), and keeping that in mind, only the ghat areas are chosen as sites here. Considering the above factors, the researcher has chosen the sites in this manner for this study.

Data was gathered by various methods; qualitative and quantitative in practicality refer to the data itself than the mode of its collection (van Thiel 2014). Primary qualitative data was acquired by telephonic semi-structured interviews of around 10 min duration with a selected few respondents from the community and key informants like consultant to the government, researcher in autonomous university, person of political influence and also residing within the site area, government officials from the concerned departments and local observer. Though primarily it was telephonic interview, other modes like Skype and WhatsApp video call options were also availed for better coordination. Two research assistants helped with the photographs and scheduling the interviews. A checklist was prepared to help the researcher in maintaining focus. Secondary qualitative and quantitative data was acquired by means of reports, etc. from the concerned government offices and the internet. Table 11.2 mentions the research strategy, method and the instruments as deemed relevant for the research.

Table 11.3 shows the unit of analysis for the research. The sampling is based on the sub-groups – community representatives from business, residential and related to the temple, organisations, etc. with saturation of information in mind. Purposive sampling by snowball method was used for the sampling. Interviews were conducted till information saturation was reached for each subcategory. Key informants

Table 11.2 Research strategy, method and instrument

Strategy	Method	Instrument
Case study	Interview	Semi-structured interviews (audio recording, notes)
	Observation	Photographs, fieldnotes
	Secondary data	Reports, etc.

Table 11.3 Unit of analysis

Sector	Categories	Subcategories	Description
Community	Business	–	Formal shops where devotees keep their shoes and buy offerings before entering the temple, formal and informal shops selling souvenirs and other items like bangles, vermilion, statues/idols, etc. pertaining to the temple, culture and faith
	Residential	Formal houses	–
		Informal settlers	Informal settlers who have a proper shelter on their own and also street dwellers
	Related to the temple	Temple administrators and workers	All levels of the temple management/ administrators/priests
Key informants	Public sector (local government) Kolkata Municipal Corporation	–	The concerned municipality
	Consultant to the government	–	Some expert with sufficient knowledge in this sector who has been consulting the government on similar issues
	Researcher	–	Someone working in similar sector
	Local observer	–	Someone from the community level who does not directly take part in such rituals but is a regular observer; religion no bar
	Politician	–	Mid-level politician of the ruling party at the state level

Table 11.4 Name of secondary data with source

Name of data	Sources
Water quality report over time	West Bengal Pollution Control Board
Environmental reports	Kolkata Municipal Corporation
Maps of river systems	
Old photographs	
Court orders	National Green Tribunal website

who had sufficient knowledge about all the sites in general were also interviewed base on purposive sampling by snowball method (van Thiel 2014).

For the secondary data collected, the same is listed in Table 11.4 with their source.

For the qualitative data derived by interviews, thick description type of qualitative analysis was done with the intention of gaining deeper understanding about the subject and the context; relevance and researcher's personal notes and observations were also taken into concern. For the secondary data, content analysis was done, that is, the researcher studied the content of the data derived in the form of documents, relevant projects, drawings, etc., and a conclusion was arrived at on the basis of the same. (van Thiel 2014) Computer-based programs like Microsoft Word and

Atlas Ti were used in coding the quotes and reports according to the operationalisation as axial coding. Microsoft Excel was used to tabulate the water data as derived and to create the charts, and Microsoft Word was used to transcribe the interviews and also helped in sorting out the data before putting them onto Atlas Ti.

Due to the richness of the knowledge gathered in this research and the depth it covered, internal validity has taken care of. For reliability and some external validity, secondary data, photographs and observation helped by triangulating the data collected. For the reliability of the research, purposive sampling by snowball method meant easily identifying the experts and the representatives which provided access to richer data and hence more reliable.

As the case study focuses more on depth than breadth of the study, it might be a little difficult to generalise for the bigger picture of the entire stretch of the Ganga or all other urban rivers for that matter. However, the research gives a general outlook of the factors in concern as the general mindset of the people practising the rituals and their connection with puja offerings, religious aspects and urban rivers are expected to be similar and be generalised upon for the entire Adi Ganga and furthermore to other urban rivers in the country. Caution was taken since snowball method of purposive sampling has a chance of biased opinions as one respondent might refer to another sharing similar outlook.

Research Findings

This section is very important as it presents the significant findings of the research. It firstly gives an insight of the sites chosen, followed by an overview of the respondents interviewed and discussed with, and finally puts forwards the analysis of the data collected in context of the research aim along with a summary of the same. This section all along discusses the factors leading to toxic and hazardous sacred waste disposal in the Adi Ganga and the contamination caused as a result. While the first part talks more about the setting, the analysis in the later part of the section will help in understanding the relationship between the variables and in turn answer the research questions.

The literature (Gatersleben et al. 2014; Gaspar 2013; Joshi and Rahman 2015; Krajhanzl 2010; Steg et al. 2014) mentions two main types of factors that work behind whether a person will do a pro-environmental action or not, and these are broadly categorised as individual or internal factors and situational or external factors. This research attempts to understand those in detail as per the case in concern. It also needs to be noted here that the urban river contamination basically refers to the types of contaminants produced by sacred waste rather than the water quality in this context because in a flowing waterbody subject to tidal effects, it is very difficult to practically segregate out the changes in water quality only due to a certain type of waste stream. However, water quality data can surely add to the knowledge on a higher level to note the changes over time for observing an overall improvement.

The main and sub-research questions are mentioned in Box 11.2 for further reference.

Box 11.2: Research Question

Main research question:

Which factors explain the disposal of toxic and hazardous sacred waste which adds to the urban river contamination in Adi Ganga, Kolkata, India?

Sub-research questions:

Which individual factors explain the disposal of toxic and hazardous sacred waste in this river?

Which situational factors explain the disposal of toxic and hazardous sacred waste in this river?

What kind of contamination happened related to toxic and hazardous sacred waste in this river?

Site Description

This section describes the canal and the three sites chosen to elaborate the case study. The setting is the city of Kolkata, the capital of the state of West Bengal in India with approximate location as 22°30′N and 82°30′E. Kolkata Municipal Corporation is the main municipality for the city. Kolkata is located in the lower Ganga basin.

Adi Ganga

Adi Ganga was known earlier as Gobindapur Creek and marked the southern boundary of Gobindapur village during the eighteenth century (Kolkata Municipal Corporation 2017). It was excavated by Edward Surman and it bore his name for some time. The nullah was deepened by Colonel William Tolly in 1773 and connected to the Circular Canal. Thereafter, it bore his name as Tolly's Nullah. In 1775, Colonel Tolly connected the Adi Ganga to the Vidyadhari, and after Tolly's renovation, the Adi Ganga remained a navigable river. However, the neglect of waterways in general and other factors such as population pressure and unplanned urbanisation caused further siltation and pollution of Adi Ganga. The present stretch of Tolly's Nullah is around 15.5 Kms with about 33 sq. km. catchment area (1 Km on either side of Tolly's Nullah). The predominant flow of Tolly's Nullah is from east to west (Garia to Hasting) and under strong tidal effect. The Tolly's Nullah flow culminates into River Hooghly. Tolly's Nullah has three tributaries, namely, Western Channel (approx. 2 Kms long), Keorapukur Khal (approximately 2 Km. long) and Chetla Boat Canal (approx. 2.8 Kms long). The existing contour or topography shows that present area is very flat with ground level varying from 2.38 m to 6.30m. The Tolly's Nullah basin at present receives water from 3 of its tributaries mentioned earlier and 74 outlets discharging directly into the Nullah. The state has planned for revival of this 15.5 Kms stretch of the Tolly's Nullah, also known as Adi Ganga channel, which starts as an offshoot of the Hooghly from Kidderpore and flows through various areas like Tollygunge, Garia, Bansdroni, Naktala and Ranikuthi area in Jadavpur constituency. The revival is to be restricted till Garia. Beyond that point, the channel has become non-existent due to encroachment. (Kolkata Municipal Corporation 2017)

Figure 11.2 shows the course of the Adi Ganga, the original flow of the Ganga in West Bengal before it meets the Bay of Bengal, River Hooghly being the current main branch of the river, as can be seen in Fig. 11.4 along with the sites chosen.

This matches with the Dutch East India Company's Governor Van de Brook's map of 1660 as seen in Fig. 11.3 where the highlighted area shows the main flow of the Ganges turn eastwards via the current Adi Ganga flow and then southwards to the Bay of Bengal, which asserts the navigation route along the Adi Ganga.

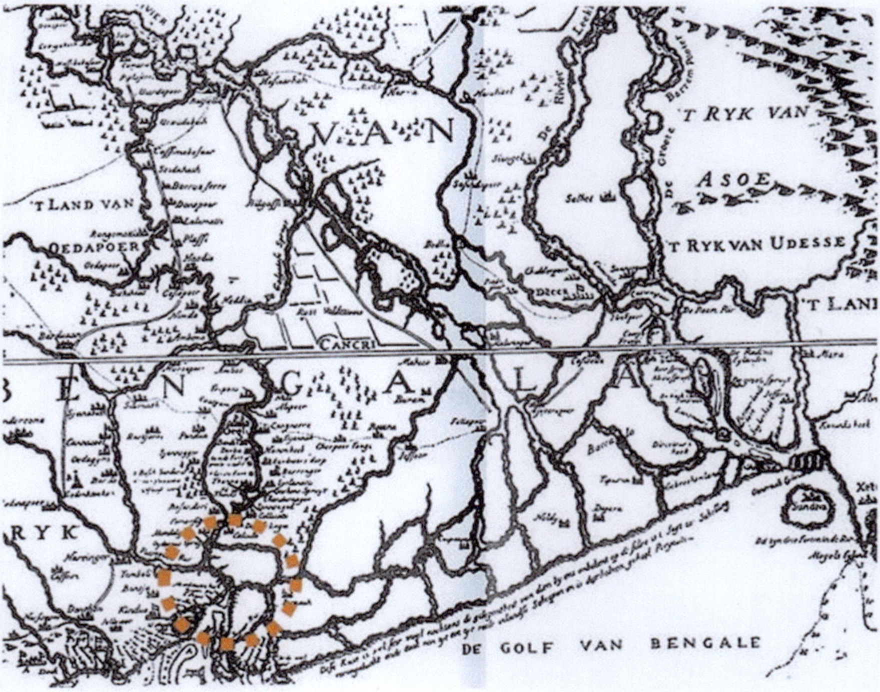

Fig. 11.3 River map of Bengal. (Source: Brook 1660)

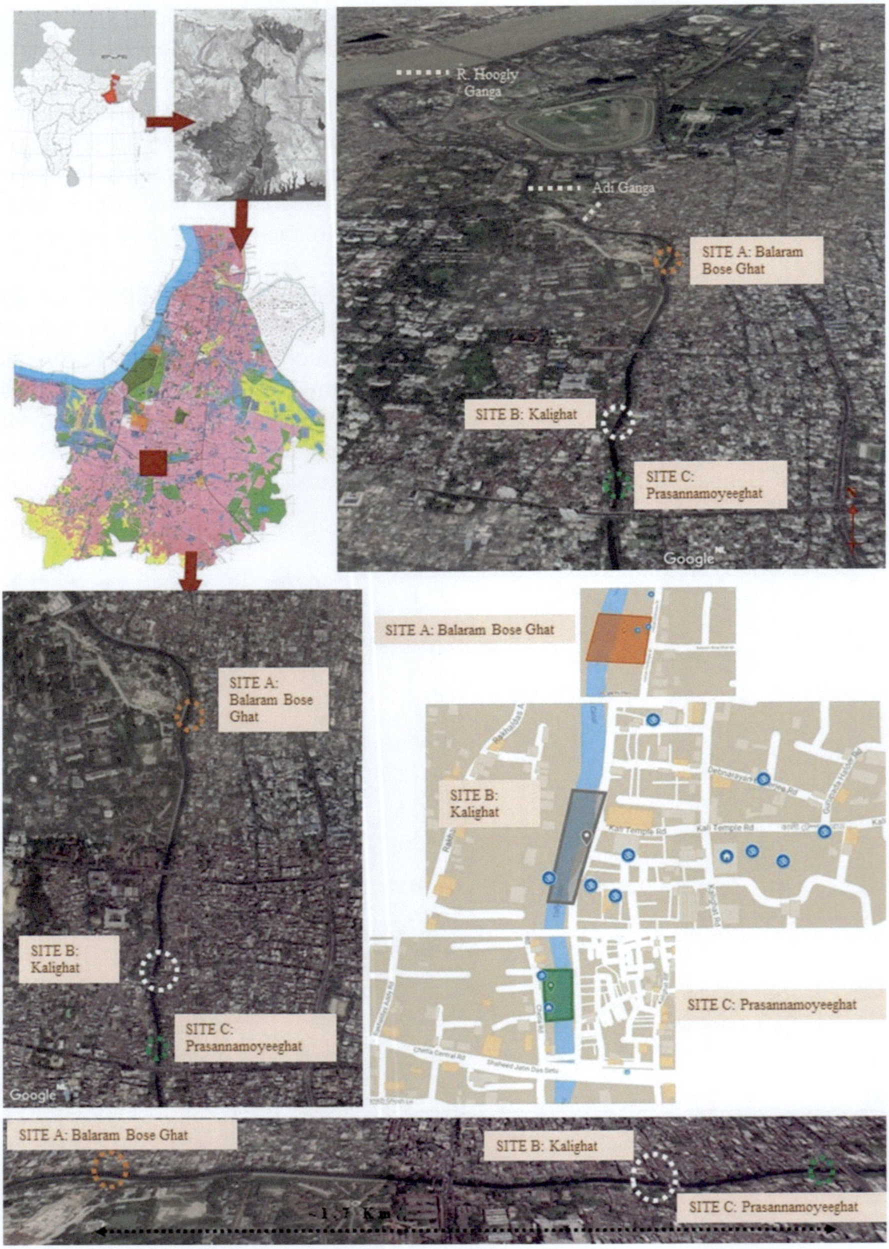

Fig. 11.4 Site map. (Source: Google Earth and Google mymaps delineated by author (2018))

Geomorphologists believed that the estuary of River Saraswati was used by maritime vessels to go upstream to reach Chinsura or Hooghly for trading purposes. Only a tidal creek, possibly navigable only during high tides, connected this estuary of River Saraswati to the Bhagirathi River. From the Hastings Point, Adi Ganga was the main flow of the Bhagirathi at that time flowing southeast. The upper part of Adi Ganga later was renamed as Tolly's Nullah when it was re-sectioned in mid-nineteenth century by William Tolly. The tidal creek connecting the estuary of the River Saraswati and the Bhagirathi River supposedly impaired easy shipment of goods leading to unloading of goods into country boats of shallower draughts during low tide to reach the main flow of the Bhagirathi for further movement upstream. It is rumoured that the Dutch re-sectioned this tidal creek for easy access of sea-going vessels up to the Bhagirathi to Chinsura, the headquarters of the Dutch East India Company. (Chakraborty 2004) This emphasised the trade route and the importance of this river.

Figure 11.4 shows the sites selected for this research: Site A, Balaram Bose Ghat which is upstream of the main pilgrimage Kalighat; Site B, Kalighat; and Site C, Prasannamoyeeghat, which is downstream of Kalighat. Ghats are the steps leading to the river and form the interface for direct human interactions connecting the land and the water. Thus, only the ghat areas are chosen for the detailed study. Figures 11.8, 11.9 and 11.10 subsequently gives the blow-up map along with the landmarks, mostly temples and significant buildings, in the Sites A, B and C in that order.

Balaram Bose Ghat has temples like Harlalka Mandir ("mandir" meaning temple in Bengali), Trinath Mandir, Shiv Mandir, Avaya Mandir, Durga Mandir and Shani Mandir dedicated to the respective gods and goddesses along with an old zamindar house known as the Mullick Bari. The opposite bank of the river being the Presidency Jail ground, there is no option for crossover in this ghat in this stretch. The area is one of the oldest in the city, Bhowanipore, and most of the residents here have been here for generations apart from the informal settlers around and mostly along the Adi Ganga. The area around this site mostly consists of middle-income group households of mostly formal houses. There are some informal settlements along the river but these do not form a slum. The current Chief Minister's residence is less than 500 metres from this site towards the direction of Site B, making this area all the more important. Also, the first "barowari" or community Durga puja is known to have started from this locality. Very near to this ghat is Patuapara, literally meaning the neighbourhood of the pot artisans, which is the second largest idol manufacturing centre in the city after Kumortuli which is in North Kolkata along the banks of the River Hooghly. Mud from Adi Ganga forms an integral part of this idol making industry which is a cottage industry by itself.

Kalighat has the highest density of temples and is the most prominent out of the three chosen sites. The significant temples here are Kalighat Kali Mandir, Kalighat Bajrangbali Mandir, Jagannath Mandir, Nakuleshwar Shiv Mandir, Jora Barir Panchanan Shiv Mandir, Radha Govinda Mandir, Jora Shiv Mandir, Shiv Mandir, Bagala Mandir, Shiv Mandir and Hanuman Mandir along with Nirmal Hriday – Mother Teresa's Home for the Dying Destitutes. This is also one of the oldest parts of the city with Kalighat Kali Mandir being the prime attraction for pilgrims here

Fig. 11.5 Dredging in progess by the Irrigation Department, Government of West Bengal. (Source: Author 2012)

leading to many shops in the area catering to pilgrims, tourists and also the locals. Kalighat being a major pilgrimage, there are also beggars around who mostly stay in slums nearby. The area is mostly inhabited by middle-income group of residents who have been living here for generations now.

Prasannamoyeeghat has the least density of temples with one main temple, Sitala Mandir, and an ashram, Bijoy Krishna Goswami Ashram. Even though this is in close proximity to Kalighat, this area has a different character compared to the other two sites in terms of its urban fabric. Located right next to the Keoratala Bridge, public transport accessibility is very good which has over time led to many apartments being built in the locality and the character of the buildings, residents and commuters have also changed over time.

Figure 11.5 shows the ongoing dredging activities in Adi Ganga as of the 3rd of December 2012.

Finally, Fig. 11.6 shows the issue as mentioned earlier at the beginning of this chapter where the 2009 beautification works can be seen mostly as futile.

Much has been said about Adi Ganga and its glorious past:

> "It had a major role in water navigation and tremendous prospect exist for growth which can directly create a mass employment in various ways. Earlier huge country boats used to ply with merchandise like rice, bamboos, tiles, bricks, sand, etc. etc. and thus many godowns and business establishments had grown on both the banks of Adi Ganga. Some still exist in decaying stage but majority had closed owing to obvious reasons like change of usages and spreading of city at distant places due to rise in population and sadly because of lack of dredging and dumping of city wastes which caused silting on the bed of Adi Ganga." – Respondent RFA3, Site A

This indeed can be tallied from Fig. 11.7.

The Pollution Abatement and Rehabilitation of Tolly's Nullah is under the National Ganga River Basin Project under the Ministry of Water Resources, River Development and Ganga Rejuvenation, Government of India. The report also talks about resettlement plans for the canal bank dwellers along with control of solid waste disposal into the waterbody and eviction of piggeries that are located along the canal and pollute the same. Walkways, greeneries, etc. have been proposed for beautification of the canal front. (Kolkata Municipal Corporation 2017)

Fig. 11.6 Banks of Adi Ganga opposite Kalighat temple in 2012 after beautification done in 2009. (Source: Author 2012)

Fig. 11.7 Bathing and boats on Adi Ganga in 1880. (Source: Bourne and Shepherd 1880)

Description of Respondents

Telephonic interviews were conducted in the month of July 2018 with the help of two main research assistants who coordinated with the respondents and also helped in locating the next respondent based on the feedback of the first. Though primarily it was telephonic interview, other modes like Skype and WhatsApp video call options were also availed for better coordination. Code names of the respondents are given which will help in easily identifying the quotes (Table 11.5).

Findings and Analysis

This section takes the findings each site at a time with a detailed map of each site followed by the photographs and the findings supported by photographs, field notes and interviews.

There are certain norms and regulations that need to be abided by. For organising a community Durga puja or Kali puja, a clearance needs to be taken from WBPCB in accordance to Environment (Protection) Act 1986 with undertaking that puja offerings, like flowers, leaves, etc., will be deposited in the designated bins, places or pits as arranged by the concerned local authorities by the banks of rivers, ponds, waterbodies, and the immersion of idol will happen as per the dates scheduled by the police and district authorities as per the case. (Government of West Bengal 2017)

Table 11.5 Category of respondents

Respondents				Site A	Site B	Site C
Sector	Categories	Subcategories	Description			
Community	Business	-	Formal shops where devotees keep their shoes and buy offerings before entering the temple, formal and informal shops selling souvenirs and other items like bangles, vermilion, statues/idols, etc. pertaining to the temple, culture and faith	2 (BA1, BA2)	2 (BB1, BB2)	1 (BC1)
	Residential	Formal houses	-	3 (RFA1, RFA2, RFA3)	2 (RFB1, RFB2)	2 (RFC1, RFC2)
		Informal settlers	This will include informal settlers who have a proper shelter on their own and also street dwellers	1 (RIA1)	2 (RIB1. R1B2)	1 (RIC1)
	Related to the temple	Temple administrators and workers	All levels of the temple management/administrators/priests	1 (TA1)	2 (TB1, TB2)	1 (TC1)
Key informants	Public sector (local government) Kolkata Municipal Corporation	-	The concerned municipality	1 (K1)		
	Consultant to the government	-	Some expert with sufficient knowledge in this sector who has been consulting the government on similar issues	1 (K2)		
	Researcher	-	Someone working in a similar sector	1 (K3)		
	Local observer	-	Someone from the community level who does not directly take part in such rituals but is a regular observer. Can preferably be from different religious background	1 (K4)		
	Politician	-	Politician of any rank as per availability who is from the locality and knows about the situation. Can be from either the ruling party or the opposition, based on availability and consent for the interview	1 (K5)		

The above respondents were interviewed and there was repetition of information after this and hence the number of interviews was kept up to this number

Site A: Balaram Bose Ghat

This is the site with medium density of temples and the upstream one (Fig. 11.8).

Adi Ganga is considered important unanimously by all, the major reason being access to holy water. A respondent also mentioned that "Once upon a time even drinking water was taken directly from this river and consumed after filtration in individual houses by sand, charcoal and alum added with a small quantity of camphor". Water from this river was also used for cleaning the roads:

> "This river is very important for the city life of Kolkata. She gives earnings to many people at the time of immersions of different pujas, selling of various commodities mainly for pujas, tea stalls, fast food stalls and the Kolkata speciality of water balls (phuchka). Porters are hired during immersions for carrying the idols to the river water for immersion. Similarly, the agglomeration of temples calls for devotees and this footfall in turn leads to food stalls and other similar businesses." – Respondent RFA1, Site A

Water sports like swimming and water polo was an essential part of this stretch but stopped around 40 years back. Only puja-related rituals happen nowadays. Collecting Gangajal for holy rituals has always been important from this river. The river has also been a major source of water at the time of emergency particularly fire breakouts. Availability of fertile clay for in-house gardening particularly for flower plantations is available here, while the soft and sticky natured clay is highly suitable for making idols and earthen pots. That is why Patuapara, one of the two havens for idol makers, is just by this river. The first "barowari Durga puja" or community Durga puja of Kolkata was first celebrated here by this ghat.

A very interesting thing came up as put forwards by the following quote:

> "The offerings are mostly flowers and fruits to the God and the idol is a representation of the God. So an idol cannot be an offering. We are offering to the idol. Like candles are being lit in front of Jesus and thus candle is the offering. The candle cannot be compared to the status of Christ. Similarly, fruits are offered to the idols. Idols are immersed after the worship is over and that is as per Hindu mythology of 'bisharjon'. The idol is not the offering to the river or anything for that matter. Suppose, an idol maker makes ten idols where seven gets sold and three remains unsold. So what happens then? To the idol maker it can be a piece of art with certain nostalgia and strive for earning but the main thing that inserts life into the idol is the ritual of 'pranpratisthan' and that marks the representation of the idol as God." – Respondent RFA1, Site A

This establishes that a particular ritual makes the idol holy and representation of God in that sense. Other than that, the idols are as such no different from any regular clay statue. This quote linked the individual factors like emotions and perceived consumer effectiveness on one side and situational factors like subjective norm and product attributes where it boils down to the subjective norm which explains the ritual when an idol is given the stature of God. This gives a very important point that it is this ritual that differentiates a normal piece of art in the form of a statue to the sole representation of God who is then worshipped.

It is also to be noted that the residence of the Chief Minister of West Bengal is in the vicinity of this ghat towards Kalighat, and that has a great impact on the cleanliness of this particular area. The municipality cleaning is quite rigorous here due to this, and they collect waste from the households and clean the roads once in the morning and once in the late afternoon, but they do not clean the river banks. Idol immersions happen here but not to that extent. A particular Zamindar family

Fig. 11.8 Site A – detailed map and satellite image. (Source: Google Earth and Google mymaps delineated by author (2018))

immerses their Durga idol in this ghat every year due to their long family traditions. Few small idols are also immersed here by the local people, like the Saraswati or Lakshmi idols after the puja in individual homes. Major immersions happen in the Baro Ganga in assigned ghats and under proper supervision of the municipality. Water has gone down in Adi Ganga making immersion difficult:

> "Firstly idols are not offerings but because you want to know, the practice is before immersion the idol odd number of people depending upon the size of the idol lift it and turn

> clockwise seven times and then immerse the same into water with face up. This resembles floating while face down means rude and disrespectful push. After the immersion, Gangajal is collected in the same earthern pot used during the puja and sprinkled on the devotees as a mark of peace and harmony amongst people." – Respondent BA2, Site A

This quote again relates to the individual factor – emotion and situational factor – subjective norm where the latter is again observed to be the prevailing factor concerning the ritual and the idols not being considered as offerings as they are supposedly the Gods themselves.

Photograph 11.1 shows the site and general characteristics of Balaram Bose Ghat. The top right picture is on the special occasion of Rathayatra. Photograph 11.2

Photograph 11.1 Site A. (Source: Observation (Jul–Aug, 2018))

Photograph 11.2 Site A – river condition. (Source: Observation (Jul-Aug, 2018))

depicts the condition of the river on the same day. Though a person is seen picking up the trash, the rest of the pictures show the waste thrown into the river. This includes flowers and leaves as in top left picture, while a major part is paper plates and bottles placed by the steps. During a puja, "Prasad" or food offered to the God is then distributed amongst the devotees which are often served in paper or thermocol plates as those can be disposed of easily.

Individual Factors Pertaining to This Site

Some people do not think that there can ever be any harmful component in puja offerings as there are considered to be obviously pure and hence being used for worship. Individual factor like values and personal norms (belief) plays a very important role in any aspect regarding sacred waste.

Habit and lack of knowledge, which are again individual factors, regarding the consequences of waste disposal and environment also add to the process of continuous throwing of the plates used for eating the Prasad by the devotees which could have easily been put in the municipal bin or, if not, the roadside from where the municipality sweeper can collect in their next round of cleaning:

> "There is a ritual at the end of the puja called 'bishorjan' which signifies the immersion of the idol by reciting mantras marking the end of the puja. This process does not necessarily

need idol immersion into waterbodies but technically the idol can be reused again and many do that as well, particularly in cases where idols are made of metal instead of clay. However, this is more of a sentiment to immerse the idol marking the end of the worship in the physical sense. Similarly, for domestic puja, I have personally no issues with putting those with the household waste as the same way we discard other fruits and vegetables but my mother will not allow that and she would insist on putting it in the water, more so because the Ganga is considered holy and we have the privilege of staying very near to it. I had tried to make her understand the logic but eventually when she persists, I throw the organic waste only in the river after segregating the plastic and we collect it in our home in a bucket and throw the puja waste once the bucket is full, say, once in 4–5 days." – Respondent RFA2, Site A

This response brings individual factors – knowledge and perceived behavioural control that enable people to make informed decisions. Simultaneously, a house is not about an individual alone. For example, with this respondent, it is the faith of the mother that keeps the process of putting the sacred waste into the river continuous, which is the situational factor of subjective norm, but it is the individual factor of values and personal norms in the case of the respondent that enables him to make a decision within his scope and segregate the waste before the disposal because at the end of the day sacred waste is thrown based on the belief that people should not walk on these as these have been offered to the Gods. Here, the complexity of a decision-making can be noticed – the respondent by virtue of individual factors realises his concern towards the environment but is forced to throw the sacred waste due to his mother's beliefs, which is situational factor. However, what he ultimately does by segregating plastic and other harmful elements, he addresses the situational factor but listens to his individual values and personal norms, and thus by drawing a balance between situational and individual factors, he does his part to reduce the harm caused to the environment to a certain extent.

Situational Factors Pertaining to This Site

That idols are not really seen as offerings is part of the subjective norm that how the society perceives this concept of offering and idol together. Throwing of domestic sacred waste into the water is also part of the subjective norm:

"For our local community puja, the committee decides on buying the idol and one person as such cannot do much and we usually end up buying the idol with lead paint as it costs almost half of the similar one without lead free paint." – Respondent RFA2, Site A

Eventually it boils down to the situational factor of price, that is, affordability of the product, and the individual factors of perceived consumer effectiveness, perceived behavioural control and values and personal norms lose to the situational factors. Within the situational factors, the eco certification, however, cannot hold any good in this case, and price overpowered all other indicators.

The usage of disposable plates is most often due to price, ease of disposal and unavailability of sustainable alternate options. Same goes for use of plastic bags for disposal of the waste. In such cases, individual factors like perceived consumer effectiveness and values and personal norms are not strong to show environment-friendly behaviour.

Net has been provided ahead of the Chief Minister's residence near this bank to restrict the major chunk of waste from the Kalighat side. So this stretch is relatively cleaner. This related to situational factor of infrastructure. This site being in the upstream area and also subject to high and low tides, the net prevents the major chunk of the waste in Site B from coming in this area. This shows what effect an intervention of political power can have. Some respondents have also mentioned that they would be happy to see similar action throughout the stretch of the river rather than just a particular stretch because of its deemed importance because the river is for everyone and felt that regular vigilance and enforcement might actually help.

Contamination Pertaining to This Site

The offerings are mostly flowers and fruits. Banana leaves, mango leaves, banyan leaves, bel leaves, tulsi leaves, "durba" – a particular type of grass etc. are important, and each deity has certain types of preferences. For example, bel leaves are for Shiva, tulsi leaves for Vishnu, etc. As seen in Photograph 11.2, disposable plates and bottles are also thrown, and even though these are not part of the ritual offering, these are still related to the rituals in a way as the devotees are served the food in these:

> "Toxicity to some extent is due to immersions out of lead based paints used on the idols. Though government has regulations but it is not properly monitored. To the hazardous, it is the usage of plastic bags and packaged drinking bottles." – Respondent RFA2, Site A

Respondents have also mentioned that puja offerings like flowers/fruits/sweets do not harm much, but throwing of those on the banks irrationally causes nuisances like getting rotten, and thus bad smell and formation of bacteria make the water bad. The usage of vermilion is great which is again bad because it contains mercury followed by alta for foot application of women:

> "Offerings are purely organic in nature like flowers, leaves, fruits etc. and hence not toxic and hazardous. However, people often knowingly or unknowingly put those in a plastic bag and the tied up plastic bag is then thrown into the river. So the plastic is the main hazardous element even though no ritual as such calls for plastic in the very first place." – Respondent TA1, Site A

Finally, abundant use of plastic carry bags and bottles creates great harm to the water:

> "Idols are not considered as offerings as offerings are made to the idols which are the physical representation of the Gods themselves. We do decorate the idols as we would dress up any normal human being as these idols or Gods become part of our family during the puja days. Clay (from the idols) gets dissolved in the water. In case of bigger idols, the decorative clothes may affect adversely to some extent. There have also been talks about lead free paints which I use." – Respondent BA2, Site A

Any specific volume cannot be calculated as this is not confined to a particular place and is also subject to tidal effects, but the respondents mentioned that it can be about 2–3 Kg. a day during peak season. Essentially peak season refers to the maximum activity irrespective of the type of the festival or the time of the year.

Site B: Kalighat

The chosen site has multiple layers of sacredness associated with it – the Ganga and the Kali Temple being a Hindu tirtha by itself, which makes it a very important site for the offerings and rituals. The offerings in the river in this stretch due to religious activities are the concerned sacred waste for this research. This mostly constitute of particular types of flowers, leaves, statues, ritual bathing of goats before they are sacrificed, etc. Around 700 goats are slain in Kalighat on the night of Kali puja (McDermott 2011). The mode and the frequency of disposal of this waste need to be observed to gain a better insight (Fig. 11.9).

"Kalighat is considered to be one of the fifty-one pithas of the Goddess. The manifestation of the Goddess here is known as Dakshinakali; her consort Shiva is

Fig. 11.9 Site B – detailed map and satellite image. (Source: Google Earth and Google mymaps delineated by author (2018))

Nakuleshwar; Vishnu in the form of Krishna, dwells under the appellation of Shyam Ray in an adjacent temple. The association of these three deities makes Kalighat a unique meeting – point for both Shakta and Vaishnav pilgrims, for ascetics as well as householders." (Chaudhuri 1990, p. 24)

All the respondents believe that Adi Ganga is important for them. This significance is very much related to their daily lives, not only from the present day but also from past reminiscences, and hence related to the individual factor – emotions:

> "I remember that up to late 60s, this river water was used for washing of roads twice a day which was fantastic as all roads used to remain very neat and clean. For the development side, it was used for navigation purpose and thus lot of businesses developed throughout the stretch of this river flowing through proper Kolkata." – Respondent RFB1, Site B

Respondents also mention that navigation along this river helped in bringing clay from the main stream of Ganga to the potters and idol makers of Patuapara nearby. Being one of the 51 shaktipeeths in Hindu mythology, Kalighat has a special significance where Gangajal taken from the Adi Ganga is used during the preparation of "bhog" or the food offered to the Goddess and also sprinkled at the time of taking the "bhog" from kitchen to temple for offering:

> "Particularly for me, it is very important as I earn my livelihood from my shop which is on the eastern side of Adi Ganga in the Kalighat side." – Respondent BB2, Site B

Essentially the river, the human connection with water and the site hold special meaning for all, be it for different or similar reasons, and enforce the individual factor of emotions.

> As said by one of the temple representatives, "the Kalighat temple used to be right on the banks of the river and this Adi Ganga used to be the original flow of the Ganga to the Bay of Bengal. This was about 400 years ago. Eventually the width and depth of the river decreased, informal settlements started coming up along the banks and also the main flow of the river had shifted to the current Hooghly river or as we now call, Baro Ganga (big Ganga) whereas Adi Ganga literally means the original Ganga."

Photograph 11.3 reveals the character of the place by means of a few old photographs from 2012 while the boat was still in working condition in Kalighat, while Photographs 11.4 and 11.5 show the current situation.

The bridge in Kalighat connecting either banks of the river in Site B, as seen in Photograph 11.7, had been recently constructed where boats used to ply previously (Photograph 11.3). Three boats were placed one after other, and due to lack of sufficient water, people used to cross over from one bank to the other simply by walking over the boats placed that way. Only during high tides, mostly during the month of Bhadro, that is, mid-August to mid-September, when water in the river is its annual high, one boat with its boatman could actually ply between the banks. The remnants of the boats can be observed in Photograph 11.4.

An interesting thing that came out was that a certain livelihood exists amongst the informal class where they pick up coins from the river bank and bed that people offer to the river as part of a ritual or for making a wish. They also get alms from the devotees, and this donation is often seen as a goodwill gesture by the devotees which will aid in their wishes coming true. This in turn opens up a whole new spectrum of livelihood in the area, and the informal settlers become an integral part of this system.

Photograph 11.3 Site B – site photographs of 2012. (Source: Observation (2012))

The character of the site with the temples, Kali Temple – the most prominent one in the area – is shown in the top left in Photograph 11.6 along with some of the other temples which bring out the character of the place. The statuette of Mother Teresa can also be seen in the same photograph, giving this site a multi-cultural perspective. The temples are of various scales and architectural styles (Photograph 11.7).

Photograph 11.8 shows few of the rituals taking places near the Kali Temple by the Adi Ganga like head shaving of a child, puja offerings set in a temporary stall and family praying together with the aid of a priest clad in red cloth.

Photograph 11.4 Site B – Kalighat bank. (Source: Observation (Jul–Aug, 2018))

Photograph 11.5 Site B – Chetla bank. (Source: Observation (Jul–Aug, 2018))

Photograph 11.6 Site B – site surroundings. (Source: Observation (Jul–Aug, 2018))

While mostly flowers, leaves and fruits are the common items used in the offerings as put forwards by all the respondents, there are also some specific offerings based on the ritual; the customs change but the items remain more or less similar:

> "For sradhh ceremony, white flowers are required while for marriages, coloured flowers are required. Also Karthik idols are not immersed but left on the banks while idols are immersed in the river." – Respondent RIB2, Site B

Banana plant is bathed for Kalabou snan during Durga puja, while goats are bathed for Kali puja. Though there is no specific time for such sacred waste disposal, Saturday is a very important day for Kali puja which makes it more auspicious and more crowded.

> "We do sradhh ceremony or last rites, thread ceremony while people also throw ashes of the deceased here mainly in Kalighat because this place is considered holy, being one of the 51 shaktipeeths of Hindu mythology and performing rituals here help attain moksha. People also collect Gangajal during high tide. The low tide water is pretty bad and is probably the wastewater of the entire city." – Respondent TB2, Site B

Photograph 11.7 Site B – access across Adi Ganga Source: Author's observation (Jul–Aug, 2018)

Bathing on special days mostly on Tuesdays and Saturdays, Sankrantis, that is last day of each month in the vernacular calendar, especially on Makar Sankranti in mid-January Mahalaya (usually in October) and eclipses is considered auspicious.

> "Our puja offering packet mostly contains flowers, garlands, sweets, candles, bangles etc. wrapped in a sal leaf packing. People also offer sarees, dhotis and gamchhas (thin handloom cloth towel)." – Respondent BB1, Site B

Another respondent also mentioned:

> "Our puja packages wrapped in sal leaves have sweet, sindoor, flowers while garlands, betel leaves, nuts, bangles, alta, incense sticks, candles, oil lamps, fruits are extra and can be taken as per choice of the devotee and also the price range varies across the options." – Respondent BB2, Site B

Photograph 11.8 Site B – rituals. (Source: Observation (Jul–Aug, 2018))

The above quotations bring out that most of the offerings are organic in nature, but it is to be noted here that as per another respondent quoted later, the bangles in the packets are often made of plastic due to the pricing factor. Also sindoor or vermilion and bangles are offered only to goddesses and not to gods.

Photograph 11.9 shows the human interactions with the river for the rituals – one shows the occasion of Bipodtarini puja, worship for keeping the evil at bay, where devotees flock to the river for ritual bath and offerings even when it was raining as can be noted from the umbrellas and river, and the other one shows ritual bathing on a regular day.

The municipality has bins on wheels which the person takes along for collecting household waste and for cleaning the roads. All these are stacked near the Kalighat temple itself as seen in Photograph 11.10.

Individual Factors Pertaining to This Site

Though lack of civic sense, values, emotions, habits and personal norms play key roles, religious belief systems that pertains to individual factors – values and personal norms–come out as one of the prime factors for sacred waste disposal. Religious ground that Adi Ganga water is holy forms the general notion.

Because the flowers, fruits and leaves and the other materials have been used in the worship, those are still considered as sacred, and people, in general, do not consider throwing the holy items together with other waste in the municipal bins an

Photograph 11.9 Site B – rituals in Adi Ganga. (Source: Observation (Jul–Aug, 2018))

option at all, and hence they offer those to the holy river, and this relates to situational factor of subjective norm, but individual factor of values and personal norms finally makes the cut here as it is after all an individual's decision.

The goddess is considered as a woman, and it is more about the association of the goddess with married women which raises the significance of these practices. So this basically gives a human touch or emotional connect (individual factor –

Photograph 11.10 Site B – municipality waste collection vans near Site B. (Source: Observation (Jul–Aug, 2018))

emotions) to the goddess which makes such rituals all the more significant to the lives of the devotees.

Some respondents also do not consider that sacred offerings can ever be toxic under any circumstance. This relates to individual factors such as perceived consumer effectiveness, lack of environmental knowledge and perceived behavioural control on one hand and situational factors such as subjective norm on the other, though the individual factors are stronger here. Same goes for the following quote:

> "These worship flowers turn holy after the worship is complete." – Respondent BB1, Site B

Sacred waste is perceived as holy and as offering and not really as waste:

> "To some extent it does because during high tide, Adi Ganga receives a lot of waste from the Baro Ganga and while most of the materials gets washed out again to Boro Ganga, whatever cannot go during low tide remains stuck here. Baro Ganga is powerful but these tributaries lack that capacity now. We don't know what happens after Baro Ganga but Ganga engulfs everything and makes it pure." – Respondent TB2, Site B

This gives a complex perspective as on one side it accepts that river should be clean, while on the other it says that Ganga has the healing power. This relates to situational factors, subjective norm and infrastructure on one hand, and individual factors, perceived consumer effectiveness, lack of environmental knowledge and perceived behavioural control on the other; lack of environmental knowledge works the strongest here.

Convenience also plays a key role as the river being close by, it becomes more convenient for them to throw all types of waste items together in the water or on the banks as that will anyway be carried away by the water which pertains to situational factor, infrastructure, and stronger individual factors, habit, perceived consumer effectiveness, perceived behavioural control and lack of knowledge about the environment.

Situational Factors Pertaining to This Site

It is to be noted here that price of the commodities given in the offerings plays an important role as well:

> "The local stalls mostly sell plastic bangles for such cases as it costs less whereas the actual material to be used is conch and coral bangles for the married women and also for the Goddess for symbolic as well as health purpose. Goddess Kali is visualised as a married woman and hence the same." – Respondent RFB2, Site B

Even though there are individual factors of perceived consumer effectiveness and (lack of) knowledge regarding the ill-effects of plastic, both amongst the buyers and the sellers, it is obvious that while some might still be conscious about the ill-effects, it is for the majority the industry survives, and for this, situational factors – price (as plastic is cheaper) and subjective norm – play a very important role. What can be observed here is that the essence of the ritual and the practice remains the same, but to keep doing that at an affordable price for access to the masses, cost of the commodity becomes a big issue. Hence bangles made out of conch shells and coral which hold the main significance in the social customs have been replaced by plastic white and red bangles consecutively. Devotees can now afford buying the same, but however, plastic creeps in the process due to the issue of affordability, which is a situational factor as found in the literature. This is also, in fact, in sync with the subjective norm, and that is about the customs and the practices which spread its roots deep into the society. It is often about carrying on the belief by one way or the other due to the significance of the same.

Lack of alternate options again becomes a crucial factor as it came out mostly from the informal settlers that they also dispose of their household waste along with sacred waste in the river as they do not have a proper system in place for disposal of household waste as the municipality collects it from the formal houses. Individual factors of values and personal norm and knowledge take a backseat, and situational factors of infrastructure and lack of alternate options take the lead:

> "Considering it to be holy and offering to Goddess Ganga. In Hinduism, all rivers and mountains are considered holy." – Respondent TB1, Site B

This again relates to both situational factor, subjective norm, and individual factor, values and personal norms, where the former becomes the main guiding role.

The complex case of the idol making industry follows mostly situational factors like price, product attributes, lack of a proper certification system, subjective norm together along with individual factors of emotions, perceived behavioural control and knowledge or lack of it, though situational factors make the main deciding factors.

Contamination Pertaining to This Site

The respondents mention that vermilion and lead-based paints used on idols, earthern potteries, oil lamps and candles may harm the waterbody to a certain extent. The rituals associated with this river happen all the year round, the exact dates and time being decided by the almanac. As from the interviews, the main rituals performed here are mostly marriage, thread ceremony, sradhh and, tarpon on Mahalaya days where one pays homage to ancestors, bathings after both solar and lunar eclipses, Chhat puja and bathing goats before sacrifice, the offerings mostly being flowers of different types, leaves of different types, fruits, earthern pots, milk, etc. Domestic sacred waste is usually put on the banks or the river water by bringing puja waste collected from home as per the convenience of the individual or on a weekly basis. The rituals usually happen all the year round, but there are some important dates as mentioned by this respondent:

> "It is being done on a regular basis like Saraswati idol immersion after the puja usually in February, Lakshmi idol immersion usually in October, Kali idol immersion mostly in November etc. along with other pujas all through the year." – Respondent RFA2, Site B

Another element that came out from the research was the use of the plastic bags for the disposal as put forwards in the following quote. Though this is not part of the sacred waste per se, this is often used for the disposal which does the harm:

> "I don't think there is anything so serious from the offerings but plastic bags are used in abundance for carrying and throwing the domestic sacred waste in the river." – Respondent BB2, Site B

Another very interesting outcome came up regarding the idol making industry and its associated toxic and hazardous components as put forwards in the following quote:

> "Plastic bags in which people throw stuff, I think, is the major hazardous element. Idols also have lead based paints but a ban was enforced on the same but it is still used. Implementation will take time. Idol making is an industry by itself afterall and is the source of livelihood of so many people, starting from people collecting the clay, making the idols, painting the same, making the clothes and accessories of the idols and so much more. Previously idols were made only of clay but due to rain the idols would get out of shape which ruined the image of worship. Eventually these people started using burnt clay and adding synthetic elements for decoration purpose like plastic beads, thermocol etc. and also so that the idol does not get ruined in case of rains. It was an innovation from the idol making industry. Thus it is very difficult to judge why something is used and why something is not used. Imagine this from the buyers' point of time? You get sponsors, spend so much money on the grand festival of Durgapuja which caters to the tourism quotient of the state as well, will you buy a simple clay idol which does not look fancy and also stands a chance of being washed away? Rains during the time of Durgapuja is not usual but not that rare as well. Apart from this, I cannot think of any other toxic waste as such." – Respondent TB2, Site B

For the volume of the sacred waste, while many of the respondents said it is very difficult to ascertain an exact value as it is a waterbody subject to tidal flow, average figure coming out of the interviews that the respondents can estimate for the sacred waste disposed in this stretch is around 50 Kg. per day during peak season, while many saw it as not much. Difference in contamination level in the river/river banks due to religious rituals can be observed in Photographs 11.4, 11.5 and 11.9 where the latter one is on the day of a festival and the contamination is noticeably higher.

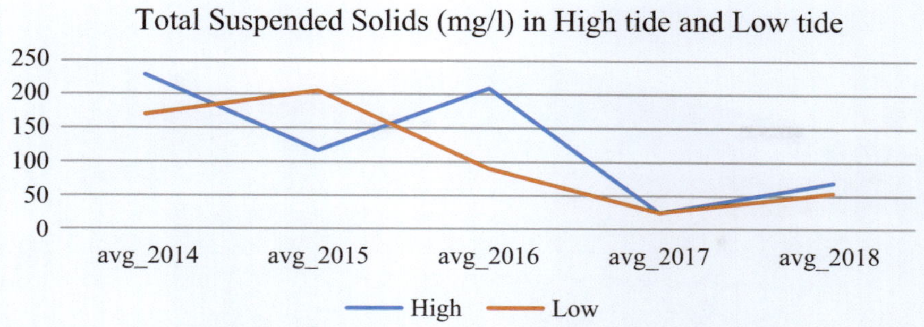

Graph 11.1 Total suspended solids in Kalighat station point, Adi Ganga in high tide and low tide over 5 years. (Source: West Bengal Pollution Control Board (2018))

This can also be noted in the photographs where the surface looks cleaner with lesser concentration of suspended materials in 2018 as compared to 2012 (Fig. 11.6 and Photograph 11.4). It is also to be observed that for the 2018 data, the total suspended solids during high tide and low tide are almost the same. This can be linked to Site A where the net in front of the Chief Minister's residence blocked the passage of the suspended solids from Sites A and B towards Site A, the latter being nearer to the main trunk of River Hooghly which induces the tidal flow in Adi Ganga. Please vide Fig. 11.4 for the location of the River Hooghly and the three sites in concern. The little difference between the readings during high tide and low tide for TSS is the movement of such suspended solids along the direction of tide in between Site B up to the net installed near Site A. It should also be noted that due to security reasons, being in front of the Chief Minister's residence, the net could not be photographed during the data collection period but has references in the interviews. Only Site B has the station point for recording water quality by the WBPCB, but as the stream experiences tidal flow, the overall same water quality is experienced by Site A and Site C as well.

A compactor machine has also been installed and has been functioning since November 2016 where the organic waste from the Kalighat temple goes along with organic waste as collected by the municipality. However, most of the respondents were not aware of this and can be associated to the individual factor – lack of knowledge (Photograph 11.11).

Photograph 11.11 Site B – compactor unit set up by KMC near Site B. (Source: Observation (Jul–Aug, 2018))

Site C: Prasannamoyeeghat

This site is further downstream of Site B:

> "In earlier days at the time of my grandmother, she used to take bath regularly. Later on, I have seen my mother also but gradually the bathing stopped because of the poor water quality." – Respondent RFC1, Site C

Prasannamoyeeghat was primarily made for personal uses of the then zamindars of this area. As such idols of worships or pujas performed in individual houses used to be immersed here. Another important part is for navigational purpose of commercial cargo – mostly rice, bamboos, tiles, bricks, jhama brick bats for soorkhee for road building, etc. led to development of a commercial hub – the Chetla Haat nearby along with other factories and godowns along the river banks. A few of these survive now:

> "It used to have a big importance earlier towards sports. Different types of sports used to happen here. Lots of swimming events used to be conducted during the high tide. Later on due to shortage of water, these got stopped and Adi Ganga virtually became a sewerage canal." – Respondent RFC2, Site C

Photograph 11.12 Site C. (Source: Observation (Jul–Aug, 2018))

Photograph 11.12 shows the overview of Site C, certain extent of beautification done in the form of a bench by the river, new buildings and also some ongoing construction but the situation is visibly not near to perfect. The main bridge right next to Site C with few informal settlers on the sidewalk of the bridge, someone disposing a packet can be observed in Photograph 11.13 along with other images of Site C (Fig. 11.10).

Holy dips are taken on various auspicious occasions. Nowadays because of insufficient water, bathing has almost stopped in this stretch. However, small-sized idols are still being immersed to some extent from pujas conducted in individual homes.

Photograph 11.13 Site C. (Source: Observation (Jul–Aug, 2018))

The usual offerings for worship are flowers, fruits and leaves. Offerings vary from ritual to ritual to some extent. For example, during Chhat puja bananas are offered in plenty, but in the case of other pujas, there is nothing specific, and all types of fruits are being offered.

Regarding the domestic sacred waste, mostly it is stored in the house for a week and then packed in plastic bags, and the domestic help disposes it off on the water or on the river bank.

Photograph 11.14 shows the faces of the idols placed on the trees; no one interviewed also knew who put those up there or the significance of the same, if any. There is also a bamboo framework of the idol that has been taken up from the water, and

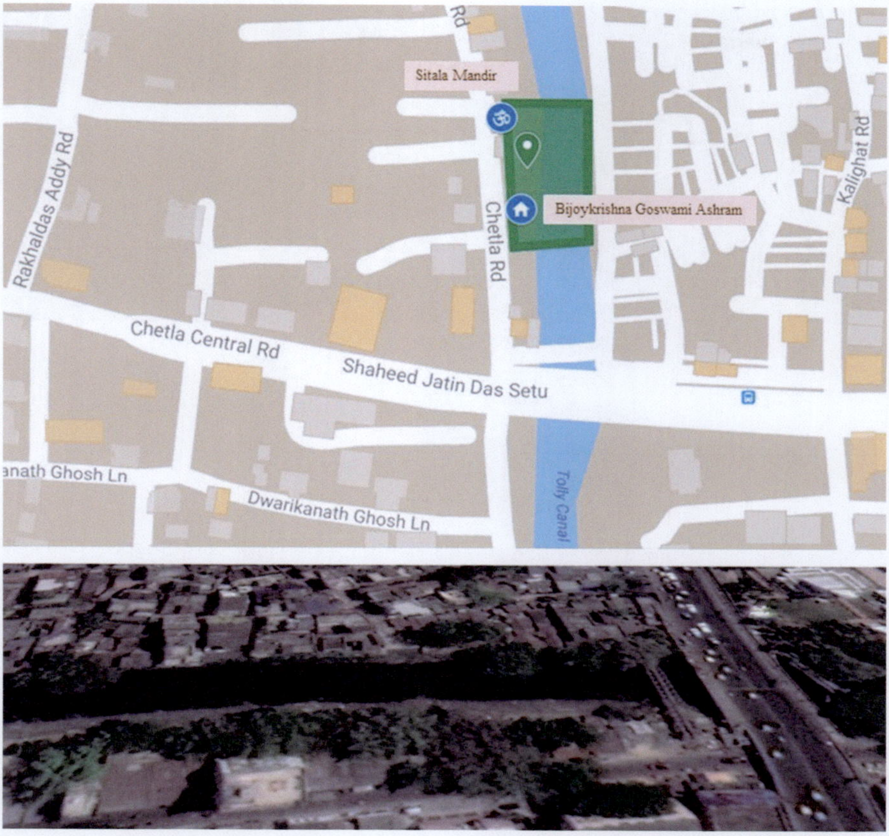

Fig. 11.10 Site C – detailed map and satellite image. (Source: Google Earth and Google mymaps delineated by author (2018))

once the other parts have been removed, the frame just stands there by the river. The plaque mentions that amount of Indian Rupees 7,50,000/= was spent on development of the Prasannamoyeeghat, and the project was initiated and completed in 2014.

The municipal vat and a municipality worker cleaning the road are seen in Photograph 11.15.

Many people come to pray in the Sitala mandir here, and also many disciples of the saint come to the Bijoy Krishna Goswami ashram regularly, mostly in the evenings for the prayer sessions. This site is very conveniently located and is less than 500 metres from the nearest metro station and less than 50 metres from the nearest bus stop making it easy for people to come here. Also many elderly people come for the evening

Photograph 11.14 Site C. (Source: Observation (Jul–Aug, 2018))

Photograph 11.15 Site C – municipality waste collection. (Source: Observation (Jul–Aug, 2018))

prayers in the ashram and, on their way back, do their regular grocery shopping from this bazaar right here under the bridge. All these together make the area very easily accessible, and this site is also the least dense one in terms of the number of temples.

At Prasannamoyeeghat, mostly sradhha and tarpan ceremonies and Chhat pujas are performed. Chhat puja usually happens in November after Kali puja. For other religious matters, people mostly prefer the Kalighat stretch. The offerings are always fruits, flowers, leaves, milk, curd, raw rice, honey, etc.:

> "Though people prefer mostly Kalighat but to avoid the heavy rush there, some people come here as well and hence this is also important." – Respondent BC1, Site C

The respondents have also mentioned about dredging happening about 5 years back, and no significant improvement has happened in the river conditions since then:

> "The overall situation is getting worse due to shortage of water in the river. Not much waste is dumped in this stretch but being a waterbody, everything is connected. This is south of Kalighat and downstream. So during high tide water from the Hoohly come to this side past Kalighat and bring the waste here as well and much is left behind when the water flows back during low tide." – Respondent BC1, Site C

Thus, it is not only about the practices in that particular stretch but is also a play of the tides, and thus the relative location also plays a very important role.

Individual Factors Pertaining to This Site

The most prominent factor coming out regarding the disposal of sacred waste into the water was because it was considered to be holy – both the offerings and the river.

That puja waste is thrown by the local residents and also by the passers-by from the bridge above shows individual factors – lack of knowledge on environment, perceived consumer effectiveness, habits and also perceived behavioural control. This is reinforced when people still put the puja waste in the river water or on the banks during low tide when there is less water in the river in spite of the municipal van being present right there.

Apparently, worshipped flowers should be put on the water only so that no one can step on those and can flow straight to the sea; thus the practice shows the individual factor of values and personal norms of the practitioners and the situational factor – subjective norm to a great extent.

The shopkeeper mentioned about her conscious effort in doing her part: "I don't throw myself and I have placed a bin in front of my shop so that people do not litter here and there". This is a very significant finding because it shows that conscious effort can be taken by anyone irrespective of the individual's position in the societal ladder or the financial status. This obviously connects individual factors – knowledge along with values and personal norms, perceived behavioural control, perceived consumer effectiveness and also habit.

It is also very interesting to note the following which exhibits a weird combination of individual factors, knowledge, values and personal norms and emotions, and situational factor, subjective norm, though the individual factors aid in the final decision-making:

> "No waste apart from puja offerings should be thrown into the holy water as that makes the river unholy." – Respondent TC1, Site C

Situational Factors Pertaining to This Site

The continuous process of waste disposal always can be related to subjective norm, and one person does because others are doing and no one is really stopping the other. It is also subjective norm that holy puja waste should be thrown only on holy water and not with other household waste in municipal bins as they will make the sacred waste impure. That sacred waste is to be put on water and not in the municipal bin, in spite of those being present right in front of the site, asserts subjective norm further. That is the reason as per the temple representative that it is "offered to the Goddess Ganga at the end of the puja to be carried to the sea and finally to the heaven".

Contamination Pertaining to This Site

Some respondents had the opinion that to some extent the formation of bacteria from rotten flowers and leaves which are part of the sacred waste exerts pungent odour, and that may be harmful.

One resident from the formal house mentioned that:

> "Other than puja offerings, small chemical factories used to drain out their effluents. At present, due to the strong control of the Pollution Control Board, direct discharge of such effluents cannot happen to the river. So the situation is improving but it will take time." – Respondent RFC2, Site C

He also mentioned that plastic bags, bottles and containers when disposed into the river are also some of the items causing harm.

It is worth noting here that everyone was aware of the tidal influence in this river. The temple representative also mentioned that it is difficult to assess the exact volume of waste in the stretch as the waste gets carried away to Baro Ganga from here. Again, there was no precise outcome on the amount of the sacred waste, but the average reply was around 5 Kg. per day during peak session:

> "People in this particular ghat do not throw much. However, some passers-by often throw discarded stuff on the go from the Keoratala Bridge here which adds up to the garbage in the stream below." – Respondent RFC2, Site C

This also relates to individual factors, habit, perceived consumer effectiveness and lack of environmental knowledge, and, to certain extent, situational factor, infrastructure, the latter being less significant, as there are waste bins by the municipality at regular intervals throughout the city and in this site as can be seen in Photograph 11.15.

Summing Up the Findings

> "Any river is important and this is more significant to us because of its values. Hinduism sees nature as God." – Respondent TA1, Site A

This quote asserts the importance of this river and marks a strong connection between individual factor, values and personal norms while marking the significance at an individual level, and subjective factor, subjective norm when it talks about a religion in general:

> "The carrying capacity of the river is going down. Dredging happens but it is not always efficient. For example, on paper it is 100 cu.m. but on ground it is 20 cu.m. Thus honest scrutiny is required during the implementation to keep in check such incidents." – Respondent K2

It is also about these factors like political will, vigilance and enforcement which hold very crucial as well. However, there is another perspective, and that is of political will and influence in materialising a project, and this is showed in the case of Site A due to the proximity to the Chief Minister's house. As a respondent had expressed his feeling quite aptly:

> "This stretch of about 500 metres is well maintained as our present Chief Minister's house is just on the bank. A net has been installed across the river so that flowing objects, particularly from the Kalighat area cannot enter this stretch. Alas! Had our Chief Minister's house been along the entire stretch of Adi Ganga, the entire river would have been clean!" – Respondent RFA1, Site A

However, political influence gives an entire dimension altogether and can be separate research by itself and hence was kept out of the main parameters of this study as mentioned in the limitations earlier. Considering that Site A showed added monitoring and cleaning work from the municipality because of the influence of the Chief Minister residing there, this point can be considered under infrastructure (situational factor). With regard to the latter quote by Respondent RFA1 of Site A, it gives a remark that political influence when used for improvement is a welcome thing:

> "Lead paints are deemed harmful but it definitely is not used anymore and a ban has been imposed on such hazardous paints. There are sufficient bins and enough of enforcement actions have been taken. There is tight security and people caught littering and throwing waste in the river is fined then and there. However, offerings come under a special category and it is very difficult to put it under waste category as sentiments and emotions of people are involved. This enforcement action will at least stop people from throwing plastic bags into the river. Ban on lead paints is another good step. Of course it will take some time and it is not an overnight solution but things will surely happen for good." – Respondent K5

The decrease in the total suspended solids in this station point, as seen in Graph 11.1, can assert to the positive influence of the enforcement as put forwards by many of the respondents.

As per order of National Green Tribunal, "Worship material like flowers, vastras (clothes), decorating material (made of paper and plastic) etc. should be removed before immersion of idols. Biodegradable materials should be collected separately for recycling or composting. Non-biodegradable materials should be collected separately for disposal in sanitary landfills. Clothes may be sent to local orphan house(s)" (Government of India 2016b, p. 2).

The court order also mentions that public awareness is very crucial and people should be made aware of the consequences of such waste disposal in waterbodies as every project has a certain portion of fund earmarked for awareness programmes, and this needs to be done for the entire state and should not be restricted to a particular area or district. The order also notes from its evidence that people throw puja waste materials into water and lake which include plastic bags as well which are otherwise used only as carry bags. (Government of India 2016b)

Another order by the National Tribunal notes the complaint filed that puja practices and idol immersion caused water pollution and adversely affect the waterbodies, and proposal was made by the applicant that "Bhumi Shodhan" can be done as an alternative to idol immersion, which means idols will be subject to water spraying on a raised platform such that the melted earth and the other remnants can be collected below the platform and reused and this will not affect the waterbodies directly. The court suggested awareness measures and advertisement via different media along with proposing the states to discourage the artisans to use toxic colours and plaster of Paris for the idols. Immersion should also be regulated as per the guidelines of Central Pollution Control Board under strict supervision. Along with this district magistrates, Pollution Control Boards, superintendents of police and other relevant enforcement bodies are to keep a check on the idol making process as well. As a note to this, WBPCB stated that plaster of Paris is not toxic and that water pollution would be minimum when removed quickly after immersion. KMC also gave the consent that CPCB guidelines are being followed and the immersion ghats get cleared by the next day. This report however mentions that it was found that all these guidelines were not being followed in reality to its maximum extent, and the authorities even found that synthetic colours with toxic heavy metals were still being used on idols. Since ancient times, idol worship has been in practice in India, but natural things like river water, milk, curd, etc. were used, and idols were also made of clay and coloured with natural agents like turmeric. Mythology and religious texts emphasised on preservation of nature. (Government of India 2016a)

That plaster of Paris is harmful has been mentioned in the literature (Giripunje et al. 2014; Bhattacharya et al. 2014; Das et al. 2012; Shrivastava et al. 2011; Dana Gopal et al. 2014), but no proof in support of WBPCB's claim of it not being harmful has been found by the author during the course of this research. These findings show the path towards the recommendation and the step forwards as discussed further in the concluding part.

Table 11.6 is a summary of the findings based on all the data collected shown site wise where increasing size of the tick symbol denotes the increase in influence (out of six levels – smallest, smaller, small, big, bigger, biggest) of that certain factor. Table 11.7 shows the most prevalent factor for each site on assimilating Table 11.6 data.

The type of contaminants is similar in all the three sites as these are essentially outcome of sacred rituals, details of which have been provided earlier as part of the study.

Table 11.6 Summary of the findings

Variables	Sub-variables	Indicators	Description	Site A	Site B	Site C
Factors affecting disposal of toxic and hazardous sacred waste (Gatersleben et al. 2014; Gaspar 2013; Joshi and Rahman 2015; Krajhanzl 2010; Steg et al. 2014)	Individual	Emotions	Feelings that make one take a decision	✓	✓	✓
		Habits	A continuous process which has become a part of life	✓	✓	✓
		Perceived consumer effectiveness	Perception regarding impact of a certain action or product	✓	✓	✓
		Perceived behavioural control	Perception of the individual's behaviour by the person himself in advance in regards to a context	✓	✓	✓
		Values and personal norms	Social, ethical and moral values and ideals of an individual which aid in taking an action	✓	✓	✓
		Knowledge	Previous learning and general awareness about a certain action or product	✓	✓	✓
	Situational	Sociocultural factors/subjective norm	The social system forming a certain image or idea about a certain action or product	✓	✓	✓
		Infrastructure	Presence or absence of infrastructure that helps in decision-making and execution	✓	✓	✓
		Price	Cost of an item; relates to the affordability of a product	✓	✓	-
		Product availability	Relates to availability of alternate options	–	✓	-
		Product attributes	The components of a certain product	✓	✓	-
		Eco certification	Marked under a certain category as environment friendly by relevant authorities	✓	✓	-

Table 11.7 Most prevailing factor based on the sites chosen

Most prevailing factor	Site A	Site B	Site C
Individual/internal	Values and personal norms	(Lack of) knowledge	(Lack of) knowledge
Situational/external	Subjective norm	Subjective norm	Subjective norm

Table 11.8 Estimation of sacred waste generated in the sites chosen

Estimation of sacred waste generated	Site A	Site B	Site C
Kg./day	2–3	50	5

It is also to be noted here that WBPCB has only one station point within the sites chosen, and as the water in the canal experiences back and forth movement due to the tides, the findings in the Kalighat station point are valid for all the three chosen sites. The total suspended solids is shown in Graph 11.1.

Table 11.8 gives an estimation of sacred waste generated in the three sites as noted from the interviews.

451 MT/day and 550 g/day per capita solid waste is generated in the entire Tolly's Nullah basin as per the reports of the municipality, and the management of the same has improved over the last few years as is evident from Graph 11.1. Waste compactors has aided in this process. (Kolkata Municipal Corporation 2017)

In the end, society and ecosystem are essentially interlinked, and without a surviving ecosystem, there is also no point of the human society and their festivities (Laborda-Pemán 2015).

Conclusions and Recommendations

This segment provides a brief overview of the purpose of the research undertaken and how the study undertaken fits in to answer the research questions. The conclusion is drawn on the basis of research results, and relevance of this study is given. At the end, some recommendations have been proposed which open doors for further research in similar area, which given sufficient time would have been really interesting to capture in this research itself or can also be an extension of this research.

The main aim of the research was to find out the factors behind the disposal of toxic and hazardous sacred waste and the toxic components associated with the disposal in urban river, and the case of Adi Ganga was taken to study the same.

The main and sub-research questions are mentioned in Box 11.3.

Box 11.3: Research Question
Main research question:

Which factors explain the disposal of toxic and hazardous sacred waste which adds to the urban river contamination in Adi Ganga, Kolkata, India?

Sub-research questions:

Which individual factors explain the disposal of toxic and hazardous sacred waste in this river?
Which situational factors explain the disposal of toxic and hazardous sacred waste in this river?
What kind of contamination happened related to toxic and hazardous sacred waste in this river?

Research Results

The results and findings have already been discussed, and here the objective is to assimilate those and put as per the sub-research questions which will in turn answer the main research question in totality.

Sub-Question 1: Which Individual Factors Explain the Disposal of Toxic and Hazardous Sacred Waste in This River?

When asked that whether the river will survive if the current state continues, while most of the respondents across the three sites mentioned that it may be difficult and much work needs to be done, a few of the respondents did mention that nothing adverse can happen to the goddess herself. This portrays the underlying strength of belief and how vital and eccentric can that be at the same time. Disposal of the sacred waste into the river has also been due to piousness or belief in most of the cases. Belief or values and personal norms are thus one of the most significant factors. The most prevalent individual factor for disposal of sacred waste in concern is values and personal norms in Site A, while for Site B and Site C, the most prevailing individual factor is lack of environmental knowledge.

In order of decreasing influence, the individual factors are listed below as per each site:

Site A – values and personal norms, perceived consumer effectiveness, emotions, perceived behavioural control, (lack of) knowledge, habits
Site B – lack of knowledge, perceived consumer effectiveness, perceived behavioural control, values and personal norms, emotions, habits
Site C – lack of knowledge, perceived consumer effectiveness, habits, values and personal norms, perceived behavioural control, emotions

Sub-Question 2: Which Situational Factors Explain the Disposal of Toxic and Hazardous Sacred Waste in This River?

It is more about a shared notion that there is no harm this way. While availability of alternative options, price and subjective norm are the main situational factors, the latter two hold more significance. It is also true that subjective norm from situational factors and values and personal norms from individual factors (Gaspar 2013) somewhat got blurred and became one, and it is most often the society which creates the norms, and how an individual then imbibes that and to what extent, imbibes or not, is based on his personal norms and his own values brought from his experiences. Price played a significant role because that reflected the affordability, be it for buying the idols; application of, or rather, the market for lead-free paints; usage of plastic bags; and multiple other factors because economy or affordability affects everyone at every level of one's lives' decisions in a society making this factor all the more important.

The most prevalent situational factor for disposal of sacred waste in concern for all the three sites is subjective norm.

In order of decreasing influence, the situational factors are listed below as per each site:

Site A – subjective norm, price, product attributes, infrastructure, eco certification
Site B – subjective norm, infrastructure, price, product attributes, product availability, eco certification
Site C – subjective norm, infrastructure

Sub-Question 3: What Kind of Contamination Happened Related to Toxic and Hazardous Sacred Waste in This River?

Contaminations are mostly organic in nature like flowers, leaves, decorative items, clothes, paint, etc. from the immersed idols, etc. which is in line with the literature (Giripunje et al. 2014; Bhattacharya et al. 2014; Das et al. 2012; Shrivastava et al. 2011).

In Site C the density of temples is less, and the entire stretch is mostly residential with some commercial godowns on the bank side making it easier for the municipality to clean; hence the volume is not very high, whereas in Kalighat, the crowd is higher, and because of Kali Temple, the number of pilgrims is also very high, and garbage or waste is also more in that stretch.

As rightly put forwards by one of the respondents, idol making and offerings related to sacred rituals are now a full profession, and its added glamour has created a market for the various elements which in turn harm the environment. It is more complex than the actual components used as it deals with a social structure in entirety and the supply chain of the items. Idol making is a cottage industry, and they strive in their own way to innovate and cater to people's needs and make the idols more attractive because at the end of day, as informed by the interviewees, idols are a representative of God and also given the human touch, as, for example, in the case of Durga or Kali, the goddesses are envisioned as married women. Durga puja is celebrated as the homecoming of the goddess with her two daughters and two sons and daughter-in-law to her parents' home for these few days, and it is this personal connect and relatability that makes Durga puja beyond just worship and turns it into a festival celebrated by everyone irrespective of religion, caste or creed. Thus it is difficult to bring a change overnight by enforcing a single law. People need to understand from deep within this connect that they feel with the festival should also be existent with the environment. Ganga surely has been given a legal identity (Pecharroman 2018), but until and unless the other factors and people's consciousness come together to save the river and environment, they unfortunately cannot speak for themselves.

As seen in Graph 11.1, the total suspended solids have been going down with time, which clearly is a good news. There is much to be done to make the planet liveable for all again in the best way possible.

Summary

As seen from the literature, there are multiple indicators of individual and situational factors, also referred to as internal and external factors in other studies, and based on the cases, these meanings associated with the indicators may change to a certain extent, and it might be difficult to have a clearcut division as some aspects may fall in the common umbrella of both the factors. This can be considered for the case of individual factor, values and personal norms, and situational factor, subjective norm, as these are very much interrelated and one is immensely dependant on the other. For example, what society says determines an individual's action, and when this same factor applies individually to a large number of people, that in turn gives rise to subjective norm (Hardin 1968). Likewise, belief or values and personal norms are an outcome of the society, but the decision of what to follow and to what extent is the choice of the individual (Richerson and Boyd 2005; Gómez-Baggethun et al. 2012).

This research uses multiple modes for data collection, and in the process, the information coming out of the interviews gets validated with observations made by the author supported by field notes and photographs along with secondary data from reports from the concerned government organisations and court orders. This provides its desired internal validity as well as external validity as respondents were interviewed from different levels of the society till saturation of information was reached.

Furthermore, the rituals described in the research are Hindu rituals and are not restricted by a geographical boundary, but rivers being a natural entity, the study can be extended to other urban rivers exposed to sacred waste disposal or rituals. It can, however, be noted that due to change of legislation, even though there are rivers and practising Hindus, such rituals cannot take place in all rivers and all countries. Thus the context and cultural background remain crucial.

There are, however, conflicting views and not a fixed strategy for pro-environmental behavioural change, but making goals for pro-environmental action and strengthening those can be a good starting point (Steg et al. 2014). So essentially, "interventions aimed at the implementation and maintenance of pro-ecological behaviours need to decrease their chances of failure. For this, they should consider not only the right facilitators and conditions – positive determinants – but also the barriers and constraints to it – negative determinants." (Gaspar 2013, p. 2972) This needs to be addressed during policy implementations where not only the factors need to be studied explicitly but the processes and strategy developments also need to be addressed carefully. (Gaspar 2013)

However, incidents have been reported in Lucknow where the concerned municipal authority had made artificial ponds for idol immersion with insulated layer to avoid leeching. Hoardings were put up requesting people to immerse the idols in those ponds instead of the Gomti river, but many people still went to the river as they found the pond water getting muddy after a while which they felt was not suitable for idol immersion. Thus part of the artificial ponds remained unused. (Times News Network 2018) Such is also observed in Bhubaneswar (Pati 2018) which imply that even though government is taking up certain steps to combat river pollution, people need to be made more aware. Being a sensitive subject, other innovative alternatives need to be worked on.

Recommendations

That paints with heavy metals are still in use clearly came up from the interviews which said the idols with eco-friendly paints were priced much higher and thus came the issue of affordability. Thus, one recommendation will be to subsidise such eco-friendly paints and incentivise purchase of such idols so that both the idol makers and the buyers are encouraged. Creating awareness and imparting knowledge on environment and the consequences of not protecting the environment is also very crucial. As is clear from this research, the mindset of the people (both subjective norm and values and personal norms) plays a very important role, and thus anything related to rituals, offerings and sacred waste becomes very sensitive.

Care needs to be taken to involve people in the decision-making process while implementing such projects so that they do not feel left out or being forced. Rather a bottom-up approach by involving citizens from all ranks of the society can aid in the process, and most of the respondents had in fact mentioned that they would like to be part of the positive change in whatever way possible, but many did not really know what to do and how to go about it, whereas two of the respondents mentioned that they are already doing whatever little they can in their own way – one by segregating the organic waste before putting it into the river water and another by putting a bin in front of her shop and encouraging people to put their waste there. Thus, a well-defined system involving all these people and encouraging many others will help. A real-life example of such a pilot project recently took off in Kolkata itself.

As already initiated in a part of the city of Kolkata (Ward number 19), collection of domestic sacred waste by the municipality in a separate van is a good stepping stone as this will also address people's beliefs that sacred waste should not mix with other waste and maintains its sanctity. The initiative was taken by the local councillor based on a public participation meeting organised where he learnt about the concern of the residents in disposing sacred waste. It was due to subjective norm and also values and personal beliefs that they throw the sacred waste in waterbodies as it is concerned auspicious and not supposed to be mixed with other kinds of waste. As the outcome of the meeting, a special battery-operated van from the municipality was decided to be brought to service which would collect only domestic sacred waste. A fine amount of Indian Rupees 250/= has been decided for putting waste of any other kind in this van. Environmentalists have also noted that eco-friendly powdered colour or "abir" can be made from the organic puja waste of flowers and leaves and hence asserting the need of such separate collection vans particularly for sacred waste. The article also states that letter seeking sanction and funding for materialising this concept of manufacturing eco-friendly powdered colour has already been sent to the Environmental Department, Government of West Bengal, for approval. (Roy 2018) Education and feeling of ownership help people care more for the environment (Frey and Berkes 2014), and it has also been noted in the literature that levying fines might not be a permanent solution whereas changes in attitude can be positive (Dobson 2007), and this initiative carefully touches upon all these (Fig. 11.11).

Fig. 11.11 Special battery-operated municipal van for exclusive door-to-door collection of domestic sacred waste. (Source: Roy 2018)

Plastic ban happens at regular intervals based on court orders all over the country, but there are issues with implementation, and there are hardly any alternative options available. Hence the city goes to a standstill for a few days and again back to square one. Currently there is a ban in place across 19 states in India on single-use plastics, and the Prime Minister has also announced complete ban of single-use plastics by 2022 in the entire country. It is also noted that the real issue is with the littering more than the mere usage. Also top-down approach does not seem to work during such implementations. Thus, care needs to be taken in implementation of any such project while providing sufficient time, cheap alternate options and education beforehand. (Jestin 2018)

There are already certain alternative options, but reach of such items needs to be explored and attended to. For example, an NGO named EnviGreen has come up with biodegradable bags that look like plastic but are made of natural starch coming from potato and tapioca and vegetable oil derivatives which are soluble in water and also edible (Singh 2016).

Alternate options to idol immersions may also be taken up by spraying water on idol while resting it on a platform (Government of India 2016a) rather than immersing in rivers and waterbodies. This will not only help reduce river contamination, but the items can also be reused or recycled. This will, however, need to counter most subjective norms, as this indicator of situational factor has been found to be extremely significant as an outcome of this research.

In recent times, a few artisans, NGOs and organisations have already initiated the concept of eco-friendly Ganesh idols made from various materials ranging from cow dung, recycled newspapers, vermicompost, etc. (Patel 2018), and some variations even have a seedling inside and a pot attached such that there is no need to immerse the idol in rivers or seas and can instead be planted in a pot wherein the water dissolves the idol and germinates the seedling (Asian News International 2018).

As mentioned earlier, Site B of the research area has a compactor unit, and organic waste from the main Kali Temple goes there. Similar units installed in other sites will surely help and also encourage everyone to segregate organic waste and put those in the compactor units. Also, two youngsters from Kanpur, a north Indian city along the Ganga, have launched a start-up to recycle floral wastes from temples to make incense sticks, soaps, etc. and also have created secured employment opportunities with benefits for provident fund, health insurance and transport for about 80 local women (Chakraborty 2018). This definitely is a good start as it recycles the sacred waste generated to make incense sticks which are again used in the worship and thus does not harm the sentiment of the pilgrims and generates employment on the other hand.

Finally, considering sacred waste as a separate waste stream for policy framework will help in dealing with the problem in a better way.

Further Research

This research has opened up many avenues for further study. The social structure of the idol making industry and the supply chain of the sacred offerings when studied deeply can open up opportunities for innovative approaches towards an eco-friendly environment and lead the path towards a sustainable planet for all. When addressed properly and cautiously, this will not only open opportunities and new thinking but will also help in making every individual more conscious and informed about the choices they make and their implications. With the importance and faith instilled that even a small conscious effort made by one person can bring about a positive change, a new wave will surely be generated. Initiative taken up by the Councillor in Kolkata regarding exclusive collection of domestic sacred waste (Roy 2018) can be considered as an appropriate pilot project which can then be extended to other parts of the city and country as well.

Also, due to limitation of time, year-round observations based on all the rituals could not be captured in this research. It would have been interesting to note the changes over time and work on all those data combined. Assessment of a pilot project would also have been interesting for deciding further course of action.

Acknowledgement The author is immensely thankful to Institute for Housing and Urban Development Studies, Erasmus University Rotterdam, Dr. Vicky Ariyanti, Prof. Somesh Sharma, Prof. Ogenis Brilhante, Dr. Raphael Smals, Dr. N. B. Basu, Mr. Tapas Choudhury, Mr. Tapas Kumar Mandal, Late Dr. Marijk Huysman, Late Mr. Robin Pal and the entire team of research assistants and respondents for their support.

Bibliography

Akintunde EA (2017) Theories and concepts for Human Behavior in Environmental Preservation. J Environ Sci Public Health 1(2):120–133. Available at: http://www.fortunejournals.com/articles/theories-and-concepts-for-human-behavior-in-environmental-preservation.pdf. Accessed 20.04.2018

Alley KD (1994) Ganga and gandagi: interpretations of pollution and waste in Benaras. Ethnology 33(2):127–145. Available at: http://citeseerx.ist.psu.edu/viewdoc/download?doi=10.1.1.566.1321&rep=rep1&type=pdf. Accessed 20.6.18

Anttonen V (2000) Sacred. In: Braun W, McCutcheon RT (eds) Guide to the study of religion. Cassell, London, pp 271–282. Available at: http://www.brontaylor.com/courses/pdf/Anttonen--Sacred.pdf. Accessed 20.6.18

Asian News International (2018) Go eco-friendly this Ganesh Chaturthi with idols that grow into trees. Hindustan Times. Available at: https://www.hindustantimes.com/more-lifestyle/go-eco-friendly-this-ganesh-chaturthi-with-idols-that-grow-into-trees/story-wWkjx6THAh39waTPK1elMK.html. Accessed 08.08.2018

Beatley T (2013) Urban rivers of life. Available at: http://biophiliccities.org/urban-rivers-of-life/. Accessed 07.07.2018.

Bell S (2013) Water supply, drainage and flood protection. In: Bell S, Paskins J (eds) Imagining the future city: London 2062. Ubiquity Press, London, pp 85–93. Available at: https://www.ubiquitypress.com/site/chapters/10.5334/bag.l/download/83/. Accessed 20.04.2018

Berkes F (2012) Sacred ecology, 3rd edn. Routledge, New York

Berkes F, Colding J, Folke C (2000) Rediscovery of traditional ecological knowledge as adaptive management. Ecol Appl 10(5):1251–1262. Available at: https://esajournals.onlinelibrary.wiley.com/doi/full/10.1890/1051-0761%282000%29010%5B1251%3AROTEKA%5D2.0.CO%3B2. Accessed 20.04.2018

Bhadula S, Joshi BD (2014) Impact of religio-touristic activities on the water quality of Ganga river and solid waste generation within Haridwar city, India. Int J Plant Anim Environ Sci 4(4):309–315

Bhatnagar A, Sangwan P (2009) Impact of mass bathing on water quality. Int J Environ Res 3(2):247–252. Available at: https://ijer.ut.ac.ir/article_52.html. Accessed 07.07.2018

Bhatnagar A, Devi P, George MP (2016) Impact of mass bathing and religious activities on water quality index of prominent water bodies: a multilocation study in Haryana, India. Int J Ecol

Bhattacharya S, Bera A, Dutta A, Ghosh UC (2014) Effects of idol immersion on the water quality parameters of Indian water bodies: environmental health perspectives. Int Lett Chem Phys Astron 39:234–263. Available at: https://www.scipress.com/ILCPA.39.234.pdf. Accessed 20.6.2018

Billore DK, Dandawate M (2015) Environmental impact of idol immersion on Kakerpura lake Mhow. Social Issues Environ Prob 3(9). Available at: http://granthaalayah.com/Articles/Vol3Iss9SE/24_IJRG15_S09_50.pdf. Accessed 20.6.18

Bourne S, Shepherd C (1880) Bathing and boats on Adi Ganga in 1880. Accessed 2018

Brook Vd (1660) River map of Bengal. Available at: https://upload.wikimedia.org/wikipedia/commons/a/a9/Van_de_Brook%27s_map.jpg. Accessed 2018

Brown EH, Atkinson BW, Wolf PO, Rodda JC et al (1976) Physical problems of the urban environment: a symposium: discussion. Geogr J 142(1):72–80. Available at: https://www.jstor.org/stable/pdf/1796027.pdf?refreqid=excelsior:074c1963bde6a69d4f6d596688415b13. Accessed 20.04.2018

Brumann C (1999) Writing for culture: why a successful concept should not be discarded. Curr Anthropol 40:1–24. Available at: https://www.journals.uchicago.edu/doi/pdfplus/10.1086/200058. Accessed 20.5.2018

Chakraborty SC (2004) The story of a river port. Kolkata Port Trust, Kolkata

Chakraborty S (2018) Ankit Agarwal & Karan Rastogi: saving the Ganga by recycling floral waste – HelpUsGreen co-founders are straddling religiosity and business. Forbes India, Available at: http://www.forbesindia.com/article/30-under-30-2018/ankit-agarwal-karan-rastogi-saving-the-ganga-by-recycling-floral-waste/49369/1. Accessed 07.07.2018.

Chandrashekhar V (2018) Dying waters: India struggles to clean up its polluted urban rivers. Available at: https://e360.yale.edu/features/dying-waters-india-struggles-to-clean-up-its-polluted-urban-rivers. Accessed 07.07.2018.

Chaudhuri S (1990) Calcutta: the living city. Oxford University Press, Calcutta

Cliff K, Wright J, Clarke D (2009) What does a sociocultural perspective mean in health and physical education? In: Dinan-Thompson M (ed) Health and physical education. Oxford University Press, Oxford, pp 165–179. Available at: http://ro.uow.edu.au/cgi/viewcontent.cgi?article=1097&context=edupapers.. Accessed 20.04.2018

Dana Gopal NM, Phebe P, Suresh Kumar EV, Vani BKK (2014) Impact of plastic leading environmental pollution. J Chem Pharm Sci (3):96–99. Available at: https://www.jchps.com/specialissues/Special%20issue3/20%20jchps%20si3%20N.%20M.%20dana%20Gopal%2096-99.pdf. Accessed 07.07.2018

Das S (2009) River of time. The Telegraph, Calcutta, India, July 05 Available at: https://www.telegraphindia.com/1090705/jsp/calcutta/story_11185529.jsp. Accessed 15.02.2018

Das P, Tamminga KR (2012) The Ganges and the GAP: an assessment of efforts to clean a sacred river. Sustainability 4:1647–1668. Available at: https://pdfs.semanticscholar.org/5e92/43901d60b51259ada3cf9c1770615f780551.pdf. Accessed 14.06.2018

Das KK, Panigrahi T, Panda RB (2012) Idol immersion activities cause heavy metal contamination in river Budhabalanga, Balasore, Odisha, India. Int J Mod Eng Res 2(6):4540–4542. Available at: http://www.ijmer.com/papers/Vol2_Issue6/DY2645404542.pdf. Accessed 20.6.18

Deb Sarkar M (2017) An urban river on a gasping state: dilemma on priority of science, conscience and policy. XVI World Water Congress, Cancun, Quintana Roo, Mexico. Mexico: International Water Resources Association. Available at: http://wwc2017.iwra.org/congress/resource/ABSID285_ABSID285_FINAL_PAPER_IWRA_MexicoConference_2017.pdf. Accessed 15.02.2018.

Dobson A (2007) Environmental citizenship: towards sustainable development. Sustain Dev 15:276–285. Available at: https://onlinelibrary.wiley.com/doi/epdf/10.1002/sd.344. Accessed 20.5.2018

Dwivedi OP (1993) Human responsibility and the environment: a Hindu perspective. J Hindu-Chris Stud 6(8):19–26. Available at: https://digitalcommons.butler.edu/cgi/viewcontent.cgi?article=1077&context=jhcs. Accessed 20.6.18

Eck DL (1981) India's "Tirthas": "Crossings" in sacred geography. Hist Relig 20(4):323–344. Available at: https://dash.harvard.edu/bitstream/handle/1/25499831/Eck_India'sTirthas.pdf?sequence=4. Accessed 22.6.18

Ellis TB (2015) Evoked pūjā: the behavioral ecology of an equatorial ritual. J Am Acad Relig 83(4):1108–1156. Available at: https://academic.oup.com/jaar/article/83/4/1108/2572359. Accessed 07.07.2018

Enserink B, Patel M, Kranz N, Maestu J (2007) Cultural factors as co-determinants of participation in river basin management. Ecol Soc 12(2). Available at: https://www.ecologyandsociety.org/vol12/iss2/art24/. Accessed 20.5.18

Findlay SJ, Taylor PM (2006) Why rehabilitate urban river systems? Area 38(3):312–325. Available at: https://onlinelibrary.wiley.com/doi/epdf/10.1111/j.1475-4762.2006.00696.x. Accessed 20.5.18

Frazier J (2012) Ritual and practice in HIndu studies. J Hindu Stud 5(1):1–9. Available at: https://academic.oup.com/jhs/article/5/1/1/2188649. Accessed 07.07.2018

Frey JB, Berkes F (2014) Can partnerships and community-based conservation reverse the decline of coral reef socio-ecological system? Int J Commons 8(1):26–46. Available at: https://www.thecommonsjournal.org/articles/10.18352/ijc.408/. Accessed 24.6.18

Frøystad K (2012) Divine intersections: Hindu ritual and the incorporation of religious others. Cosmopolitan Civil Soc J 4(2):1–21. Available at: https://epress.lib.uts.edu.au/journals/index.php/mcs/article/view/2589. Accessed 07.07.2018

Gaspar R (2013) Understanding the reasons for behavioral failure: a process view of psychosocial barriers and constraints to pro-ecological behavior. Sustainability 5:2960–2975. Available at: www.mdpi.com/2071-1050/5/7/2960/pdf. Accessed 07.07.2018

Gatersleben B, Murtagh N, Abrahamse W (2014) Values, identity and pro-environmental behaviour. Contemp Soc Sci 9(4):374–392. Available at: https://www.tandfonline.com/doi/full/10.1080/21582041.2012.682086. Accessed 07.07.2018

Gavriletsa S, Richerson PJ (2017) Collective action and the evolution of social norm internalization. Proc Nat Acad Sci U S A 114(23):6068–6073. https://www.ncbi.nlm.nih.gov/pmc/articles/PMC5468620/. Accessed 20.04.2018

Giripunje MD, Fulke AB, Meshram PU (2014) Effects of idol immersion on water quality and Tilapia fish in Futula, Gandhisagar and Ambazari lakes of Nagpur, India. SpringerPlus 3. Available at: https://www.ncbi.nlm.nih.gov/pmc/articles/PMC4234744/pdf/40064_2014_Article_1359.pdf. Accessed 20.6.18

Gómez-Baggethun E, Reyes-García V, Olsson P, Montes C (2012) Traditional ecological knowledge and community resilience to environmental extremes: a case study in Doñana, SW Spain. Glob Environ Chang 22(3):640–650. Available at: http://icta.uab.cat/etnoecologia/Docs/[120]-Gomez-Baggethun%20et%20al_GEC12.pdf. Accessed 20.04.2018

Goody J (1961) Religion and ritual: the definitional problem. Br J Sociol 12(2):142–164. Available at: https://www.jstor.org/stable/pdf/586928.pdf?refreqid=excelsior%3A04af629bb4219c9397 0400cb6a5a81ad. Accessed 21.6.18

Government of India (2011) Population by religious community. Government of India, New Delhi. Available at: http://www.censusindia.gov.in/2011census/C-01.html. Accessed 07.07.2018

Government of India (2015) Joydeep Mukherjee vs The Chairman Pollution Control Board & Ors. Government of India, New Delhi. Available at: http://www.indiaenvironmentportal.org.in/files/Adi%20Ganga%20pollution%20NGT.pdf. Accessed 15.02.2018. Eastern Zone Bench, Kolkata

Government of India (2016a) Ambar Nath Sengupta vs State of West Bengal & Ors. Kolkata, Eastern Zone Bench

Government of India (2016b) Subhas Datta vs State of West Bengal & Ors. Eastern Zone Bench, Kolkata

Government of West Bengal (2003) Rapid estimation of major wastewater discharges of River Hoogly between the stretch of Palta to DhanKheti Khal. West Bengal Pollution Control Board, Kolkata

Government of West Bengal (2016) Annual flood report. Government of West Bengal, Kolkata. Available at: http://www.wbiwd.gov.in/uploads/ANNUAL_FLOOD_REPORT_2016.pdf. Accessed 15.02.2018

Government of West Bengal (2017) Undertaking. West Bengal Pollution Control Board, Kolkata

Government of West Bengal (2018) Water quality data. West Bengal Pollution Control Board, Kolkata. Available at: http://emis.wbpcb.gov.in/waterquality/showwqprevdata.do;jsessionid=8160607615AEB5A6A40322D3E79FE17E. Accessed 07.07.2018

Gregory KJ (2006) The human role in changing river channels. Geomorphology 79(3):172–191

Hardin G (1968) The tragedy of the commons. Science 162(3859):1243–1248. Available at: http://science.sciencemag.org/content/162/3859/1243/tab-pdf. Accessed 07.07.2018

Haseena M, Malik MF, Javed A, Arshad S et al (2017) Water pollution and human health. Environ Risk Assess Remed 1(3). Available at: http://www.alliedacademies.org/articles/water-pollution-and-human-health-7925.html. Accessed 07.07.2018

Hawkins G (2006) The ethics of waste: How we relate to rubbish. Rowmann and Littlefield Publishers, Inc., Oxford. Available at: https://books.google.nl/books?hl=en&lr=&id=2zHXheETyjQC&oi=fnd&pg=PR7&dq=rituals+waste&ots=JaMM2Bfyln&sig=y712XmGF4pr3oQ3aIcQHYci4RHw#v=onepage&q=rituals%20waste&f=false. Accessed 20.6.18

Jestin P (2018) India's plastics ban faces implementation hurdles. ICIS News, Mumbai. Available at: https://www.icis.com/resources/news/2018/07/19/10242867/india-s-plastics-ban-faces-implementation-hurdles/. Accessed 07.07.2018

Joshi Y, Rahman Z (2015) Factors affecting green purchase behaviour and future research directions. Int Strat Manag Rev 3:128–143. Available at: https://www.sciencedirect.com/science/article/pii/S2306774815000034. Accessed 07.07.2018

Keesing RM (1974) Theories of culture. Ann Rev Anthropol. Available at: http://kodu.ut.ee/~cect/teoreetilised%20seminarid_2009%20s%C3%BCgis/1_seminar_KULTUUR_29.09.2009/text_1.pdf. Accessed 20.5.18

Kolkata Metropolitan Development Authority (2005) Vision 2025 Perspective Plan of CMA: 2025. Kolkata Development Authority, Kolkata. Available at: https://drive.google.com/file/d/0B83sc0sJSLaXbFRMaFhCWl9sd2c/view. Accessed 20.02.2018

Kolkata Municipal Corporation (2009) Graded list of heritage buildings. Kolkata Municipal Corporation, Kolkata. Available at: https://www.kmcgov.in/KMCPortal/downloads/Graded_List_of_Heritage_Buildings_Grade_I_IIA_IIB.pdf. Accessed 15.02.2018

Kolkata Municipal Corporation (2017) Pollution abatement and rehabilitation of Tolly's Nullah: detailed project report. 1. Kolkata Municipal Corporation, Kolkata

Krajhanzl J (2010) Environmental and proenvironmental behavior. School and Health 21:251–274. Available at: http://www.ped.muni.cz/z21/knihy/2011/36/36/texty/eng/krajhanzl.pdf. Accessed 07.07.2018

Kroeber AL, Kluckhohn C (1952) Culture: a critical review of concepts and definitions. The Museum, Massachusetts. Available at: http://www.pseudology.org/Psyhology/CultureCriticalReview1952a.pdf. Accessed 20.05.2018

Kumar D (2017) River Ganges – historical, cultural and socioeconomic attributes. Aqua Ecosyst Health Manag 20:8–20. Available at: https://www.tandfonline.com/doi/pdf/10.1080/14634988.2017.1304129. Accessed 20.5.18

Laborda-Pemán M (2015) Wall, Derek (2014) The commons in history. Culture, conflict and ecology. Int J Commons 9(1):466–468. Available at: https://www.thecommonsjournal.org/articles/10.18352/ijc.565/. Accessed 24.6.18

McDermott RF (2011) Revelry, rivalry, and longing for the Goddesses of Bengal: the fortunes of Hindu festivals. Columbia University Press, New York. Available at: https://books.google.nl/books?id=ggBeH_lmUu8C&pg=PA306&lpg=PA306&dq=kalitirtha+kolkata&source=bl&ots=KxM9MMoWQl&sig=JYrD8p7g0c8Z2zeEagZoP4ZnvBc&hl=en&sa=X&ved=0ahUKEwiJ_-Tn2ufbAhVFuRQKHSYrDXcQ6AEIXDAI#v=onepage&q=kalitirtha%20kolkata&f=false. Accessed 20.6.18

Mitra D (2017) River of no return. Outlook India, 31 July. Available at: https://www.outlookindia.com/magazine/story/river-of-no-return/299138. Accessed 15.02.18

Natarajan K (2013) Review essay: Diana L. Eck, India: a sacred geography. Harmony, Random House, New York, 559 pp., Rs 599, ISBN 978-0-385-34708-2. Strategic Analysis, 37 (3), pp. 366-371. Available at: https://www.tandfonline.com/doi/pdf/10.1080/09700161.2013.782665?needAccess=true. Accessed 20.6.2018

Park C (2004) Religion and geography. In: Hinnells J (ed) Routledge companion to the study of religion. Routledge, London. Available at: http://www.lancaster.ac.uk/staff/gyaccp/geography%20and%20religion.pdf. Accessed 20.6.18

Patel T (2018) Cow dung to clay: this Chaturthi, go green with these 6 beautiful eco-Ganeshas. The Better India, Cow Dung to Clay: this Chaturthi, Go Green with These 6 Beautiful Eco-Ganeshas! Available at: https://www.thebetterindia.com/158925/ganesha-chaturthi-eco-friendly-green-idols/. Accessed 08.08.2018

Pati A (2018) Odisha: civic body makes ponds for immersion, devotees choose river. The Telegraph. Available at: https://www.telegraphindia.com/states/odisha/idol-immersion-worry-stays/cid/1669916. Accessed 08.10.2018

Pecharroman LC (2018) Rights of nature: Rivers that can stand in court. Resources 7(13). Lidia Cano Pecharroman

Rai B (2013) Pollution and conservation of Ganga river in modern India. Int J Sci Res Public 3(4). Available at: http://www.ijsrp.org/research-paper-0413/ijsrp-p1634.pdf. Accessed 20.6.18

Rennie BS (2007) Mircea Eliade and the perception of the sacred in the profane: intention, reduction, and cognitive theory. Finnish Soc Stud Relig 4(1):73–98

Richerson PJ, Boyd R (2005) Not by genes alone: how culture transformed human evolution, Chicago, University of Chicago Press. Available at: http://avalonlibrary.net/ebooks/Peter%20Richerson%20-%20Not%20by%20Genes%20Alone%20How%20Culture%20Transformed%20Human%20Evolution.pdf. Accessed 20.04.2018

Roy P (2018) Municipal van to collect domestic puja waste directly from houses. Ei Samay, Kolkata

Sadhguru (2017) The importance of rivers. Available at: https://isha.sadhguru.org/rally-for-rivers/importance-of-rivers/. Accessed 07.07.2018

Sado-Inamura Y, Fukushi K (2018) Considering water quality of urban rivers from the perspective of unpleasant odor. Sustainability 10. Available at: www.mdpi.com/2071-1050/10/3/650/pdf. Accessed 20.6.18

Samakov A, Berkes F (2017) Spiritual commons: sacred sites as core of community-conserved areas in Kryrgyzstan. Int J Commons 11(1):422–444. Available at: https://www.thecommons-journal.org/articles/10.18352/ijc.713/. Accessed 24.6.18

Sampat P (2009) The river Ganges long decline. WorldWatch 9(4). Available at: http://www.gangaaction.org/book-review/the-river-ganges-long-decline/. Accessed 20.6.18

Sayers J (2002) China's mother river scolds her young: Modernization and the National. Transformations, 5 Available at: http://www.transformationsjournal.org/wp-content/uploads/2017/01/janesayers.pdf. Accessed 20.5.18.

Shafaghat A, Ghasemi MM, Keyvanfar A, Lamit H et al (2017) Sustainable riverscape preservation strategy framework using goal-oriented method: case of historical heritage cities in Malaysia. Int J Sustain Built Environ 6(1):143–159. Available at: https://www.sciencedirect.com/science/article/pii/S2212609016300474. Accessed 20.5.18

Shrivastava N, Mishra PK, Bajpai A (2011) Impact of idol immersion activity on the water quality of Machna river, Betul, Madhya Pradesh, India. Int J Chem Sci 9(4):1637–1646. Available at: http://www.tsijournals.com/articles/impact-of-idol-immersion-activity-on-the-water-quality-of-machna-river-betul-madhya-pradesh-india.pdf. Accessed 07.07.2018

Singh T (2016) These 'lastic' bags are actually made of potato and tapioca – and can become animal food on disposal. The Better India. Available at: https://www.thebetterindia.com/77202/envigreen-bags-organic-biodegradable-plastic/. Accessed 08.08.2018

South Asian Forum for Environment (2017) Adi Ganga & wetlands: heal the broken link for a flood resilient Kolkata. Available at: http://safeinch.org/uploads/Report%20-Symposium%20on%20Adi%20Ganga%20&%20Wetlands%202017.pdf. Accessed 20.02.2018

Stausberg M (2017) The sacred, the holy, the numinous – and religion: on the emergence and early history of a terminological constellation. Religion 47(4):557–590. Available at: https://www.tandfonline.com/doi/full/10.1080/0048721X.2017.1377889?src=recsys. Accessed 20.6.18

Steg L, Bolderdijk JW, Keizer K, Perlaviciute G (2014) An integrated framework for encouraging pro-environmental behaviour: the role of values, situational factors and goals. J Environ Psychol 38:104–115. Available at: https://www.rug.nl/about-us/who-are-we/sustainability/greenoffice/cursus/myth2/anintegratedframeworkforencouragingpro-environmentalbehaviour.pdf. Accessed 07.07.2018

Stengs I (2014) Sacred waste. J Objects Art Belief 10(2):235–238. Available at: https://www.tandfonline.com/doi/abs/10.2752/175183414X13990269049482?journalCode=rfmr20. Accessed 20.6.18

Sutton MQ, Anderson EN (2010) Introduction to cultural ecology. 2. Altamira, Plymouth

Suyanto E, Soetarto E, Sumardjo, Hardjomidjojo H (2014) Analysis on local wisdom-based green community participation supporting the developmental program at the green city Purwokerto Indonesdia: a case study of household waste management. Int J Res Earth Environ Sci 2 (5)

The British Medical Journal (1905) Indian sanitary reports. Br Med J 1(2311):851–853. Available at: https://www.jstor.org/stable/pdf/20284758.pdf. Accessed 24.6.18

Times News Network (2017) NGT orders central agency to locate missing Adi Ganga links. The Times of India. Available at: https://timesofindia.indiatimes.com/city/kolkata/ngt-orders-central-agency-to-locate-missing-adi-ganga-links/articleshow/56841715.cms. Accessed 20.6.18

Times News Network (2018) Artificial ponds lie unused, idols immersed in Gomti. The Times of India. Available at: https://timesofindia.indiatimes.com/city/lucknow/artificial-ponds-lie-unused-idols-immersed-in-gomti/articleshow/65879992.cms. Accessed 20.09.2018

United Nations (2018) World urbanization prospects: The 2018 revision – key facts. United Nations, New York. Available at: https://population.un.org/wup/Publications/Files/WUP2018-KeyFacts.pdf. Accessed 07.07.2018

van Thiel S (2014) Research methods in public administration and public management: an introduction. Routledge, New York

VishnuRadhan R, Zainudin Z, Sreekanth GB, Dhiman R et al (2017) Temporal water quality response in an urban river: a case study in peninsular Malaysia. Appl Water Sci 7:923–933. Available at: https://link.springer.com/content/pdf/10.1007%2Fs13201-015-0303-1.pdf. Accessed 22.6.18

Watkar AM, Barbate MP (2014) Impact of idol immersion on water quality of Kolar river in Saoner, dist. Nagpur, India. Int Res J Environ Sci 3(3):39–42. Available at: http://www.isca.in/IJENS/Archive/v3/i3/8.ISCA-IRJEvS-2014-046.pdf. Accessed 07.07.2018

Wirtz K (2009) Hazardous waste: the semiotics of ritual hygiene in Cuban popular religion. J R Anthropol Inst 15(3):476–501. Available at: https://www.jstor.org/stable/pdf/40541695.pdf. Accessed 22.6.18

Yin RK (1981) The case study as a serious research strategy. Sage 3(1):97–114. Available at: http://journals.sagepub.com/doi/pdf/10.1177/107554708100300106. Accessed 20.6.18

Zhao YW, Yang ZF, Xu F (2007) Theoritical framework of the urban river restoration planning. Int Soc Environ Inform Sci 5:241–247. Available at: https://www.grad.unizg.hr/_download/repository/theoretical-framework-of-the-urban-river-restoration-planning.pdf. Accessed 20.5.2018

Chapter 12
Spatial Evaluation Supporting Sustainable Tourism Development in Riverine Global South

Shreeja Ganguly, Mainak Ghosh, and Joy Sen

Abstract Global South is a very enriched region being home to a number of famous, ancient and fertile riverine ecosystem. In light of present economic development, a significant part of the riverine Global South is the combined region of South, Southeast and East Asia, where several large rivers like Mekong, Yangtze, Irrawaddy, Zhujiang, Red River, Ganges, Brahmaputra, etc. are situated. Many of these rivers' lower valley especially their mega-deltas once inhabited the early civilisations which still continues to this day.

These locations have a typical characteristic, which include abundant rainfall, large rivers, a high population density and often on the lower side of the economy, and for that reason they attract a lot of tourists from all over the world. The inhabitants of this region also welcome such tourism as a way out of the economic problem though often with the absence of any prior study to judge its suitability.

On one hand, the World Tourism Organization (UNWTO) envisaged river-based tourism in Southeast Asia as an emerging activity with positive contribution to regional economic and community development; on the other hand, negative effect of uncontrolled tourism often destructs the very asset on which it was created.

Therefore, sustainability of tourism has to be ensured from the very start especially in environmentally sensitive areas like the riverine ecosystems of Global South. Accurate identification of the tourism nodes is that start and a thoroughly developed method of spatial evaluation is required for that.

In this chapter, the Gangetic riverine delta of West Bengal State, India, which is characteristically a typical example of the riverine system of Global South, is studied for the purpose, and a strategy of spatial analysis is demonstrated through a method of using multi-criteria weighted matrix to reveal the different potential areas for tourism.

S. Ganguly (✉) · M. Ghosh
Department of Architecture, Jadavpur University, Kolkata, West Bengal, India

J. Sen
Department of Architecture & Regional Planning, IIT Kharagpur, Kharagpur, West Bengal, India

© Springer Nature Switzerland AG 2020

M. Ghosh (ed.), *Perception, Design and Ecology of the Built Environment*, Springer Geography, https://doi.org/10.1007/978-3-030-25879-5_12

Keywords Sustainable tourism · Weighted matrix · Global South · Spatial evaluation · River-based tourism

Introduction

Water and Ancient Civilisations

> "Water is a unique resource, appearing in different forms and shaping all physical and biological processes. Known for its fluidity and celebrated for changing its shape and taking new forms, water also plays a special role in social and cultural representations, and these values and norms are reflected in the ways water is perceived, used, governed, and treated. Civilization has historically flourished around rivers and major waterways." (Lahiri-Dutt 2014)

Historically, civilisations tend to crop up around water, especially river. As nomadic hunter-gatherers settled into non-mobile sedentary societies, they began relying on agriculture more than hunting, developed surpluses of crops and slowly started to grow. Initially the river helped in agriculture. Later with the rise of trade, river also carried significance for trade and transport and as source of water for industries. For example, the famous Indus civilisation forms in Sindhu river valley, ancient civilisation of Egypt grew on both sides of river Nile, Mesopotamia (cradle of civilisation) was situated between the major rivers Tigris and Euphrates, and Rome was also founded on the banks of the Italian river Tiber.

Many of the old cities which are still running are formed along a river. In fact, the oldest part of the town is the part around river. Large cities and metropolises like London, New York City, Paris, Chicago, Montreal, Rotterdam, Buenos Aires, Shanghai, Tokyo and Hong Kong owe their initiation and accomplishment partly to their easy accessibility via water and the consequential growth of trade.

Rivers of Global South

Global South is home to a number of famous, ancient and fertile riverine ecosystem, enriched with agriculture, ecological and heritage all at once. The lower river valley especially their mega-deltas, both coastal and inland, is always concentrated with population.

The ancient civilisation of Egypt formed around Nile River, which is the longest river in the world. The Nile originates in the high mountains of Central Africa, passes through 11 different countries, takes the north-eastern path to finally end at the Nile River Delta and empties into the Mediterranean Sea. Due to the presence of surrounding desert and climatic location, the agriculture had to adjust to its climatic cycle and depended foremostly on annual flood. Due to this short-lived seasonal incident, the irrigation network was robust.

Zambezi, another African river, runs through a region that is often seen as under-developed with significant mineral resources, three UNESCO World Heritage resources and five Ramsar sites protected by the Convention on Wetlands of International Importance, which includes the Zambezi Delta. These sites are important for tourism, a sector which is constantly gaining importance for the national economy (Jensen and Lange 2013). The Zambezi is originated at highland in northern Zambia and flows in and out of the country up to its delta in Mozambique in the south, journeys into southeastern Angola and then briefly touches the northern parts of Namibia and Botswana, before dropping over the side of Mosi-oa-Tunya (Victoria Falls), which is shared by Zambia and Zimbabwe and then develops into Lake Kariba, controlled by the Cahora Bassa dam built in 1978 in Mozambique before finally ending in the Indian Ocean. More than 85% of the basin population lives in Malawi, Zambia and Zimbabwe (Jensen and Lange 2013).

The Amazon of South America ranks as the largest (in terms of average water discharge) as well as the second longest river of the world and the longest of the rivers of South America. Its source lies at the Andes, while its course crosses national and international borders through seven countries of the continent, including Brazil, Ecuador, Bolivia, Peru, Guyana, Venezuela and Colombia, finally joining the Atlantic Ocean, running most of its course through rainforests.

There are so many other significant rivers and their famous deltas in the region, especially within South, Southeast and East Asia that, together, this combined region can truly be called a 'land of large rivers'.

River Deltas of South, Southeast and East Asia

Geologically, a delta is an irregular progradation of the shoreline directly fed by a river.

River deltas are generally formed on the lower parts of rivers, where the current of water spreads out and slackens, with the deposition of silts and sediments into the area of lower depths of water, i.e. wetlands, floodplains, marshes, etc. The deltas are divided into two types: coastal deltas near the sea-river mouths and inland deltas where the river deposits in floodplains and marshes (Saito, 2000).

Most of the large deltas are distributed in Asia. According to Saito (2000), three major reasons can be assigned for the occurrences: (i) presence of large rivers with sources in the high mountains of the Himalayas and the Tibetan Plateau along with having a high sediment yield. This supports the fact that 70% to 80% of the world's sediment flux are contributed by the rivers of Southern Asia and Oceania during their journey, (ii) high rate of rainfall due to the monsoonal climate and (iii) the unchanging or slightly sinking sea level over the last 6000 years.

In light of present economic development, a significant part of the Global South, in this perspective, is the combined region of South, Southeast and East Asia, where a number of large rivers form huge deltas.

For example, Mekong River, which is a transboundary river in Southeast Asia, that runs through six countries, forms a riverine ecosystem with the Mekong Delta – one of largest deltas in Southeast Asia, located at the mouth of the Mekong River. The earliest recorded civilisation on the Mekong was found in this delta. In present day, river-based cruises and tourism activities are one of the key livelihoods for the inhabitants, with many major attractions and destinations like Angkor Wat, Vat Phou, Luang Prabang, etc. located either on or near the river. But there are opinions that haphazard and rapid construction of infrastructure to promote tourism has caused severe damage to ecosystems, community life and local people making them vulnerable to exploitation.

Similarly, the Yangtze River measuring 6300 km is the world's third longest river and flows through 11 provinces of the People's Republic of China (PRC) before entering the East China Sea at Shanghai. It plays a very significant role in the history, culture and economy of the PRC and flows through a wide range of ecosystems binding with it the long history of tourism of the surrounding places along the Yangtze.

Similar situation prevails in almost all riverine ecosystems of the region like the Chao Phraya River of Thailand, Irrawaddy or Ayeyarwady River of Myanmar, Zhujiang or Pearl River of China, Red River that flows from Southwest China through Northern Vietnam to the Gulf of Tonkin, Yellow River of China or the Ganges and Brahmaputra River basin in India.

Connections Through Rivers

In absence of advancement in land transport in the past, heavy materials were easier to carry by water. As a result, almost all the old rivers have been used as a mode of interstate or intercountry transportation or as trade corridor. Since many of the mentioned rivers are international in nature, in the process, these regions have also developed significant reciprocal sharing of cultural thoughts and beliefs along with trade materials.

Rivers like Nile in Egypt are not navigable yearlong. Therefore, the same large ships which transferred goods and grain barges travelled the length of Egypt, during the flood season, simultanously collecting the taxes. The dependency on river for daily activity of common people is much less in Egypt. But that is not so, in the case of the easy navigable rivers in South, Southeast and East Asia.

Mekong and its basin rivers and channels have historically been the main means for transporting people and goods to villages on the banks and are still the only method for reaching remote communities (World Tourism Organization (UNWTO) 2016). Similarly, due to dilapidated condition of road in several Southern poor countries, travel by car is difficult as well as expensive. So the Lao people use the river and its tributaries to travel by boat to different parts of the country. Stargardt (1973) gave ethnographic evidence that Khmer traders and/or Chinese traders who had called at Khmer ports used the inland waterways of isthmian Thailand for their ship.

Socio-economic Perspective

The primary livelihood of people inhabited on these deltas are agriculture, farming or fishing; the change in livelihood is very much in sync with the nature of river alongside.

As per UNWTO (2016), 60% of the Mekong River basin's population relies on agriculture for food security and their livelihoods. Paddies cover most part of Mekong's basin. While most rice farming takes place in Northeast Thailand (low yield) and the delta (high yield), cassava, sugar cane, soybean and maize are also grown during the rains, with riverbank vegetable gardens cropping up when the high river waters recede in November. Also, during the dry season, the exposed floodplains (grasslands) are used for grazing livestock. Fishing provides another livelihood for some, which accounts for up to 25% of the global freshwater catch. The Mekong Delta is called as Vietnam's 'rice bowl' because the rice farming there supplies over half of the total rice production for Vietnam (World Tourism Organization (UNWTO) 2016).

Many least developed countries (LDCs), which are mostly situated in the Global South, are among the largest producers of fish in the world. Six of the top 16 producers of fish from inland waters are LDCs (United Nations 2018).

Tourism as a Way Out: Riverine/River-Based Tourism in Global South

Most of Asia's population lives on these riverine zones and deltas. Numerically, more than 50% of the world's population lives in Asia, and a significant majority of Asia's demography lives in deltaic areas (Saito 2000). As per Saito (2000), these locations are typically gifted with copious rainfall and large rivers and have a high population density and often economically subordinate to their global north counterparts. Due to abundance of natural resources and inexpensive costs, they attract a lot of tourists from all over the world. The inhabitants of this region also encourage such tourism as a source of income often with the absence of any prior study to judge its suitability.

On one hand, the World Tourism Organization (UNWTO) envisaged river-based tourism in Southeast Asia as an emerging activity with significant potential for growth and positive contribution to regional economic development, community-based tourism (CBT) and poverty alleviation; on the other hand, negative effect of uncontrolled tourism often destructs the very asset on which it was created.

Requirement of a Framework to Identify Destinations of Sustainable Tourism

Sustainability of tourism is an important factor in appropriate urban and regional growth due to its highest multiplication factor. But in an uncontrolled way, it can destroy a place especially environmentally sensitive area. Accurate identification of the tourism hub becomes the first criteria to ensure sustainability of the tourism network. This is a crucial part to understand the maximum capacity of the region and design circuit.

'Ecotourism' and 'sustainable tourism' are constantly used alternatively nowadays as a marketing term. However, according to the International Ecotourism Society (TIES) – the biggest and oldest ecotourism society in the world, which formed in 1990, ecotourism is purposeful travel to natural areas to understand the culture and natural history of the environment, taking care not to modify the integrity of the ecosystem along with creating economic opportunities by making conservation of natural resources helpful to local people.

River-based ecotourism is a combination of river tourism and ecotourism or nature tourism. It can also be a particular sector of nature-based tourism in which the various natural places, situated in a river valley, and cultural places, which are depended on the river, are developed in the light of ecotourism. All the rules and regulations are applicable here with the spatial boundary limited by the river, as this is the main natural reserve here. Emphases are given on using the river as a potential resource like transport, recreation and irrigation but in an eco-friendly manner. Sustainable tourism is dependent upon the attractive power of the destination's primary resources.

According to UNWTO there are numerous factors that affect tourism use of rivers, including spatial (length, width, seasonality, location and navigability); political (local, state, national, legislative and riparian); management (planning, catchment, resource allocation); biological (variety of species, fishing impacts); industrial (irrigation, manufacturing, sewerage, human consumption and hydroelectricity); recreational use (swimming, boating, diving, fishing); transportation (passenger, pleasure cruise); accommodation and environmental (toxicity, salinity). To make it sustainable, these must be carefully considered in selecting the destinations and developing river-based tourism plans.

Spatial Analysis Supporting River-Based Tourism

Multi-criteria weighted matrix is an application of multi-criteria decision-making (MCDM). According to Hwang and Yoon (1981), when there are several often usually conflicting criteria, MCDM refers to 'screening, prioritizing, ranking or selecting the alternatives' based on human intelligence, understanding and out of a finite set of substitutes.

The method was used by Ian McHarg for his Staten Island, New York City Project in 1968, where he had to evaluate which lands are suitable for conservation, recreation, residential and commercial uses. The objective of the approach was to incorporate resource values, social values and aesthetic values in addition to the normal criteria of physiographic, traffic and engineering considerations.

He used this method as it employs an explicit, rational methodology. Any other person accepting the method and evidence is likely to reach the same conclusions as those demonstrated. This is in direct contrast to the bulk of planning, where criteria are often obscure and covert. This permits the most important improvement in planning method, which the community can employ its own value system.

For this research, to keep the process simple, weighted sum model (WSM) is used as the basic module for developing the matrix. According to Gayatri and Chetan (2013), in decision theory, the WSM is the simplest and still the widely used multi-criteria decision analysis (MCDA)/multi-criteria decision-making (MCDM) method for evaluating a number of alternatives in terms of a number of decision criteria. This method is appropriate for group decision matrix, handles multiple criteria and does not involve complex mathematics.

The relative importance of the considered criteria is represented by the attached weights. These weighted multi-criteria are then developed in a matrix that can be applied on the geographical area.

For each of the major data categories, a number of smaller parameters are selected and evaluated, and a value is assigned to each factor. For certain land uses, the maximum condition will be preferable; for others the minimum will have the highest value. For each proposed land use, there will be certain parameters of greatest importance, and these can be preferred for highest weightage. Moreover, there will be a ranking importance, and so factors or parameters can be arranged in a hierarchy. This value of the parameter is then multiplied with the value of each criteria and summation obtained.

Multi-criteria Weighted Matrix

The weighted matrix is formed from evaluating different parameters of the area. The methodology followed for the study is explained here graphically (Fig. 12.1).

The broad framework of study is developed based on the initial objectives formed in first stage of the study. Then it is detailed out subsequently from objectives to sub-objectives to some sets of parameters. Each parameter is then expanded to give rise to some variables. These variables are basically the information gathered through various primary and secondary data inventory process. In fact, the questionnaire format for the survey procedures was largely decided by the linking of these variables to the basic framework of sustainability for tourism.

Overview of the Broad Study Area

Gangetic riverine delta of West Bengal State, India, which is characteristically a typical example of the riverine system of Global South, is studied for the purpose, and a strategy of spatial analysis is demonstrated through a method of using multi-criteria weighted matrix to reveal the different potential areas for tourism. The main river of West Bengal, Bhagirathi/Ganges, serves as the lifeline of this state.

Within West Bengal, the lower Gangetic River Plane, more precisely, Kolkata Metropolitan Region (KMR), is taken as the study area. The study emphasizes the Inner Metropolitan Region (IMR), whereas KMR is treated as the influencing area.

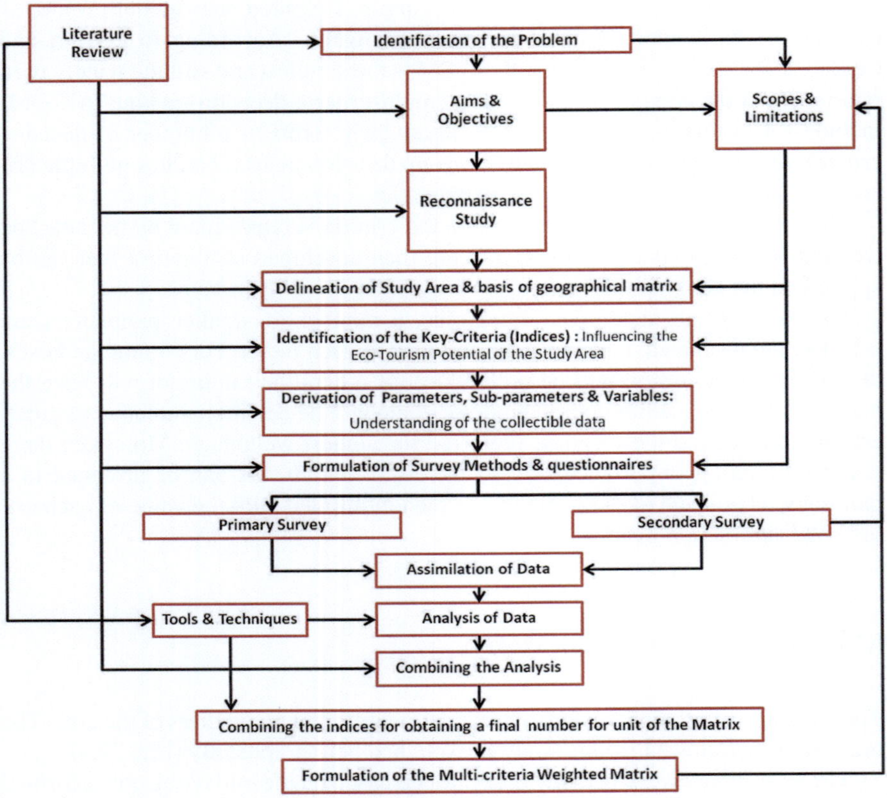

Fig. 12.1 Methodology for the research. (Source: Author)

KMR is a boundary in planning terms first mentioned in 'Perspective Plan for Calcutta: 2011' by the State Planning Board (SPB) in 1990 (KMDA 1990) and later echoed in 'Perspective Plan of Kolkata – 2025' (KMDA 2005). The area of KMR is 23185.8 sq. km according to census, 2001.

The initial study has been done graphically, mainly from literature review, reconnaissance and satellite images to demarcate the study area. The factors considered for the formulation of the graphical matrix are navigable inland network, proximity to river, canopy coverage, wildlife, scenic beauty and presence of existing tourist places.

Thus, a river-influenced core area is delineated as shown by the red dotted line in Fig. 12.2. This study as an example of typical Global South will focus on this area primarily.

For the purpose of pre-survey analysis, the delineated area is divided in three parts – the top, middle and bottom zone – on both banks of river as shown in Fig. 12.2. The main nodes, which in this case are 3 in number, are also identified in each zone as shown in Fig. 12.2, namely, the confluence of Bhagirathi River and

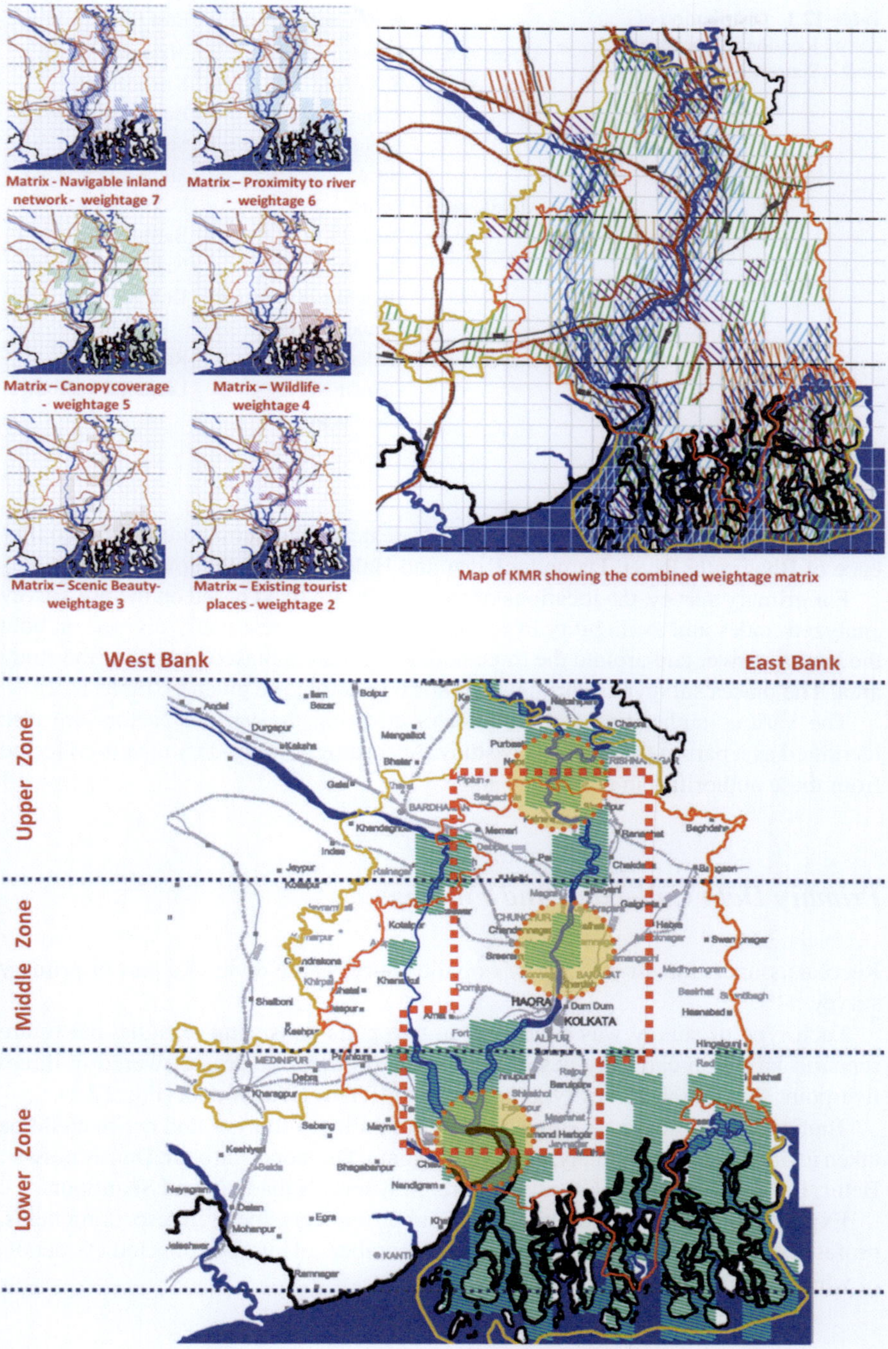

Fig. 12.2 Pre-survey framework matrices & area delineation. (Source: Author)

Table 12.1 Distribution of survey locations along river bank. (Source: Author)

	West Bank	East Bank
Upper zone	Katwa	Mayapur
	Nabadwip	Krishnanagar
	Kalna	Bethuadahari WLS
	SobujDwip (on river)	Shantipur
	Nirol	Fulia
Middle zone	Belur	Dakshineswar
	Bandel	Barrackpore
	Bansberia	Naihati
	Hooghly – Chinsurah	Kolkata
	Chandernagore	
Lower zone	Gadiara	Raichak
	Geonkhali	Diamond harbour
	Garchumuk	

Jalangi River in the north zone, the middle node near Chandernagore and the confluence of Bhagirathi River, Damodar River and Haldi River in the lower zone.

For primary survey, the locations of the surveys are fixed based on the pre-survey analyzed nodes and availability of tourists. The places are equally divided on both the banks of river and around the three nodes to get an unbiased picture of the study area. The places surveyed, starting from north to south, are given in Table 12.1.

The various authorities and parameters affecting the various factors are also identified as a part of the pre-survey study. Accordingly, secondary data is collected from these authorities in the survey stage.

Primary Data Collection and Findings

Reconnaissance, tourist and expert's opinion surveys are done as a part of primary survey.

Each type of survey was conceived in two parts consisting existing and future scenario based on carefully designed structured questionnaire, answered in three-five-point scale, each survey type yielding a different level of data (Fig. 12.3).

Tourist opinion survey is subjected to availability of tourists and hence could be taken in only 11 places, namely, Raichak, Gadiara, Diamond Harbour, Dakshineswar, Belur, Bandel, Bansberia, Chandernagore, Mayapur, Nabadwip and Shantipur.

Expert opinion survey was collected from prominent experts in respecting fields, professionals, professors, etc. The total 128 numbers of survey collected the details of which are given in Table 12.2.

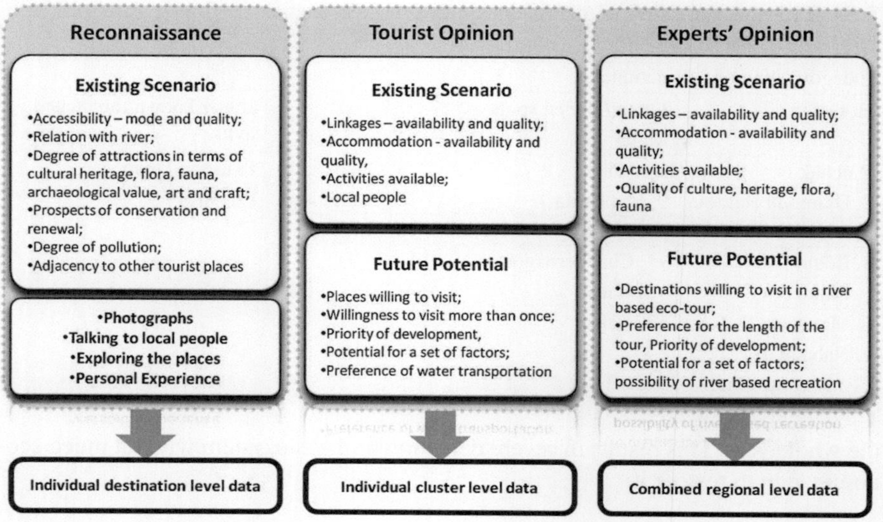

Fig. 12.3 Showing the points considered for different surveys and type of data collected through them. (Source: Author)

Table 12.2 Details of primary data collected. (Source: Author)

Survey type	Number of survey
Tourist opinion	128 (Approx. 10 in each location)
Expert opinion	15
Reconnaissance survey	19

Reconnaissance Survey Findings

During reconnaissance survey, based on the tourist influx, the destinations can be broadly classified in three categories as shown in Table 12.3, namely, hotspots, latent/hidden spots and lesser known or threatened spots.

Hotspots These are the most thronged places around Kolkata. Due to the proximity from the city and lack of accommodating infrastructure, most of these places are treated as daytrip destinations, resulting into severe damage to the spots in spite of huge influx of revenue.

Latent/Hidden Spots These places have potential to attract tourist all year round but at present are thronged by tourists, only during a particular season or festival. These can be further divided in major and minor spots according to the volume of incoming tourists.

Lesser Known/Threatened Spots These places though are very attractive in terms of culture or natural beauty and are very less known to common tourists. Hence these are at present used as picnic spots for a particular season and lie abandoned in

Table 12.3 Categorization of destinations based on reconnaissance survey. (Source:Reconnaissance Survey by Author)

Categorization of destinations			
Hotspots	Latent/hidden spots		Lesser known/threatened spots
	Major	Minor	
Raichak	8. Gadiara	14. Geonkhali	18. SobujDwip
2. Diamond Harbour	9. Bansberia	15. Garchumuk	19. Nirol
3. Dakshineswar	10. Hooghly – Chinsurah	16. Shantipur	
4. Belur	11. Chandernagore	17. Fulia	
5. Bandel	12. Katwa		
6. Mayapur	13. Kalna		
7. Nabadwip			

the whole year. This results in severe depletion of the ecosystem without much economic gain to restore it.

Primary Survey Findings

Table 12.4 shows the weightage calculated for each place surveyed in purpose of visit. To consolidate findings from primary surveys for better understanding of the current situation:

- More than 60% visitors are over 35 years.
- Highest percentage between 62 and 48 years.
- Upper node caters elderly visitors on an average.
- Short of young generation visitors because of absence of activities and maintenance.
- Table 12.4 shows upper node is more popular for pilgrimage purpose, but it also has a very close score for leisure – high potential for river-based leisure tourism.
- Pilgrimage is most prominent in upper part – gradually transforms into leisure activities and scenic beauty-oriented tourism downwards.
- Heritage potential also gradually reduces downwards – the highest at Nabadwip and Mayapur – but is not well utilized.
- Huge percentage of tourist are day-trippers – resulting into very less or no gain for the local economy generation.
- During season time, mass day-trippers cause destruction of ecology and heritage.
- For visiting places like Shantipur where no infrastructure is present, Nabadwip acts as a hub.

Table 12.4 Weightage obtained for each place surveyed in purpose of visit. (Source: Primary Survey done by Author)

Location of survey	Purpose of visit							
	Leisure	Sight-seeing	Heritage seeing	Pilgrimage	Packaged tour	Business	Education	Seeing friend/ family/ relative
Shantipur	6	0	2	7	0	0	0	0
Nabadwip	2	2	4	5	0	0	0	1
Mayapur	15	0	4	14	0	0	0	1
Upper node	23	2	10	26	0	0	0	2
Chandernagore	2	2	3	1	0	1	1	4
Bansberia	3	0	1	6	0	0	0	0
Bandel	3	1	2	1	0	0	3	1
Dakshineswar	9	0	2	12	0	0	0	0
Belur	7	1	0	0	0	0	0	0
Middle node	24	4	8	20	0	1	4	5
Raichak	8	2	0	0	0	0	0	0
Gadiara	6	1	0	0	0	0	0	0
Diamond Harbour	1	3	0	0	0	7	0	0
Lower node	15	6	0	0	0	7	0	0

- In Bandel and Dakshineswar, day tour is prevalent – mainly for two reasons – proximity to main city and lack of quality accommodation and other tourist facilities.
- Rowing/cruising comes as an alternative option as no facilities are present.
- Visiting cultural site has highest score in terms of tourist interests as per survey.
- Highest preference is for Nabadwip and Mayapur; next is Dakshineswar and Belur.
- Among the lesser known, Gadiara and Raichak are preferred.

Secondary Data Collection and Findings

Several secondary data are collected from various authorities involving both government and private organizations. The details are given in Table 12.5. Apart from that a lot of data are collected through authorized websites and discussion with experts. Both structured and non-structured interview are planned for experts.

With the help of the collected secondary data, two mosaics are formed.

Table 12.5 Showing details of secondary data collected. (Source: Author)

Reports/data collected	Concerned authority
Location and list of identified heritage places	State archaeological department; heritage commission
Location and list of wildlife park and sanctuaries	West Bengal Forest Department
Details of the fauna	
Location and list of identified tourist places; tourist influx details	West Bengal Tourism Development Corporation
Master plan, vision – 2025, Kolkata	KMDA
Five-year development plan for environment, wetland, urban amenities and heritage within KMA Study on passenger ferry service on river Hooghly Urban development strategy for West Bengal	
District census handbook	Department of Publication
Maps	Various sources

Physical-Demographic-Infrastructure Mosaic

For the first mosaic physical, demographic and infrastructural conditions within KMR are compared with places outside KMR, in terms of density, decadal growth rate, unit road length, rural road, forest area and employment scenario as shown in Fig. 12.4.

The findings consolidated from the comparative study are as follows:

- The highest density is in core city where difference from the next order is too large.
- Density of South 24 Parganas is less due to the presence of huge forest area (Sunderbans).
- The east bank is denser due to presence of Kolkata than the west bank, creating huge difference in level of development.
- Disparity in decadal growth rate is present of more than 10% between the two banks of river in the lower region, which shows the disparity is much larger in the southern side.
- Decadal growth is least in the core city indicating that it is almost saturated; hence any new development should be outward.
- Total road length is maximum in South 24 Parganas though in Kolkata, per unit area is highest.
- Road length per unit rural area is highest in KMR part of Howrah followed by Hooghly, and minimum length of rural road per unit area is in Nadia, which points out the unplanned growth of Nadia.
- Except small sanctuaries like Bethuadahari, the whole KMR has marginal forest belt.
- The work participation rate is almost similar for the whole KMR region.

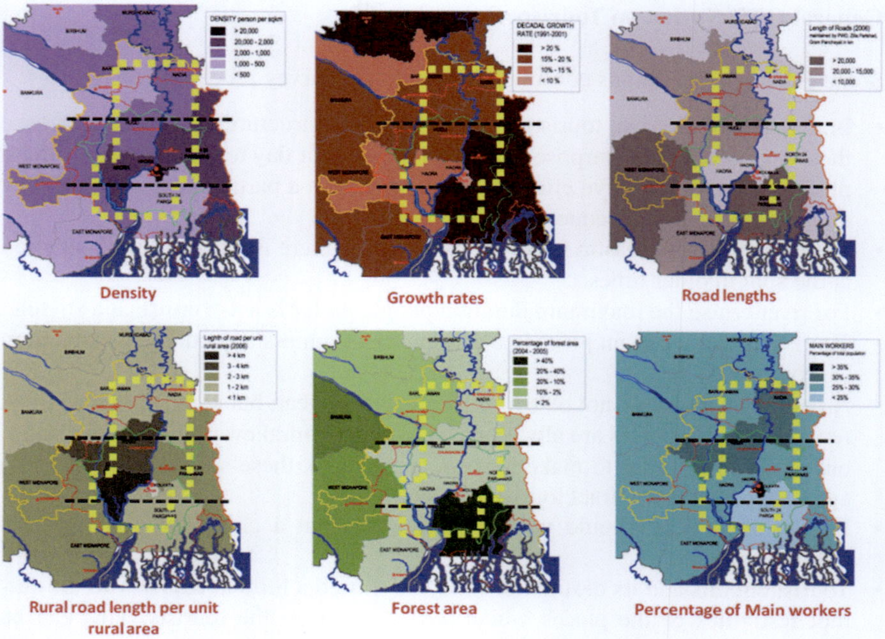

Fig. 12.4 Physical-demographic-infrastructure mosaic. (Source: Author based on Secondary data)

- The maximum employment is seen in Nadia apart from the core city area – this region also has a huge tourist influx, second after Kolkata – which emphasizes the importance of tourism industry in employment generation.

Sociocultural Mosaic

For the second mosaic, an inventory of different art, craft, fairs and festivals of different localities within KMR is carried out. These traditional antiquities are to be promoted as attractions for the individual destinations. The different forms are:

Music: Apart from the folk songs, the important types are Kirtan, Baul Gaan, Tagore tradition, musical instruments, etc.

Dance forms: Each district and festival is attached to a distinct dance form. The most famous forms are Gambhira, Tusu and Bhaduriya Saila performed during Makara Sankranti, Chhau of Puruliya District, Jhumar and Ashariya Jhumar, Raibense of Birbhum district and Dhanuchi dance forms.

Fairs promoted for tourism: Some fairs are already promoted by tourism department of West Bengal. These are Kenduli Mela, District Birbhum; Vasanta Utsav, District Birbhum; Chaitra – Baishakhi (allied to Charak), Gangasagar Mela – District S 24 Parganas, etc.

General Observation on Tourism Scenario

Some general observations consolidated during the survey procedure are:

- In majority of the river tourism spots, proper infrastructure is not present; hence these are used for the purpose of seasonal picnics or day tours. In this way these places suffer the negative effect of mass tourism in a particular season, but they do not gain enough revenue to rectify them.
- Winter is the peak season for ecotourism unless there is some cultural occasion at the spot in other times.
- For river cruise the maximum duration people prefer is a 4-h cruise at a stretch.
- Tourists prefer a certain place in a certain season unless any other new attraction is provided.
- A particular spot may not be active for the whole year, but some places within a few kilometer radiuses are always having some cultural events or other throughout the year. In order to make tourism sustainable, these spots should be managed collectively to attract tourists all year round.
- By destination we should not mean one spot but a cluster of spots usually interrelated.
- Tourist circuits and its destinations may be shifted a little to adjust with the current festivities of the places which are attractive to the tourists. This can be rotated throughout the year to keep tourist influx constant.
- There are a huge number of heritage buildings, some of which are very unique in nature. Some of them can be reused adaptively, particularly in tourism sector. These will also reduce an enormous cost of rebuilding some infrastructure from new.

Moderated Delineation of Study Area for Calculation

The delineated area is divided into a matrix of 20 km x 20 km grid as in Fig. 12.5, to make the calculation easier. Each grid is then evaluated in 3-point scale for each of the parameters or sub-parameters based on the various data collected.

Each parameter also has a certain weightage, which is obtained from expert opinion survey. Then combining all the parameters according to a set of logic, four indices are generated: one for each of the grid and, thus, four weighted matrices. Each matrix highlights the potential zone in its aspect.

Figure 12.6 shows the method of combination of all the data from primary, secondary and reconnaissance surveys by virtue of which four sub-indices are formed, each containing several weighted parameters.

Here the grid of the matrix is kept at a broader level, for exercise purpose. If data is available at a very minute level, the grid can be divided further to form a more intricate matrix.

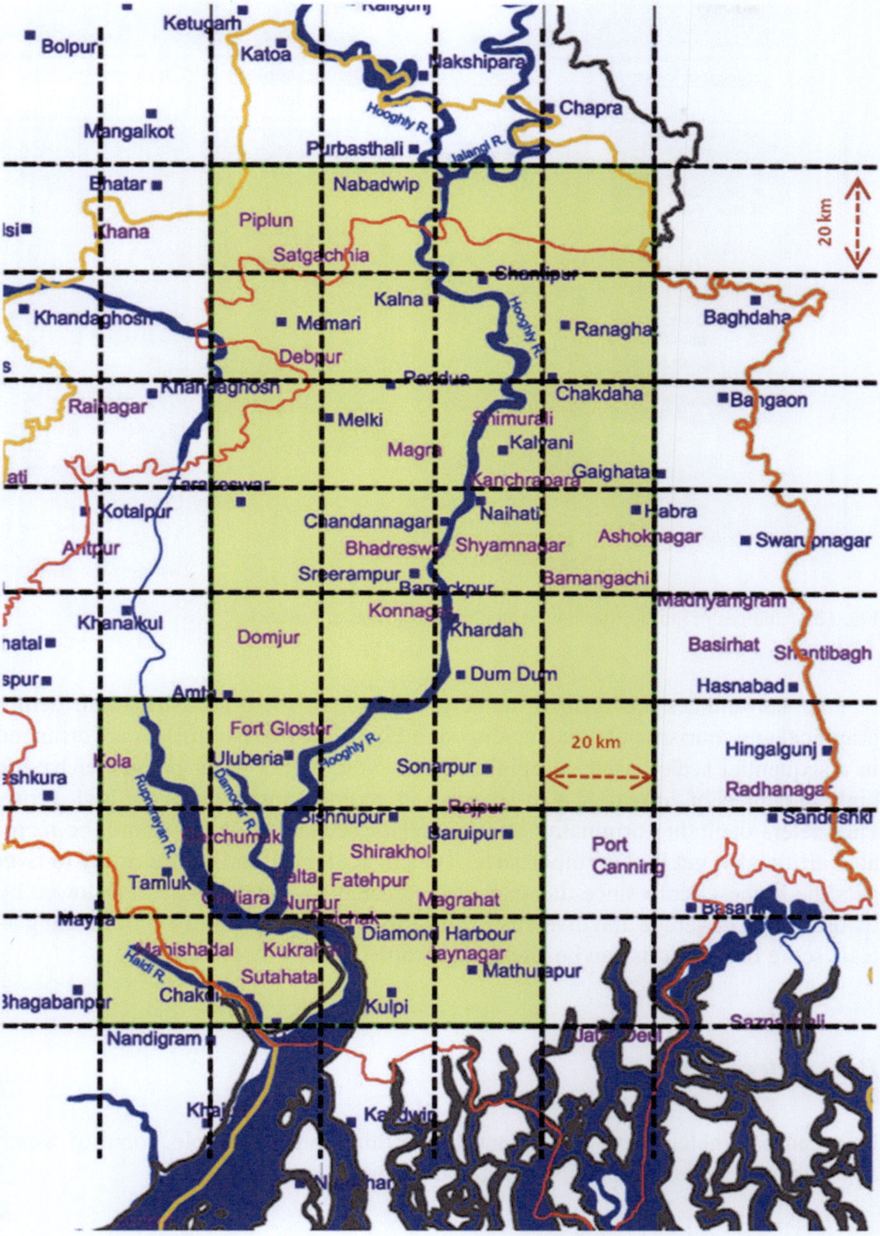

Fig. 12.5 Moderated delineation of study area. (Source: Author based on Secondary data)

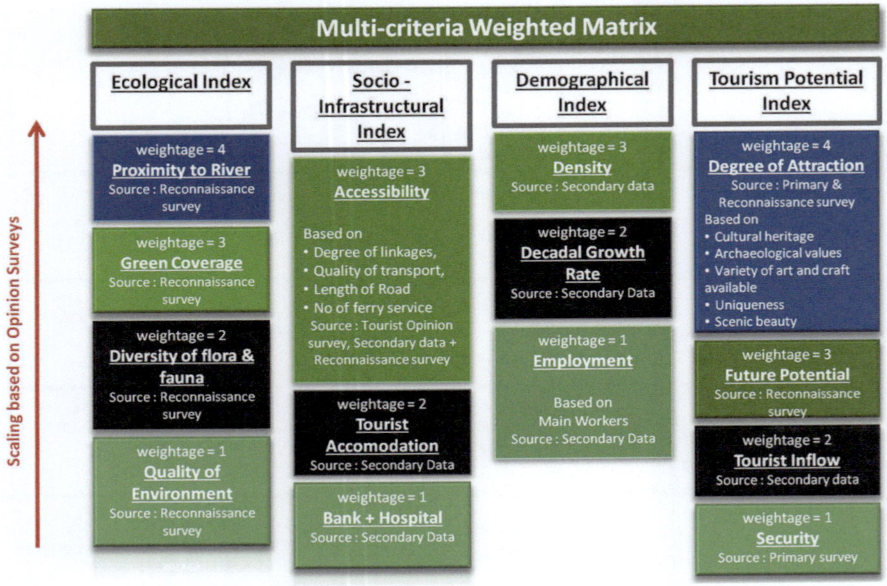

Fig. 12.6 Indices in multi-criteria weighted matrix. (Source: Author)

Four sub-indices are formed, namely, ecological, socio-infrastructural, demographical and tourism potential, as shown in Fig. 12.6. The parameters are arranged in a sequential scale based on opinion survey; the higher-order parameter having highest degree of importance in river-based ecotourism hence has a high score. Parameters depicting originality in attributes or scenarios, which cannot be recreated artificially, get higher importance. Thus in ecological index, proximity to river gets the highest score since this cannot be created by any means and followed by natural green coverage and diversity of flora and fauna. Quality of environment gets least score because this can be altered by proper policy and guidelines.

Parameters (P_i)

Fourteen parameters have been identified in this typical example, some of which also have several variables. These parameters are:

1. Proximity to river (P_1)
2. Green coverage (P_2)
3. Diversity of flora and fauna (P_3)
4. Quality of environment (P_4)
5. Accessibility (P_5)
6. Availability of tourist accommodation (P_6)

7. Availability of bank and hospital (P_7)
8. Density (P_8)
9. Decadal growth rate (P_9)
10. Employment (P_{10})
11. Degree of attraction (P_{11})
12. Future potential (P_{12})
13. Tourist inflow (P_{13})
14. Security (P_{14})

All the parameters are marked in 3-point scale.

For parameter like accessibility and degree of attraction, each of them has several variables, since these are more complex to define by one aspect. These two parameters are explained in detail as examples. The valuations of the other parameters are done in similar fashion.

Two parameters are explained in detail as examples: accessibility (P_5) and degree of attraction (P_{11}).

Accessibility ($\mathbf{P_5}$)

The variables under this parameter are:

- Degree of linkages
- Length of road
- Quality of transportation
- Ferry service

Combined accessibility matrix is the rounded off average of the four abovementioned variables to retain the 3-point-scaled character for further analysis as shown in Fig. 12.7.

Degree of Attraction ($\mathbf{P_{11}}$)

The variables under this parameter are:

- Quality of cultural heritage along river
- Diversity of flora in the region
- Diversity of wildlife in the river sanctuaries
- Scenic potential of the river
- Archaeological value of the region
- Variety of art/craft available
- Quality of product
- Presence of any uniqueness

The variables are first combined with the help of 3-point scale, and then the combined data is taken for further analysis.

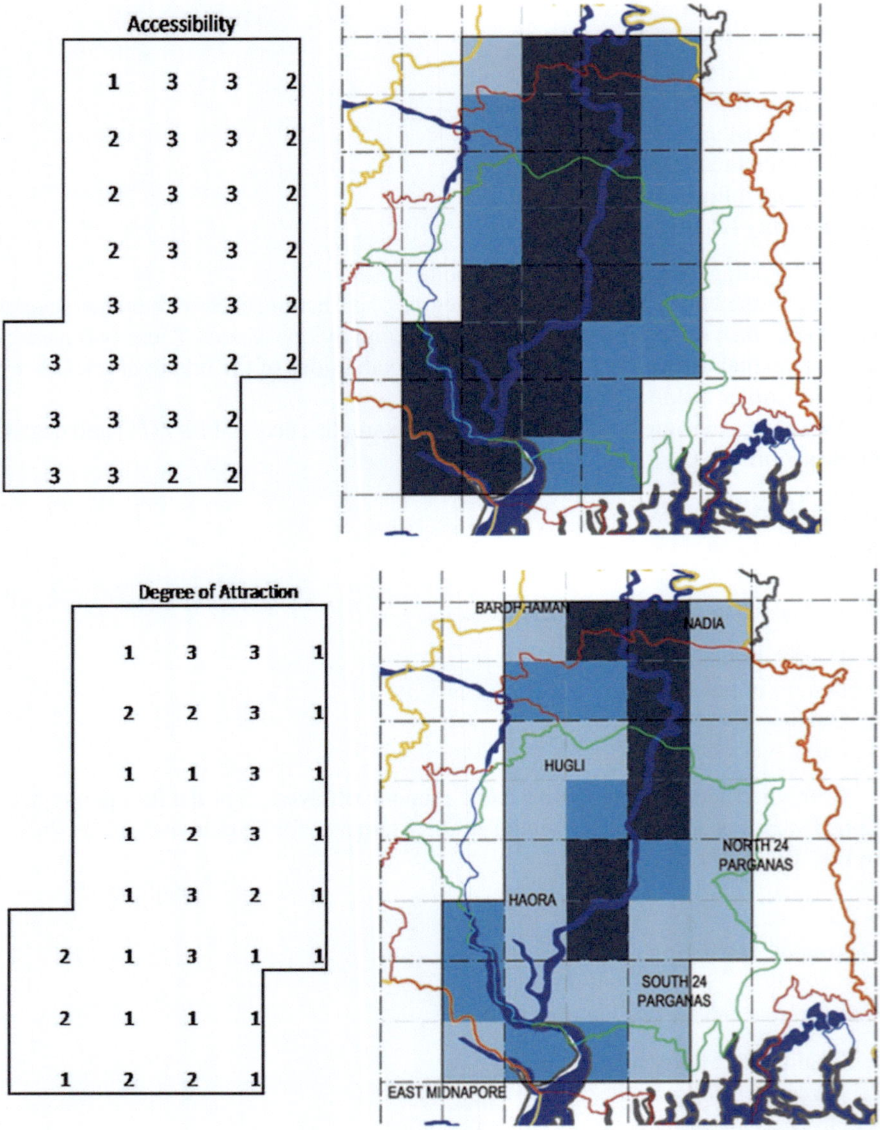

Fig. 12.7 Matrix: accessibility and degree of attraction. (Source: Author)

Based on the survey questionnaire, an index denoting the degree of attraction of a place is calculated considering both existing infrastructure and future potential.

From the result it is seen that the places with highest degree of attraction, i.e. the places with highest point, i.e. 3 (green) in Table 12.6, are already very populated and have a huge tourist influx.

Table 12.6 Degree of attraction of the spots. (Source: Author based on Primary Survey)

Location of survey	Score of degree of attraction	For converting to 3-point scale formation of matrix
		Input: 1 for greater than or equal to 67 percentile, 2 for less than 67 but more than or equal to 33 percentile, 3 for less than 33 percentile
Raichak	10.0	1
Nurpur	9.0	1
Gadiara	12.0	1
Geonkhali	13.0	2
Garchumuk	14.0	2
Diamond Harbour	13.0	2
Lower node	**11.8**	
Chandernagore	18.0	3
Hooghly-Chinsurah	14.0	2
Bansberia	16.0	3
Bandel	14.0	2
Naihati	14.0	2
Barrackpore	14.0	2
Kolkata	19.0	3
Dakshineswar	19.0	3
Adyapith	13.0	2
Belur	15.0	2
MIddle node	**15.6**	
Shantipur-Fulia	18.0	3
Nabadwip	16.0	3
Mayapur	16.0	3
Krishnanagar	18.0	3
Katoa	15.0	2
Kalna	15.0	2
SabujDwip	12.0	1
Bethuadahari	11.0	1
Upper node	**15.1**	

Places with point 1 like Raichak, Gadiara, Nurpur, SabujDwip and Bethuadahari have a low degree of attraction to tourists as these places lack the necessary tourist infrastructure and due to lack of awareness among tourists. These places have a very high nature-based tour potential.

Places with point 2 have a medium degree of attraction as the potential of these is not fully utilized in tourism sector. The existing infrastructure is less, but they have a rich cultural background.

Sub-indices (I_m)

Here four sub-indices (I_m) are considered. These are:

1. Ecological index (I_1)
2. Socio-infrastructural index (I_2)
3. Demographical index (I_3)
4. Tourism potential index (I_4)

Now, the formula to calculate sub-indices (I_m) is as follows:

$$I_m = \frac{\sum_{i=1}^{n} (P_i \times W_i)}{\sum_{i=1}^{n} W_i}$$

Where,

I_m are the calculated sub-indices, here m is 1 to 4.
P_i are of the parameters from where the sub-indices are calculated.
W_i are of the weightages that each parameter carries for calculation of sub-indices.
i is the number of parameters involved, where the maximum value of i is n.

Ecological Index (I_1)

For the purpose of explaining the detail scoring process and how the graphical representation is derived from that process is explained below for the ecological index (Table 12.7 and Fig. 12.8).

Table 12.7 Steps to calculate ecological index

Parameters P_i	Proximity to river P_1	Green coverage P_2	Diversity of flora and fauna P_3	Quality of environment P_4
Input values for parameters	3 for excellent			
	2 for good			
	1 for poor			
Weightage W_i	W_1=4	W_2= 3	W_3= 2	W_4= 1
Index I_m	**Ecological index:** $I_1 = \frac{\sum_{i=1}^{4} (P_i \times W_i)}{\sum_{i=1}^{4} W_i}$			
Simplified index I'_m	**Simplified ecological index (I'_1)** Input 3 for range more than or equal to 2, 2 for range more than or equal to 1.5 to less than 2, 1 for range less than 1.5			

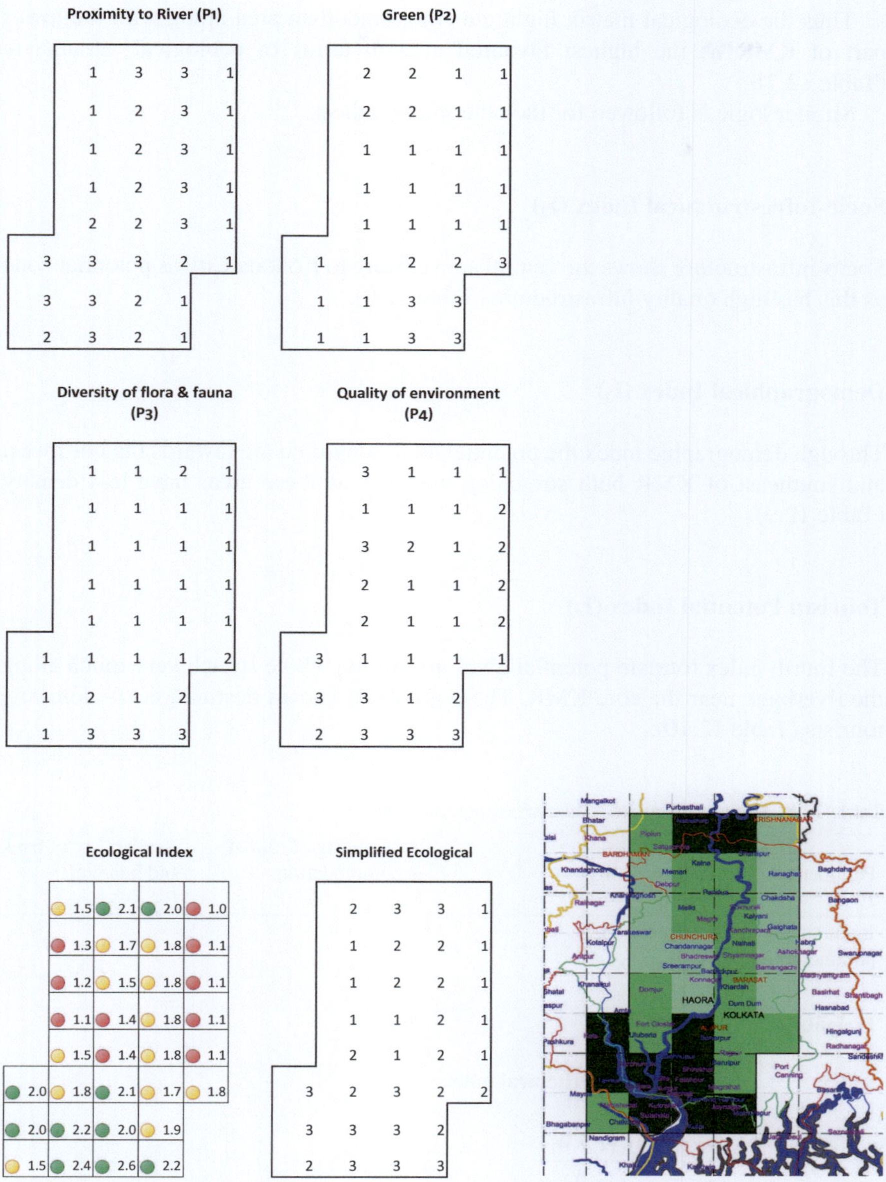

Fig. 12.8 Parameter matrices under ecological indexed matrix. (Source: Author)

Thus the ecological matrix highlights the Sunderban area along with the lower part of KMR as the highest favoured area in terms of ecological parameters (Table 12.7).

Similar logic is followed for the other three indices.

Socio-infrastructural Index (I_2)

Socio-infrastructure shows the central area closure to Kolkata City as potential zone as this has high quality infrastructure (Table 12.8).

Demographical Index (I_3)

Through demographic index the potential areas identified are towards the northwest and southeast of KMR both stretching outward, as these areas have less density (Table 12.9).

Tourism Potential Index (I_4)

The fourth index tourism potential gives almost the whole stretch very much along the riverbank near the core KMR. These are more known destinations to domestic tourists (Table 12.10).

Table 12.8 Steps to calculate socio-infrastructural index

Parameters P_i	Accessibility P_5	Availability of tourist accommodation P_6	Availability of bank and hospital P_7
Input values for parameters	3 for excellent		
	2 for good		
	1 for poor		
Weightage W_i	W_5=3	W_6= 2	W_7= 1
Index I_m	**Socio-infrastructural index** $I_2 = \frac{\sum_{i=5}^{7}(P_i \times W_i)}{\sum_{i=5}^{7} W_i}$		
Simplified index I'_m	**Simplified socio-infrastructural index (I'_2)** Input 3 for range more than or equal to 2, 2 for range more than or equal to 1.5 to less than 2, 1 for range less than 1.5		

Table 12.9 Steps to calculate demographical index

Parameters P_i	**Density** P_8	**Decadal growth rate** P_9	**Employment** P_{10}
Input values for parameters	3 for excellent		
	2 for good		
	1 for poor		
Weightage W_i	W_8=3	W_9= 2	W_{10}= 1
Index I_m	**Demographical index:** $I_3 = \frac{\sum_{i=8}^{10}(P_i \times W_i)}{\sum_{i=8}^{10} W_i}$		
Simplified index I'_m	**Simplified demographical index (I'_3)** Input 3 for range more than or equal to 2, 2 for range more than or equal to 1.5 to less than 2, 1 for range less than 1.5		

Table 12.10 Steps to calculate tourism potential index

Parameters P_i	Degree of attraction P_{11}	Future potential P_{12}	Tourist inflow P_{13}	Security P_{14}
Input values for parameters	3 for excellent			
	2 for good			
	1 for poor			
Weightage W_i	W_{11}=4	W_{12}= 3	W_{13}= 2	W_{14}= 1
Index I_m	**Tourism potential index**: $I_4 = \frac{\sum_{i=11}^{14}(P_i \times W_i)}{\sum_{i=11}^{14} W_i}$			
Simplified index I'_m	**Simplified tourism potential index (I'_4)** Input 3 for range more than or equal to 2, 2 for range more than or equal to 1.5 to less than 2, 1 for range less than 1.5			

Simplified Sub-indices (I'_m)

These simplified sub-indices are a manifestation of calculated sub-indices (I_m).

Since the calculated sub-indices have a range of decimal numbers, simplified sub-indices (I'_m) are obtained to simplify the process of calculation by further categorizing sub-indices by a range based on the distribution of dataset. In this case, a highest number of 3 for range more than or equal to 2, 2 for range more than or equal to 1.5 to less than 2, and 1 for range less than 1.5 has been assigned.

Combined River-Based Ecotourism Potential Index (I)

Finally, overlapping the abovementioned four matrices, a combined multi-criteria weighted matrix is found out as in Fig. 12.9, which gives the potential areas for river-based ecotourism in overall scenario. This shows the most potent places in darkest colours to the least potent in lightest colours.

Since the delineated area is previously divided into smaller units of 20 m square, the final aim is to find the combined river-based ecotourism potential index I for each one of the smaller units with the help of the following formula:

$$\mathrm{I} = \frac{\sum_{\mathrm{m}=1}^{\mathrm{z}} \mathrm{I}'_{\mathrm{m}}}{\mathrm{z}}$$

Where,

I is the combined river-based ecotourisn potential index.
I'_m is the simplified sub-indices, measured from calculated sub-indices (I_m). Each sub-indice has a set of parameters from where they are calculated.
m is the maximum number of simplified sub-indices considered, the maximum value of m is z. Here z is 4 since four simplified sub-indices (I'_m) are considered. These are:

1. Simplified ecological index (I'_1)
2. Simplified socio-infrastructural index (I'_2)
3. Simplified demographical index (I'_3)
4. Simplified tourism potential index (I'_4)

In other cases more number of indices can be considered, and the subsets can be created accordingly.

The combined matrix reveals the characteristic of the deleniated zone in the light of one of the four indices as well as in the combined light of the four.

In this example, the ecological matrix highlights the Sunderban area along with the lower part of KMR as the highest favoured area. Socio-infrastructure shows the central area closure to Kolkata City as potential zone as this has high quality infrastructure. Through demographic index the potential area identified is towards the northwest and southeast of KMR both stretching outwards, since these areas have less density. The fourth index tourism potential shows almost the whole stretch very much along the riverbank near the core KMR. These are more known destinations to domestic tourists. Finally, in the combined multi-criteria weighted matrix, it is seen that the two confluences of Bhagirathi, the first one with Jalangi near Nabadwip

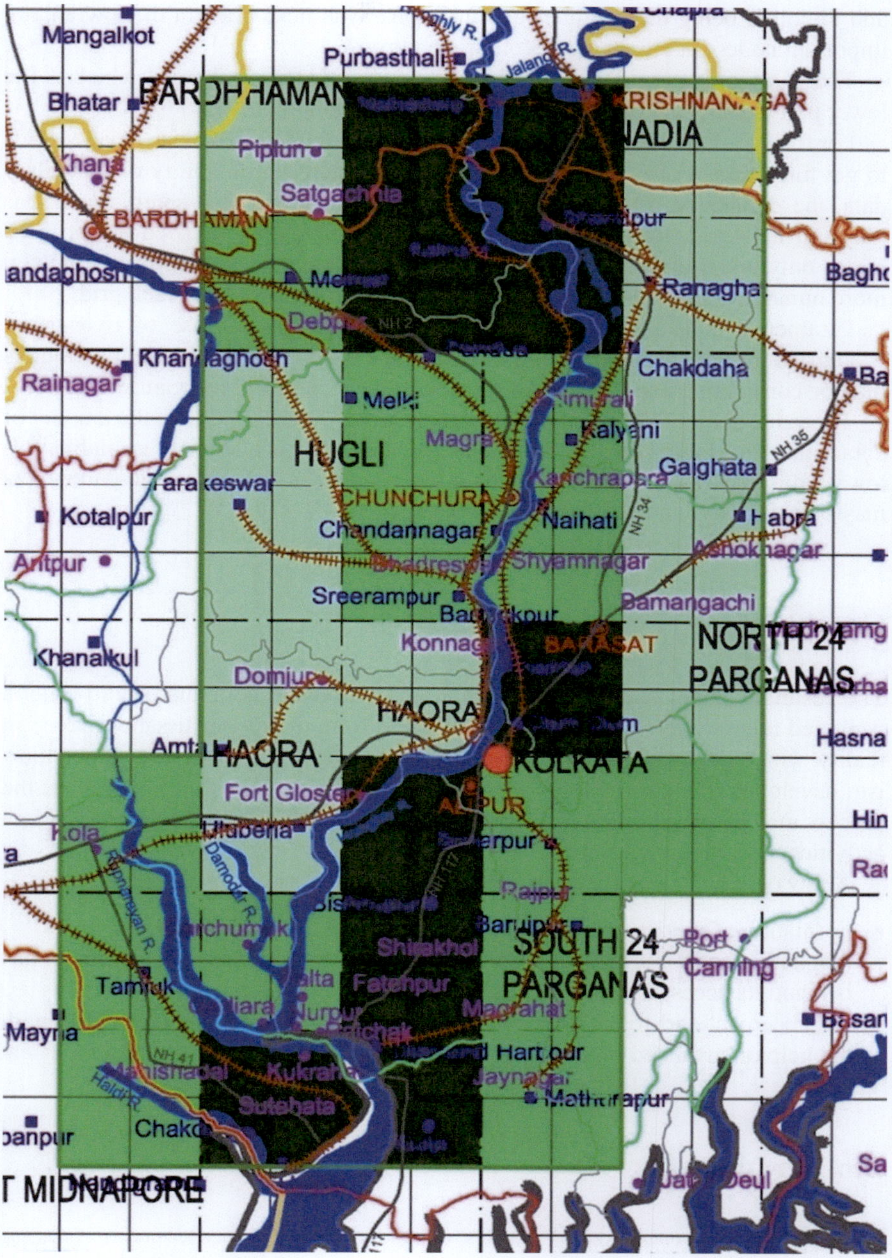

Fig. 12.9 Spatial representation of combined matrix. (Source: Author)

and the other being the confluence with Rupnarayan near Gadiara in Howrah, are important nodes for river-based ecotourism purpose.

The study area considered here is lower part of Gangetic Bengal which is the lower part of the State of West Bengal situated in India.

For further minute studies, the target area can be divided into much smaller grids to get fine-grained results. Due to time constraint and unavailability of secondary data, the delineated area is divided in broad grids, where 2–3 sample locations may fall in 1 grid. Reconnaissance and primary surveys are carried out at these nodes only and considered for the entire grid. Where data availability is greater, more number of nodes has to be considered, preferably 1 node for each grid.

Parameters considered in this study are specific to the case of river-based ecotourism of Global South only. For other type of tourism, additional parameters may be considered, and the process have to be rechecked and reorganised.

The value associated with the parameter is largely associated with the amount of data available. Hence, for very remote areas where data is not easily available, following this process may be difficult. But for those areas, the concerned parameters may be changed to match the available dataset.

Conclusion

The resultant nodes found in the typical case from the combined multi-criteria weighted matrix coincide with the nodes found from the reconnaissance survey in reality, which implies that this technique can be replicated in other river-based tourism development area of Global South. For other type of tourism or location, the process may be enhanced on case-to-case basis to predict nodes of potential tourism activities in an unknown zone.

A few outreach areas for further research on this broad subject can be:

- Formation of tourist circuits and identifying the gap in infrastructure
- Application of combined multi-criteria weighted matrix in areas other than tourism and its necessary modification
- Enhancement and refinement of combined multi-criteria weighted matrix with the help of advanced statistics

Bibliography

Aabo D (2006) Sustainable tourism realities: a case for adventures service tourism. Capstone Collection. 650

Anon (2012) Destination wetlands: Supporting sustainable tourism. Secretariat of the Ramsar Convention on Wetlands (Ramsar, Iran, 1971); World Tourism Organization (UNWTO), Gland/Madrid

Anon. (2018). Mapsofindia.com. [Online] Available at: https://www.mapsofindia.com. Accessed 20 May 2018

Dey A (2016) goUNESCO. [Online] .Available at: https://www.gounesco.com/heritage-kumartuli-kolkata/. Accessed 14 Nov 2018

Gayatri SV, Chetan SM (2013) Comparative study of different multi-criteria decision-making methods. Int J Adv Comp Theor Eng (IJACTE) 2(4):2319–2526

Hall CM, Härkönen T (2006) Lake tourism: an introduction to lacustrine tourist system. In: , Lake tourism: an integrated approach to lacustrine tourism systems. s.l. Channel View Publications

Hwang C-L, Yoon K (1981) Multiple attribute decision making. s.l. Springer, Berlin/Heidelberg

Jafari J, Xiao H (2016) Encyclopedia of tourism. s.l. Springer

Jensen KM, Lange RB (2013) Transboundary water governance in a shifting development context: New Development Finance, Development Spaces and Commitment to Cooperation: A Comparative Study of the Mekong and the Zambezi River Basins, s.l.: Danish Institute for International Studies

KMDA (1990) A perspective plan for Calcutta, 2011, Kolkata. State Planning Board, Development & Planning Dept., Govt. of West Bengal, Calcutta

KMDA (2005) VISION 2025 "Perspective Plan of KMA: 2025". Kolkata Metropolitan Development Authority, Kolkata

Lahiri-Dutt K (2014) Beyond the water-land binary in geography: water/lands of Bengal re-visioning hybridity. ACME 13(3):505–529

Moutinho L (2011) Strategic management in tourism. CABI, Wallingford

Prideaux B (2009) River tourism. Cabi, Wallingford

Ritchie J, Crouch G (2009) The competitive destination. CABI, Wallingford

Saito Y (2000) Deltas in Southeast and East Asia: their evolution and current problems. Joint Conference on Coastal Impacts of Climate Change and Adaptation in the Asia-Pacific Region, Kobe, Japan, pp 185–191

Samanta D (n.d.) Planning for ecotourism in Sunderban, s.l.: s.n

Shaw R, Thaitakoo D (2010) Water community. In: Community, environment and disaster risk management. Emerald Group Publishing Limited, s.l., pp 1–13

Shoval N, Raveh A (2004) Categorization of tourist attractions and the modeling of tourist cities: based on the co-plot method of multivariate analysis. Tour Manag 25(6):741–750

Stargardt J (1973) Southern Thai waterways: archaeological evidence on agriculture, shipping and trade in the Srivijayan Period Southern Thai waterways: archaeological evidence on agriculture, shipping and trade in the Srivijayan period. Man, New Series, March, 8(1), pp. 5–29

Terencia NM (2018) Community based tourism and development in third world countries: the case of the Bamileke region of Cameroon. Int J Soc Business Sci 12(1):26–30

United Nations (2018) Forging a path beyong borders: The Global South by UNCAD. s.l., United Nations Publications

World Tourism Organization (UNWTO) (2016) Mekong river-based tourism product development. World Tourism Organization (UNWTO), Madrid

Zacharias P, Manalel D, Jose P, Salam A (2008) Back water tourism in Kerala: challenges and opportunities. IIMK, s.l.

Chapter 13
Between Mountain and River: A Vernacular Settlement-Architectural Concept in Indonesian Archipelago

Indah Widiastuti

Abstract The architecture between mountain and river discusses about the concept of indigenous vernacular villages in Indonesia, as a representational character of settlement-architecture in insular Southeast Asia. It typically demonstrated a cultural landscapes formed by villages with row of houses standing on stilt with overwhelming roof designs, shaded by volcanoes, and fronted by paddy field and a plain that stretch from mountain feet toward river. Mountainous terrain, volcanoes, dendritic rivers, and maritime background generally characterize the ecology of archipelago in tropical belt.

Paddy farming is the main agricultural livelihood, marked by granary as basic model to elaborate building's typology and settlement arrangement. Compounds of houses and granary in various spatial compositions mark the arrangement of settlement in a landscape. The design of houses shows resemblance or expansion from granary structure. More than needs and commodity, paddy and rice are also conceptual sources of many social values, symbols, and mechanism, by which social stability maintained. The wholesome of ecology, settlement and architectural design, and social and communal values forms a conceptual character of cultural landscape of paddy farming society in Southeast Asian archipelago.

The archaic concept of village in Indonesia is called "wanua" – an Austronesian term that means a land and home. Typically, several wanuas develop a confederation which hierarchically comprises of core village(s) or the earliest settlement, peripheral villages, further expansion of the settlement, and migrant land or settlements outside the cultural boundary. These villages laying on hinterland and migrant land ecological-wise demonstrated network of villages, organized in various operations, such as kinship, temple network, and water management. Some cases are drawn to illustrate the settlement concepts, which are the Nagari settlement with the rumah gadang residential architecture of the Minangkabau in West Sumatra

---

I. Widiastuti (✉)
History, Theory and Criticism of Architecture – School of Architecture, Planning and Policy Development, Institut Teknologi Bandung, Bandung, Indonesia

© Springer Nature Switzerland AG 2020

M. Ghosh (ed.), *Perception, Design and Ecology of the Built Environment*, Springer Geography, https://doi.org/10.1007/978-3-030-25879-5_13

and the desa concept with the Uma residential architecture of the Bali Aga people in Bali.

Keywords Nagari · Banua · Minangkabau · Bali Aga · Wanua · Settlement-architecture · Jero Kahyangan · Mountain and river

Introduction

Indonesia is the biggest archipelago in the world, with approximately 13,000 islands, lying on the zone of equator-tropical Southeast Asian archipelago, the meeting of three geotectonic plates, and the "Pacific Ring of Fire" volcanic chain. Vernacular settlements or villages in Indonesia substantially appear as groups of habitable sphere laying on mountains feet or slopes, with rivers intertwining down and irrigating the agricultural land. Visually volcanoes shade behind the paddy field, arranged in tiers, down to the river. The architecture generally stands on stilt with sophisticated carpentry and ornaments, topped by overwhelming roofs, providing a traditional idyllic image of vernacular indigenous agriculturist living. During the occupation of Dutch colonial administration, this image was dramatized, as an exotic and orientalist Asian villages – *Mooi Indie* (the beautiful India – Dutch). After independence, this image persisted and became a stereotypical image of Indonesian cultural landscape. The aim of this article is to bring a glimpse about vernacular settlement-architecture in Indonesia as distinct architectural and dwelling traditions that emerged in the tropical and volcanic archipelago. The term vernacular implies that the settlement-architecture refers to commoner's habitation with architectural traditions that sustain through time and undergo continuous development and transformation.

Settlement-Architectural Concept

Defining vernacular settlement in Indonesia is often problematic, because of the following situations: (1) dwelling includes physical and mental activities that may not accurately represent each other, due to the high mobility of the people and interconnections among villages; (2) the social-cultural sphere of a settlement could extend beyond their formal spatial boundary; (3) formal boundaries such as gates and fort are not too known; and (4) the local notion of inhabitants and functions cannot be easily understood in modern functionalism sense. Human is not always the center of importance, comparing to rice, the ideas of divine mothers and ancestor's spirit.

It has been also a general situation that architecture – as a modern functional concept – often fails to provide a holistic frame to explain the phenomenon of dwelling and habitation that includes the traditional pre-modern settlements. Hence, many scholars working in architecture and anthropology tried to explore various alternative definitions and considered better equipped with methodological properties to embrace the complex dimensions of a dwelling sphere. For the same intention, Nold Egenter (1996) coined the term "settlement-architecture," to define settlement as a horizontal unity where human, habitable places, and tectonic tradition developed. The concept promotes a holistic understanding about human settlement through the emergence of its dwelling culture, from the elementary structure to the higher and the highest order and from the rudimentary to its more sophisticated and well-refined design. It implies that a settlement as a concept of habitation has its architecture. An architectural concept can appear as an inter-scale emergence of coherence that simultaneously frames architecture, settlement, human, and other inmates on multiple scales and levels of spatial narrative, projected on a horizontal landscape. Egenter termed it as toposemantic. In the same spirit, Domenig (2014, p. 9) proposed "Spatial Anthropology" to address a field of study devoted to human behavior projected on a horizontal spatial organization. In this stance, Habraken's theory about territorial control "Control Hierarchies in Complex Artifacts" (1987) suits to explain the notion of territory. The sense of boundary is then a continuous production and reproduction of a controlled sense of territory until the development meets a steady balance. The definition of settlement in this chapter refers to this dynamic concept of space and projection of a long accumulative human process in spatial form with common genetic codes in varied scales of habitation.

The Living Culture Between Mountain and River Through Time

Domenig (2014, pp. 57–64) underlined the importance of mountain for the ancient Indonesian as a place where the deceased ancestors were buried. Mountains are also container of rich natural deposit beneficial as an asset of agriculture and other aspects of livelihood. The chain of volcanoes in Indonesia has made the rice culture in Indonesia or insular Southeast Asia (*sawah*) more specific than mainland Southeast Asia. This made volcanoes sources of inspirations and value for pragmatic, spiritual, and economic orientation. Indeed, the mountain became more associative with a place of the spiritual higher place (Wales 1959, in Ahmad 2010, pp. 49–53). Scattered in islands and remote places in Indonesia are local myths typically depicting spiritual power and ancestors as descending from the sky to mountain peak before spreading down the slopes to lower land.[1]

Although in many places local myth rarely elaborate rivers in their narrative, compared to mountains, the toponymical names of rivers and the use of local term

[1]The Bugis and Toraja in Sulawesi Island acknowledge the ancestor(s) *To Manurung* descending a coconut-like mountain peak. The mountain is said to evolve from ocean. The Minangkabau in Sumatra Island mentioned traveling ancestors by boats reaching an egg-shape mountain peak. The memory of ocean migration and prehistoric glacialization may have lend base to the plot.

that means "water" or "river" to call "a settlement" proved its importance (Asnan 2016, pp. 23, 43–45). Through the river system, ancient culture of Dongson from South China penetrated the archipelago's higher land around 2000 BC and introduced the tradition of rice culture and domestic buffalo, along with double burials, ancestral worship, bronze drums, streamline decorations, menhir stone crafts, and tiered pyramids for religious symbols (Munoz 2006, pp. 33–35). The Minangkabau people of West Sumatra in general believe that their original ancestor came from the highland slopes of Marapi Mountain. However, their narrative of origin was full of tales of the traveling ancestors across the oceans and rivers. Given maritime and insular-migrations traditions, Indonesian men traveled overseas, and female managed their farmland. As follows, homeland, female, and residential buildings are commonly associated with one another. The Indonesians address their homeland *tanah-air* (the land and water, Indonesian), which is again associative with feminine character – *pertiwi* (*pritvi*, Sanskrit). The primordial memory of water was articulated in the symbolic totems like fish, serpent, and amphibious creatures or to a more advanced organization of water management. Subak irrigation in Bali is the signature of Indonesian paddy farming (*sawah*) indigenous irrigation.

The tradition of rice farming developed upon prehistoric horticultural layers. The old memory of horticulturist society was transmitted to the present social conscience in a form of vegetal memory and tree totem as foci of a village. Rice culture established significant meaning for rice barns, manifested physically in grain storage box designs, granary building, and granary-based settlements. Renewing the ancient concept of divine mother was the rice goddess (Dewi Sri), adopted from the character of Uma, Shiva's companion from Hindu-Buddhist traditions. The nature of habitations in Indonesia stands on the prehistoric templates of forest and river ecological substratum (Jarzombek 2013, p. 249). In the beginning, people dwelled around the accessible living source, in caves or in groups of sheds. They maintained their dwelling sphere and differentiated it from the unconquerable place. The cave painting depicted various complex of living, hunters and gatherers, anglers, natural elements, and the realm of spirits. Jarzombek (2013, pp. 116–148) explored that conical shelters with a central pillar or tipi-like form were typical forms found in many places in the world. The conical structure in Indonesia was explored by Domenig, and he concluded that the structure was the earliest genotype of what later would be developed into hipped-gable roof architecture (1988, pp. 9, 83, 103).

River, the feminine characters, dynamic joints, and compressive construction like boat structures constitute unique architectural characters. Outlining Southeast Asian and Thai architecture, Jumsai in his book the *Naga* (1995, pp. 11–13) elaborated the occurrence of oceanic-riverine-mountainous characters epitomized in the figure of *naga* (a serpent in the Sanskrit language) – as a reflection of cultural subconscious of the Southeast Asian, shaped by the primordial memories of glacialization and prehistoric migration. He implied that liquid and oceanic primordial archetype lend base to the aquatic artistic ornamentation and cosmography. Amphibious-tectonic principles characterize the construction in Southeast Asia. They are articulated in typical structures standing on stilts; structural tensioning and weaving characters enable dynamic structure of the overwhelming roof, constructed without nails.

"Wanua," Granary, and Residential Architecture

The conceptual model of settlement in Indonesia is identified as an archaic concept of a hamlet, called *wanua* (Santoso 2008, p. 9; Ahmad 2010, pp. 207–209). *Wanua* is a generic Austronesian phoneme found among the indigenous people in ancient Southeast Asia, mainland and insular, Pacific, Oceania, and up to Madagascar, which means the "inhabited sphere," land, country, place, or village – *wanua/banua/manua/bonua* (Fox 2006, p. 9). The smallest unit of *wanua* comprises rows of individual houses, facing a main common east-west orientated boulevard, and its supporting function, like assembly hall, pounding shed, granaries, and worshiping compound. The dwelling unit alternatively appears as single-hall houses, house compounds, or longhouses where people live together. Jarzombek underlined the architecture of Dayak ethnic on Kalimantan Island, Mentawai Island, and Nias Island in Indonesia, and the architecture of Andaman, India, keeps the remaining traces of dwelling culture based on horticulture civilization. In later period the adoption of Indian culture yielded new Sanskritized words and value, for *wanua* such as *huta* among the Bataknese in North Sumatra Batak (modified from *kuta*-Sans), nagari of the Minangkabau in West Sumatra, or *desa* in Java (taken from *desham*-Sanskrit). Local traditional agriculture found its design – *sawah* – and was managed in various ways, such as pond and irrigated canal and rain-fed paddy farming in Sumatra Island, subak in Bali, and centralized intensive irrigation in Java. The agriculturist society generally demonstrates communal solidarity of network villages. Their cohesiveness is generally legitimized by common myth, kinship, rituals, water consolidation, and overall common cosmological orientation toward mountain and river.

By intra-village network, several *wanuas* establish a confederation, comprising hierarchically one or more village(s) of origin, the mainstream village where the bulk of inhabitant lives and peripheral villages or the extended villages. Further are migrant lands (*rantau*), usually outside the confederation or overseas. There are various ways of establishing horizontal network or confederation, as follows:

1. The confederation arranges the settlement units in hierarchical order according to distance and attachment to the origin and yields concepts such as the "village of origin," mainstream or inner villages, and of the extended village at the perimeter or outer village where modern lifestyle is allowed to some degree.
2. The confederation arranges horizontally the member villages according to clusters representing a community with specific social roles, i.e., aristocratic, administrative, spiritual, and commoner's clusters.
3. The confederation is segmented socially or spatially or both at once based on clans or matriclans and its extended kin.

Agriculture living is marked by elaborate constructions of granaries. Granary contains more than functions but spiritual value. Under the idiom of Dewi Sri (rice goddess), instead of "stored in granary," paddy "returns home" to her kin folks in their granaries. The construction is generally a box structure made of wood and bamboo, resting on four high stilts, with anti-rat disc installed at the joint of box structures and the stilts. The elaborate complex ornamentations on granary could carry symbolic significance. Granary appears in various architectural forms and construction: granary-gazebo, where the granary leaves underneath space for sitting; granary-in-the-house, where the house keeps a particular room for granary; granary house, where the granary is an integral part of the house (see Fig. 13.1 about Three residential architectures and granaries). It lends base to Sato's claim (1991) that the residential architecture in Indonesia is conceptualized as granary dwelling.

Josef Prijotomo outlined how human in the tropic needed more shelter to shade than to protect and need more a limbus of exterior-interior than a closed and rigid space. Tropical spatiality tends to obscure the spatial definition of interior-exterior, as well building and landscape – a conceptual blend that Prijotomo called *bangunan dalam kebun* or "built form in a garden" (2018, p. 21). Spatial continuity between inner living space, landscape, and rice and landscape is the character of indigenous architecture in Indonesia and Southeast Asia. House, granary, and common space/hall are arranged in various composition, such as (1) standing in rows on common space among residential area, (2) separated clusters outside the residential clusters, (3) standing each in front of the houses, and (4) within each house compound, in particular, orientation.

In Toraja in Sulawesi Island, granary is metaphorically called "husband" of the "house" as wife. Granary is always designed with elaborate ornamentation, full of symbolic value, and delicate craft. Domenig (1980, 2005) shows how the structural evolution of the residential architecture of the Toraja in Sulawesi Island, the Batak, in Sumatra Island, and South Japan developed from the earlier conical tipi-like shelter, after which it developed into granary structure with slanting walls and bent-roof, before later it developed into a residential structure.

Residential architecture is typically a free layout wooden building centered in a hearth, with an enclosed place for in-house granary and inner room for the elderly. They generally stand on stilts, with a central post, log ladder, and tiered floor. Indeed, the residential structure slightly has a common form with granary.

As most architecture in the earthquake-prone area, the architectonic traditions incorporate dynamic joints using peg and pen without nail – "shaky architecture" (Prijotomo 2018, p. 73). Jumsai claimed architecture with dynamic and tensile joints as specific characters of Southeast Asian architecture and call it aquatic architecture. The incorporation of bamboo is due to its light and tensile characters (Jumsai 1989, pp. 72–76). The conical and huge structure of the roof is associated with the mountain image.

Fig. 13.1 Three residential architectures and granaries. (**a**) *Uma* and *Lumbung* of the Sasak in Lombok Island; (**b**) *Imah* and *Leuit* of the *Sundanese* in West Java, Java Island; and (**c**) *Tongkonon* and *Alang* of the *Toraja* in South Sulawesi, Sulawesi Island. (Author 2017)

Cosmography of Mountain and River

An urge to assure the stability of the sedentary living technically, socially, and spiritually led to efforts for legitimating dwelling process and construction through some spiritual consecration and schematic interactional concept to install human and their habitation in nature and access to natural sources and through cosmography and various forms of rituals. Hence, dwelling place conceives meanings more than its functionality but a process of initiation and cohabitation with nature. Residence and settlement kept collective memories of the ideal landscape and its future projections. The familiarity to mountain, forest, and rivers is sanctified and represented in conceptual references, such as sacred grove and ornamentation based on foliage of the earlier horticulture plants. Allegories like a mountain as "nail of the land" and abode of ancestors inspired the residential architecture with an overwhelming roof design while recollecting visual memory about the mountain. Animal totems, inspired by amphibious and reptile animals (crocodile, lizard), marked the essence of space underneath as place for the underworld. Along with roof and living space, it creates the tripartite symbolism. All can be narrated in legend and myths. In Java Island, the concept of residential architecture acknowledges inner sacred space that recollects archetypical memory of granary, which is currently left empty and utilized as an ancestral or inner room. The trace of the Javanese granary is still traceable in the vernacular house in Sumbawa Island and other islands eastern ward of Java. The memory of the tree made a central pillar, or the first installed pole became the center of rituals in the house. The fire utilized for hearth preserve food, and the ashes preserve the wooden pole. The complex roles of residential architecture drew the emergence of distinct skill mastering architectural traditions, craft, and knowledge in a holistic manner. Priest-builder artisan thence appeared as *undagi* in Bali, *uragi* in Sulawesi, *tukang* in Malay, or *dulah* in Sunda.

The influence of Indian culture on traditional cultures in Indonesia appears as conceptual appropriation and adoption by the locals than influence as a strategic maneuver to legitimate the existing local power and to advance the corporate estates, including territorializing the habitable sphere. The diffusionist idea of Indianization in Southeast Asia has received many challenges from many scholars. Southeast Asian world had been a sophisticated civilization when Indian culture arrived (Coedes 1968, p. 9). The Hindu-Buddha cosmology rather appropriated to advance the existing vernacular cosmography with the Sanskrit language, Vedic paradigm, therefore allow the appropriation of linga-yoni concept to advance the megalithic traditions into monuments and funerary tower or *candi*. The Sanskrit words like *kuta* and *desa* alter the archaic word *wanua*. Wiryomartono (1995, pp. 22–61) argued that even the concept mosque and *alun-alun* (local urban square) was a result of continuous transformation from prehistoric menhir into candi during Hindu-Buddha period and later mosques during the introduction of Islam (Poerwono 1995, pp. 2–10).

The *meru* concept advances the idea of volcanoes as the spiritual center of the people and villages (axis-mundi) and reference for monuments and religious buildings, temples, residential structure, and even mosque. The serpent "*naga*" was appropriated to advance the indigenous amphibious animal characters, symbolizing the unpredictable mysterious characters of nature. It became the representational alteration for archaic totemic animals, including rivers, ocean, and nymph. The coincidental oceanic characters of the Southeast Asian archipelago setting with the cosmographic concept of *meru* and *jambudvipa* triggered Sumet Jumsai to a speculated possibility that the *meru* concept may have referred from Southeast Asia but refined in South Asia (Jumsai 1989, p. 13).[2]

In a sum, the concept of universal divinity is introduced through Hindu-Buddha to advance the local belief system, bringing new ways in conceptualizing architecture, imbued by symbolic ornamentation and designs. Architecture became then a representational object than pragmatic craft. In Java, the palace incorporates the concept of *meru*, *nagara*, and *jambudvipa* for planning and designing city and palaces. The Buddhist temple Borobudur took after mountain figure and depicted the complexity of human life along the circumambulatory galleries. The Hindu Balinese architecture is full of symbolic ornaments and heavenly figures. The installation of Pagaruyung kingdom in West Sumatra referred to the design of *rumah gadang* architecture but bigger in size and more symbolic in nature.

Two Examples of Confederation Settlements

There are two settlement-architectures elaborated here. They are (1) the Minangkabau's *nagari* settlement and *rumah gadang* architecture, in West Sumatra of Sumatra Island, situated along the highland of Merapi Mountain socially characterized by society practicing matrilineal kindship, and (2) the Bali Aga's settlement and architecture's model, in Bali Island, situated on the highland of Bangli Mountain, characterized by cultural landscape governed by temple network (Reuter 2005, p. 7).

The Bali Aga case represents the architecture of the southeastern groups of islands of Indonesia. The Minangkabau case represents most architecture in Western Indonesia. The myth of origin in both places acknowledges narratives of the traveling ancestors overseas, arriving on a local mountain wherein they reside and dwell. Both Bali Aga and Minangkabau acknowledge the traveling ancestors in their legend, generally traveled and dropped on several places, and initiated villages before the final destinations. In real, indeed the traditional villages developed from few huts into hamlets and later villages, before they distributed in various sectors of the villages. Land, in the perspective of both ethnics, is female. In Minangkabau, they

[2]Given many conical volcanoes in insular Southeast Asia, the coincidence triggers Jumsai (1989, p. 12) to speculate that the concept of jambudvipa was a more recent concept and possibly a conceptual refinement of what had existed in pre-Hindu Southeast Asian. The reverse origin Indic concepts are also stated by many linguistic scholars, like Sylvain Levi, Jean Przyluski, and Bloch in (Bagchi 1929). Pramar (2005, p. 13) also observes that certain substratum of Indian architecture shared characters with the Asiatic people in the East V.S. Pramar.

acknowledge the sequential evolvement of huts (*taratak*), into hamlet (*dusun*), village (*koto*), and confederation villages (*nagari*). In the case of Desa Bayung Gede, village (*desa*) evolved from the grouping and development process of farming huts into hamlets (*pondok*).

Nagari of the Minangkabau in West Sumatra

Minangkabau is an ancient ethnic in Sumatra Island – the second largest island in Indonesian archipelago, administered under West Sumatra Province. Barisan Mountain – stretching along the northwest to southeast of the island divides the island into the west fertile highland facing Indian Ocean and eastern plain running to Malacca Strait. Various ethnics in Indonesia, including the Malay world, Sulawesi, and other small islands of the archipelago, claim Minangkabau as their ancestor. They are prominent with their paddy farming tradition, occasional migration, matrilineal kinship, communal living, and strong Islamic value.

Nagari is a traditional concept of village confederation of the Minangkabau people. The spatial expansion of nagari rooted on the expansion and dispersion of matrikin (*kaum)* and the corporate descent group or matrilineal clans *(suku). Suku* constitutes *nagari's* subgovernance. Each *suku* comprised of matriclans *(kaum)* living in a cluster of house compound called a *kampung. Kampung* cum *kaum* appears to be the smallest social-spatial unit of a *nagari*. A *nagari* must have minimum of four main matrilineal clans *(suku).* Its foci sits in a clan assembly or *surau. Surau* is an education-dorm and house for the male members of the clans or kins member. The location of suraus is generally near riverfront facing the vast paddy field, along with other facilities, such as open-air assembly (*pamedanan*), bathhouse (*tapian mandi*), and cemetery (*pandam pakuburan*).[3] The roof ridge orients to Merapi Mountain.

Intersecting the clan network of a nagari is, village sectors, called *jorong*. Physically a *jorong* appears more like a hamlet in a territorialized area, comprising house compounds (*kampung*) of the matriclans (*kaum*) from various matrilineal clans (*suku*). It appears as a unit settlement on a mound, enclave, or plane in a landscape. In 1979, the modern government of Indonesia took alteration of traditional governance into national administration by making *jorong* the administrative unit. It consequently disrupted the solid governance of *suku* and the cultural integrity of the *nagari* with its matrilineal clans (*suku*). However, it was restored in 1992. Minangkabau is one among few vernacular villages in Indonesia, whose concept sustain in modern administration in Indonesia.

A *kampung* or house compound consists of the ancestral house (*rumah gadang*), *house* (rumah), lumbung (*rangkiang*), and farmland (*sawah*) (see Fig. 13.2b, c about The settlement-architecture of Nagari Rao-Rao). A *kampung* along with paddy field

[3] Complete elaboration of this organization is in my earlier paper (Widiastuti and Vedamuthu 2009, pp. 20–25).

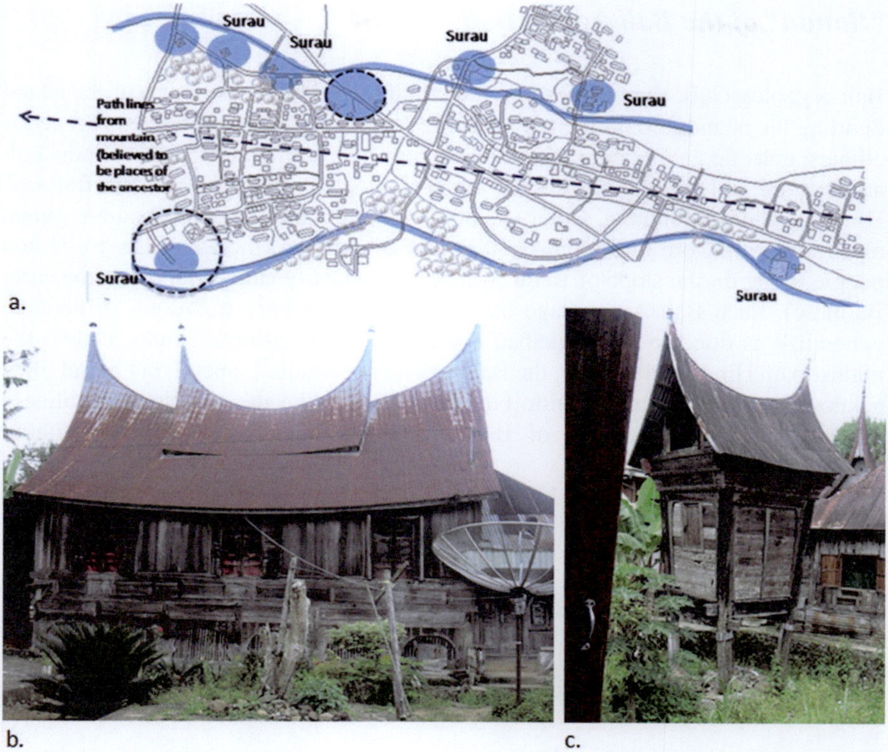

Fig. 13.2 The settlement-architecture of Nagari Rao-Rao, West Sumatra. (**a**) Site plan of a *nagari*, (**b**) *rumah gadang* (Marikin-house), and (**c**) *rangkiang* (granary). (Source: Author 2017)

farmland (*sawah*) is a feminine sphere and entity. Only female members of the mother lineage are allowed to enter the ancestral house or "big house" (*rumah gadang*) and a house (*rumah*). Traditionally, a kampung is more than a house compound for a matrikin member but also for paddy. The rumah gadang contains rooms (*bilik*) dedicated for a married female members. Men theoretically have no space at home; they dwell in a dorm-house (*surau*). *Rangkiang* stands as freestanding granary storage, adorned by elaborated ornamentations.

Rumah gadang (big house) or ancestress house and *rangkiang* or granary are important architectures with buffalo-horn-shape roof, wooden box structure standing on stilts, construction about nail as salient architectural features. Customarily, a *rumah gadang* is centered toward a central post (*tiang tuo* or "old post"). It was the first pole to be installed during construction. Orientation toward Mountain Marapi is reflected in the edge of ridge oriented toward the mountain.

Migration tradition allows men to venture beyond the villages to earn a living and prove his prowess, and economically the migrant-land concept (*rantau*) enables the nagari to extend its social and spatial entanglement outside its cultural boundary and yielded a network that bound regions of the inland (*darek*), seashore (*pesisir*), and migrant land (*rantau*).

"Banua" of the Bali Aga in Bali

Bali is geologically characterized by mountain ranges in northern part of the island dividing the island into northern part facing Java Sea, dominated by heterogeneous ethnics, eclectic architecture, and the southern part, dominated by the Hindu Bali architecture and people. In general, the cosmological orientation of the Balinese Hindu is Agung Mountain. Other smaller and more remote mountains are regional axis mundi for more local communities and sub-ethnics, including the pre-Hindu people living on the slope of Batur Mountain, popularly called Bali Aga (the early Balinese). Each Bali Aga village has their own customary traditions, difficult to generalize as done to Minangkabau and mainstream Balinese Hindu. Unlike the mainstream Hindu Bali people, the Bali Aga do not conduct funeral, but burial; they acknowledge social classification but not caste. Unlike the mainstream Balinese architecture, the architecture of Bali Aga is more archaic and plain without ornaments.

To elaborate further I would bring the case of Desa Bayung Gede as the case of the Bali Aga village in Kintamani region (see Fig. 13.3 about The

Fig. 13.3 The settlement-architecture of Desa Bayung Gede, Kintamani, Bali. (**a**) Boulevard of Desa Bayung Gede, (**b**) *bale pepingitan* (house), (**c**) *jineng* (Granary), and (**d**) merajan (family shrines) at the rear of *pekarangan* (house compound). (Source: Author 2013)

settlement-architecture of Desa Bayung Gede, Kintamani). It is considered the origin of many Bali Aga villages in Kintamani district over Batur Mountain highland and slopes. Like most Balinese village, Desa Bayung Gede has three village temples *kahyangan tiga* (the three heavens), arranged along a boulevard, constituting an imaginary axis running from mountain to ocean, called *hulu-teben* (upstream-downstream) axis (Adiputra et al. 2016). As they do not practice the funeral, they had a cemetery on the *teben* end (downstream). The solidarity unit is sub-desa sectors, each bearing a specific communal role or stratum in the village level. The Bayung Gede village acknowledges a council of women called *Jero Kahyangan* (council of heaven) who are responsible about the ancient shrine, tegalan suci (sacred yard). They are wives of the village leaders and dignitaries. No village affair and plan can proceed without their consent.

However, the wholesome concept of settlement-architecture of the Bayung Gede is a wider *desa*, or a confederation, called *banua*. *Banua* is a confederation village with its elementary settlement unit – *desa* (village) (Reuter 2005, p. 19). There are several *desas* spread over the plain slopes of Batur Mountain ranges, in Bangli district. Each village has a common cosmological orientation toward an ancient temple complex in a village, on the summit of Batur Mountain of Kintamani district. This temple complex binds all these dispersed *desas* in a connected network or confederation – the *banua.*

The Desa Bayung Gede network was initiated by the establishment of simple residential hamlet in the agricultural land, called *pondok*. As it developed into a compound and later clusters, the pondok would soon be established into new *desa*, marked by the erection of temples, that refers to Desa Bayung Gede or temples as their origin. Through time, Bayung Gede formed a firm solidarity network with desas of common origin. The sphere of network is called *gebog satag*. The *desas* member of *gebog satag* must attend to the temple of origin at Desa Bayung Gede. Along with additional nine desas, bound by myth of common origin in different sections of the slopes, the *gebog satag* and the other accompanying desas formed a *banua*. Temple network bound the banua solidarity and provides a social-spiritual field for the *gebog satag*'s member village and banua.

A *desa* or a village physically consists of collections of house compounds (*pekarangan*) in rows facing common pathway boulevard. A house compound or *pekarangan* in Bayung Gede consists of a granary-gazebo (*jineng*) with sitting space underneath generally for males outside the house compound, in common with one or more houses (*bale pepingitan*), kitchen (*paon*), and family shrine (*merajan*) (Laskara 2008, in Widiastuti et al. 2017). A pekarangan is the smallest social-spatial units that build a desa. An array of pekarangan unit stand along the village lanes, standing back to back with another rows facing the corresponding lanes. *Merajan* (family shrines) keeps relics that represent connectivity of the pekarangan with desa, gebog satag's, and *banua* (Fig. 13.3a, b, about The settlement-architecture of Desa Bayung Gede, Kintamani). This row stretched from the highland where temple complexes situated, ran down the boulevard until it ended in cemetery or setra, beyond which is the river.

Discussions

Through time, mountain and river had stayed with the native Indonesian as geo-physical facts and schemata and establish an idyllic image about the ideal home landscape. Tropical climates, earthquake hazard, the tectonic soil, and volcanic mountains shaped the vernacular settlement-architecture in Indonesia. Earthquakes had trained the traditional society with dynamic compressive construction and space structure, in wooden-bamboo tectonic.

Spatially, the archaic settlement in Indonesia was *wanua*. Socially, it could be stretched beyond the boundary, organized in a network, and made into confederation villages, expandable to migrant land. The land and house were feminine and female. In a more advanced sedentary living, supported by agriculture, the memories and perception of the old living culture were reorganized in a cosmology, which was advanced by the introduction of a new belief system. *Meru, nagara, and jambudvipa* system were maintained in Hindu-Buddha belief systems, and introduced the concept of divine king. Later, the influence of Islam transformed the basic assumption about the divine king, as supreme leader and protector of its people and ecosystem. The archipelago ecosystem allows a short distance and slopes between mountain and ocean in between which rivers are running to support villages that emerged and grew along the flow.

Banua and *nagari* are examples of tropical settlement-architecture in Indonesia that appear as confederation village's concept archipelago. Throughout history, these settlement-architecture systems relatively sustain while adopting new values that allowed them to expand and cope with the wider community of global network in the Southern Ocean.

A residential compound in both banua and nagari and most confederation villages are more than the house of human but also paddy, with its guardian inmate, the paddy goddess, Dewi Sri. Given matrilineal kinship, the residential architecture of the Minangkabau appears as house compound (*kampung*) and centered in an ancestral house called *rumah gadang* (big house, Minangkabau). Given cognatic kinship, the residential architecture of the Bali Aga people appears as small house compound (*kampung*) accommodating house of a nuclear family (pekarangan), the inner house *(bale pepingitan*), granary (*jineng*), and family shrine (*merajan*).

For banua and nagari, mountain and river are more than a natural object or sources on their landscape but a primordial image about the universe and how humanity is situated within and dwelling space. Orientation to nature, as reflected in cosmography, orientation and conceptual spatializing indicated attachment of the settlement concept, with nature wherein mountain and river are the important cultural idioms. Currently, these architectures underwent modern advancement incorporating more durable material, bigger size, and more sophisticated ornamentation, but the architectonics characters without nails survived in many places. Rapid urbanization and tourism industrialization that follow soon inevitably change the local perception about their own dwelling culture. It is the modern lifestyle that transforms the idealization of ecology that led to ecological detachment and caused the disappearing mental image of nature as part of human life, including the lost sensibility toward mountains and rivers. Currently, under modern term "cultural landscape," the notion of the mountain and river was reduced into mere "natural sources," "cultural representations," and "irreversible asset." The meaning of paddy also turned away from a vegetal inhabitant of a house into inmates of a food chain.

Bibliography

Adiputra T, Sudaryono S, Wiyono D, Sarwadi A (2016) Konsep Hulu-Teben PADA Permukiman Tradisional Bali Pegunungan/Bali Aga di Desa Adat Bayung Gede Kecamatan Kintamani Kabupaten Bangli, vol 37, no. 1, Forum Teknik, Bali

Ahmad JJ (2010) The Javanese perception of landscape. Universitas Trisakti Press, Jakarta

Asnan G (2016) Sungai dan Sejarah Sumatra. Penerbit Ombak, Jogjakarta

Coedes G (1968) The Indianized States of Southeast Asia (trans: Crowning SB, ed. Fela WF). East West Center Press, Honolulu

Domenig G (1980) Tektonikimprimitiven Dachbau. Materialien und Rekonstruktionzum Phanomen der auskragenden Giebel an alien Dachformen Ostasiens, Sudostasiens und Ozeaniens. Einarchitekturtheoretischer und bauethnologischer Versuch. ETH-Honggerberg, Zurich

Domenig G (2005) Consequences of functional change granaries, granary-dwellings, and of the Toba Batak. In: Schefold R, Nas JP (eds) Indonesian houses: traditions and transformations in vernacular architecture. NUS Press, Singapore, pp 61–97

Domenig G (2014) Religion and architecture in Premodern Indonesia – studies in spatial anthropology. Brill, Leiden

Egenter N (1996) Anthropology of habitat and architecture. In implosion[online]. Available at http://home.worldcom.ch/negenter/. Accessed 25 Mar 2018

Fox J (2006) Chapter 1: Comparative perspectives on Austronesian houses: an introductory essay. In: Fox J (ed) Inside Austronesian house- perspectives on domestic designs for living. ANU Press, Canberra, pp 1–28

Habraken NJ (1987) Control hierarchies – control hierarchies in complex artifacts. In: Proceeding of international congress on planning and design theory. Available at https://www.habraken.com/html/downloads/control_hierarchies.pdf. Accessed 25 Mar 2018

Jarzombek M (2013) Architecture of first societies: a global perspective. Wiley, Hoboken

Jumsai S (1989) The Naga – cultural origin in Siam and the West Pacific. Oxford University Press, Oxford

Munoz PM (2006) Early Kingdoms of the Indonesian Archipelago and the Malay Peninsula. Editions Didier Miller, Singapore

Pramar VS (2005) A social history of Indian architecture. Oxford University Press, New Delhi

Prijotomo J (2018) Prijotomomembenahi Arsitektur Nusantara. Wastu Laras Grafika, Surabaya

Reuter TA (2005) Custodian of Sacred mountain: budayadanmasyarakat di pegunungan Bali (trans: Zainuddin AR). University of Hawai Press, Honolulu

Santoso J (2008) Arsitektur-kota Jawa: kosmos, kultur & kuasa. Jakarta, Centropolis

Sato K (1991) Menghuni Lumbung: Beberapa Pertimbangan Mengenai Asal-Usul Konstruksi Rumah Panggung di Kepulauan Pasifik. J Antropol Indones 49:31–47

Sato K (2015) Introduction to the manifestation of Indonesian wooden architecture. Available at http://www.sumai.org/asia/refer/its2015.htm. Accessed 25 Mar 2018

Widiastuti I, Vedamuthu R (2009) A comparative study of vernacular settlement and dwelling culture: a case study in Kerala, South India, and Minangkabau in Sumatra, Indonesia. Int J Interdisciplinary Soc Sci Ann Rev 4(2):15–30

Wiryomartono ABP (1995) Seni Bangunan dan Seni Bina Kota di Indonesia. Gramedia, Jakarta

Widiastuti, Syamsul AP, Dwijendra NKA, Laskara GW (2017) Perubahan Arsitektur Tradisional Hunian Desa Bayung Gede, Bangli. In: Proceeding "National Seminar of Arsitektur dan Tata Ruang (SAMARTA)". Denpasar, Indonesia.

Chapter 14
Reflection on Rhetorics, Appropriate Building Materials, and Domestic Utilities Towards Reduction of Housing Costs in Africa: A Case of Tanzania

Moses Felician Moses and Livin Henry Mosha

Abstract Africa is facing housing deficit in terms of social infrastructure and number of dwelling units, basically because of overdependence on industrial building materials, technology and commercial energy. Provision of ideal housing includes adequate shelter, safety, sanitation, sewerage, security, privacy and utilities. This chapter gives a reflection on Tanzania's housing national strategies and approaches in solving housing deficiencies. Tanzania is one of the countries in Africa. It is in the eastern part of the continent endowed with vast savannah plains with lakes, rivers, valleys and mountains including Mount Kilimanjaro and Serengeti national parks. Its current population in 2018 is estimated to be 59 million people with a density of 67 people per square kilometre in a tropical climate. The chapter highlights on the use of appropriate building materials such as soil-cement interlocking blocks and sisal concrete roofing tiles and alternative domestic utilities including solar energy and rainwater harvest as an attempt to reduce urban housing cost. The concept of reduce, reuse and recycle of housing resources is emphasized. Reducing is to achieve housing needs by decreasing housing amounts without compromising living standard, reusing is to repeat consuming housing resources instead of throwing, and recycling creates and uses new housing resources from old ones instead of trashing. Architects, engineers, spatial planners and others have a great and important role to play in reducing urban housing cost as the result of critical thinking and innovative professional services deserved to be provided to housing stakeholders especially in the Global South.

Keywords Housing · Building materials · Utilities · Housing costs

M. F. Moses
Tanzania Buildings Agency (TBA), Songea, Tanzania

L. H. Mosha (✉)
Department of Architecture, School of Architecture, Economics and Construction Management (SACEM), Ardhi University (ARU), Dar es Salaam, Tanzania

© Springer Nature Switzerland AG 2020
M. Ghosh (ed.), *Perception, Design and Ecology of the Built Environment*,
Springer Geography, https://doi.org/10.1007/978-3-030-25879-5_14

Contextualizing Housing in Tanzania

Housing supply has never met its demand in developing countries especially in Africa. For instance in 2011, it was estimated that there was a demand of 60 million new house units in Africa to accommodate the rapidly growing urban population. This huge demand for housing complicates national efforts of construction and supply of adequate housing to be a fundamental human settlement issue in all African countries (UNCHS 2001). Despite the fact that the term *housing* is popular to many, it appears that there is no general agreed definition. In 1962 Ad Hoc Expert Group convened by the UN Secretary General stated that "Housing is not a 'shelter' or 'household facilities' alone, but comprise a number of facilities, services and utilities which link the individual and his family to the community, and the community to the region in which it grows and progress". In 1970, another Ad-Hoc Group of experts on housing and urban development concluded that in the fulfillment of social needs, housing plays both direct and indirect role, and both roles are decisive. In its direct role, housing serves as a place where individuals become capable of experiencing community and privacy, social well-being and shelter and protection against hostile physical forces and disturbances. It serves social services such as social-cultural education, recreation, intercourse, sports, social welfare and health protecting services, shopping and transportation (UN 1970). In 1975, at the interregional seminar on the social aspects of housing, it was stated that the concept of housing is more than merely a physical shell. Housing encompasses all the ancillary services and social services, which are necessary and essential to human well-being. Therefore, social services create an integral part of the housing concept and should receive equal or greater attention than the housing unit itself (UN 1975). Housing can be described in broad terms as inclusion of houses, flats and other housing typologies, as well as infrastructure services, the whole residential neighborhood and public spaces. Housing is not just about dwellings, but it includes issues in the localized location, social aspects and domestic utilities. Housing should include facilities and services that affect the site on which a house is built such as roads, drainage channels, water, waste disposal and security of tenure. Housing is therefore a composite complex that provides a heterogeneous mix of services including shelter, safety, sanitation, security, privacy, utilities (such as commercial energy and water), total environment and social culture (Mosha 2011).

Tanzania is considered to be a housing window in this research to explore and expose various housing phenomenon and efforts to establish the magnitude of housing potentials and challenges. Tanzania is a country located in East Africa in the region of African Great Lakes. It is bordered by Kenya and Uganda to the north; Rwanda, Burundi and the Democratic Republic of the Congo to the west; and Zambia, Malawi and Mozambique to the south. The country's eastern border is formed by the Indian Ocean. The country's name comes from the union of Tanganyika which is the mainland territory and Zanzibar islands. These two states united in 1964 and formed the United Republic of Tanzania (URT). Tanzania has a total area of 947,300 km^2 of which land area is 885,800 km^2 and water covered area is 61,500 km^2 (DGW 2016).

The population of Tanzania is increasing at a rate of about 2.7% out of which urban population is growing at a rate of 6% annually (Mosha 2017). Hence, the housing issues become more challenging because population increase is not directly proportional to people's income. In the period between 2002 and 2012, the population of Tanzania increased by 30% from 34.4 million to 44.9 million (URT 2016), and it is estimated to be 59 million by the end of 2018. At the continental level, it is estimated that the population of Africa will reach two billion people by 2040, where the level of urbanization is projected to rise from 40% in 2010 to 50% in 2035 and 58% in 2050 (UN-HABITAT 2014).

Dar es Salaam is Tanzania's largest and richest city and a regionally important economic hub. It is the capital of the Dar es Salaam regional administrative province and consists of five local government areas or administrative districts, namely, Kinondoni, Ilala, Temeke, Ubungo and Kigamboni. Dar es Salaam population increase was 5.6% per year from 2002 to 2012. It is the third fastest-growing city in Africa after Bamako in Mali and Lagos in Nigeria. Dar es Salaam is the ninth fastest in the world. According to the population and housing census of 2012, Dar es Salaam accounts for 10% of the total Tanzania Mainland population, whereby it had a population of 4,364,541 people in 2012. Its population in 2017 was estimated to be 5,502,000. Dar es Salaam is close to the equator and warm throughout the year. The city experiences generally tropical climatic conditions of *hot and humid weather* throughout the year. Annual rainfall is approximately 1100 mm, and in a normal year, there are two rainy seasons: *the long rains* in April and May and *the short rains* in October and November. These varying climatical conditions are part of housing issues and challenges in the context of limited resources, poverty and inadequate infrastructural services.

Poverty in Africa is the greatest housing challenge. The World Bank defines poverty by using the bench mark of persons, living cost of less than 1 US$ per day (Mosha 2017), but Tanzania uses the national poverty line to determine poverty. URT (2014) shows that in 2007, percentage of population below the basic needs poverty line in Tanzania was as high as 34.4 (Fig. 14.1). Furthermore, according to the Household Budget Survey (HBS), the national average food poverty percentage was 11.6; rural had 13.5%, and Dar es Salaam had the lowest of the food poverty line of 1% (Fig. 14.2) followed by other urban areas with 8.7% of people who are below food poverty line. This shows that the economic situation of Dar es Salaam is better compared to other urban areas because it is a major economic hub in Tanzania. Today, the world is struggling to ensure that the world population has access to decent housing facilities. Housing as a human basic need should be available to all people. However, the current situation shows that the problem is getting worse as the world population increases especially in the urban areas. In 2010 Africa's population exceeded one billion though majorities were dwelling in rural areas. It is projected that by 2025, half of African population will be urban dwellers (UN-HABITAT 2010). This situation of rapid urbanization as experienced in Dar es Salaam is now acting as a driving force towards worsening housing situation in many developing countries. However, this article does not focus on rural housing context where the majority of African population dwell. Mosha (2017) indicated

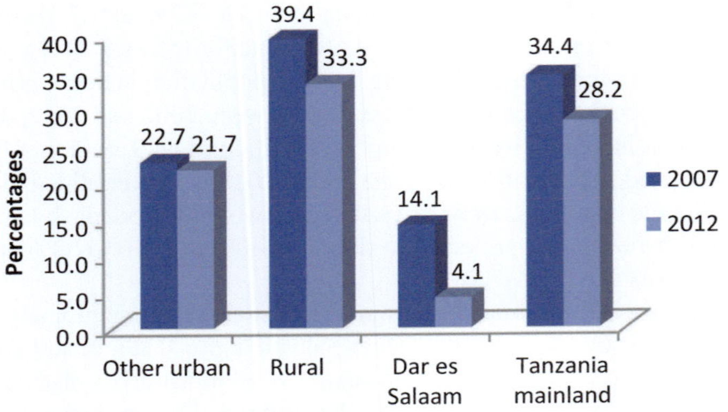

Fig. 14.1 Percentage of population below the basic needs poverty line in Tanzania mainland. (Source: Mosha 2017)

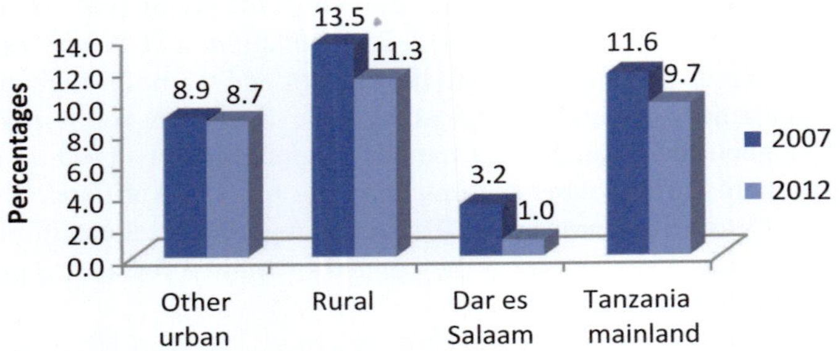

Fig. 14.2 Percentage of population below the food poverty line in Tanzania mainland. (Source: Mosha 2017)

that 79% and 70% of sub-Saharan Africa and Tanzania population, respectively, were rural dwellers in 2017 with average housing condition as shown in Fig. 14.3 and as shown in Table 14.1 expressing use of industrial and/or non-industrial materials in housing. There is a need therefore of critical thinking on possible alternative means and approaches to reduce construction costs to enhance affordable housing concepts.

The current housing deficit in Tanzania is estimated to be more than three million units, and it is growing at a rate of 200,000 units per annum. The problem is more pronounced in urban areas where according to statistics, the urban population has grown from 14.8% in 1980 to 37.5% in 2005. However, there is limited information on how to reduce construction cost of housing as an attempt to provide affordable housing. Lack of knowledge on the effective utilization of both appropriate building materials and utilities has led to unresolved economic burden to residents. This has been a result of overdependence on conventional building materials, technology and commercial energy sources. There is a need to take steps towards provision of

Fig. 14.3 Example of rural housing condition in Tanzania. (Source: 2018 Field Survey in Misungwi Rural Village – Tanzania)

Table 14.1 Household percentages showing usage of industrial and non-industrial building materials in Tanzania

Construction material	Dar es Salaam			Other urban area			Rural area			Tanzania mainland		
	2001	2007	2012	2001	2007	2012	2001	2007	2012	2001	2007	2012
Floors												
Nonmodern	7.6	9.7	3.5	38.8	38	31.6	87.5	84.4	80.1	74.8	68.2	60.6
Modern	92.4	90.4	96.5	61.1	61.9	68.4	12.5	15.6	20	25.2	31.8	39.4
Walls												
Nonmodern	11.5	9.9	2.9	61.7	49.4	32.1	83.3	78.2	66.9	75.3	66.0	51.8
Modern	88.5	89.9	97.1	38.3	50.6	67.8	16.7	21.9	33	24.7	34.1	48.2
Roofs												
Nonmodern	1.8	2.8	0.8	16.3	15.4	9.5	68.7	58	45.2	56.4	44.4	32.3
Modern	98.2	97.1	99.2	83.7	84.6	90.5	31.2	42	54.8	43.6	55.6	70.3

Source: Mosha (2017)

improved housing and related social facilities in both rural and urban areas, especially for low-income groups. Housing professionals are striving to remedy challenges of unplanned urbanization and to undertake necessary urban and rural planning. Particular efforts must be extended to uphold affordable housing concepts in both private and public housing programmes on a self-help basis and through cooperatives, utilizing as much as possible local available building materials, technology and labour.

Different Housing Definitions and Concepts

For the purpose of this study, the term *house* is defined as a structure serving as a dwelling unit for one or more persons, especially for a family. Therefore a *house* as structure for housing has the following basic components: floors, walls, roofs and its surrounding environment. A house is required to act as an entity to keep bad weather from the users of that particular house unit. A house is one of the basic human needs that also act as security and treasured fixed asset as stated in number of human theories of needs.

According to National Housing Building Research Agency (NHBRA), a good house should have the following features: Structurally stable – house stability will depend on the building materials as well as the construction techniques used. If the construction technique used is poor, then, even if the building materials used are strong, the house will have an inferior quality. The house must be durable; capable of protecting inhabitants and their properties from bad weather; able to meet the requirements of all activities which are likely to take place within or outside the building, including cooking, washing, eating and playing; and must meet environmental and human hygiene rules. Human health is very important to all, and hence, a good house should, at all times, be capable to meet health and environmental requirements and capable of meeting privacy and cultural and traditional aspects of inhabitants. These are important because there are some traditional groups in Africa, Tanzania in particular, where parents and their in-laws are restricted to use common washrooms. Likewise for other ethnicities, males and females should not take meals on one table at a time! A good house must be able to meet these cultural values (Mosha 2011). A good house should be attractive in its form. It should have a good visual appealing to both users and passer-by. House possession is a treasured investment and has got its value. A good house should be at the level of investment status which can be saved as collateral and can be sold at a cost worth it and therefore be a source of income.

For the past 53 years after Tanzania's independence, the government of the United Republic of Tanzania has been struggling to ensure that adequate, decent and affordable housing is accessible to all citizens. As an attempt to provide adequate housing facilities which are accessible by the majority of Tanzanians, the government has set out different strategies and action plans to solve the housing deficit in the country. These strategies include formation of the Tanzania Housing Bank (THB), establishment of the National Housing Corporation (NHC), establishment of the Building Research Unit (BRU) now the National Housing and Building Research Agency (NHBRA) and initiated different low-cost housing campaigns to address housing deficit in the country. National Housing Corporation is responsible to provide housing facilities to Tanzanians in a manner that aims at solving the problem of shortage of housing facilities in the country. NHC has the following roles: to construct houses for sale; to facilitate availability of building materials and components; to do business as building contractors, planners or consultants; to rent out and manage houses or properties built by the corporation and those acquired by the government; and to carry out other activities related to construction of residential houses or other public buildings.

Appropriate Building Materials and Technology in Housing

Use of appropriate building materials has a huge cost implication to any building construction. This aspect is the most crucial in controlling and managing construction cost. The choice of appropriate building material can reduce or increase cost of housing and its services. The choice of building materials can positively affect the affordability of a particular housing unit to a person or group of persons at a certain locality. Appropriate building materials should not be subjected to long distance of transportation. The materials that are specified by designers for walls, roofs, foundations, ceilings and other building elements should be selected wisely so as to establish cost implication on housing delivery. For instance, residential building constructed at Kigamboni housing estate is constructed using appropriate building material of soil-sand interlocking bricks which do not need application of mortar or plaster (Fig. 14.4).

The National Housing and Building Research Agency (NHBRA) is a government housing executive agency, which was launched in 2001 under the Ministry of Lands, Housing and Human Settlements Development. This agency has the following main functions: to conduct research on building materials and technology; to collaborate with central and local government authorities, non governmental organizations (NGOs), community-based organizations (CBOs), development partners

Fig. 14.4 Kigamboni residential house built with soil-cement interlocking bricks. (Source: Moses 2015)

and individuals in the formulation and training of grass-root building construction and production teams/brigades; to promote human resources (i.e. technical, financial and managerial) of all actors involved in housing delivery and human settlements development; to ensure that planning, legislation, building regulations, standards and other controls are in place; and to promote the production and use of appropriate local and affordable building materials. Therefore, the agency is focused on training, research and demonstration on how to build strong residential houses with respect to low cost; to conduct intensive research and to make sure the outcome of research on housing is available on time; to give out education to all; and to create knowledge about quality building materials and technology which are found in a particular context. Figure 14.5 shows NHBRA's appropriate wall material of soil-cement interlocking bricks which are produced by a pressing machine (Fig. 14.6) that can make four different types of bricks without cutting. The machine can produce a full brick, a three-quarter brick, half brick and or quarter brick. Dimension of one full soil-cement interlocking brick is 100 × 150 × 300 mm. Table 14.2 provides comparative dimensions of soil-cement interlocking and non-interlocking cement sand bricks. The second alternative appropriate building material is the sisal concrete roofing tiles (Fig. 14.7) which can be made into two different shapes. Table 14.3 provides dimensions of these tiles. House construction cost is reduced significantly when these two alternative building materials are used. Several non-governmental organizations (NGOs) are assisting development of innovative and appropriate materials and technologies in Africa. The International NGO Practical Action (formerly Intermediate Technology Development Group – ITDG) has greatly assisted in increasing housing affordability through changes in low-cost building materials and technologies (Majale 2005).

Fabrication of sisal concrete roofing tiles is achieved by mixing sand, cement and sisal fibre into a mould of 8 mm thick. A workshop table of appropriate technology (Fig. 14.8) produces vibration by the use of car-cell which can be used up to 7 days at a rate of 10 h a day. When vibration is done, the tile paste mixture is placed on a plastic sheet and transferred into other moulds to get the shape of a desired tile. Drying of tiles is done into two main stages, namely, indoor drying which takes 24 h,

Fig. 14.5 Interlocking bricks. (Source: Field study at NHBRA)

Fig. 14.6 Interlocking bricks pressing machine. (Source: Field study at NHBRA)

Table 14.2 Interlocking and cement sand blocks laboratory data

Item	Soil-cement interlocking brick	Cement sand block
Dimension	100 mm × 150 mm × 300 mm	120 mm × 230 mm × 450 mm
Average weight	6.5 kg	2.4 kg
Expected wall thickness	150 mm	120 mm

Source: Moses (2015)

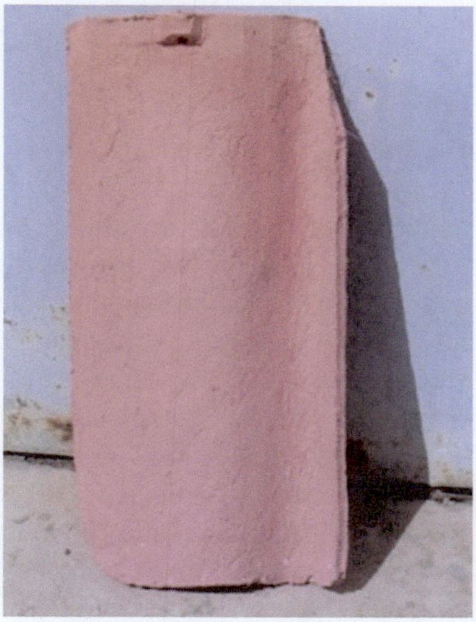

Fig. 14.7 Example of sisal concrete roofing tiles. (Source: NHBRA Field Survey)

Table 14.3 Quantity and specifications of sisal concrete roofing tiles produced per 50 kg of cement

Tile	Size (mm)	Thickness (mm)	Weight (g)	Number of tiles per 50 kg of cement bag	Surface area to be covered (m^2)	Number of tiles (per m^2)
Type 1	250 × 500	8	3553	95	7.60	13
Type 2	300 × 495	8	2993	75	8.25	10

Source: Moses (2015)

Fig. 14.8 Sisal reinforced tiles workshop. (Source: NHBRA Field Survey)

and it is then deepen into outdoor water pond for 28 days. After that, tiles are ready to be used. There are two types of tiles that can be produced in different sizes, shapes and weights but in the same thickness. When production of tiles is carefully done, 50 kg bag of cement can produce 95 tiles of type 1 which can cover roof surface of 7.60 sqm. However, the same bag of cement can produce 75 tiles of type 2 which can cover 8.25 sqm roof surface as shown in Table 14.3.

Appropriate disappearing beam formwork (Fig. 14.9) is a special beam formwork that upon its application, it becomes part of the structure. It is a precast beam formwork made from sand-cement mixture (1:6), and it is initially prepared in soft timber formwork. This type of beam formwork is used for ring beams of a wall structure (Fig. 14.10). Procedure of making appropriate disappearing beam formwork is as follows: Timber formwork is prepared with exact dimensions as the structural wall and height of the ring beam; mix sand and cement at the ratio of 1:6; sand used must be of good quality as that of plastering. Cast the mixture of cement, sand and water into prepared beam formworks and leave it for 2 days to harden, and remove timber formwork within 3 days. Watering should start on the next day and can be done up to 7 days. Figure 14.11 is a construction model which shows use of

Fig. 14.9 Appropriate disappearing beam formwork. (Source: Moses 2015)

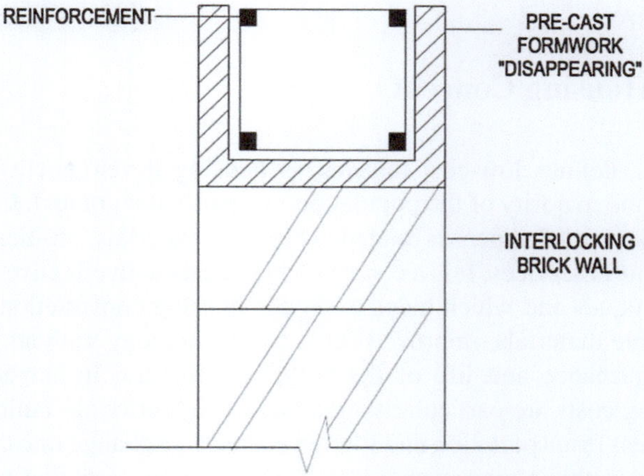

Fig. 14.10 Cross section of a wall to illustrate use of appropriate disappearing beam formwork. (Source: Moses 2015)

soil-cement interlocking brick wall, disappearing beam formwork and sisal concrete roofing tiles on timber truss members. This model provides great knowledge to builders and clients who would wish to use these appropriate building materials and technologies in housing. Construction cost of a structure of soil-cement interlocking brick wall, roofed with sisal concrete roofing tiles, is reduced by 40% from that of a same structure but of sand-cement block wall, roofed with corrugated iron sheets of 30 gauges.

Fig. 14.11 Construction model showing: soil-cement interlocking brick wall, appropriate disappearing beam formwork and sisal concrete roofing tiles on timber truss members. (Source: 2018 Survey at NHBRA)

Low-Cost Housing Concept

UN-HABITAT defines low-cost housing as housing development which can be accessible by the majority of the population in a particular context. Cost of housing is indicated by the total expenses related to building materials, utilities, maintenance and infrastructural services. It is a concept which deals with effective budgeting and building techniques and which helps to reduce housing construction cost by using locally available materials, improved skills and technology without sacrificing the strength, performance and life of the building structure. In sub-Saharan Africa region, housing costs are particularly high due to high prices in building materials, professional fees, transportation and foreign currency exchange rates. About 80% of the population in this region cannot afford construction costs that meet minimum housing standards (Haregewoin 2007). Consequently, UN-HABITAT and few dedicated individual countries devoted their efforts in thinking ways and means of reducing construction cost. Furthermore, UN-HABITAT (2001) indicates that in 1981 the government of Tanzania started a low-cost housing scheme in Dodoma, the new capital of Tanzania, to provide technical and financial support to cooperatives as well as typical house plans, site supervision and construction guidelines. A huge misconception exists that low-cost housing is suitable for only sub-standard works and they are constructed by utilizing cheap building materials of low quality meant to meet poor people's housing demand. Proper management of resources can influence reduction of construction cost of any form of housing to carter for the needs of all income groups. You (2007) assert that there is no universally agreed measure of what constitute "affordable housing" or "low-cost housing" but certainly there are three common measures to all, namely, house price-to-income ratio, house rent-to-income ratio and the residents income assessment. Chapinduka and Cloete (2007) gives an example of Malawi that only 35% of the urban population is able to access finance from formal sector and less than 16% of households in major urban areas can afford construction of an average house. This signals that majority in Africa cannot afford to have an average decent house as stipulated by the UN-HABITAT standards.

United Nations Habitat estimates that 1.1 billion people live in inadequate housing conditions in urban areas. The right for adequate housing is fundamental human right, enriched in various international human right treaties and instruments and applying equally to all people of the universe, but millions of households across Africa, in particular the poor and women-headed households, lack tenure security (UNCHS 1997). In many cities of developing countries, more than half of the population lives in informal settlements, without security of tenure and in conditions that can be described as life threat. Everyone has the right to own property alone as well as in association with others. Everyone has the right to a standard living, adequate housing, food, clothing, medical care and necessary social services and right to security in the event of unemployment, sickness, disability, widowhood, old age or other lack of livelihood in circumstances beyond his or her control. We recognize that access to safe and healthy housing is essential to person's physical, psychological, social and economic well-being and should be a fundamental part of our urgent actions for more than one billion people without decent living conditions in the world. As part of the strategy of the Tanzania Government to provide low-cost housing, the National Housing Corporation in Tanzania has launched a famous project commonly known as "My life my house" in 2012 with the target of constructing 5000 house units. Few of these residential units are already built. Kigamboni Housing Scheme at Kibada in Dar es Salaam is a good example as shown in Fig. 14.4.

Sustainable Construction Towards Reduction of Housing Cost

Main housing utilities in this discussion are water and electricity supply. Commercial energy sources in developing countries are normally inadequate. See, for instance, condition of poor people struggling for water shown in Fig. 14.15, unreliability of power supply due to undersupply and frequent power shedding in Tanzania and elsewhere with similar situations of poverty (Ozolins 2015). People use water for cooking, washing, flushing toilets, gardening, washing cars and so forth. At night, electricity is needed to light residential houses. The same electricity is used to operate different electronics in bedrooms and lounges. Electricity is also used to run mechanical cooling appliances like ceiling fans and air conditioners. However, human beings are able to perform different domestic activities without the presence of commercial utilities in their places of domicile. Bennetts et al. (2003) express that sustainable construction is how a built structure responds to the effects of human activities on the environment, sociocultural and economy of a particular context. For the purpose of comparative study, we can see domestic utilities have direct cost implication on housing. Therefore, proper measures should be taken so that the consumption of different utilities in residential buildings is used at a minimum cost in a reductive manner when needed. For instance, use of smart bulbs allows light to be on only when needed. Lighting security lights during night times should be on only when needed. Sustainable construction sometimes referred as "intelligent buildings" and/or "digital building" uses automated power and lighting control systems which bring about energy saving and efficiency (Watson 2011; Sinopoli 2010).

If the cost of housing is to be kept as low as possible, undue reliance cannot be placed on mechanical means of controlling the indoor environment, but housing must be designed to carefully consider the climate and weather to promote human thermal comfort. Climatic factors, namely, winds, temperature, rainfall and humidity, have to be considered. The building design is supposed to address all calamities that are likely to jeopardize minimum housing cost and maximum human comfort. Sustainable construction emphasizes on the effective use of utilities and building materials not only in cost reduction but also environment consideration. Appropriate orientation of a house can reasonably contribute to human thermal comfort of house users. For instance, orienting bedrooms to the eastern side of the house in East Africa is professionally correct than when they are oriented to the western side where the afternoon solar radiations are maximally absorbed in bedrooms and re-emitted during sleeping time. The utilities consumption within the building should be analysed carefully and properly consumed only when needed. This practice reduces the cost of energy on housing significantly. Well-thought utilization of utilities and building materials can lead to minimum wastes during construction and in the course of living. Housing facilities have to be part of the environment so as to minimize the use of mechanical means to perfect indoor thermal condition. This includes the use of passive measures. In this particular discussion, the type and size of window openings may increase or reduce the cost of power consumption in housing. For instance, sliding aluminium windows which allow only half of the window to open may require air conditioners in Dar es Salaam because of the harsh warm and humid weather conditions. A simple thermal performance analysis conducted to compare first floor window in Fig. 14.12 with its openable aluminium window panel shutters showed double performance level to that of sliding aluminium window glass panel of the ground floor which opens only 50%, hence reducing passive thermal cooling by 50% of that of the first floor window. Residents complained of uncomfortable passive thermal cooling and requested the landlord to introduce openable aluminium window shutter panels especially for rooms without two-side exterior windows. When the thermal performance of a window is poor, it necessitates application of mechanical means of cooling, which in turn have a rise in housing cost to achieve the reasonable human thermal comfort inside the house.

Renewable energy sources have constant supply and are reliable, and their cost is less compared to other commercial sources. Water from DAWASCO (Dar es Salaam Water and Sanitation Company) is the most reliable. Electricity main source

Fig. 14.12 Passive thermal cooling through sliding and openable aluminium window shutters. (Source: 2018 Field Survey)

Table 14.4 Water and electricity sources at Boko National Housing Corporation Estate in Dar es Salaam

Source of water	Number of houses	Source of electricity	Number of houses
Bore hole	0	TANESCO	15
DAWASCO	11	Solar power	0
Water vending trucks	0	Generator	0
Water vending points	4	Other sources	0
Rainwater	0		
Total	15	Total	15

Source: Moses (2015)

in Tanzania is TANESCO (Tanzania Electricity Supply Company) which is a sole source of commercial power in Tanzania. DAWASCO users pay between 15,000 and 200,000 Tanzanian shillings (TZS)[1] water bills per month if there is a constant supply. Table 14.4 shows different water and electricity sources in Tanzania. Water vending (Fig. 14.13) is also practised at 200 Tanzanian shillings per 20 l bucket.

The main concern is how we can plan to incorporate renewable sources such as rainwater harvesting in housing design as shown in Fig. 14.14. Among reliable and cost-effective water sources found during field survey is from rain which plays a great role towards cost reduction in housing. However, major challenge associated with rainwater collection is *how to collect* and *how to store*. Rainwater harvesting exercise needs gutter systems to collect and storage tanks for later use. Rainwater is

[1] Exchange rate: 1 US$ = 2226 TZS as of 25 January 2018

Fig. 14.13 Peoples searching for water. (Source: Moses 2015)

Fig. 14.14 Rainwater harvesting. (Source: Moses 2015)

free and pure. Once harvesting infrastructure is set, the harvest can be performed at no extra cost. For example, it was observed during field survey that one resident in Dar es Salaam had an underground tank with a capacity of 25,000 l, while his water consumption per day is 200 l. This means these 25,000 l can last for 125 days which is equivalent to more than 4 months!

Electric and thermal energies are the two components of solar energy systems, which are useful in reducing housing cost. Photovoltaic (PV) panels as shown in Fig. 14.15 are used as roofing materials and produce electricity. Solar thermal systems in Fig. 14.16 produce heat energy. Installation charges may be expensive but eventually turns to be cheap in the course of use because there is no fuel cost needed, and it has low operational and maintenance cost. Photovoltaic solar panels can be integrated into walls and roofs of residential houses and commercial buildings and reduce consumption of electric load from national power system. Tanzania is within the tropical zone where the sun is in abundance throughout the year. The persisting problem of carbon dioxide emissions can be minimized by the use of solar energy (Mosha 2006).

Fig. 14.15 Use of solar energy in buildings in Mbweni – Dar es Salaam. (Source: 2018 Survey)

Fig. 14.16 Use of solar thermal energy system for heating in Mbweni – Dar es Salaam. (Source: 2018 Field Survey)

Reduce, Reuse and Recycle (3Rs) of Construction Resources Towards Reduction of Housing Cost

Reduce means finding way to decrease or lessen the amount of anything that its absence has no harmful to the performance or attainment of intended objectives. In resource management principles, reduce means using fewer resources in the first place. This is the most effective of the 3Rs. It can be the hardest because it requires letting go of some of the resources we love, but they might be absent and affect nothing. However, we do not need to let go completely or all at once. "Reduce" is a comparative terminology. It says cut back from where you are now, hence "use only when necessary". When resources are not wisely used, they can be termed as being "abused". A resident must choose electronics and appliances that are energy-efficient; control consumption of water at home; waste less energy on lights and equipment; avoid overly packaged goods; use products made from post-consumer recycled materials, especially paper and washroom tissue; turn off lights when you are not using them, this shall save energy; avoid disposable or "use once only" items; and use rechargeable batteries. Reducing will promote conservation efforts and decrease landfill with wastes. Reducing results contribute to less pollution and cleaner environment. It also helps conserve natural resources. Therefore conserving resources, housing dwellers have to reduce unnecessary use of resources.

Reuse means finding way to use things repeatedly instead of throwing them away. We can reduce the purchase of new products by reusing some of them. Before disposing anything, we should consider whether it has life left in it and if it can be used to serve other purposes even after few modifications rather than buying new ones. Reusing methods include repairing damaged items, donating items to less affluent people or finding another way to use them. For example, water used in one activity can be suitable for use in another activity different from the first one. For instance, water used for laundry can be reused to flush toilets. For example, if 50 l of water is used for general cleaning, this same water is reused to flush toilets per day, then the amount of water needed will be 1500 l in a month and 18,000 l in a year. This amount of water is saved if the concept of reusing laundry water is practiced for the purpose of flushing toilets! Figure 14.17 is another example of reusing waste water from soak-away pit in gardening. There is a lot of saving in water bills if water for gardening is tapped from soak-away pits instead of the commercial main water supply source.

Recycling is converting something old to something new. This process involves making new products out of old products. This means potential landfill waste becomes a new product. We can protect the environment by buying products that contain recycling instructions after the first use. Many of the things we use in our everyday life are recycled (Langston et al. 2004). New plastic bottles can be recycled from old ones and can be filled with sand and become construction building material in place of cement sand blocks. Recycling is a process, which uses recovered materials to make new products, and the value of recycling lies primarily in the fact that all products derived from used materials.

Fig. 14.17 Reusing waste water for gardening. (Source: Field Survey 2018)

Conclusion

This article discussed and provided housing rhetoric, appropriate building materials and domestic utilities in the context of Global South in Africa with an in-depth case study of Tanzania. It has been observed that majority of people are poor in Africa; hence poverty is the major housing challenge. In Tanzania, for example, 34.4% of her population are below basic needs poverty line, and they definitely cannot afford adequate housing. Rural communities in Africa are more than 70% of the entire population and certainly cannot afford to buy industrial building materials. Furthermore, 21% of urban population in Tanzania live below basic needs poverty line and cannot afford decent housing units. Even in cities and towns, it is only 48% and 70% of urban population can afford to purchase industrial materials for walls and roofs, respectively. This research promotes use of soil-cement interlocking blocks and sisal concrete reinforced tiles as appropriate building materials for walls and roofs, respectively, to reduce housing cost of dwelling units. Application of renewable energy for lighting and heating as well as upholding principles of "3R", namely "reducing", "reusing" and recycling housing products has significant cost reduction in the construction and utilization of housing resources.

References

Bennetts H, Radford A, Williamson T (2003) Understanding sustainable architecture. Spon press, London
CAHF (2010) Housing finance in Africa: a review of some of Africa's housing finance markets, 2010 year book. Centre for Affordable Housing Finance in Africa, South Africa
Chapinduka N, Cloete C (2007) Lack of affordable housing finance in Malawi's major urban areas. Prop Manag 25(1):54–67
DGW (2016) Dar es Salaam glass works. Unpublished Manuscript, Dar es Salaam
Haregewoin Y (2007) Integrated housing development programmes for urban poverty alleviation and sustainable urbanization the case of Addis Ababa. In: Proceedings of the ENHR International Conference on Sustainable Urban Areas, Rotterdam
IPS (2018) Poverty and housing are common things which out to feature at Davos meeting. In: Tanzania National Newspaper, Daily news, ISSN 0856 – 3812, No. 12,122, Thursday, January 25, 2018
Langston A, Abbott L, Lewis V, Kellett M (2004) Early childhood. In: Fraser S, Lewis V, Ding S, Kellett M, Robinson C (eds) Doing research with children and young people. Sage, London
Langton G, Ding G (2001) Sustainable practices in the built environments, 2nd edn. Heinemann, Oxford
Majale M (2005) Urban housing and livelihood: comparative case examples from Kenya and India. In: Haamdi N (ed) Urban futures: economic growth and poverty reduction. ITGD Publishing, Rugby
Moses F (2015) Architectural considerations towards cost-effective housing in Tanzania – a case of effective utilization of domestic utilities and building materials in Dar es Salaam. Dar es Salaam: Unpublished B. Arch. Dissertation, Ardhi University
Mosha L (2006) Realities of energy efficiency in shelter adequacy in Tanzania. In: Workshop proceedings promoting renewable energies in Africa. International Solar Energy Society (ISES), Dortmund, pp 3–14
Mosha L (2011) Architecture of African dwellings policy influence towards rural-urbanism in Tanzania. LAP Lambert Academic Publishing GmbH, Saabrucken
Mosha L (2017) Questioning the concept of affordable housing – to who in Tanzania. Int J Recent Sci Res 8(5):16866–16871
NHBRA Brochure (2018) Soil cement construction for low-cost housing. Unpublished Brochure, Dar es Salaam
Ozolins P (2015) Sustainability and scarcity: a handbook for green design and construction in developing countries. Routledge, New York
Sinopoli J (2010) Smart building system for architects, owners and builders. Elsevier, Burlington
UN (1970) Report of the Ad Hoc Group of Experts on housing and urban development, Series no. 63.IV.1. United Nations Publications, New York
UN (1975) Social programming of housing in urban areas, Series no. E.71.IV.10. United Nations Publication, New York
UNCHS (1997) The Istanbul declaration and the Habitat agenda, United Nations Centre for human settlements. Nairobi, HABITAT
UNCHS (2001) Synthesis of national reports on the implementation of the Habitat agenda in the Africa region. Special Session of the General Assembly for an Overall Review and Appraisal of the Implementation of the Habitat Agenda, New York
UN-HABITAT (2001) Shelter co-operatives in Eastern and Southern Africa: contributions of the co-operative sector to shelter development. Nairobi, UN-HABITAT
UN-HABITAT (2010) The state of Africa cities. Nairobi, UN-HABITAT

UN-HABITAT (2014) Sustainable building design for tropical climates: principles and application from Eastern Africa. Nairobi, UN-HABITAT

URT (2014) Tanzania mainland household budget survey main report 2011/2012. Government Printers, Dar es Salaam

URT (2016) Tanzania demographic and health survey and malaria indicator survey 2015–2016. National Bureau of statistics Publication, Dar es Salaam

Watson A (2011) Digital buildings – challenges and opportunities. Adv Eng Inform J 25:573–581

You N (2007) Making the market work for pro-poor urban housing. 1st Asia-Pacific Housing Forum, Singapore

Chapter 15
Design, Form, and Ecological Characteristics of the Traditional Cunda Houses in Anatolia

Ayten Erdem

Abstract Cunda, a small island in the Aegean Sea, was incorporated into the Ottoman Empire in the fifteenth century. It is home to an array of natural, archaeological, and urban site areas with its monasteries, churches, mills, stores, and houses protected in large measure. The traditional settlement is situated in the southern tip of the island. The urban architecture created by the "Rum" population is magnificent, especially in the shore environs. The two- and three-story houses with their stone walls, monumental doors, balconies, projections, stone consoles, and iron supports are especially remarkable. Most of the neoclassical style houses have an enclosed hall reached by two doors: one door an entrance to depots known as "storerooms" and the other into the house proper. In the past, the inhabitants of these houses who engaged in the production of such goods as grapes, wine, olives, and olive oil used these depots for storage of their products. This tradition was continued in the same way by the new home owners after the exchange of populations. This study represents a research and documentation of the architectural formation of these protected Cunda houses.

Keywords Alibey İsland · Cunda · Grid fabric · Neoclassic houses · Sarımsak stone

Introduction

Our present time is one in which we are witnessing the rapid destruction and disappearance of both our natural environments and energy resources. Upon the recognition that the kinds of artificial environs being created to replace these nature-attuned developments are resulting in an alienation of man with nature, developers are now using the concept of "sustainable environments" in their designs of urban spaces that can be perceived as "green" and ecologically balanced. Current efforts also serve to underline the importance of these traditional settlements, which themselves were

A. Erdem (✉)
Architectural Faculty, Restoration Department, Yıldız Technical University, Istanbul, Turkey

© Springer Nature Switzerland AG 2020
M. Ghosh (ed.), *Perception, Design and Ecology of the Built Environment*, Springer Geography, https://doi.org/10.1007/978-3-030-25879-5_15

based on such ecologically attuned conceptualizations. The principle characteristic of the homes found in traditional settlements is that they are architectural products that emerged without architectural support. Categorized as "vernacular" or "regional" architectural architecture, such traditional constructions reflect responses to such factors as the unique geographical, climate, water availability, and soil of the area and especially to the abiding culture of the area and the kinds of construction materials easily available. Unfortunately for Turkey, this country has been late to recognize the importance of the kinds of architecture that emerged through generations of shared knowledge and experience and were built by master builders who utilized basic tools and materials. Although Turkey is located in the northern hemisphere, it has many historical and ecological settlements that they have still continued traditional architecture, lifestyle, production techniques, and culture such as global south countries.

This study thus chose to focus its interest on Cunda Island, one of Turkey's unique traditional settlements and one that is made up of especially beautiful natural, archaeological, and urban preservation sites. This settlement is composed of structures that, while inclusive of the Ottoman Empire, originate from a culture that differs from the culture of Ottoman Turkish Muslims and that reflect the Empire's Orthodox Greek population.

Cunda Island lies in Anatolia along the shores of the Aegean Sea to the south of the Gulf of Edremit. Administratively attached to the Province of Balıkesir and the County of Ayvalık, Cunda is the largest of the 22 islands belonging to the Ayvalık archipelago. While termed an "island," Cunda is actually a peninsula that is narrowly accessible by both land and sea. Located in the Mediterranean climate region, its traditional settlement is situated in the southern reaches of the island, giving it a climate that has both mild temperatures and is protected from the winds. Hot and dry in the summer months, Cunda is cooled by western and southwesterly breezes (Balkan 1997, p. 71). In 1989 the whole of the traditional settlement was certified as an "urban preservation site." Its present economy depends on fishing and tourism and on the production of olives, olive oil, and wine.

History of the Settlement

Excavations carried out in Cunda and its environs have uncovered artifacts stretching back to the Bronze Age (Beksaç 1999, p. 10). In the fifth century BCE, the Greek historian Herodotus referred to Cunda and the other islands in the archipelago as Eflooronisos/Heflooronnesoi/the Hundred Islands (Bayraktar 1998, p. 11). It was during this period that the ancient city of Nesos (whose ruins are still visible today) was founded on the eastern side of the island. This city continued to flourish during the Hellenistic, Roman, and Byzantine empires, but through time it gradually declined and was abandoned. It was replaced in the tenth century AD by a new city called Moshanisia or Moshonesi, meaning "fragrant island" (Atilla and Öztüre 2004, p. 70). In 1430 the region fell under Ottoman dominion, at which its Muslim population began referring to the settlement as "Cezire-i Yunda/Yunda Island." The name "Yund" which is a now an obsolete Turkish word meaning "stray stallion," or "herd of horses," was first used by Piri Reis (Tekin 2002, p. 47). Over time the name Yunda underwent a phonetic change, becoming Cund and Cunda.[1]

From the mid-seventeenth century on, Ayvalık and Cunda served as some of the most important of the Greek-speaking enclaves on the Aegan Sea (Akın 2005, p. 18). Thanks to a special inducement provided by the Ottoman authorities, the region enjoyed a partial immunity benefit (Bayraktar 1998, p. 6). In the nineteenth century, industrialization and developments in sea trade led Ayvalık to a position of some importance among Mediterranean cities (Akın 2005, p. 22). It is certain that these developments also left a mark on the Cunda Island settlement. In 1862 Cunda was classified as a municipality separate from that of Ayvalık and in 1908 it was reclassified as a sub-district/county (Yorulmaz 2004, p. 158). Both Cunda Island and Ayvalık were occupied by Greek forces during the years spanning 1919–1922. In the aftermath of Turkey's War of Independence, the Lausanne Peace Treaty of 1923 included stipulations for an exchange of minority populations between Turkey and Greece.[2] A referendum voted on by the local population of Cunda in 1952 resulted in the administrative independence of Cunda from Ayvalık. In 1980, the island was officially renamed "Alibey," honoring the Turkish commander who freed the island from its occupying forces during the War of Independence (Bayraktar 1998, p. 17). Today the "Historical Preservation Site" on Cunda includes 551 examples of civil architecture and 18 structures and sites of religious, cultural, and natural importance, all of which have been deemed worthy of preservation.

Characteristics of the Traditional Fabric

Cunda represents an apt example of a small Greek/Orthodox settlement with its churches, mills, shops, and houses (especially those constructed in the second half of the nineteenth century) all bearing traces of antique period architecture. The characteristics of both its street and general island layout, along with its architectural features, differ significantly from those of Ottoman cities that are more reflective of Turkish/Islamic integrations, and do so in both architectural form and in choice of materials. In place of the organic layouts of Turkish towns and cities with their winding streets, varying and differing layouts of land utilizations, and wooden framed and courtyard houses with multiple facades, Cunda reflects a town of grid layout, rectangular lots, and two facade stone-walled houses with backyards. One

[1] The word Cunda is of Italian origin and is a maritime's term signifying the tip of a horizontal object, the staff or a ship, or the staff upon which the ship's sail is hung (Akın 2005, p. 112). It remains unclear why this area was given an Italian originated name.

[2] The Greek-speaking inhabitants/Rums of Ayvalık and Cunda were resettled in mainland Greece, while the Turkish-speaking inhabitants of the islands of Crete and Lesbos were settled in Ayvalık and Cunda (Bayraktar 1998, p. 17).

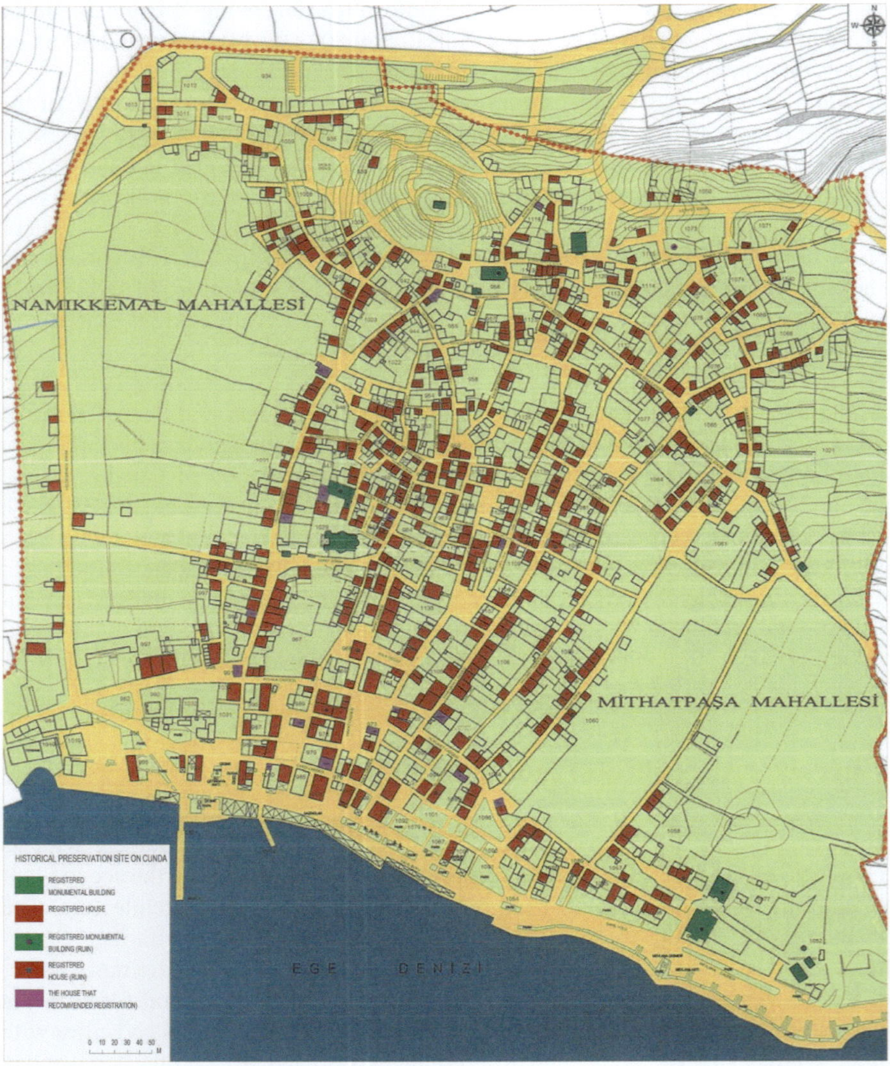

Fig. 15.1 The map of Cunda settlement

Cunda settlement sits on both the flat lands that line the shore and on sloped lands that move further inland (Fig. 15.1).

The highest point of the settlement perches on a hill *Âşıklar* Hill. It is believed that the island's earliest settlement was situated on that hill. Today this windy site is home to a number of edifices, including wind mills, the churches of St. *Iannis* and *Panagia* and the latter's secondary buildings, and the Greek Girls' School. The southern slope of the hill is home to the one *Taksiyarhis* Church, which is located

Fig. 15.2 A view of island from south (Aegean Sea)

nearest to the shore and the Greek Boys' School and administration buildings which are in its immediate environs (Fig. 15.1).

At present, the traditional settlement has two main neighborhoods: the commercial center located along the southern shores of the sea with its one- and two-story stone shops and the harbor itself. Among the primary and most monumental edifices in this neighborhood—none of which exceed two stories and therefore do not obstruct the silhouette of the overall view—are the Taş Coffeehouse, the Home of the Despot (Figs. 15.2 and 15.3), and the city's sole mosque, the *Hamidiye* Mosque (Fig. 15.18).

Leading up from the shore, the main part of the settlement reflects a grid plan of parallel and vertical streets and crossroads. As one nears the *Âşıklar* Hill and *Taksiyarhis* Church, these grid pattern roads give way to a more organic fabric of winding roads and pathways. These narrow streets are lined with stone, row houses that face the street in front (Fig. 15.4). Only a handful of homes are located in the middle of large yards (Fig. 15.5). The settlement's green areas consist of the gardens in the backyards of the houses. A few examples of stand-alone houses with both back and side yards can be found on some of the corner lots. Among those streets that have preserved their traditional features, we can name Halk, Namık Kemal, and Cumhuriyet avenues (Figs. 15.6, 15.7, and 15.8). These avenues, which run down to the coast, are home to some of the most beautiful examples of Cunda homes in the shore area, homes which display both the strength and the wealth of their inhabitants. The homes in the vicinity of *Âşıklar* Hill are smaller and more modest in proportions (Fig. 15.9).

Fig. 15.3 A different view from south

Fig. 15.4 Row houses, Selamet Street

Fig. 15.5 A stand-alone house, Selamet Street

Fig. 15.6 A view of Cumhuriyet Street

Fig. 15.7 Selamet Street view from south

Monumental Buildings

Especially noteworthy are the church buildings of the traditional settlement. While in the past Cunda was known to have seven churches, remaining today are the church of one *Taksiyarhis* (1873) and the *church of St. Iannis* and *Panagia* (Yorulmaz, 2004, pp. 167–168). The *church of Taksiyarhis* ranks the most magnificent of all of the buildings on the island and also serves as one of its primary symbols (Figs. 15.10 and 15.11). Constructed of stone and brick, this large church reflects the enclosed Greek cross plan and is of neoclassical style with three naves and one main dome. When it lost its congregation, the church remained unused for a number of years. The static cracks occurred on the walls, vaults, and dome of the church due to the 1945 and 1999 earthquakes. Repaired in earthquake in 2013–2014, the building today serves as a museum. A fountain known as the *Aşağı Çeşme* (Fig. 15.12) is situated on the southern corner of the *Taksiyarhis* and facing *Şeref Street*. Ranking as the best example of the settlement's fountains, it is especially appreciated for its stone workmanship and ornamentation. The small and rectangular *church of St. Iannis* (Figs. 15.13 and 15.14) is located on *Âşıklar Hill*. Both it and its adjacent windmill fell into disuse and disrepair after the population exchange, but both

Fig. 15.8 Vertical Street to the sea

structures were repaired in 2007. Today used as a municipal library and cafe, their hilltop location leads them to serve as the crown of the settlement and completing elements to the settlement's silhouette. The *Panagia Church* is in ruins today (Fig. 15.15).

The *Taş Coffeehouse* is the best known and most eye-catching of all the shops in the commercial district (Fig. 15.16). This one-story structure of neoclassical style has a semicircular arched door, and its windows are framed with pink-colored casings made of stone quarried from the nearby district of *Sarımsak* (garlic) and locally known as *Sarımsak stone*. The building is assumed to have been constructed during the second half of the nineteenth century.

Yet another Greek/*Rum* building is the *Despot House* (1862), a stone-walled, two-story plus basement house situated along the settlement's shore to the south (Fig. 15.17). Upon the death of the despot (1877), the home was purchased by the Ottoman authorities and used first as an administration building and later as an orphanage (1921). In the early years of the republic, the building was used as both a dormitory and a primary school but was abandoned to its fate in 1980 (URL-1). Today efforts are being made to plan for its eventual restoration and repair. The square and single dome *Hamidiye Mosque* was built in the Ottoman period in 1905 to serve the small number of *Turk* and Muslim population of the settlement

Fig. 15.9 Modest houses on Aşıklar Hill

Fig. 15.10 Taksiyarhis Church, before the last restoration

Fig. 15.11 Taksiyarhis Church, after 2014 restoration

(Yorulmaz 2004, p. 178). Completely restored in 2016, the mosque is again open for worship (Fig. 15.18).

Traditional Houses

The traditional houses of Cunda generally consist of adjacent row houses situated at the front of a rectangular lot (Figs. 15.19 and 15.20). The houses located on Lovers' Hill and its environs are oriented in such a way that they face the sea, rather than the street or the yard. The area to the immediate north of the commercial district along the southern shore is home to the settlement's finest homes built on spacious plots of land (Fig. 15.21). As the streets stretch upward toward the hill and its environs, the size of the plots grows smaller, and the houses themselves become more modest in architectural stance (Fig. 15.22).

These traditional Cunda houses are, generally speaking, either two- or three-storied. Extant modes are of two floors +ground floor, two floors + mezzanine floor, one floor+ground floor, and single floor houses (Figs. 15.19, 15.20, 15.21, and 15.22). One of the doors opens onto the ground floor or the *mağaza*, which is the

Fig. 15.12 Aşağı Fountain, Şeref Street

Fig. 15.13 St Iannis Church, before the restoration, Aşıklar Hill

Fig. 15.14 St Iannis Church, after the restoration

Fig. 15.15 Panagia Church, Aşıklar Hill

Fig. 15.16 Famous Taş Coffeehouse, on the shore

Fig. 15.17 Despot House, on the shore area

Fig. 15.18 Hamidiye Mosque, on the shore area

Fig. 15.19 The row houses, Bahar Street

Fig. 15.20 A house on the corner

Fig. 15.21 A characteristic house on the shore area

Fig. 15.22 Modest houses in Cunda

regional term for a storeroom, while the other door leads to the entry hall of the house (Figs. 15.23 and 15.24). Some houses have only one entry door, and access to the *mağaza* is also accessed from this door (Fig. 15.19). In some instances the ground floor is slightly below street level, and it is here that the storerooms, used to house large earthenware jugs that hold such products as olives, olive oils, grapes, etc., are located. Most often the kitchen, laundry, cistern (Figs. 15.25, 15.26, and 15.27), or water well are also located on this lower-level ground floor. The cisterns are filled with rainwater that flows from roof pipes to these containers. These ground floors open both to the street and to the garden of the house. When the house has been built on a steep slope, this ground floor is located at basement level. The ground floors and upper levels of some three-story homes are sometimes connected via a service staircase, while some of the houses have a side door leading off the main entrance hall that connects to this level. Most of these houses have a back garden that is protected from view with a high wall (Fig. 15.28). These gardens are used to serve as spaces for a house, toilet, kitchen, laundry, etc. Some of these gardens also have cooking hearths and water wells.

The mid-level and top (main living) floors of these houses were used as the household's actual living and sleeping quarters. In addition to the connecting hall (the *sofa*) and living rooms (Figs. 15.29, 15.30, and 15.31), the mid-level floor could also include a kitchen. A balcony was included in the floor in which most living functions were carried out. The height of the floors increased as they moved

Fig. 15.23 A house with two doors, Şeref Street

Fig. 15.24 "Mağaza"/storeroom, Ali Kesebir house

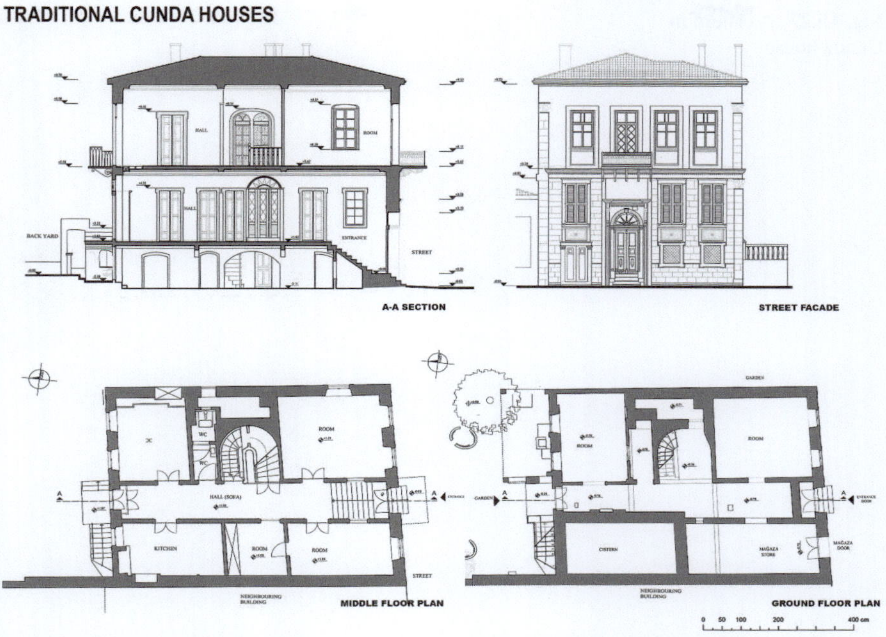

Fig. 15.25 Plans of ground floor and middle floor, the section and street facade of a traditional Cunda house on Ayvalık Street

Fig. 15.26 The kitchen in the mid-level floor Müjdat Soylu house

Fig. 15.27 A cistern in Cunda house

Fig. 15.28 Back gardens of Cunda houses

Fig. 15.29 The narrow hall provides access to the rooms

up from ground to the highest level. While the ground floors generally ranged from 1.90 to 2.00 m in height, the floors used for living could be 3.5 m or more tall. In three-story houses, the main door generally led to the mid-level floor, while the main door led to the ground floor in two-story homes. Because many of the homes were built on sloped parcels, two-story houses appeared from the street side to be of one story (Fig. 15.32), while from the garden, it was evident that it was a two-story structure. In these kinds of houses, the main door opened onto the living floor, which was at street level.

Plan Characteristics

The plan categorization of the Cunda houses is determined by the upper level "actual living space floor," while the other floors were laid out in a way that corresponded to this plan arrangement. In the region of Cunda, the "sofa," the space which serves both as the determinant of the plan type and from which the individual rooms are accessed, could be situated either in front of the rooms, in between the rooms at the

Fig. 15.30 Another narrow hall "sofa"

corner of the rooms, or in the middle of all of the rooms and was designed as a protected space (Fig. 15.33). In other words, the plans of the living room floors were categorized according to their sofa layout as an inner, outer, outer and inner or L-shaped, corner, and central sofa-type house. Just as the sofa could be vertical to the street, as a buffer between the street and the backyard, the sofa space could also be parallel to the street. The sofas of most of the Cunda houses are relatively small, making them areas to access rooms, rather than serving as living spaces. In these houses one of the rooms was always larger than the others, and it was this space in which family members could gather and sit together. This room faced either the street or the backyard.

The most commonly encountered type were the houses with a rectangular, "inner sofa" that extended from one end of the house to the other (Erdem et al. 2007, p. 82). In this plan, the rooms were accessed from the long ends of the sofa, while the balcony and staircase were accessed from the short ends. A maximum of four rooms could be accessed by the long sides of the sofas. The sizes of the rooms varied, but, generally, doors were also used to interconnect the two large back-to-back rooms. In some examples of this plan type, the rooms on either side of the sofa that

Fig. 15.31 A view of the room

face the street are built to project over the street itself. The sofa that is between these rooms ends in a balcony that projects slightly further than the room projections.

The plan of house in which the inner sofa ran parallel to the street was most often utilized in houses situated on lots that had narrow street facades. In this form, one room faced the street, while second room faced the backyard. Along with those examples that included two rooms to the front and two rooms to the back, for a total of four rooms, there are also examples of a single large room facing the street that was used as a formal living room, or *salon*, thus providing an example of a three-room plus sofa plan.

In contrast to the rectangular sofa plans, the removal of one of the back rooms facing the yard resulted in an L-shaped sofa (outer and inner sofa are together). Yet another frequently seen type is the corner sofa model. This arrangement utilizes a sofa with a staircase situated off one of its corners, while rooms were accessed from the other three corners. In this type, the sofa could be on the street side or on the side facing the yard.

In the "outer, enclosed sofa plan," the sofa is situated along either the street or yard side or parallel to the street and side. In the central sofa plan type, the sofa is situated in the middle of the rooms. In these plan types, the sofa or one of the rooms facing the street or backyard may sometimes include a balcony.

Fig. 15.32 Single-story house from the street level

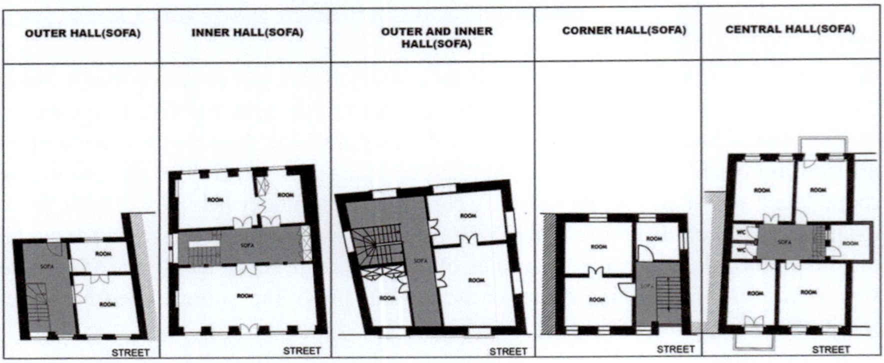

Fig. 15.33 Plan types of the traditional Cunda houses

Facade Characteristics

Most of Cunda's two- or three-story row houses have two facades, one facing the street and the other facing the yard (Figs. 15.34 and 15.35). If the house is situated on a corner, or in the middle of the garden, the number of facades will, of course, increase (Fig. 15.36). While the street facades of the Cunda houses appear to be

Fig. 15.34 Street facades of the houses, Selamet Street

Fig. 15.35 Back facade of the Cunda house

Fig. 15.36 Corner house, Şeref Street

symmetric, they generally are of asymmetric arrangements (Fig. 15.37). Despite that, there are also examples of homes laid out with a symmetrical understanding (Fig. 15.38). There are also examples of adjacent row houses that share the exact same symmetry (Fig. 15.39). It may be assumed that these houses were most likely the property of close family members and were built to be twin houses.

The general features of these facades are best reflected in the houses' own entrance/street facades (Figs. 15.34 and 15.37). The main entry doors comprise the most important identifying features of the street facades (Fig. 15.40). Generally speaking, the doors are set back within a protective niche providing these entries with a sense of decorum. In most cases, yet another door opens from this street facing facade and is usually situated to either the left or right of the entry door. This second door leads to the house's storeroom "*mağaza*" (Fig. 15.41). Smaller in dimension than the entry door, this "*mağaza*" door most always sits at street level. The formal appearance of the entry door is often further enhanced with the addition of wrought iron railings and iron bracket-supported balconies (Fig. 15.37).

Door and window casings, pilasters reflecting architectural styles of the antique period, and projections supported by brackets often made of the local pink stone referred to as "Sarımsaklı stone," served enliven the painted and plastered facade

Fig. 15.37 Asymmetric facade, Ayvalık Street

wall surfaces. There are also examples in which all faces or basement faces are left unplastered (Fig. 15.42) or the street facades are completely or partially sided with Sarımsaklı stone (Figs. 15.43 and 15.44). While not used as frequently as balconies, there are also examples of homes with projections supported by iron brackets or carved stone consoles.

Less often seen, but evident nonetheless, are homes with both projections and balconies. Ground floor windows tend to be square in share, while upper floor windows are rectangular (Fig. 15.45). The houses are secured with the use of wrought iron bars covering the ground floor windows and wood or metal covers or shutters used to secure windows on upper floors.

The side exterior house walls are constructed of rough stones and are usually left unplastered (Fig. 15.35). The cornices are sometimes used between the floors, while pilasters are used at corners, at facade connecting points, and around or two sides of the door niche as accents (Figs. 15.34 and 15.37). The facades are finished off with moldings that lead up to the eaves and with roofs covered with Turkish-style tiles. Despite the fact that the row houses are adjacent to one another, some of the houses have slanted roofs, while others have gable roofs and no eaves.

Figs. 15.38 and 15.39 Symmetric street facades

Doors The magnificent entry doors especially seen in the homes of wealthy inhabitants are usually situated above street level and are protected within a niche and accessed by a low flight of stairs (Figs. 15.46 and 15.47). Sometimes one or two of these steps extend into the street. Depending on whether the house has a basement or not, this staircase can be made up of five or six steps. Besides the houses with entry niches, there are also examples of houses with entry doors at the same level as the house facades (Fig. 15.45). The spaces above the entry and balcony doors have small windows that allow light to enter the interior. The entry doors are double-leaf doors, while the "*mağaza*" door leading to the storeroom is a single- or double-leaf door constructed of wood. The lower portion of the entry door leafs has a platform table, while the upper part has the iron bar-secured window that can be opened. Some of the entry doors have a plaque providing the date of the house construction attached over the door. Almost each door of every house in Ayvalık and Cunda is adorned with a door knocker carved with an image of one or other of the ancient Greek gods and goddesses (Fig. 15.48).

Figs. 15.38 and 15.39 (continued)

Windows The windows of the Cunda house are encircled by pink Sarımsaklı stone casings (Fig. 15.49). Simple brackets/cornices have been placed above the upper casings. The exterior facades of the rooms and sofa windows are rectangular in appearance, while arched windows have been used in the interior spaces. The windows opening into the ground floor "*mağazas*" are square in shape. Room and sofa windows are either sliding or casement in type. Whatever the type, all of the room or sofa windows are protected from outside effects with the addition of wood or metal coverings or shutters (Figs. 15.43 and 15.56). The storeroom/*mağaza* windows are protected with a local version of a covering of oblique, wrought-iron square bars (Fig. 15.47). Along with these commonly seen features, there are also examples of facade, the windows that are arched (Figs. 15.23 and 15.52) or that rise in a triangle, and the ground floor windows are also protected with steel bars.

Balconies Street-facing balconies are generally situated in such a way that they extend over the entry doors, thus providing an eaves for the entryway (Figs. 15.43 and 15.50). Most commonly seen are balconies that lead from the sofa; these wood-floored balconies are generally held in place with wrought or cast iron supports and sometimes stone consoles, and their rails are also made of wrought or

Fig. 15.40 The main entry door

cast iron. In some cases, both stone consoles and iron supports are used together (Fig. 15.64). The solid balcony doors, with their door jambs, are topped with a small window designed to allow light to seep into the connecting room. While most balconies overlook the street, there are also examples of balconies that look onto the back or side yards at the main living floor level.

Projections Rectangular shaped, room projections that lead off from the living rooms are used as a means to provide better viewing of the street or to increase the views of two rooms. These projections are supported by wrought iron supports or stone consoles (Figs. 15.51 and 15.52). Sometimes homes with wider fronts often have projections at their corners. In the middle of the facade, these kinds of homes have a balcony that extends even further over the street than the extension. Narrow-faced houses may have a single extension situated at either end of the facade (Figs. 15.53 and 15.54). In addition to these types, there are also examples of corner situated homes with projections on both front and side facades (Fig. 15.44).

Fig. 15.41 The mağaza/storeroom door

Construction Characteristics

Cunda houses are constructed by using a traditional masonry system based on the use of both stone and brick. The 50–60 cm exterior walls of these houses rest on a rough-stone foundation system. These walls (also of rough stone) continue all the way up to the roof (Fig. 15.55). In some cases, the walls are of rough stone up to the ground floor and then, after the ground or mid-level floor, change to a brick wall construction (Fig. 15.56). Also, in some cases wooden posts are placed between the brick or stone walls at determined intervals (Fig. 15.57). The dividing walls are of either wooden framing or single brick-type construction (Fig. 15.58). Lathes are nailed into the wooden framed walls and then plastered over. The dividing walls on the upper floors and the staircase leading up from the ground floor are carried by the stone dividing walls, stone pillars, arches, and vaulted flooring laid in certain parts (Figs. 15.24 and 15.58). In the construction of the rough stone walls, generally speaking, the corners of the house walls are made up of large-cut stone blocks, while the rest of the wall is built by placing smaller pink-colored, rectangular stones among larger rough stones (Figs. 15.59 and 15.60). The stones used in the construction include the black-colored local stones found on the island, the pink Sarımsaklı

Fig. 15.42 Unplastered facade

stones, and sometimes the yellow-tinted "tulip bulb" stones (Fig. 15.61). The Sarımsaklı stones are mostly used to frame door and window ways, in pilasters and in house projections supported by consoles. The street-facing facades of some of the houses have been covered with Sarımsaklı stone, while there are also a few examples of houses whose rough stone walls are built entirely of these pink, local stones. Mud plaster was used to bind the stone and brick walls (Fig. 15.62), and the surfaces of these walls have been covered with mud mortar and after lime-based plaster on both the exterior and interior. In later constructions it is seen that mud plaster was replaced with lime plaster as a binding element.

Because the houses are constructed of stone and brick materials, there are only a few examples of houses with projections leading over the street or the yard, and those projections are supported by stone consoles (Figs. 15.52 and 15.53). Also, examples of wood frame houses with projections are extremely rare (Figs. 15.51 and 15.54). In contrast, the frequently seen balconies are seated on iron framework or iron beams supported by iron supports and, in some cases, by stone

Fig. 15.43 A facade covered with Sarımsaklı stone, Selamet Street

consoles (Figs. 15.63 and 15.64). Balconies are either floored with wood or with marble. The window and entry door openings are framed with brick arches (Fig. 15.65); however, the windows on the exterior facades are rectangular in shape due to their casings.

While the flooring used between the levels of the house is usually wooden boards nailed onto wooden supporting beams, some of the ground level floors are seen to rest on arches and vaults (Figs. 15.66 and 15.67). While the ceilings of the ground floor areas are made of exposed beams, the supporting beams in the ceilings of the upper floors are almost always concealed with wooden boards (Figs. 15.68 and 15.69). However, there are also some examples of ceilings whose beams have been covered with lathes which have then been concealed with a plaster coating (Fig. 15.70).

Fig. 15.44 Plastered facade, Selamet Street

Fig. 15.45 Square windows of ground floor, rectangular windows of upper floors

Figs. 15.46 and 15.47 Main doors protected within a niche

Figs. 15.46 and 15.47 (continued)

Fig. 15.48 The ancient Greek god knockers

Fig. 15.49 Windows with pink Sarımsaklı stone casings

Fig. 15.50 The balcony over the main entry door

Figs. 15.51, 15.52, 15.53, and 15.54 Different projections on street and side facades with iron supports and stone consoles

Figs. 15.51, 15.52, 15.53, and 15.54 (continued)

Figs. 15.51, 15.52, 15.53, and 15.54 (continued)

Figs. 15.51, 15.52, 15.53, and 15.54 (continued)

Fig. 15.55 Exterior walls are made up of rough stone

Fig. 15.56 Sometimes exterior walls of the upper floors are made up of brick

Figs. 15.57 and 15.58 Main and dividing brick walls

Fig. 15.57 and 15.58 (continued)

Figs. 15.59 and 15.60 The rough "Sarımsak" stone walls

Figs. 15.59 and 15.60 (continued)

Fig. 15.61 The walls are built with local stone

Fig. 15.62 Use of mud mortar as a binding element

Fig. 15.63 Balcony is seated by iron supports

Fig. 15.64 Balcony is supported by stone consoles and iron supports

Fig. 15.65 Inner view of windows of houses

Fig. 15.66 The wooden flooring with wooden beams

Figs. 15.67 and 15.68 The ceilings of the ground floors

Figs. 15.67 and 15.68 (continued)

Figs. 15.69 and 15.70 The ceilings of the upper floors

Figs. 15.69 and 15.70 (continued)

Conclusion

Cunda Island ranks as an Anatolian settlement with very unique natural, archaeological, and urban preservation sites. The settlement is in harmony with the island's topography and does not hinder the natural landscape. While the settlement is protected from the strong winds, it is also open to the gentle breezes that blow from see. The traditional architecture of the island ensures that each house can take advantage of the positive aspects of the sun and the sea. Most of the construction in the settlement is based on masonry systems and utilizes the kinds of nature building materials available in the local environs. The settlement has been designed and built in such a way that it respects its inhabitants, does not physically pollute the area, harmonizes with its natural surroundings, does not cause damage to the flora and fauna of the island, and contributes to the health, comfort, and safety of its populations.

Stone makes up the primary building material of the island's traditional constructions. This stone absorbs heat, is water-resistant, is recyclable, and requires relatively low amounts of energy in its extraction. Also, the wood used is very little of wasted in its applications, can be reused, and offers its users a sense of warmth, both physically and psychologically. The soil used as a binding material is a natural material that is both readily available and inexpensive. This soil is also used in the production of jugs and pots used to store olives and olive oil.

The architectural design of these traditional houses reflects a harmonious use of dimension and shape in the windows and doors that have an exterior exposure. This design and utilization have positive effects on the building's heating, cooling, and lighting needs. Additionally, such features as the proportions, rhythm, materials, and colors of the buildings' windows also are psychologically soothing to the home's inhabitants. The storerooms/*mağaza*s provided on the ground or basement levels of the houses assist in the storage of the products, and requirements of the house and the wells and cisterns aid in the collection of rainwater, thus preserving such valuable water resources.

In addition, even though constructed years earlier, the traditional architecture of the Cunda settlement allows for their continued use, thus injecting these structures with features of sustainability. In order to ensure that this unique feature is bequeathed to coming generations, the traditional constructions in the preservation area need to be subjected to a process of detailed research so that any future negative interventions can be prevented. Any repairs made to these structures need to be conducted by using local material and in strict accordance with scientifically based restoration criteria. Nontraditional interventions must be prevented.

Bibliography

Akın B (2005) Kentli Ayvalık. Graphis Matbaası, İstanbul

Atilla AN, Öztüre N (2004) Adabeyi: dünden bugüne adım adım Ayvalık. Öztüre AŞ Kültür Yayını, İzmir

Balkan A (1997) Ege kıyı yerleşmelerinde kentsel mekanların analizi Ayvalık ve Cunda Adası örneği. PhD thesis, Mimar Sinan Üniversitesi Fen Bilimleri Enstitüsü, Kentsel Tasarım Yüksek Lisans Programı, İstanbul

Bayraktar B (1998) Osmanlı'dan Cumhuriyet'e Ayvalık tarihi. Atatürk Araştırma Merkezi Yayını, Ankara

Beksaç E (1999) 1998 Yılı Balıkesir İli, Ayvalık, Gömeç, Burhaniye ve Edremit ilçelerinde pre-prohistorik yerleşmeler yüzey araştırması. Kültür ve Turizm Bakanlığı 17. Araştırma Sonuçları Ankara 24–28 Mayıs 1999, (2), pp 231–280

Erdem A, Özakın R, Yergün U (2007) Ayvalık (Balıkesir) Alibey/Cunda Adası kentsel mimarlık envanteri 2005–2006. TÜBA Kültür Envanteri Dergisi/J Cult Invent Mayıs 6:77–99

Tekin Ş (2002) Cunda Adası'nın Adları. Tarih ve toplum dergisi 217:43–50

URL-1.: Cunda.net/tarihi-yerler/despotun-Evi. Access 20 Apr 2018

Yorulmaz A (2004) Ayvalık'ı Gezerken. Dünya Yayınları, İstanbul

Chapter 16
Regionalising Contemporary Architecture in a Case of Global South: Masjid Raya Sumatra Barat in West Sumatra

Feni Kurniati

Abstract We usually perceive architecture as an object. A constant object in which we, architects, put a message into: a message that brings out the reason of its existence per se. This has been happened for years up to the moment. Pyramid, for example, was built to deliver the message of eternality, while Pantheon was for the mightiness. Even in the practice of regionalism, however, where the context and people aspects should be fully considered, architects still treat architecture as an object.

Nevertheless, I believe that architecture is not merely an object. It is a very dynamic process, particularly in relation to regionalism in which people and culture interactions are embodied. To show that architecture is a process, not mere an object, is the purpose of this paper. This paper discusses the process of regionalising Masjid Raya Sumatra Barat, the Great Mosque of West Sumatra, Indonesia which was built in 2007.

In this sense, architectural regionalism is seen as a process, where discussion between society and design can be occurred. The process has started since the idea of having a provincial mosque was initiated, to the present time, after it was built and occupied by the people. I understand that the social process of regionalism in Masjid Raya emerges in two ways. First is formulating design's references which best represents the local. This is important due to its task: to be the platform of regional representation. Thus, understanding the essence of local culture and how the society sees themselves should be carefully considered by the experts. Second, socialising the work is another key to succeed. Continuous socialising by the experts is an attempt to legitimise the mosque as part of the community due to its distinctive and controversial visual qualities compared to the surroundings.

Keywords Regionalism · Contemporary architecture · West Sumatra · Regionalising process

F. Kurniati (✉)
History, Theory and Criticism of Architecture, School of Architecture, Planning and Policy Development, Institut Teknologi Bandung, Bandung, Indonesia

© Springer Nature Switzerland AG 2020
M. Ghosh (ed.), *Perception, Design and Ecology of the Built Environment*, Springer Geography, https://doi.org/10.1007/978-3-030-25879-5_16

Positioning Regionalism

We usually perceive architecture as an object. A constant object in which we, architects, put a message into: a message that brings out the reason of its existence per se. This practice has been conducted up to the moment for years. Pyramid, for example, was built to deliver the message of eternality, while for the contemporary case, was built to bring out the spirit of certain party. In its development, when the message is derived from where building is built, it contains territorial characteristics which are chosen to best represent the region. And in architecture, we acknowledge it as an approach to highlight regional characteristics and potencies: known as regionalism.

Regionalism gives architecture an opportunity to bridge the long debatable gap between built environment (buildings architects design) and the users (for whom the architecture is designed). In this sense, architects need to read the context thoroughly and explore possibilities of key factors within the context, to finally reproduce it into design as a regional image. This formal translation is expected to create a familiarity as well as to invite an attachment from the people with the building. This is part of image-making process intentionally conducted to create a particular image of the region and further to establish a certain identity of the place.

Searching for an identity has been a primary focus in many regions, particularly in developing countries, due to the pressure of independence movement. This is a phenomenon which grew largely fuelled by architects who are concerned about their national and cultural identity and dreamed of establishing a more authentic regional voice for their architecture, as a method in defining and declaring their nations (Steele 1997, p. 226). This has been a long trend in many developing countries, including, not limited to, Egypt, Jordan, India, Mexico, and Indonesia, which then characterised their post-independence development.

This phenomenon then has marked an early stage of development of the country, when the question of self was 'discussed' within the society. Although regionalism is not exclusively emerged from post-independence movements, it has become a strategic attempt and trusted method to establish political and cultural roots for the nation. This regional-based identity, on the other hand, cannot be separated from the global issue of universalism brought by International Style and Modernism. For southern developing countries, which regionally have strong traditional characteristics, however, this became a threat. The awareness of the loss of regional identity resulted from this cultural globalisation which was initially fuelled by architects to take a mindful part in the development.

Although regionalism is commonly perceived as a tool to create an identity, it is not necessarily limited to this understanding. Regionalism itself has a long history regarding image-making and the search for identity. It has various understandings and can be implemented in a wide range of practical cases in architecture. In this sense, it is important to briefly revisit the theoretical discourse of the term itself.

A Brief Look on Theoretical Development

This part does not necessarily follow the chronology of regionalism's theoretical development. This attempts to highlight some important insights experts contribute to the concept of regionalism, which also includes the practices followed later. Further, this also offers a discussion of those definitions and relations to help us construct an established understanding of it. In this writing, regionalism particularly refers to architectural regionalism.

From its literal meaning, regionalism is derived from a situation in which systems, administrations, and economy, culture or politics are more regional rather than central. In the context of architecture, therefore it can be seen as an approach to regionalise those elements by using architectural designs or elements to be able to create a certain regionally based identity. The notion of regionalism itself is attached to the territorial attribute embodied in particular regions. This definition tries to summarise physical and nonphysical determinants that exist within the region, including climate, topography and land use pattern, living culture, organization, finance, language, materials and architectural styles.

Furthermore, although regionalism itself is almost impossible to define solely, Canizaro (2007, p. 19) offers a definition that regionalism can be seen as a process of establishing connections between new works and pre-existing local and regional characteristics. This definition tells us that there are two poles which draw one and another in regionalism: the past and the present. This might bring up a question in mind: How do we define the term of 'established connection' between the two? How do we contextualize the past into the present? And perhaps, how do we use this definition in the practical realm?

Practically speaking, using regionalism as an approach in architecture is usually applied through employing the strong visual impressions of the regions from the past to the present needs. The most found practices of architectural regionalism are commonly perceived to recover some semblance of cultural identity by cultivating visual aspects and attempting to return to vernacular expression (Steele 1997, p. 226; Canizaro 2007; Salama 2014). Not only does it try to highlight a formal expression of the local but also attempts to establish an attachment (or at least a familiarity) of the people with the building.

At some point, it is acceptable that people easily relate to visual impression, especially to aspects shared within the community as a cultural characteristics, in this case is vernacular architecture. However, this may lead to simplification of the meaning of regionalism itself. Along with its development, on the other hand, focusing on visual expression seems to bring regionalism practices to its own limitation. These attempts start to bring out discussions with various opinions: some perceives visual representation as simplifications in understanding a region, as well as an exaggerating way to assert an identity, while others see it as an effective way to build a familiarity (hopefully an attachment) of the people and the building it represents.

Learning from the Experts: Some Insights

In the era of modern architecture, concepts of regionalism as an attempt to contest local identity by bringing back local architecture expressions, according to Steele (1997), are not the favoured one. Regionalism seems bounding too closely to the past, therefore does not fit in into the present. This insight resulted from the favouritism of 'sophisticated' architectural forms developed in Western world. Moreover, modern architecture was more a thread to the local identity and diversity rather than an achievement of architecture development.

Within this phenomenon, concerns towards discourse of regional identity increased. Along with its development, awareness of relation between architecture and the environmental sustainability stimulated the thoughts on regional potencies. This consequently brought out the discussion of environmental friendly based identity for regions. This might be then understood as the start of regionalism in architecture.

Regarding this, one of notable architects who speak for local and environmental-rooted identity is Hassan AL-Fathy. He graduated from King Fuad University, which is now known as the University of Cairo. His ideology leads him to disagree with modernism that universality is not the solution to all problems. On the contrary, he believes that the solution to the problem (especially) for his country is hidden within the people (Steele 1997).

Having a profession as an architect helps him to contribute to his country to find its identity through architecture based on local potentials. It can be seen from his project on school in Talkha, a small city near the river Nile. He explores knowledge existing within the site and attempts to discover possibilities out of it. His focus is on approaching solution from local potentials, including aspects of material, architecture and system of construction developed within the community. Although some critiques arise regarding his work, particularly criticising on the way Hassan Al-Fathy using certain form of its local architecture as transplanting (literally reuse the form), his emphasis on his work is his sincere intention to demonstrate how architecture should have been able to develop appreciation towards its local community through empowering its people and benefiting its local sources.

Despite this debate, the spirit of regionalism continues to evolve. There are two notable young architects to continue Hassan Al-Fathy's dreams, they are El-Wakil and Rasem Badran. While El-Wakil used explicitly local Arab traditional form as a design philosophy in his career as an architect in urban era, Rasem in later generation used a quite different approach by optimising relations between social-cultural, environmental, and morphological factors, as well as recent technology. This approach helps Badran to bridge the gap between local and vernacular architecture which is perceived as the past and modern architecture which is understood as a representation of the now. Badran manifested this in designs by exploring values and philosophy within the architecture which is then applied in to the contemporary context (Steele 1997). This is known as an important step in the development of architectural regionalism due to its ability to cultivate the past and adapt it into the present or, even more, reflect to the future.

Unlike those architects in Egypt land who introduce us the concept of regionalism from practical realms, Kenneth Frampton presents it from the perspective of a theorist. Frampton is one of the key figures in regionalism with his famous thoughts on critical regionalism elaborating in the writing titled 'Critical Regionalism: Modern Architecture and Cultural Identity' (2007). Frampton in addition argues that regionalism has two opposite latent potentials within it. On the one hand, regionalism relies on the past particularly its culture and tradition as the root of one's identity, while on the other hand, a region should move towards the future, where modernism takes place altogether with science, technology and power and politics which mostly are apart from the past. In this sense, Frampton argues that culture would be irresistant if it is faced towards the movement of modernism.

In addition, regionalism gives us an insight of ways to position culture and attributes of the past in the context of modernisation in order to sustain its cultural core values. From this perspective, we can see regionalism as means to highlight and establish the main characteristics. In other words, it shows that regionalism does not necessarily utilise locality to create universality; however, it connects the past (culture) and the present (modernisation particularly in influence of science, technology, political power developments). As one of the key figures in regionalism, Kenneth Frampton tries to put an established emphasis that regionalism should be able to contextualise culture and local history as products from the past into the present, the modernity, where technology becomes the centre of development. In fact, the next question may arise in mind is about *how*: how is this abstraction appropriately manifested in practical, formal or technical worlds?

When we come to this question, there might be no right or wrong, nor there is a single appropriate answer. Every practical discourse, every architect, works with their own definition of regionalism. And there are always insights to look out and to dig deeper beneath the surface. While Hassan Al-Fathy perceives regionalism as cultural approach to empower and lift people welfare, Zumthor gives us a different insight about regionalism. In 1996, through his work of Thermal Bath Vals in Swiss, he opens up people perspectives around the world who are concerned about local potentials and characteristics rooted within the region. To some extent, this can be seen as an act of exploring local sources to reveal its regional identity. Although Zumthor is not necessarily known as regionalist, not explicitly he present himself as one, his works can provide us a new insight to understand the term regionalism per se. His regionality is reflected in how deep he understands and appreciates the place. He believes that place is the heart to create a soul in architecture, as Zumthor (2006, p. 42) points out that architectural design should be drawn both on emanations of contemporary life and genuine vibration of the place.

In his work, he exercises geological condition of the site to show its true characteristics: he brings the light and at the same time presents the darkness into the building through deep exploration on openness of the building. He stimulates human sensories through the light, geological texture, shades, reflection on water and smell and sound of nature, which eventually result in a boundary-less spatial experiments (Zumthor 2006; Armand 2011). Thus, this shows that regionalism is not necessarily related to cultural movement only, it can cover a broader discourse, including environmental existence, in this case is geological aspects of the site.

This discussion so far has given an insight that regionalism can be very varied. It depends on who the actor is, where it is placed and when it is constructed. This may help to at least understand that regionalism can also rely on another aspect such as sociocultural condition of the given region. For more developed country, practices of regionalism can be very different from those in less developed country. This then may lead us to look out on the next aspect of regionalism which is regions motive. In this writing, I would like to focus on regionalism practice in a developing country, particularly in Indonesia. A short and may be an unchronological story of regionalism would be laid out in this writing to show how sociopolitical and sociocultural aspects around the region contribute to formulate the form of regionalism practices taken.

Regionalism in Indonesia: A Discourse on Motives

In Indonesia, the cognition on regionalism may firstly took a place in the era of colonial expansion, particularly at the end of colonial era. By referring to Widodo's periodicity of cognition of regionality (2007), there are four periods of architectural development in Indonesia: (1) proto-modern period, context for modernisation; (2) early modern period, the transplantation of European typology; (3) recent modern period, climatic adaptation and cultural accommodation; and (4) hybridization and the maturity of Indonesian architecture. According to such periodisation, Indonesian regionalism movement takes place in the period three, which is the 'recent modern period' around year 1940 when VOC governance were oppressed by the local people. In this era, Indonesia as a nation starts to need a declaration of the unity which led to searching on self-identity. In this context, regionalism was driven by the motive based on pride, 'which behaves like nationalist' (Berry 2007). In another words, independence becomes an important momentum in the search for identity.

Regarding this condition, architecture may start to take place in built development in Indonesia. Widyarta (2012) furthermore argues that after independence, there are discussions focused on local characteristics conducted by Netherland architects in Indonesia including Thomas Karsten, Maclaine Pont and V.R. van Romondt. They attempted to find concept of Nusantara through exploration of local climate, material and skills. In this era, climatic adaptation and cultural accommodation were understood as an act of nationality which meant that local people, Indonesian, had started to appreciate their own regions' characteristics to be able to accommodate the contemporary needs of the nation. This then led to an awareness of values which have been there for them for so long, what they have been living and what they have been believed. In this stage, environmental-regional self-awareness and self-appreciation after the long and hard colonialism in Indonesia may become the motive driven by architects to build nationalism. This motive then be transformed and developed into regional identity in the next stage of defining what Indonesian architecture is.

Therefore, regionalism needs architecture to materialise vision about what it looks to be Indonesia. The motive to gain recognition for regions became so important in this period. An environmentally based approach which was used in considerable ways before nevertheless was gradually replaced. For some, this environmental-based approach in regionalism seems quite prosaic, and sometimes regionalism is minimally interpreted as a response to the local climatic conditions or specific topography only. It is minimally adaptive and acclimatising (Berry 2007, p. 19). This insight brings regionalism a new brand in architecture, which is more like an antithesis of the previous: formal-based approach.

In addition, architecture becomes a powerful tool to translate and crystalise chosen regional-based characteristics into formal and visual aspects. This phenomenon has taken a lot of attentions from architects, who are concerned about regional and cultural identity and dreamed of establishing a more authentic regional voice for their architecture, as a method in defining and declaring the identity (Steele 1997, p. 226). In this sense, architects need to read the context thoroughly and explore possibilities of key factors within the context, to finally reproduce it into design as a regional image. In fact, formal-based approach becomes one of the fruitful approaches to create such a familiar identity for people.

In the context of Indonesia, a country with thousands of cultures and traditions, this phenomenon was hugely stimulated by the government decentralization programme in 1999 during the reign of new order. This programme brought a measure of autonomy to every cultural region in Indonesia to build the capacity to carry out local planning and to promote local economic growth (Green 2005). Not only did the practice of decentralization bring the power to local, it increased intraregional competitiveness among regions in Indonesia as well. Every region attempted to explore its distinctive characteristic as an image of the region. Consequently, architecture plays an important role to create certain image through the built environment by referring to some semblance to cultural identity which mostly refers to its formal aspects. This particular practice cannot only be found in the capital city of Indonesia but also throughout the islands due to the programme of decentralisation.

Understanding Regionalism in the Context of *Masjid Raya* West Sumatra

West Sumatra is located in Sumatra Island, western part of Indonesia. This province is originally known as Minangkabau Island, named after its indigenous people who inhabit the area: Minangkabau people. Minangkabau people are known as the community with matrilineal system to live with. Despite this kinship system as one of the characteristics, Minangkabau has various specific cultures and tradition, ranging from its vernacular architecture 'gonjong', its local culinary, woven clothes, traditional ceremony and many others.

In its development, Minangkabau is one of regions which accepted major effects of municipal by Law Number 6 year 1990 by central governance. Obligation to follow this law brings a new face to the built environment in West Sumatra. Every governmental building in West Sumatra uses *gonjong* on its top: not only governmental buildings but also other public and private buildings, including banks, market and mosque. In fact, the programme of governmental decentralisation in 1999 contributes to present more 'gonjongs' in the region due to the increasing number of new buildings constructed. In this era, every region started to compete to declare their power over others through progressive physical development. Consequently, architectural regionalism becomes tools to fulfil this regional political motive: how to create local-rooted identity and at the same time bring the image of power and modernity.

Likewise in other regions, this phenomenon in West Sumatra brings out controversy among scholars. Although some argues that this approach is one way to represent the strong sense of Minangkabau in the development, others perceive it as an exaggeration use of 'gonjong', which was then called as *gonjongisasi* (Nas 2007). This discussion may lead architects to do deeper explorations on formal aspects of local architecture to result in innovative and creative designs of 'gonjong' as a representation of Minangkabau. In the last 20 years, West Sumatra has initiated various architectural competitions to find a local-rooted modern building for the area. One of significant building which has been built and occupied is the grand Mosque, called Masjid Raya Sumatra Barat. It was completed in 2005, firstly constructed in 2007 and officially launched in 2015, although its construction process continues up to this moment.

Masjid Raya Sumatra Barat is the first provincial mosque built in the region. It is located in the capital city of West Sumatra, Padang City (see Fig. 16.1). Using its provincial name gives the mosque an important role in the region: it is expected to represent the province and its people. Regarding this, it is obvious that the idea of having a provincial mosque in West Sumatra has requested great intentions from various parties within the region, including person as individual, communities and even government. Each actor plays an important role in different parts of regionalism process in West Sumatra, particularly in the case of Masjid Raya.

Having the word 'regionalism' followed with 'process' shows that regionalism in case of Masjid Raya is a huge attempt of discovering and highlighting the essence of local culture and identity. In this sense, regionalism cannot be seen as merely an architectural design or building. It is far beyond mere formal and visual dimensions. This insight allows us to roughly trace every signs of regionalism in the long and dynamic process of presenting Masjid Raya into the site as it is now.

Furthermore, if it is allowed to point out the beginning and the end of regionalism process in the case of Masjid Raya, it would be long before the design is produced and far after the building is constructed. It may show us that regionalism has come to a stage the moment when the idea of having the great provincial mosque was initiated. This was requested by the former governor, Gamawan Fauzi, which was initiated from the event of Vice President Jusuf Kalla visitation to the province. After the visitation, the governor said that West Sumatra, having known with its Minangkabau

Fig. 16.1 *Masjid Raya* of West Sumatra

culture closely tied to Islam, is very unfortunate not to have a grand mosque in the region.

Masjid Raya was designed by Rizal Muslimin, the winner of architectural design competition for West Sumatra mosque. The competition was conducted in 2007, one of the biggest architectural competitions in Indonesia. This was expected to encourage Indonesian architects from various backgrounds to participate in the competition. While searching for good design from this competition, it also intended to capture wider public attention towards its presence. In this sense, it is acceptable that regionalism in Masjid Raya could not be seen merely as a product, but it is a process: a process to capture and highlight one's image or identity.

Regionalism in Masjid Raya: A Process of *Regionalising*

When we take a closer look, there are various stories that come along with the process of bringing the mosque into the site. Those stories are from different actors who are involved in constructing the mosque, Masjid Raya. Thus, when we try to put each of those stories into a trajectory, we can see that the process of regionalism is made of three parts. Firstly, it is the process of initiating the regional mosque. In this case, actors who play an important role are the governance as well as the committee of the competition. Within this period, conceptualising process is conducted by both actors to formulate certain image, hopefully an identity, for the region. This results in an abstraction for the region which then is the guidance for the competition, as stated in the competition brief. One of the local philosophies for Minangkabau people is *Adat Basandi Syara', Syara' Basandi Kitabullah* (adat is based on Islam, and Islam is based on Koran). This philosophy is one of the core cultures in Minangkabau. Having it stated in the competition brief becomes a significant step in the process of regionalising the mosque for West Sumatra due to its role for the next part in the regionalism process.

Secondly, regionalising is signified by the concepts brought by the architect into his design. In this step, it is important to reveal and understand which concepts the architect chose to represent the region and, perhaps, why. Within the case of Masjid Raya, concepts crystallised by Rizal Muslimin, the architect, into the design are based on a deep reading on the physical and natural environment as well as characteristics of Minangkabau people. Having studied the local philosophy *Adat Basandi Syara', Syara' Basandi Kitabullah*, he decided to represent this into physical and technical dimensions of the building. Rizal Muslimin perceives this philosophy as an approach to bring adat and Islam side by side. He explores local architecture, local curving patterns, its custom clothing and its built environment surrounding the site and how those are combined with the Islamic rules and values.

The architect tries to bring local and Islamic values altogether in his design. He investigates the *gonjong* roof of Minangkabau and explores the visual imaginary of an event of Prophet Muhammad moving Hajr Aswad stone in mecca (see Fig. 16.2a, b), and both are exercised to discover design possibilities combining the two. Both are juxtaposed and iterated altogether to finally produce the design which comprises both Islam and adat at the same time. Another regionalising attempt strived by the architect is environmental aspect. Having the tropical climate, he benefits the concepts of solid and void into the building so that the building can breathe and provide a good thermal comfort for praying without any additional air conditioner. While providing cross-ventilation for the building, geometrical combination of solid and void applied on the building also brings aesthetic patterns of shade and shadow into the praying hall. This value generates the sacred atmosphere within the building which hopefully helps people to engage more into their worships.

Furthermore, the architect also regionalises his building by applying local curving to facilitate the need of solid and void on the building. By referring to the philosophy, he cultivates the geometrical pattern of the building shape to enable him to combine local curving with Islamic calligraphy on it. These carvings are quite distinctive because it is on the majority of the building cover.

Besides those formal and technical benefits from using vernacular shape of *gonjong as well as* local curving patterns, there is a strong reason why they are applied in the building. Cultivating those formal and visual aspects is an approach to create a familiarity within the people towards the building. He argues that those are shared and stored in the memory of Minangkabau people as part of their identity. In this sense, although form, dimension and technology aspects of the building are relatively different and novel to the surrounding as well as to its local people, he successfully delivers a message that this building somehow still relates and is cultivated from themselves.

Fig. 16.2a Adaptation of *gonjong roof* on *Masjid Raya*

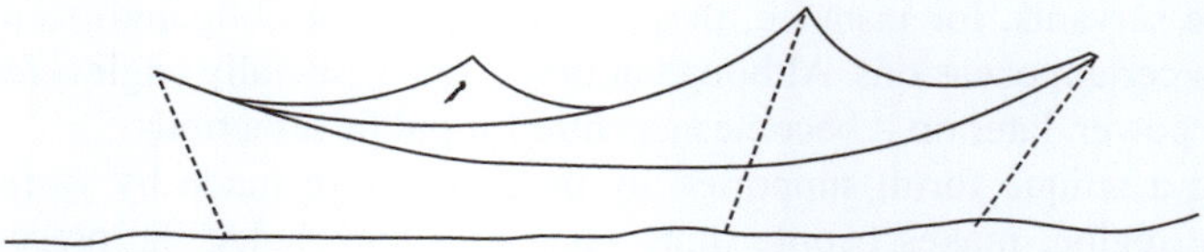

Fig. 16.2b Illustration of the exemplary story of moving Hajr Aswad

Thirdly, this familiarity approach in the design can be seen as a starting point of establishing a connection, if not an attachment, between people and the building. This building is expected to be able to represent the people by itself, as a part of the people. While it does not necessarily guarantee an attachment between people and the building, this may become a good point to start developing such thing. The familiarity may lead to the sense of belonging. At this point, the socialising phase of the regionalism has begun since the conceptual formulating regionality within the competition.

Although socialising overlaps with the initiating process, the step of launching the final design of the mosque to public through various media ranging from newsletters to government websites and billboards displayed around the city still attracts major attentions from the local people. In this sense, there are two major occasions involving public in the process: visual introductory and physical engagement. Visual introductory is the preliminary introduction of Masjid Raya to the public by means of 2Ds visualisation. In this case, the fact that the mosque is produced from one of the biggest contests in the country gives an advantage to publicising actions, not only to whom it is designed but also to a wider society. This also takes a benefit from aspects of innovation and novelty of the design.

Furthermore, as a preliminary introductory to the mosque, it takes first responses and social dynamics from the society. During this period, it was vulnerable due to unpredictable responses which might arise from people. It is likely because this is the first time for the public to experience such a distinctive and massive building within the region. Accordingly, the adat and Islamic scholars together with governance carefully observe social situations. And when it is needed, they give a legitimation towards a certain issues which arise from the society. Fortunately as Minangkabau people who have a culture that greatly trust and respect leaders, various social thread related to the mosque can be prevented before it gets worse. This control happens for years, since the first visual publication was released until the construction get started and finally finished. During this long period, social control has always been the consideration.

When the construction gets started, its physical appearance starts to communicate with the public. The communication gets stronger as the form revealed. Along with this, wide range of acceptances may follow as an accumulation from the visual introductory period as well. In the case of Masjid Raya, although its distinctive physical appearance can be conceptually or formally traced from the cultural elements, its presence in the centre of the city does not necessarily attract public everyday life in significant ways. To see this mosque as the biggest and the most notified building in the city yet does not get many interactions with citizens needs persuasive actions. In this context, local government together with Islamic scholars and the adat persuades people in many ways to have direct experiences with the building. For public servants, for instance, they are asked (if not obligated) to pray in the mosque in certain occasions. Although at first it seems socially engineered by those who have power, later on it becomes a centre for public attentions.

Having a unique form, supported by positive image made by various parties within the region, makes people more open to acknowledge the presence of the mosque. Furthermore, once it was photographed, it will not only spread quickly to the locals but also reach out to people around the country. In fact, as a competition product, this status can strengthen its position as one attractive building in the region. In broader context, this building can be one of tourism attractions within the region. Its local yet innovative form can be one of the benefits the building has to attract tourists to the region. In this sense, people acceptance can be developed, and designs understanding can be evolved over time. To sum up, without repression, power holders in the region can benefit their positions to encourage people to first get to know the building, then to interact with it and at last to acknowledge the new image it brings into the city. This process keeps going on to the point no one can predict. One thing for sure, regionalism process will never be constant because it simultaneously depends on its sociocultural settings.

Having regionalism explored in such ways, it shows that there are attempts needed to execute in order to bring out specific regional characteristics into its architectural identity. On the other hand, the process of regionalism in Masjid Raya also plays significant role in creating the image of the region, Minangkabau. From those, we can specifically see that regionalism is a dynamic process: not only a process of bringing together of concepts, strategies, techniques, ideologies between the two opposite poles, globalisation and localism and modernity and tradition but also a process of providing an understanding of the act of regionalism as well as facilitating the public to experience the regionality per se. Due to social changes it may bring, the process of regionalising may take some time, and also several socially based attempts need to be taken during the process.

Bibliography

Armand A (2011) Arsitektur yang lain: sebuah kritik arsitektur. PT Gramedia Pustaka Utama, Jakarta

Ashraf M. Salama, (2014) Interrogating the Practice of Image Making in a Budding Context. International Journal of Architectural Research: ArchNet-IJAR 8 (3):74

Berry W (2007) The regional motive. In: Canizaro VB (ed) Architectural regionalism: collected writings on place, identity, modernity, and tradition. Princeton Architectural Press, New York

Canizaro VB (ed) (2007) Architectural regionalism: collected writings on place, identity, modernity, and tradition. Princeton Architectural Press, New York

Frampton K (2007) Modern architecture: a critical history, 2nd edn. Thames & Hudson Ltd, London

Green K (2005) Decentralization and good governance: the case of Indonesia, Munich Personal RePEc Archive

Nas PJM (2007) The past in the present: architecture Indonesia. KITLV Press, Leiden

Steele J (1997) Architecture today. Phaidon Press, London

Widodo J (2007) Modern Indonesian architecture: transplantation, adaptation, accommodation and hybridization. In: Nas PJM (2007) The past in the present: architecture Indonesia. KITLV Press, Leiden

Widyarta MN (2012) Mencari arsitektur Indonesia yang utama pada masa akhir kolonial. In: Pusat Dokumentasi Arsitektur Indonesia (2012) Tegang bentang: seratus tahun persfektif arsitektural di Indonesia. Gramedia Pustaka Utama, Jakarta

Zumthor P (2006) Thinking architecture, 2nd edn. Birkhauser Architecture

Chapter 17
Spatial Growth of Religious Architecture: Case of Indian Temples

Soumen Mitra and Mayukh Ch. Sadhukhan

Abstract: As a part of South Asia, Indian Hindu temples played a significant role in shaping sociocultural, economic, and political development for Indian societies in spite of varieties, externalities, and religious turmoil across space and time. The journey started with a single-cell development in Gupta period (300 AD) of central India and obtained ultimate maturity as a temple city during Nayaka period (1700 AD) in south India. Evolution of temple architecture in India has become complex due to multidimensional developments considering the styles, movements, geography, economy, and externalities.

Across space and time, spatial growth of Indian temples has been observed as significant but non-uniform corresponding to the changes in axial dimensions, in spite of strict regulatory instructions of Vedic texts. The trend of development, both in spatial (horizontal) and time-based (vertical) format, has been derived mathematically through curve fitting techniques. For running the experiment, database of 47 temples have been finalized from around 80 samples and synthesized in a master chart considering their dimensional features. Degree of interconnectivity and correlations among linear trends has been calculated from the most appropriate curve(s) from a set of linear, polynomial, exponential, and logarithmic systems for both macro and micro (zonal) level. Impacts of geography, socioeconomic strata, externalities, regulatory measures, and level of patronage on the trends in spatial growth were discussed at length.

Keywords: Indian Hindu temples · Spatial growth · Curve fitting · Central tendency · Standard deviation · Skewness · Correlation

---

S. Mitra (✉)
Department of Architecture, T&RP, IIEST Shibpur, Howrah, India

M. C. Sadhukhan
Department of Architecture, Om Dayal College of Architecture, Howrah, India

© Springer Nature Switzerland AG 2020
M. Ghosh (ed.), *Perception, Design and Ecology of the Built Environment*, Springer Geography, https://doi.org/10.1007/978-3-030-25879-5_17

Introduction

The civilization of Indian subcontinent has been considered as one of the oldest in the world that has continued its rich legacy for centuries. Currently, the subcontinent is one of the vital parts of "Global South". Geographically, the subcontinent is spread near and around the equator in the eastern longitudinal hemisphere. It encompasses the nations of India, Bangladesh, Pakistan, and Sri Lanka. Across time, the rich heritage and culture of Indian subcontinent influenced a larger hinterland in northwest (Afghanistan), northeast (Nepal, Bhutan, and Tibet), and Southeast Asia (comprising Myanmar, Cambodia, Laos, Vietnam, Malaysia, and Indonesia). The long cultural traditions of Indian subcontinent have been regionalized in the said countries and amalgamated with their own traditions (Kossak and Watts 2001).

Since centuries, Indian subcontinent primarily possessed a stronghold of religion-based societies. There was not a single religion. After the downfall of Indus Valley Civilization (around 1500 BC), Hinduism got its prominence during Vedic era. Around 400 BC, two parallel religions, Buddhism and Jainism, came into existence. Through passage of time, Jainism got practiced by few specific groups of people due to its hard dictums. As a comparison, Buddhism was more popular first in India and later was transferred to Nepal, Tibet, China, and Korea through trade and cultural exchanges. All three religions tied the society through their worship centers. For Buddhism, the centers were chaitya, vihara, stupa, pagoda, and temples, while for Jainism, it was rock cut gumphas and later as temples. In Hinduism, the predominant worship centers were temples. Hindu temple architecture could be distinguished from Jain and Buddhist temples, Jain being introvert and elegant in interior and Buddhist being devotional uses of monks (Fletcher 1992).

This chapter unfolds some issues on Hindu temple architecture, confined in India, as a part of larger continent. Undoubtedly, numerous small and large temples were built in India across time and space (provinces). Parallel running of Jainism and Buddhism, Hinduism was revived during 300 AD under patronage of Gupta kings of central India. During the revival of Brahmanism (synonymous but inferior to Hinduism), India experienced extreme turmoil, and the country was ruled by various Hindu kings in fragmented provinces. There were political disturbance due to frequent wars among the kings. However, the society all through was dictated by the rules of temple and the kings. Patronization of kings accelerated the growth of temple architecture. Temples were not only treated as worship centers of commons and royals but were also established as the center for education, culture, politics, and social functions. The invasion of Islam during thirteenth century, reduced the pace of growth for Indian temples and began installing mosques, maqbaras, garden, and pavillions; however, the Hindu architecture again reshaped its spatial configurations for more 500 years. Evidences prove that from a single cell of worship, it culminated to temple cities in various provinces of India during a span of more than 1400 years. But, as societal and political points of views, there arises a few significant queries. Did the societal turmoil and changes influenced in spatial configuration of Hindu temples? Was there any influence of space and time in designing

Hindu temples in India? If both the answers are "yes", *what was the exact quantitative manifestation of growths for Indian temples?* The researchers are almost silent. Here the authors opened up the discussion on the said issues and have tried to analyze the spatial configurations of Indian temples, considering a set of appropriate samples, using mathematical tools to identify the specific changes in space designs, and having tried to establish an overall idea to connect the architectural design and sociopolitical strata.

Indian Temples: A Brief Overview

Concept of Indian Temples

The word "temple" is derived from the Latin word *templum* that means a sacred precinct (Vardia 2008). Originally it is a sacred place for prayer and sacrifice for spiritual attainment. Globally, the term is commonly used since ancient and classical civilizations. Therefore, India, in function of a temple, was not a pioneer. But, it demonstrated a different philosophical framework in shaping the place of worship. These are the most characteristic artistic expression of Hinduism providing a focus for both the social and spiritual life of the community it serves (Michell 1988). Overall, the Hindu temples are the unuttered soul of India that has been playing consistent role in shaping the sociocultural, economic, and political development for medieval Indian societies.

Traditionally the Hindu temples follow the rules of ancient texts like *Sthapatya Veda, Brihat Samhita, Manasara, Mayamatam,* and many more. The design principles and guidelines were usually followed by the architect group known as Viswakarma in northern India and Maya in southern India. Astronomy, orientation, and sacred geometry played vital role in structuring Hindu temples irrespective of place and time (Kak 2002). The initial philosophy of temple architecture in India was "Vastu-Purusha-Mandalam", where vastu is referred to built space, purusha being a male, and mandala is a system. Hinduism believes that the universe is created, destroyed, and recreated in an endless series of repetitive cycles, where Brahma is the creator, Vishnu the sustainer, and Shiva the destroyer. This trinity combines to form Parameshwar (Purush), the Supreme Being who is the manifested form of the whole of infinity (Trivedi 1993). In Hinduism, several gods were created, both male and female, as a specialization of various natural elements and material/spiritual demand of worshippers.

Spatial Arrangements and Evolution of Spaces

The purest form of an Indian Hindu temple is a square on plan, conceptually a terrestrial format, created from a circle, the accepted celestial form in Indian subcontinent. The interface of celestial and terrestrial forms connects the invisible gods and the commons in the core sanctum. The square chamber is conceptualized as an embryo, the womb chamber, which is the source of life (Brown 1976). The Indian term of the chamber is garbhagriha. The square chamber houses an idol of a god, as preferred by the local royals and commons, accessed by a single and narrow door placed centrally to a wall, preferably in east. At inception stage, the chamber was devoid of any window for avoiding any distraction in prayer. However, the concept was criticized by many researchers due to creation of a dark and mystic environment inside without considering the comfort of the space. But, on the contrary, Hindu societal system talks about salvation for attaining the "moksha," i.e., gaining the ultimate and leaving all earthy possessions. The idol is placed centrally on "Brahmasthana", the energy receiver, and a circumambulatory path is kept. This was probably from a Buddhist origin that was adapted permanently. The three-dimensional feature of the square temple was initially a cube characterized by either a flat roof or a conical shikhara, the spire. The apex of the shikhara was a place to fix the weapon of the deity placed in sanctum. Due to presence of multiple gods in Indian Hindu society, such symbolism helped to distinguish among various categories of temples.

With due course of time, compartments were plugged with the garbhagriha as a small assembly hall or "mandapa". A room or a verandah, as the assembly, was added, preferably in eastern direction of the temple. Such orientation or easting was common for Indian temples. Initially the size of mandapa was developed to accommodate a few worshippers; however, gradually it took a shape of another big square or rectangular covered room. To differentiate these two spaces, i.e., garbhagriha and mandapa, a spire was included over the garbhagriha. In early matured stage of temple architecture in India, the mandapa was further expanded and fragmented as various types of mandapa for various groups of people. More often these mandapas were termed as ardha-mandapa, maha-mandapa, and mukha-mandapa based on their functions. Therefore, the temples in early matured stage took a form, a narrow rectangle in proportion. During this phase, the artisans were ordered to develop large temple compounds to accommodate large number of people. The function of worshipping only was transferred into an institution of academics, a center for important discussions, a zone for cultural development (visual arts like dance, drama, songs, etc.), a seat of political power, and lastly, the recreational and ceremonial hub of the settlements. During 1000–1200 AD, the magnificent temples in India came into prominence both in northern and southern part of India. As a fashion statement, gigantic temples were continued to be constructed across provinces of India in simultaneous time frames. The fragmented powers of India flourished and nourished temple architecture in their own styles. Onward, the size of temple increased drastically along with its compound comprising of various activity zones and smaller temples. The temple campus in later stage accommodated large numbers of populations and also permanent activity zones. Such activities include residential buildings of priests, teachers, devadasis (temple prostitutes), shops, schools, and other commercial activities. Freehand development of temples was not easy after 1200 AD. The fragmented societies of India attracted the external forces to capture the motherland. After several trials, powerful Islamic rulers finally invaded the northern India during 1200 AD. The Islamic rulers started demolishing Hindu and Jain temples to spread their power and started making Islamic buildings. The fragmented society of Hindu people receded and took protection in making temples. Thus, the spatial dimensions changed over the time, suddenly or slowly depending upon the situation.

Evolution of Indian Temples Across Space and Time

After the downfall of Indus Valley Civilization in the northwestern part of India, the period of Vedas became prominent. The Vedic period considered the worship of natural powers without placing any idol. Evidences of Vedic architecture is hardly found in India; however, some important texts explain the systems of society, politics, building science, and spiritualism. Gradually, the Hinduism became faded as Buddhism was nourished by Emperor Ashoka in third millennium BC. The Hellenic influences from northwest and vernacular influence from Gangetic Plain shaped the building morphology of Mahayana Buddhism (Brown 1976). Parallel to Buddhist belief, the Hinduism again revived under the powerful Gupta rulers. Chronologically, Gupta era can be treated as the inception of Hindu temples in earliest form.

The life of Hindu temple architecture spanned over 1400 years, starting at 300 AD and ending around 1700 AD. In a vast country like India, architecture of temples was not uniform. The nonuniformity was supplemented by the truth that the country was ruled by many rulers at a single span from fragmented regions. The ruling powers were fragmented and staggered. However, all the powers, at a specific time frame, obeyed Hindu religion and patronized construction of temples. On an average, every ruling era spanned around 300 years. In some provinces, multiple genres of powers were observed time to time. Therefore, Indian temple architecture got its excellence with space-time combinations. Figure 17.1 explains the locations of major temple architecture in the map of India.

The earliest temples were observed during Gupta period around 300 AD in around central India. Gupta temples (300–600 AD) were probably the smallest-sized temples that housed one idol, male or female. The temples were made by stone or brick, with or without spire. While Gupta was at its mid phase, another power, Chalukya (450–750 AD), grew at Karnataka, one of the southern states of India. The contribution of Chalukyan architecture was to finalize the plan forms of temple architecture and also to develop two different styles of envelopes. The first style was adopted by north India which was named as Nagara order, and the second style was carried to south India in the name of Dravida style.

Appropriate Dravida style was generated under patronage of Pallava kings (600–900 AD) in present state of Tamil Nadu. They contributed in shaping the elevational layers with a concept of sharanga (six parts of a body). Therefore, the Chalukyan and Pallava kings shaped the three-dimensional manifestation of Indian temples. Further, the batten was shifted to Chola kings of Tamil Nadu. They expanded the temples both horizontally and vertically. The crest of Indian temple architecture in southern part was definitely achieved by Cholas during 1000 AD. Chola kings handed over their power to Pandyans. They introduced multiple enclosures and multiple gateways of a temple for expansion of activities as well as protection of the

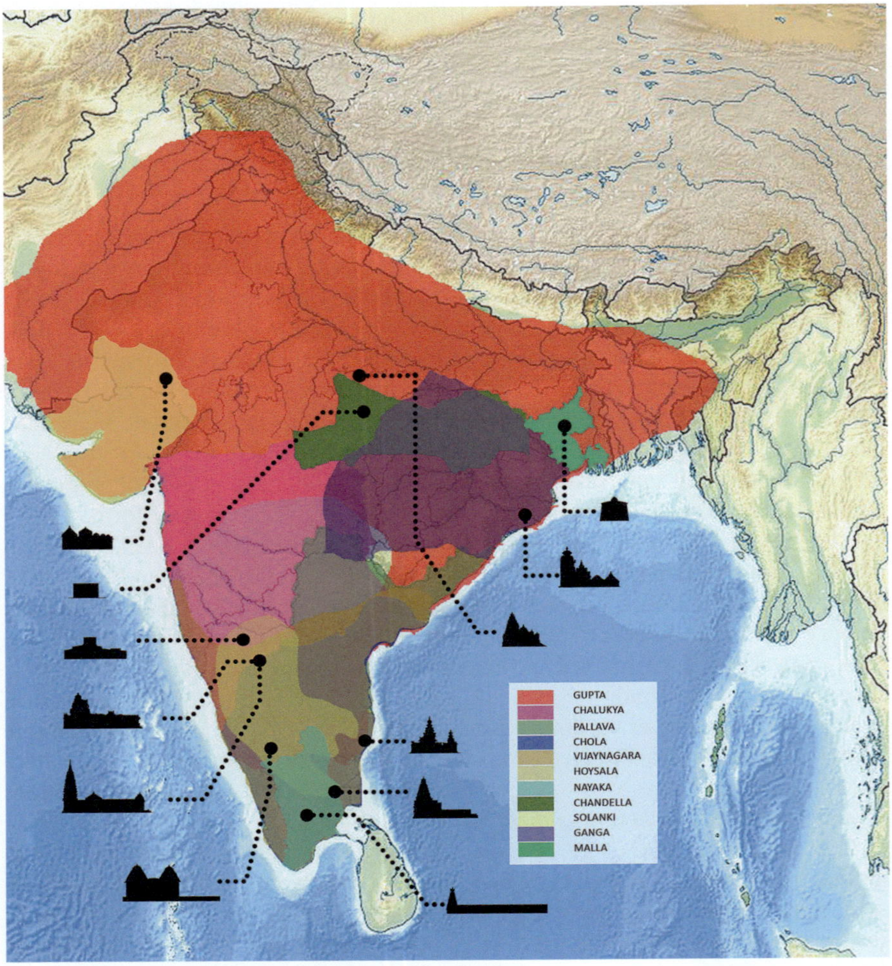

Fig. 17.1 Locations of major dynasties in Indian temple architecture

temple. In thirteenth century, a major part of south India was captured by Islamic rulers, and therefore, the then Vijayanagar kingdom of Karnataka compelled to restrict the temple dimensions. Also, they introduced musical columns made by stone for both cultural issues as well as protection issues from any attack. So, Vijayanagar dynasty exhibited a squeeze in dimensions for sociopolitical reasons. The Islamic rulers encroached the states of Karnataka and Andhra Pradesh but couldn't penetrate the state of Tamil Nadu. This helped a lot in making the terminal temple architecture in Madurai during 1700 AD in the form of a temple city.

The Nagara style was distributed over five major areas – Odisha, Khajuraho, Gujarat, Bengal, and Rajasthan. No such continuity and uniformity, like Dravida order, was observed in Nagara styles. Temple architecture in northern India was probably egoistic and based on own regional styles. Nagara architecture in different provinces was different. In Odisha, monumental buildings are observed. In central India, the sizes of temples were almost half of Odisha temples, and they considered local climate and protection in shaping the temples. Gujarat temples are microscopic and prone to details. In Bengal, due to unavailability of stones, the temples were made of terracotta tiles, more in human scale and more like a residential setup. In Rajasthan, both Hindu and Jain temples are observed that those confirm the pattern of central Indian temples of later stage.

Figure 17.2 (Evolution of temple plans) and Fig. 17.3 (Evolution of temple elevations) demonstrate the plans and elevations of selected temples as a sample representation along with varied scales. Both figures explain the figure-ground relation of the built forms. Emphasis has been given on the shapes and not on iconography. Both plan and elevations exhibit an overall shape and structure of the systems. For each temple, time span of construction has been marked. Variability in scales explains the changes in profile. The spatial growth has been elaborately discussed later.

Glimpses of Literature

Studies on spatial growth of sacred architecture in connection to sociopolitical changes across significant time duration in a country are scarce. Some studies are observed on town morphology, some focus on geometry of sacred built forms, and some identify spatial versus cultural connections. Good numbers of studies are available for church and mosque architecture in terms of geometry and orientations. Budi (2004) worked on Javanese mosques to review and match the previous theories with them. Izumida (2006) discussed on historical development and variety of the mosques that were built before Second World War in Southeast Asia and southern China. Ahmadi (2013) researched on scale of mosque as a visual communication and pedestrian access by the commons. He applied his theory for Safavid mosques in Arab. Garcia and Belmonte (2014) have done extensive works on finding orientation of churches in Mediterranean region.

Few studies on sacred architecture are observed in India and mostly in southern part of India. Dutta and Adane (2014) and Sardar and Kulkarni (2015) both have studies on Indian temples in the light of fractal geometry. Ely (1997), as a faculty of art, highlights the ornamentations and iconographic issues in Hampi City of Karnataka state, south India. She has analyzed how the Hindu architects have added in particular ways to the creative language of the corporeal, architectural, mythical, and sculptural. Kavilkar and Deshmukh (2014) again worked on the same city on the importance of form and space both for sacred cluster and royal cluster. Panjikaran and Vedamuthu (2013) have made the objective of his study to understand the spatial organization of the indigenous church architecture of Kerala and to evaluate the changes in spatial organization during the sixteenth to seventeenth centuries. He identified a quantified relation between functional organization and

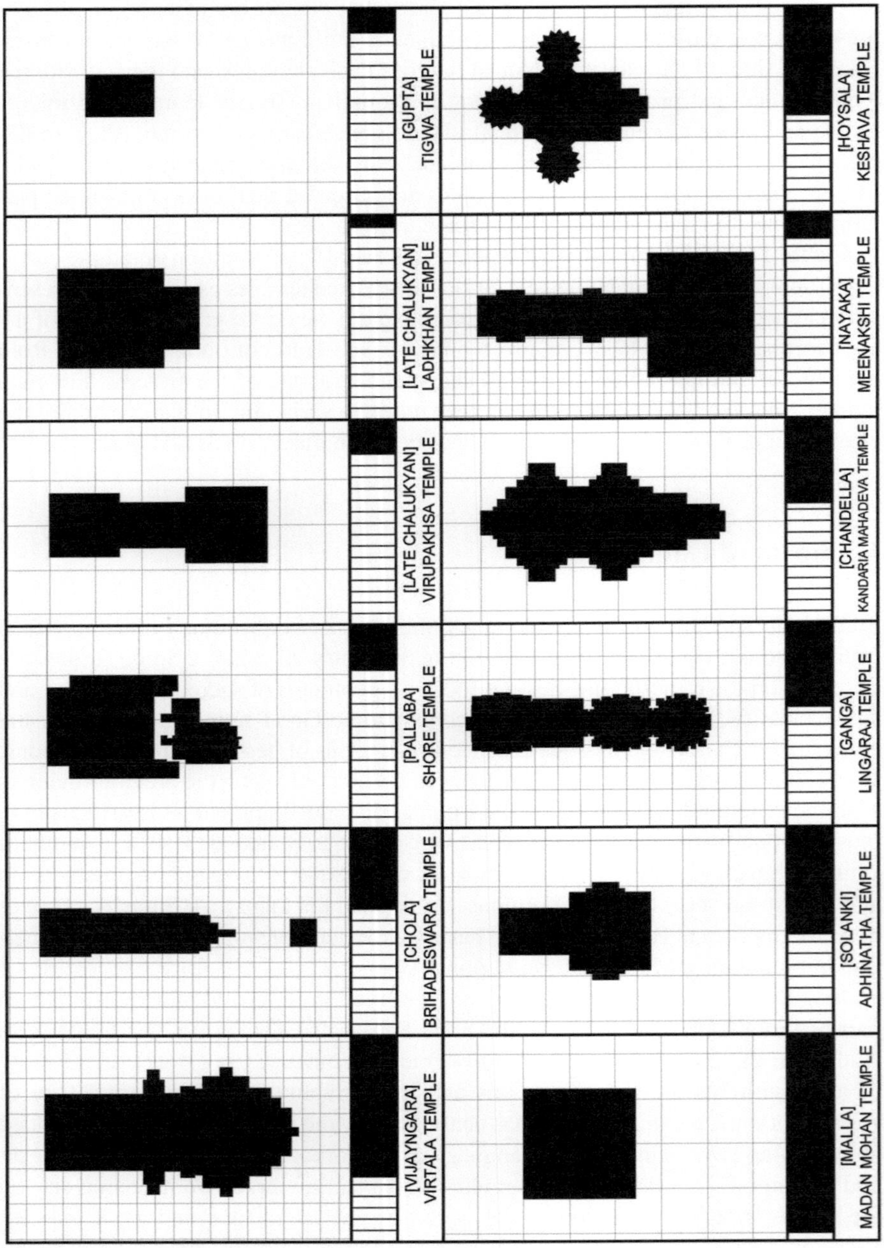

Fig. 17.2 Evolution of plan forms for Indian temples

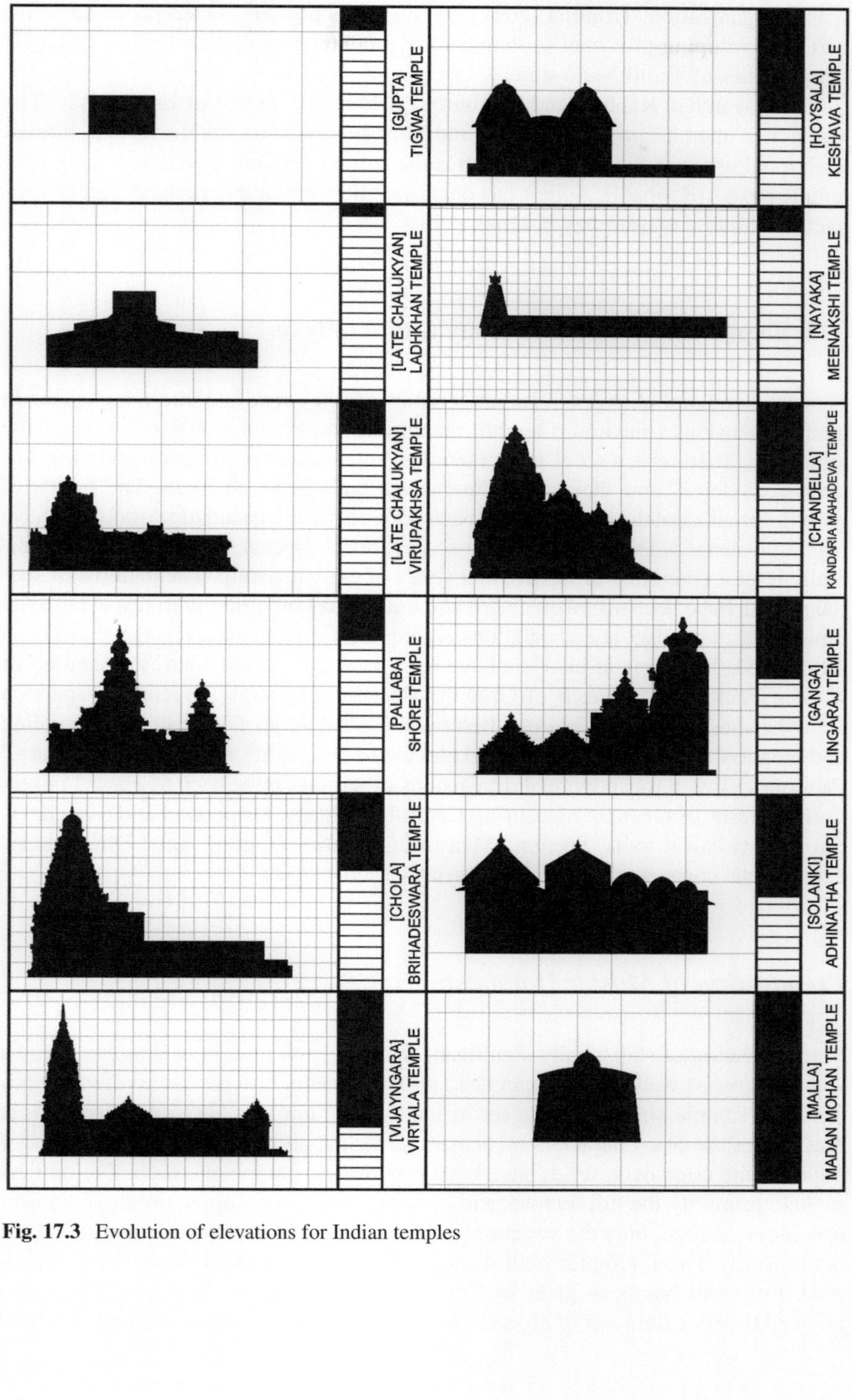

Fig. 17.3 Evolution of elevations for Indian temples

spatial organization. Trouillet (2017) demonstrates how Hindu temples take active part in developing proximal settlements by producing centrality. He has taken six temple cities of Tamil Nadu state.

The research of Raphael and Kasthurba (2015) is the closest to this article – "The research is intended with the need of understanding built environments in a selected traditional urban core of Kerala which is the cultural capital as well as a traditional temple town" (Raphael). Author has quantified how intangible heritage can be connected with the morphology of temple cities.

The Experiment: Methodology and Analysis

Patterns of spatial changes in Indian temples can be identified through changes in axial dimensions (maximum length, maximum width, and height of apex) of the structures. Therefore, a set of Indian temples, representatives of space and time, has been considered, and their axial dimensions have been obtained. The trends of growth for all axial dimensions have been derived based on statistical tools and have been correlated with parameters like sociocultural aspects, political scenario, climatic issues, economic influence, and styles of craftsmanship. The patterns of spatial growth have been tested on macro and micro level timelines with respect to axial dimensions, and also, the correlations among intra-dimensional changes have been interpreted. The steps of analysis have been explained in sections "Preparation of Master Database," "Patterns of Growth: Measuring Central Tendency, Dispersion and Skewness," and "Measuring Relations among Axial Dimensions." Sampling and creation of database have been explained in "Preparation of Master Database." Pattern of growth for individual dimensions of temples have been discussed in section "Patterns of Growth: Measuring Central Tendency, Dispersion and Skewness" through statistical tools. Section "Measuring Relations among Axial Dimensions" explains the correlations among the variables across ages.

Preparation of Master Database

Since selection of statistically significant samples from numerous temples in India is a sensitive as well as herculean task, a set of sieving criteria has applied to optimize the sample size. First, the set of mainstream temples have considered those that reflect the true characters of Nagara, Dravida, and Vesara styles. Temples of regional and composite styles have been kept aside from this research. The outliers include primarily the hill temples and also the temples of Kerala, north, and northeast India. Second, only the structural temples have been included in the list and not the rock cut. Third, temples built during 300 AD and 1800 AD have been considered. Emphasis has been given on "dynasty-based temples," and therefore, it has allowed the time duration of around 1400 years and discarded the modern Hindu or

Jain temples built after colonization of European traders. Fourth, for any dynasty, samples from early, medieval, and later temples have been considered avoiding repetition. Fifth, information on the temples has been collected from secondary sources, i.e., books, journals, and web. Sixth, except early stage, most of the temples are located within walled enclosures. Even multiple temples are located in single enclosure. However, dimensions of only the temple envelopes have been considered instead of temple compounds. Seventh, temples both living (active in worshipping) and nonliving (heritage precinct) have been taken into account.

Initially a set of 80 temples was listed; however, the dimensional aspects of some temples could not be found from secondary sources. Finally, a list of 47 temples has been prepared for experiment and analysis. It may be observed that some renowned temples of India could not be accommodated in the list due to repetition of dimensional aspects. In the absence of data, the dimensions of some important temples have been acquired from approximate measurements through CAAD software. In some cases, exact timeline of construction was not available. Therefore, a tentative timeline (AD) has been considered.

The master chart comprises ten columns, namely, (1) serial number, (2) name of the temple, (3) dynasty (era, i.e., patronage group), (4) city or settlement, (5) state (province), (6) timeline (AD), (7) region, (8) length (m), (9) width (m), and (10) height (m). Timeline indicates the year of construction or completion, as mentioned in written sources. Region indicates the broad geographic areas (north, east, west, south, and central) in India. The database is provided in Table 17.1.

Patterns of Growth: Measuring Central Tendency, Dispersion, and Skewness

Trends of spatial growth have been tested in macro levels considering all data from sample. The test relied on three factors – arithmetic mean value, standard deviation, and skewness. Mean provides a value of central tendency of all values in one. Standard deviation is defined as a quantity expressing by how much the members of a group differ from the mean value for the group. Skewness is defined as the degree to which a statistical distribution is not in balance around the mean. As a tendency, skewness may be positive or negative. The curve of positive skewness exhibits greatest value of mean followed by median (exact mid value in distribution) and further followed by mode (maximum occurrence of similar dimensions). Positivity in skewness happens when a few values in the set are much larger than others. The reverse incident occurs for negative skewness. Calculations of mean, standard deviation, and skewness have been done on a Microsoft Excel 10.0 platform. The exercise has been performed in two hierarchies – (a) considering all databases as listed and (b) timeline-based database. All the tests have been performed individually on lengths, widths, and heights of the sample temples.

(a) Calculations based on overall sample data

Table 17.1 Master list of selected temples

Sl	Name	Dynasty	City	State	Timeline	Region	Length	Width	Height
1	Parvathi T.	Gupta	Nachana	Madhya Pradesh	350	Central	4.80	4.57	9.00
2	Vishnu T.	Gupta	Tigawa	Madhya Pradesh	425	Central	6.00	6.00	4.50
3	Lad Khan T.	Chalukya	Aihole	Karnataka	450	South	24.00	15.00	6.00
4	Bhitargaon T.	Gupta	Kanpur	Uttar Pradesh	450	North	14.33	10.97	20.80
5	Bhumara T.	Gupta	Satna	Madhya Pradesh	500	Central	9.09	4.00	12.00
6	Dashavatara T.	Gupta	Deogarh	Uttar Pradesh	525	North	16.90	16.90	22.00
7	Durga T.	Chalukya	Aihole	Karnataka	550	South	25.60	11.00	9.10
8	Ekambareswarar T.	Pallava	Kanchipuram	Tamil Nadu	600	South	60.00	25.00	59.00
9	Parasurameswar T.	Ganga	Bhubaneswar	Odisha	650	East	15.24	6.70	13.70
10	Papanatha T.	Chalukya	Pattadakal	Karnataka	680	South	27.50	12.10	11.60
11	Vaital Deul	Ganga	Bhubaneswar	Odisha	700	East	5.50	7.70	10.70
12	Shore T.	Pallava	Mamallapuram	Tamil Nadu	715	South	15.00	10.00	17.00
13	Kailasanatha T.	Pallava	Kanchipuram	Tamil Nadu	705	South	14.50	13.20	55.00
14	Virupaksha T.	Chalukya	Pattadakal	Karnataka	740	South	27.50	20.10	13.70
15	Mallikarjuna T.	Chalukya	Pattadakal	Karnataka	740	South	39.62	11.00	13.41
16	Vaikuntha Perumal T.	Pallava	Kanchipuram	Tamil Nadu	850	South	30.00	20.00	20.00
17	Chausath Yogini T.	Chandela	Khajuraho	Madhya Pradesh	885	Central	31.40	18.30	8.00
18	Pampapati Temple	Vijayanagar	Hampi	Karnataka	900	South	50.00	30.00	25.00
19	Koranganatha T.	Chola	Srinivasanlm	Tamil Nadu	949	South	7.60	6.00	15.24
20	Mukteswar T.	Ganga	Bhubaneswar	Odisha	950	East	13.70	7.62	10.67
21	Viswanath T.	Chandela	Khajuraho	Madhya Pradesh	1000	Central	26.50	14.02	26.00
22	Chaturburj T.	Chandela	Khajuraho	Madhya Pradesh	1000	Central	25.90	13.41	28.00
23	Jaganath T.	Ganga	Puri	Odisha	1000	East	93.00	24.40	65.20
24	Jagadamba T.	Chandela	Khajuraho	Madhya Pradesh	1025	Central	70.00	50.00	31.50
25	Brihadisvara T.	Chola	Tanjore	Tamil Nadu	1035	South	65.00	35.00	66.00

26	Lingaraj T.	Ganga	Bhubaneswar	Odisha	1040	East	70.00	35.00	55.00
27	Kandariya Mahadeva T.	Chandela	Khajuraho	Madhya Pradesh	1050	Central	31.00	20.00	31.00
28	Sun T.	Solanki	Modhera	Gujarat	1064	West	24.40	15.30	12.00
29	Kalesvara T.	Hoysala	Kukkanur	Karnataka	1100	South	20.42	11.28	11.28
30	Somnath T.	Yadava	Patan	Gujarat	1100	West	70.00	35.00	47.24
31	Rajarani T.	Ganga	Bhubaneswar	Odisha	1105	East	20.73	11.60	20.80
32	Chennakeshava T.	Hoysala	Belur	Karnataka	1117	South	54.25	47.55	12.00
33	Airavatesvara T.	Pandya	Darasuram	Tamil Nadu	1166	South	12.00	12.00	59.00
34	Jain T.	Hoysala	Lakkundi	Karnataka	1191	South	28.35	10.67	12.80
35	Mahadev T.	Hoysala	Ittagi	Karnataka	1200	South	36.58	18.29	12.20
36	Konark Sun T.	Ganga	Koanark	Odisha	1250	East	96.00	68.30	70.00
37	Pattabhirama T.	Vijayanagar	Hampi	Karnataka	1250	South	50.00	25.00	10.00
38	Kesava T.	Hoysala	Somnathpur	Karnataka	1268	South	26.50	25.30	9.14
39	Ananta Basudev T.	Ganga	Bhubaneswar	Odisha	1278	East	36.60	9.75	21.40
40	Vitthalaswami T.	Vijayanagar	Hampi	Karnataka	1500	South	60.00	25.00	8.50
41	Rasmancha	Malla	Bishnupur	West Bengal	1600	East	24.50	24.50	12.50
42	Meenakshi T.	Nayaka	Madurai	Tamil Nadu	1600	South	50.00	22.50	9.00
43	Shyamray T.	Malla	Bishnupur	West Bengal	1643	East	11.40	11.40	10.70
44	Jor Bangla T.	Malla	Bishnupur	West Bengal	1655	East	11.80	11.70	11.00
45	Radha Laljiu T.	Malla	Bishnupur	West Bengal	1658	East	12.30	12.30	9.50
46	Madanmohan T.	Malla	Bishnupur	West Bengal	1698	East	12.20	12.20	13.00
47	Radhashyam T.	Malla	Bishnupur	West Bengal	1758	East	12.50	12.50	10.00

Table 17.2 Overall data at macro scale

	Length (m)	Width (m)	Height (m)
Mean	31.71	18.51	22.15
Standard deviation	23.03	12.76	18.42
Skewness	1.15	1.92	1.51

The sample data explains that smallest dimensions of temples belong to Gupta dynasty, the earliest era (350–650 AD) of temple installations, while the largest sizes are occupied by Ganga dynasty in Odisha state (750–1250 AD), and that belongs to the mid-phase of Indian temple architecture. From the list, the smallest length, width, and height are observed in different temples of Gupta era, namely, Parvati Temple (4.8 m in length), Bhumra Temple (4 m in width), and Tigawa Temple (4.5 m in height), respectively. The greatest dimensions of all axes have been occupied by the Sun Temple at Konarka from Ganga dynasty. The hikes in axial dimensions from Gupta to Ganga are 20 times, 17 times, and 15.5 times for length, width, and height, respectively.

Mean values of length, width, and height from samples are 31.71 m, 18.51 m, and 22.15 m, respectively (Table 17.2). The data indicates the proportion of length/width is almost 2:1. The reasons may be apprehended in two ways – first, lack of knowledge in making larger openings which compelled the artisans to focus on plugging smaller compartments in unidirectional way. The second one may be derived from the first one to arrange all gatherings in a longitudinal way to conceptualize the philosophy of layered movements with specific restrictions of caste systems. Heights definitely played a vital role to make the people aware of its grand existence. 22.15 m as a mean value and 18.42 m as standard deviation explain a moderate variability in temple heights, specifically the height of the crest. Skewness for all parameters exhibits a surprising result. All depict positive skewness, i.e., mean happens to be more than median and mode. The least skewness is observed in lengths, while the largest is found for width.

(b) Calculations based on time-split data

In second level, similar exercise has been done based on split timelines of Indian temple architecture. The 1400-year history of Indian temples accommodates a number of ruling powers in simultaneous timelines. Individual timelines exhibit varied durations, lowest being 100 years for Chandela dynasty of central India and highest being 900 years in Bengal. However, the temple architecture of Bengal was never continuous, and therefore, consideration of 900 years may not be wise. The timeline-based grouping has been done considering the average duration of eras (Table 17.3). The value is 297 years, excluding Bengal and 321 including Bengal. Therefore, an average of 300 years duration has been considered for grouping. Ultimately, the 1400 years of Indian temple architecture has been distributed over 5 timelines, i.e., 300–600 AD (T_1) (8 temples), 600–900 AD (T_2) (10 temples), 900–1200 AD (T_3) (17 temples), 1200–1500 AD (T_4) (5 temples), and 1500–1800 AD (T_5) (7 temples).

Calculations on mean, standard deviation, and skewness have been performed for each 300 years slab (Table 17.4). Without using any mathematical tool for assessing spatial changes in dimensions, the timeline-wise data of mean, standard deviation, and skewness for length, width, and height provides a notional idea for the evolution.

Table 17.3 Timeline of dynasties

	Start AD	End AD	Duration years
Gupta	350	650	300
Chalukyan	450	750	300
Pallava	600	900	300
Chola	900	1150	250
Pandya	1100	1350	250
Vijayanagar	1300	1565	265
Nayaka	1600	1750	150
Odisha	800	1250	450
Khajuraho	950	1050	100
Rajputana	800	1100	300
Gujarat	941	1311	370
Hoysala	1050	1300	250
Bengal	800	1700	900
Average			321.92

Table 17.4 Timeline-based database

Timeline	Parameters	Length (m)	Width (m)	Height (m)
300–600 AD 8 temples	Mean	20.09	11.68	17.80
	SD	17.87	7.16	17.84
	Skewness	1.87	0.80	2.17
600–900 AD 10 temples	Mean	25.63	14.91	18.81
	SD	13.33	7.16	13.62
	Skewness	0.33	0.99	2.48
900–1200 AD 17 temples	Mean	39.38	21.60	30.35
	SD	25.57	13.83	20.32
	Skewness	0.71	0.95	0.75
1200–1500 AD 5 temples	Mean	53.82	30.67	23.81
	SD	26.81	22.06	26.36
	Skewness	1.08	1.69	2.03
1500–1800 AD 7 temples	Mean	19.24	15.30	10.81
	SD	14.34	5.64	1.49
	Skewness	2.16	1.26	0.45

Experiments on Central Tendency (Mean)

(a) Changes in arithmetic mean values

The means of lengths during T_1–T_5 are 20.09, 25.63, 39.38, 53.82, and 19.24 m. The average dimensions indicate a sharp growth during 300–1500 AD; however, a drastic downfall is observed in the last lap. Likewise, the average values of width are 11.68, 14.91, 21.60, 30.67, and 15.30 m. Definitely, the average values of width have changes similar to length being dimensions almost half of it. The change in heights shows a deviation from lengths and widths. Till 1200 AD, heights of temples exhibit a sharp upward turn, while afterward, it sharply falls. Table 17.5 explains the actual changes and percentages in dimensions.

Figure 17.4 explains the trends in means of length, width, and heights. The trend graphs exhibit two pivotal lines. First one exists around 1200 AD and the second one at 1500 AD. Before 1200 AD, all three axial dimensions increased. Among three, lengths had the highest dimensions from its inception, followed by height and width. At around 1200 AD, the average dimensions of length, height, and width became 39.38 m, 30.35 m, and 21.60 m, respectively. During 1200–1500 AD, the heights of temples started diminishing, while the horizontal dimensions got increased in length (36.67%) and width (42%). Height decreased 21.5% during the timeline. The phenomenon can be interpreted based on sociopolitical issues. Till 1200 AD, the Hindu, Buddhist, and Jain systems in India were able to rule uninter-

Table 17.5 Change of means in time gaps

Changes in intertimeline	Ch L	%	Ch W	%	Ch H	%
Change in T_1–T_2	5.54	27.56	3.23	27.65	1.01	5.68
Change in T_2–T_3	13.75	53.67	6.69	44.85	11.54	61.34
Change in T_3–T_4	14.44	36.67	9.07	42.01	−6.54	−21.55
Change in T_4–T_5	−34.58	−64.25	−15.37	−50.11	−12.99	−54.58

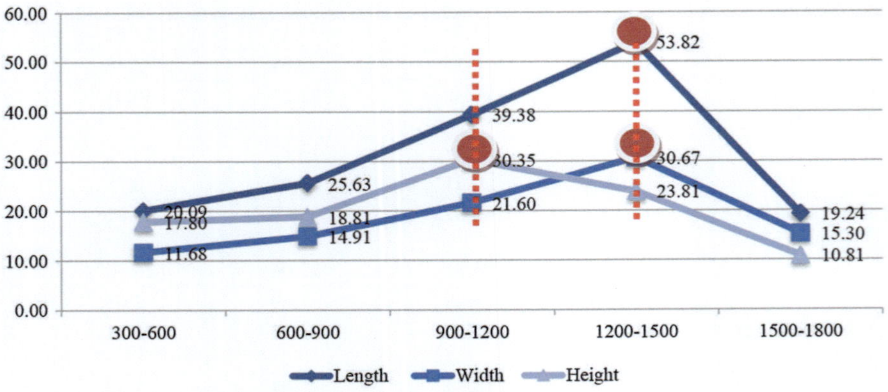

Fig. 17.4 Time-wise changes in means of axial dimensions in Indian temples

ruptedly, while they lost the power after invasion of Islam. Historically, India experienced the Islamic rule since 1200 AD with the startup by Qutbuddin Aibak, the first emperor of Slave dynasty. Evidences proved that the setting of the Islamic rule in and around Delhi, the capital of India, was possible by destructing and demolishing thousands of Hindu temples. Therefore, it can be interpreted that the heights of the temples, after Islamic invasion, started decreasing for protection of the Hindu and Jain wealth as well as to timid the visibility.

Average spatial database in 1500 AD indicates highest value of temple length as 53.82 m followed by temple width 30.67 m and height as 23.81 m. After 1500 AD, all the dimensions started decreasing. The percentages of decrease exhibit 64.25%, 50.11%, and 54.58% for lengths, widths, and heights, respectively.

(b) Suitable curve fitting for means

The trends in mean values for each axial dimension have been tested on five mathematical equations of regression, and the best-fitted curve has been identified. As output, both the equation and "coefficient of determination," i.e., R^2, have been tested on how close the equation is to its database. The tests have been done on the same excel software platform. Five types of curves include (a) exponential, (b) linear, (c) logarithmic, (d) polynomial of second degree, and (e) index. Table 17.6 explains the equations and R^2 values for all axial dimensions of mean considering the Timeline T_1–T_5.

Surprisingly, the result indicates that the polynomial equations of second degree exhibit highest R^2 values for all axial dimensions. The R^2 values are 0.596, 0.614, and 0.754 for lengths, widths, and heights, respectively (Fig. 17.5). This may be assumed that the heights can be explained best, followed by widths and lengths. For widths, all equations show a short spread of R^2 values, exponential being the lowest (0.281) and polynomial being the highest (0.614).

Experiments on Standard Deviation and Skewness

(a) Standard Deviation

There exist two pivotal lines for standard deviation at T_2 and T_4. For all axial dimensions, standard deviations decreased at T_2, while it increased till T_4 (Fig. 17.6). The trend exhibits a fall at the first phase, rise in the second phase, and a steady fall in the last phase. Surprisingly, the standard deviation and arithmetic mean show a difference in trend at the first phase. The mean is increasing, while the standard deviation is decreasing in the phase. Therefore, it may be apprehended that the axial dimensions from T_1 to T_2 increased significantly, while the differences in intra-axial values became stabilized.

At T_4, the standard deviations of length and height both acquire extreme upper value. This phenomenon indicates rapid horizontal and vertical development during 1200–1500 AD, but not uniform for all temples. The data explains that some of the temples showed extreme spatial developments, while some others don't, and hence,

Table 17.6 Best-fitted curves for axial dimensions

	Length		Width		Height	
Curves	Equation	R^2	Equation	R^2	Equation	R^2
Exponential	$Y = 23.93e^{0.0065x}$	0.054	$Y = 12.16e^{0.126x}$	0.281	$Y = 24.13e^{0.07x}$	0.097
Linear	$Y = 2.650x + 23.68$	0.080	$Y = 2.299x +11.93$	0.233	$Y = -0.879x + 23$	0.033
Logarithmic	$Y = 9.39\ln(x) + 22.64$	0.162	$Y = 6.838\ln(x) + 12.28$	0.333	$Y = 0.29\ln(x) + 20.59$	0.000
Polynomial (2)	$Y = -5.61x^2 + 36.73x - 16.08$	0.596	$Y = -2.486x^2 + 17.21x - 5.474$	0.614	$Y = -3.292x^2 + 18.85x -0.034$	0.754
Power	$Y = 22.64x^{0.263}$	0.140	$Y = 12.35x^{0.378}$	0.410	$Y = 20.86x^{-0.08}$	0.020

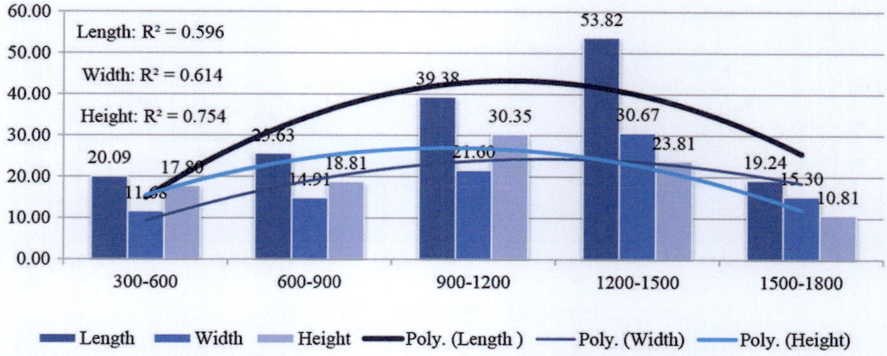

Fig. 17.5 Best-fitted curves for axial dimensions in Indian temples

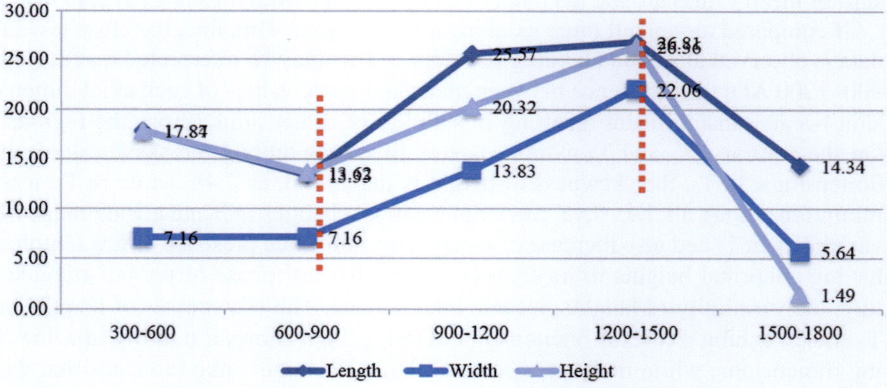

Fig. 17.6 Time-wise changes in standard deviations in Indian temples

the standard deviation differed. One of the striking features for standard deviation is observed in the last phase. Values of all axial dimensions sharply decreased in the range of 1.49–14.34. Till T_4, standard deviation of widths indicated the lowest growth. But, the trend exhibits lowest value of standard deviation for heights, followed by widths and lengths.

(b) Skewness

Trend of skewness for axial dimensions during T_1–T_5 doesn't exhibit any uniform character. Figure 17.7 indicates that skewness in width across time holds the most stable trend within lowest range. The values are 0.80 in T_1, 0.99 in T_2, 0.95 in T_3, 1.69 in T_4, and 1.26 in T_5. Trend of skewness for length has a little more spread from 0.33 at T_2 to 2.16 at T_5. It also indicates a downward turn in the first 600 years; however, it grew gradually till the end. Apparently, the most erratic trend happened

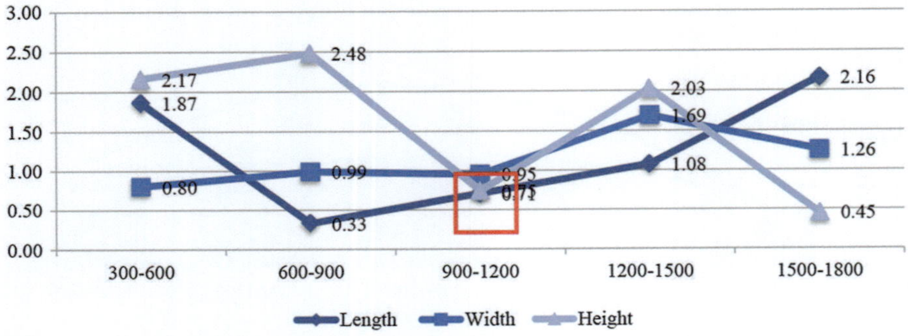

Fig. 17.7 Time-wise changes in skewness of axial dimensions in Indian temples

for heights. In alternate time slots, it shows rise and fall which indicates continuous shift of mean values around the mode. The ranges vary from 0.75 in T_3 to 2.48 in T_2.

If compared among all three axial measures for each Timeline, the closeness of data is observed at T_3 and all being less than 1. This may be interpreted that during 900–1200 AD, the difference between mean and mode values of each axial dimension became insignificant resulting in stability of dimensions across the regions. On the contrary, T_2 and T_5 exhibit drastic difference among the skewness of all dimensions. In T_2, the skewness of height is maximum, as 2.48, while in T_5, it is minimum among all, i.e., 0.45. Such dimensional changes indicate a wide range of variability in T_2 and also increase of mean value due to the presence of few temples having abnormal heights in its context. During the last phase of temple architecture, the variability of heights became insignificant. The skewnesses of lengths in T_2 and T_5 exhibit a reverse phenomenon. During T_2, it shows a stability in achieving dimensions, while in T_5, it becomes 2.16. Such large value indicates that the difference between mean and mode in last stage is largest, which provides an impression on instability in lengths of temples across geographic locations. The samples indicate smaller-sized group of temples in Bengal and large-sized few temples in Tamil Nadu.

Observations and Inferences

Table 17.7 explains the changes in axial dimensions for all parameters across every time slot.

(a) Observations on length

Lengths of Indian temple gradually changed following a best-fitted polynomial curve of second degree ($Y = -5.61x^2 + 36.73x - 16.08$). The fall at the last phase (1500–1800 AD) is drastic, as the installation of larger temples, preferably in south India was almost stopped, while group of smaller temples were observed in Bengal as a blend of Hindu and Islamic characters. Such temples of Bengal under Malla

Table 17.7 Changes in axial dimensions for all parameters across every time slot

Changes in intertimeline	Length			Width			Height		
	Mean	SD	Skewness	Mean	SD	Skewness	Mean	SD	Skewness
Change in T_1–T_2	5.54	−4.54	−1.53	3.23	0.01	0.18	1.01	−4.22	0.31
Change in T_2–T_3	13.75	12.24	0.38	6.69	6.67	−0.04	11.54	6.70	−1.73
Change in T_3–T_4	14.44	1.24	0.37	9.07	8.23	0.74	−6.54	6.04	1.28
Change in T_4–T_5	−34.58	−12.47	1.09	−15.37	−16.42	−0.44	−12.99	−24.87	−1.57

dynasty were designed by brick or laterite blocks and cladded with terracotta tiles, and they never could be compared with monumental temples similar to other Indian provinces. The changes in standard deviation strengthen the pattern of spread across geography in specific time frames. In the two midphases, the values increased significantly. This implies that large spreads of lengths were available in Indian temples during these two phases. However, standard deviation drastically dropped down in the first and last phase with an indication of stabilized length values proximity to means.

(b) Observations on widths

Change in mean values for width in first three timelines is almost uniform, and there is a drastic downfall in mean for the terminal timeline. Standard deviation during the third phase is highest (8.23 m) as a gradual increase from the first phase. From the database, it can be interpreted that the artisans could have technological limitation to experiment with the size of widths, and therefore, both growth of widths and its spread couldn't be compared with other two axial dimensions. However, the change of mean can be suitably fitted within a polynomial curve ($Y = -2.486x^2 + 17.21x - 5.474$).

(c) Observations on heights

Change in heights exhibits a different pattern than lengths and width. After significant rise in dimensions, it started falling much earlier than lengths and widths, probably to restrict the visibility of the temples toward Islamic people. During T_2–T_3, average difference of mean value of heights in sample temples is maximum 11.54 m. Difference in standard deviation is almost half of lengths during the mid phase. In the mid phase, the deviation of heights is not more than 6 m. It may be apprehended that horizontal expansion was easier and safer compared to changes in heights. The best-fitted curve, i.e., $Y = -3.292x^2 + 18.85x - 0.034$, provides the best R^2 value among all.

Measuring Relations Among Axial Dimensions

This section (Table 17.8) explains the ratios and correlations among the axial dimensions for both overall sample data and time-split data. Considering all sample data, the primary ratio of length/width/height is 1.7:1:1.2. This may help a person, who is unaware about Indian temples, to visualize a proportional three-dimensional format of average Indian temple. Across time, as specified in this article, the proportions change. On plan it always exhibits a rectangle of narrow linear pattern and height being closer to the length. The height is not uniform, and it exhibits only the crown over garbhagriha.

In different time durations, as mentioned in Table 17.8, the proportion of length and width is almost same except the last phase. The data indicates that heights also

Table 17.8 Interrelations among axial dimensions: ratios and correlations

	Ratio			Correlation coefficient		
	L	W	H	L vs W	W vs H	H vs L
Overall	1.7	1	1.2	0.84	0.41	0.50
T_1	1.7	1.0	1.5	0.89	0.80	0.85
T_2	1.7	1.0	1.3	0.77	0.14	−0.10
T_3	1.8	1.0	1.4	0.79	0.36	0.62
T_4	1.8	1.0	0.8	0.88	0.87	0.83
T_5	1.3	1.0	0.7	0.80	0.03	−0.37

maintain a uniform proportion with the other dimensions till T_3. Suddenly, the height reduces at T_4, though the proportion of length and width remained the same.

Correlation measures the pattern of mutual changes in values for two variables. The coefficient of correlation (r) ranges between −1 and + 1. The positive value indicates that when the values of a particular variable kept in ascending or descending order, the other variable also exhibits the same pattern, and it is reverse for negative correlations. Correlation coefficient 0 indicates that there exists no relation between two variables.

The best correlation coefficient is observed for length and width as 0.84, followed by length and height as 0.50 and lastly width versus height 0.41. It is evident that height of the temple couldn't be correlated with lengths or widths in suitable measures. However, the correlation-based split timeline shows better result than the average. The best measures of correlations for all three components are explained in T_1 and T_4. T_1 represents the initial timeline of temple constructions, in which all three axial dimensions were in formative stage. Neither the lengths nor the heights could be made monumental. Hence, the restricted development in widths couldn't be far away from lengths and widths. The temples of Gupta and early Chalukyan era exhibit high correlations as provided in Table 17.8. T_4, i.e., 1200–1500 AD, exhibits the postclassical era of temple architecture. In this particular time frame, the proportions among heights, widths, and lengths became stabilized across regions. Temples of later Ganga, Solanki, Hoysala, and Vijayanagar dynasties, irrespective of varied sizes, exhibit similar proportions resulting in high correlations. The most surprising result is obtained from T_2, i.e., 600–900 AD. Values of correlations for length-width, width-height, and height-lengths are 0.77, 0.14, and 0.10, respectively. It may be interpreted that Indian temples under various dynasties exhibit various stages of growth in the same timeline. This was supplemented by material, local climate, and patterns of craftsmanship, and therefore, uniformity in axial dimensions was hardly available. During the phase of T_2, the data indicates a negative as well as insignificant value of coefficient in terms of height versus length. T_5 also represents similar situation as of T_2.

Conclusions

This article studies the spatial evolution of axial dimensions of Hindu Indian temples during 300 AD–1700 AD. The study finally concludes a set of rationale, observations, and realities in a nut shell. A wide range of spatial developments are observed for Indian temples across time and spaces. From a core cubical chamber in 300 AD ended with a temple city during 1700 AD. Increased temple-centric activities and dependence, social and political pivots, material and climate, and wealth of ruling parties dictated the shaping and developments of temples in India. All the three axial dimensions, i.e., lengths, widths, and heights, of temples increased gradually; however the paces of changes were not uniform. The study indicates that the lengths of temple increases more freely, followed by heights and lastly widths. Lengths were kept increasing for accommodating additional functions in longitudinal directions. The increment got saturated after 1500 AD as the craftsmen focused on developments of larger temple campus intensely. During preinvasion of Islam, the power of Hindu kingdoms was exhibited through increasing heights of temples, and therefore, after 1200 AD, in true sense, the heights started diminishing in constructions. The utter necessity and dependence of temples in one hand, and protection of the temples in another hand, resulted in decreased heights of temples. The visibility and grandeur of temples were defocused in Islamic ruling in India at first. But, the width, as per longitudinal zoning concept, could not be increased drastically. In quantitative analysis, as done in the study, all the growths are best fitted in polynomial curve of second degree with varying R^2 values. The rise, crest, and fall for length are maximum. The curve never intersects the other curves. But, the polynomial curves for heights and width intersect in later half of temple constructions. Heights became diminished more than the width. The proportion and correlation also exhibit better results in time-based temple analysis. Therefore, as a whole, the article captures the spatial changes across time and region based on individual dimensional changes and also their interrelations. The patterns of changes across timeline reflect the dominance of sociopolitical scenario followed by wealth of rulers, materials, and climate. Similar experiments may be carried out for other religious and royal structures across the globe and time.

The study, in spite of having a few drawbacks like selection of sample size and consideration of nonmainstream temples, highlights some relevant issues regarding the connectivity and dependence of sociopolitical scenario and sociocultural setups on shaping temple architecture. The study may be extended to other parts of South Asia for a rational comparison between India and the other selected countries.

References and Bibliography

References

Ahmadi M (2013) The effect of the scale of mosques and the relationship with passenger access in Islamic cities in the Safavid age. Sustain City VIII(2):989–999

Budi BS (2004) A study on the history and development of the Javanese mosque. J Asian Arch Build Eng 3(1):189–195. https://doi.org/10.3130/jaabe.3.189

Dutta T, Adane V (2014) Symbolism in Hindu Temple architecture and fractal geometry – 'thought behind form'. Int J Sci Res 3(12):489–497

Ely B (1997) The spatiality of Hindu temples, southern India. J Asian Arts Soc Aust 6(4):10–11

Fletcher B (1992) The history of architecture. CBS Publishers and Distributors, New Delhi

Garcia CG, Belmonte JA (2014) Sacred architecture orientation across the Mediterranean: a comparative statistical analysis. Mediter Archaeol Archaeom 14(2):95–113

Izumida H (2006) A study on development of mosque architecture in Southeast Asia. 19th IAHA Conference, Manila, Philippines

Kak S (2002) Space and cosmology in Hindu temple. Vaastu Kaushal: international symposium on science and technology in ancient Indian monuments, New Delhi, November 16–17

Kavilkar RD, Deshmukh R (2014) Sacred architectural 'form' and 'space' as symbols of political ideology and authority: a case study of sacred Centre of Hampi. J Indian Inst Arch, November

Kossak M, Watts EW (2001) The art of South and Southeast Asia. The Metropolitan Museum of Art, New York

Michell G (1988) The Hindu Temple: an introduction to its meaning and forms. The University of Chicago Press, Chicago/London

Panjikaran S, Vedamuthu R (2013) An evaluation of spatial organization of the church architecture of Kerala during the sixteenth to seventeenth centuries. J Inst Eng India Ser A 94(2):123–130

Raphael JP, Kasthurba AK (2015) Social logic of morphological evolution of temple towns of Kerala, India: understanding the cultural influence on the spatial configuration of temple towns. Int J Chem Environ Biol Sci 3(3):208–215

Sardar D, Kulkarni SY (2015) Role of fractal geometry in Indian Hindu temple architecture. Int J Eng Res Technol 4(5):532–537

Trivedi K (1993) Hindu temple: models of a fractal universe. International seminar on Mayonic Science and Technology, Madras, pp 243–258

Trouillet P (2017) Hindu temples and development of localities in Tamil Nadu (South India). In: Denis É, Zérah M-H (eds) Subaltern urbanization in India: an introduction to the dynamics of ordinary towns, Exploring urban change in South Asia series. Springer, Delhi, pp 311–334

Vardia S (2008) Building science of Indian temple architecture. A Master's thesis, Universidade do Minho, Portugal

Bibliography

Acharya PK (2011) Architecture of Manasara. Bharatiya Kala Prakashan, Delhi

Brown P (1976) Indian architecture (Buddhist and Hindu periods), seventh reprint. D. B. Taraporevala Sons and Co. Pvt. Ltd., Mumbai

Kramrisch S (2002) The Hindu temple (vols. I & II). Motilal Banarsidas Publishers Private Limited, Delhi

Lorenzetti T, Scialpi F (2011) Glimpses of Indian history and art reflections on the past, perspectives for the future. Proceedings of the international congress Rome, 18–19 April 2011

Chapter 18
Co-existence: Migrated Settlement Redefining Cultural Heritage – A Case from Bangladesh

Mohaimeen Islam, Huda Mohammed Faisal, and Md. Tawhidur Rashid

Abstract This chapter explores the interplay and cultural relationship between a migrated settlement at Chandanaish in Bangladesh and their surrounding natural setting, where the prevailing biodiversity immensely influenced their dwelling pattern. It tries to argue that their co-existence on such diverse topography has an innate value to observe and study their historical transformation as a subject of cultural heritage. In scope of this chapter, four types of mud houses of the migrated settlement, named 'Dherr', located in the Shopnonagar village, were studied, through on-site photographic exploration, analysis and observation into tangible and intangible aspects with reference to its ecology and built environment. Chandanaish has a rich culture of mud houses with hills and forest all around, which harbors a variety of wildlife including large mammals like the Asian elephants. However, sometimes elephants attack the village for food which force, the villagers to design mud houses in a way that can prevent elephant attack. Such adaptation in dwelling pattern allowed co-existence among built environment, human and nature. Currently, clusters of mud houses with courtyards have formed the character of the 'Dherr' settlement, where customs as well as wisdom of Bangladeshi people have been transmitted to this locality as a form of a cultural heritage. This chapter is an attempt to investigate in detail the habitation of the settlement which derived from adopted co-existence of human, built environment and nature, to redefine cultural heritage of this place.

Keywords Co-existence · Cultural heritage · Mud house · Dwelling pattern · Migrated settlement

M. Islam (✉)
Assistant Professor, Department of Architecture, Bangladesh University of Engineering and Technology (BUET), Dhaka, Bangladesh
e-mail: mohaimeenislam@arch.buet.ac.bd; mohaimeen.islam@gmail.com

H. M. Faisal
Department of Environmental Design, Graduate School of Design, Kyushu University, Fukuoka, Japan
e-mail: fhuda51@gmail.com

Md. T. Rashid
Architect & Managing Partner, INFORME DESIGN GROUP (IDG), Dhaka, Bangladesh
e-mail: ar.tawhidurrashid@gmail.com

© Springer Nature Switzerland AG 2020
M. Ghosh (ed.), *Perception, Design and Ecology of the Built Environment*, Springer Geography, https://doi.org/10.1007/978-3-030-25879-5_18

Introduction

Elephant Attack in Chandanaish Causes Severe Damage!

A troop of wild elephant came down from the nearby terrain and committed a brutal attack in the Kanchannagar area of Chandanaish upaozila in Chittagong on Saturday midnight (Fig. 18.1.) Eight of the households were severely damaged due to this incident. They ruined banana and jackfruit garden on a land of 4 acres of area. The victims are Jibon Das, Milon Das, Dulal Das, Sumon Das, Robi Das, Mukul Das and Vajan Das. People who witnessed inform that a troop of around 18 elephants came down in Kanchannagar area and ran this disaster.

...Though people of most of the families escape to a safe location, hence eight of these families are now staying under open sky after this severe loss. Those who witnessed informs there were 16 of the elephants in the troop on the day when the incident occured who actually entered into the adjacent village of the Karunarchara Tea State. And ate up the stored dry rice, boiled rice, salt and the banana tree around the houses... (source: Local Daily)

Rural settlements are the places where rural inhabitants live and engage in agricultural production, which is the main defining factor of human habitation formed by the interaction of local community with the natural, economic, social and cultural environments (Li et al. 2015). Because they are the main carrier of ethnic culture (i.e. religion, beliefs, customs and traditions, languages, food, arts and values), the core function of rural settlements has been integrated into the culture of their

Fig. 18.1 Elephant attack in Chandanaish

Fig. 18.2 Bangladesh map

ethnicity and is reflected in the spatial characteristics of rural settlements (Segun 2012). There is a community living in the village named 'Dherr Para' near Chandanaish upazila in the district of Chittagong, south-east part of Bangladesh (Fig. 18.2). The 'Dherr', as they have been called for hundreds of years, migrated mainly from Birbhum-Bakura, north-eastern part of India during the time of British Empiricism in the subcontinent. They were hired as workers for the tea estates under British companies. After the independence of India and Pakistan, they remained in that place, and they continued their existence after liberation war in 1971 under the supervision of the National Tea Estate of Bangladesh. They were accommodated by the Tea Estate Authority, and they lived there by their own interest of survival.

The village was later named 'Shawpno Nagar'. According to some sources, the ancestors of the villagers migrated here 150 years ago. Migrated people need a transitional time to settle down. This migrated community feels lack of confidence and security since after so many years of their being in this territory, they are yet not well accepted as an equal beholder of this land. In addition to this dilemma, very often the present villagers have to search for a shelter to save their life during the elephant's arrival in their village. For this reason, they still feel insecure and unstable in spite of having own living space. Their sociocultural and socio-economic development lack behind for the rapid repetition of this undesired situation.

However, they feel an urge for their community development. They find aspiration of life from surroundings and response from their basic instincts. Most of their learnings and experience are from nature. They don't fight against the nature; rather they live coherently with it. They are frequently attacked by elephants from adjacent forests, but they have never attacked back. They believe in co-existence.

The formation and evolution of geographical characteristics of such rural settlements are not only influenced by natural factors, economic factors and policy but also by various factors such as ideology, ethics, religious beliefs, agricultural production and cultural customs (Gude et al. 2006; Polat and Olgun 2004; Sanjay 2007; Shmueli 1980). Under these circumstances a certain community that belongs to no place, having no identity, no field of their own to cultivate and no water for fishing, confronts a social phenomenon to inhabit with other community (mainly Muslims) and has developed a symphony of mutual understanding for survival. They are just caretakers and guards of assets of others that are worth millions, but they do not get sufficient benefits. The local people, either Muslim or Hindu, treat them as slaves but often give them responsibility of their gardens or estates. The only reason that they are entrusted with so much responsibilities is because they are honest and trustworthy. Moreover, in addition to the consequences of frequent elephant attack from adjacent hill side contributes even more to the regular dilemma of their lifestyle.

Effectively all of these issues define a unique pattern of their livelihood which were explored and analyzed from an architectural point of view and discussed in this chapter. The analysis primarily focuses on the pattern and character of their settlement with the question in mind whether their traditional living pattern – the cultural heritage of this community – is redefined. They have established a settlement using our regional materials and elements. And that other issues (e.g. natural forces and elephant attack) play a role to shape their living pattern becomes a concern to consider. This chapter puts light on the architectural character that is shaped under such reality of their existence and endeavoured to investigate how a cultural heritage can be redefined under diverse state of co-existence with built environment and nature. Here we aimed to emphasize the historical value of the living pattern of the 'Dherr' community by investigating into their habitation, building construction techniques of their mud houses with timber structures and other related social and political factors that contribute in realising their existence as part of a cultural heritage, in response to international regulations (ICOMOS 1964, 1975, 1999a, b).

Shawpno Nagar village **Location** Shawpno Nagar (adjacent to Kanchannagar Tea Estate), P.O. – Kanchannagar, Union: Kochuai, Upazila: Potiya, District: Chittagong, Bangladesh.

Geography and Nature

The area is fully covered by green paddy fields (Fig. 18.3). There were also tea estates, but these are transferred in garden of guava, litchi, pineapples, etc. Sugar cane is also cultivated in some places. There is a lake of fresh water in the far distance of the village. The forest is wildered by the leopards, wild birds and wild pigs. In the winter the migrated birds color the lake and green fields. But the increasing brick field industries have created noise pollution and are damaging the peace of the place.

Since the community does not belong to this land, they encounter many social, cultural and economical limitations. In addition to this, for the location of their village near forest and hill, frequent elephant attack is an unavoidable incident in their life. The houses almost got hidden beyond the green lash of the area. The landscape fabric of vast crop fields is enclosed and guarded by the sloppy terrain. The earth is

Fig. 18.3 View of Shawpno Nagar village from Google maps

pretty stiff; therefore landslide is not frequent here. Hence during monsoon the pathways become really muddy and slippery.

Population and Lifestyle

It is unknown how many families were brought in during those times. A member of one of these villagers' family informs probably 50 families of around 500 people were living under the supervision of the tea state. But at present around 30–40 families with approximately 200 members are living in this place. Some families with their members migrated to the Banshkhali Tea Estate for better living and work opportunity, many years ago. The National Tea Estate has been out of business in 1984, and from then they are yet surviving with immense miseries and hardship. They are living as guards and daily labourers of the Guava Estate, Litchi Garden, sugar cane field, and hilly cultivation (locally called 'Jum') and caretakers of domestic animals of other community (mainly Muslims). They earn money by cleaning the gardens, carrying wood and wood stocks, plucking the fruits from trees and other physical labour. The forest and hills are also a great source of their livelihood. Their houses are usually made of mud and bamboos.

Land Ownership and Administration

DC Notice! 30 Tea Labour Families Under Fright of Eviction!

Hathazari Representative || 30 labour families of Lot Elahabad Tea State in Chandanaish upazila are in utter pressure from the Chittagong DC office since they have been ordered to evacuate their territory. In 1947 after the partition, Indian National Tea State established an expanded tea state between Kochuai and Kanchanbad union in Srimai area of Potiya. The Indian authority brought hundreds of workers at that time to work as labourers in the tea gardens. They allotted the workers in small huts built in hilly terrains inside the garden area. Then after the liberation war in 1971, the authority abandoned the place and gave permission to the workers to have legal rights (written) over the lands and hills of the state. And from then on, the people were continuing their family life by working as daily labourers and chopping woods from the hills. Recently 30 migrated families of around 200 members have been upset with a notice of eviction from the additional district commissioner (tax) signed on behalf of the DC office of Chittagong. If they are ejected, they would have no place for shelter. They are now under immense fear by the notice. On the other side, a local group is imposing power to uproot the families intended to have hold on those government properties. Already this enforced group has captured huge area of the tea state and some ponds illegally. And their future target remains on the few leftover areas where now settle the poor families (source: Local Daily).

Previously they had a little grip on the farm lands, but now they only occupy the area on which they have settled since liberation. Rest of the farm lands are controlled by the people of Kanchannagar area. Recently they have taken hold of the two of the ponds after negotiation. Hence, they get no benefit from the crop lands around their territory being an inferior community.

They have body of administration called 'Panchayat'. They have a sardar (leader) whose job is to take care of the whole community from all kinds of accidental or incidental occurrences. This sardar is nothing like what we perceive of people like him. He is good in nature and sincere to his people.

Education and Profession

No education was promoted or provided for them in any level of schooling by the authority or even government. Local inhabitants are not interested about their education. Even they were also not aware of the need for education of their children. They think it is better to work in fields rather than to waste time on study. But things are changing now. The National Tea Estate has been out of business in 1984, and from then the tea garden workers are living as guards and daily labourers of the Guava Estate, Litchi Garden, sugar cane field and hilly cultivation (locally called 'Jum') and caretakers of domestic animals of other community (mainly Muslims). They earn money by cleaning the gardens, carrying wood and wood stocks, plucking the fruits from trees and other physical labor. The forest and hills are also a great source of their livelihood. The male and adult boys of the family go to the forest hill to chop wood sticks that they sell in the nearby market. They also engage in repairing household furniture, roofing sheets, etc. And the females mainly rear children, cook food, cater animals and engage to repair the cracks of mud walls. In most cases they do the foundation work of the mud houses, and the males do the rest of the covering and roofing part.

Co-existence: People and Occupation

The people of the community suffer from an inferiority complex. They lack modern facilities. The domestics and family members stay under the same shade which promotes a situation of co-existence. Apart from earning for sheer survival, they hardly can imagine for future planning. Mostly the males are involved in collecting wood from the forest to supply for household activities and also to repair houses for a very little fee (Fig. 18.4). Being a Hindu-based community, they are again in the conflict of racism and mostly dominated by the Muslim people of Kanchannagar

Fig. 18.4 People and occupation – Shawpno Nagar village

village. They are deprived of social security. Being a migrated community, they basically copied local custom for sake of existence.

Food and Facility

Their prime food is rice. They have habits of taking tea with puffed rice and some cakes or biscuits. They eat vegetables which are available in nature such as kochu, potatoes, Kolmi shak, Palang shak, etc. Because of their poverty, they cannot buy frequently to eat. Again they are habituated in hunting wild deer and pork. Apart from all these, recently they have planted sagoon tree along the pathways which they will sell after some years of growth. They are even farming strawberry to improve economy.

The community is deprived of proper health and sanitation system. There is scarcity of pure drinking water in this village. They now collect water from a well erected during British period. Hence it becomes unusable during the monsoon. The only pond they have in the territory is currently used for daily household wash and also for bathing purpose. In most of the houses, there is no proper sanitation system. They either use open latrine or just simply go with the natural process.

Marriage, Religion, Festivals and Myths

People never marry them even though they are of the same caste of Bengali Hindus. They can marry among themselves, and interesting is they arrange marriage with the inhabitants of Banshkhali Dherrs who migrated from their community. Hence at present it is seen that though the girl cannot marry outside of their community, the male can choose their bride of same religion but from other village.

The Dherr (as they are called) basically belong to low caste of Hindu religion, and they live in very low standard because of their slave classes who are considered 'untouchable' by the Brahmin concept of Hinduism. Their usual titles are 'Ray', 'Das', or 'Bose'. They worship 'Kali Devi' and 'Manasa Devi'. The temple of Kali is located at the top of a small terrain. In most of the houses, a small chamber to worship Hindu gods/goddesses is evident. Though most of the families belong to Hindu caste, recently a Muslim member of the family who used to be a guard married a Muslim woman and is expanding his family alongside this community. He maintains connection with the people of Kanchanangar, and therefore an event of clash is frequent between these two religious groups.

They cannot afford big celebration during the festival of their religion, but within a small arrangement, they pray and sing religious songs and 'Poonthi' (legends and mythical ballad) with their crafty musical instrument 'the dhol/the mandira' and 'flute'. They have a small arrangement on top of the hill with this musical ballad. In the month of Srabon, they celebrate Monosha Puja, and in Kartik they celebrate Kali Puja. During those times they play Holi (festival of colours) in the temples. Apart from these they enjoy family gathering and other festivals organized by people of the adjacent village.

The place is historical and the events like tea estates, gardens are articulated by the interest of colonial authority. So many stories bloomed on that tree: The Fountain of an Old Lady (Burri'r Kup), Rest House, the Temple of Maa Kali, Death/Murder of a Guar, the Tomb of Punjabi (Pakistani), The Bridge of World War II, Mill House, Manager's Daughter's Epitaph, etc.

Children of the Area

No education was promoted or provided for them in any level of schooling by the authority or even government. Local inhabitants were not at all interested about their education. Even they were also not aware of the need for education of their children. They thought it is better to work in fields and in households rather than to waste time on study. But things are changing now. Since they are a small group of people, they are more intimate to each other.

Two B.B.A students, Rafiqul Islam and Wasim Uddin, of B.G.C. Trust University at Biddanagar area in Kanchannagar came forwards to flourish the light of education among the children of this poverty caste community. They established Momtaz Memorial Institute in 2006. The classes started in a bamboo-made small house. They managed all necessary books and stationaries for more than 30 students bearing expenses from their own pocket. The local people inform though they lack proper school infrastructure, the school tries to provide high-quality education. They have asked for government approval for a school building. At present the children help their parents in household activities. But they find their separate time for schooling and play (Fig. 18.5).

Fig. 18.5 Household activity by children

Fig. 18.6 Elephant attack

Farmer Killed in Elephant Attack!

(Blog by Dhrubajyoti Hore, 8 June 2010 at 7:48 pm)

*One farmer from our neighboring village Kanchannagar was killed by a wild elephant in Shawpno Nagar village area (*Fig. 18.6*). The deceased is Ali Munsi who used to come here everyday as he has a little vegetable field just beside our school* *area. At noon, he was grabbed by an angry elephant when he was trying to drive away the herd together with some other farmer to protect their crop field. Friends, you know about the frequent invasion of wild elephants in Shawpno Nagar and the adjacent area which is actually much closed to a deserted forest of Bandarban. From last few days, a group of around 15 elephants have been roaming around the village. At noon they got very near to the crop field and became face to face with the anxious farmers. And then the killing took place after tug of war.*

In this area, wild elephants come down from the hilly forest five to six times every year. Off and on they do some damage to the field and houses. But long after, this time they took a human life. Last similar incident happened some 7 years ago. Actually the elephants invade due to scarcity of food in their own forest which is already ruined up by people. Such situation compelled them to come out and seize crops what our poor farmers grow after a year-long hard work. And then 'fight for food' becomes obvious between these two marginalized species. Sometimes poor people and sometimes poor animal kill one another. This is the paradox of poor's life in such area!

Should we accuse elephants for this? Or should we point the finger at the inconsiderate, unjust deeds of human which happen due to the reluctant government policy and measures in such issues?

When the elephant troop is near the village area, the workers can hear them marching at least from 1.5 miles apart. And it is a psychological signal system through which the villagers get aware of their upcoming hazards. Within these 30 years of time, there were frequent attacks, and usually the elephants attack the houses at night. These hungry animals search for dry salt and rice stacked inside the household. And with intention of finding food, they just break through the soft earth mold walls. Fire and wire mesh fencing is no option to tackle these incidents.

Settlement Pattern

The settlement is the functional grouping and distribution of people. Kirk H. Stone defined settlement as the distribution of buildings by which people attach themselves to the land. It is not incidental. Rather it is a process of time, space and culture (Stone 1965). In Bangladesh, the process of rural habitation is traditional and mingled with the origin of agriculture, and the settlement in this area dates back to the remote past. It is deeply rooted in this subcontinent's past history (Bashem 1959). The process of village formation has been continuous due to the facts of population increase and the movement of people among and within the villages. There are both inward and outward forces at work (Anas 1954). The settlement of Shawpno Nagar village is nucleated and linear along with the edge of hills. The houses are arranged in a way that they assume a regular village shape which gives nucleated and linear pattern to the settlement. Different physical, cultural and local conditions play a great role in shaping distinctive forms and patterns of settlements (Vecco 2010). The houses are compact, and Shawpno Nagar village has been developed on the communication route of the villagers. This type of settlement is common in Chittagong Hill Tracts. Also local natural condition like elephant attack guided the villagers' way of living. The villagers stay in a group so that they can face the elephant attack all together to save the village from destruction.

Dwelling Scheme: Group Settlement

Shawpno Nagar village bears the cultural heritage of mud houses of Bangladesh. The selection criteria of cultural heritage have changed; while, initially, the historic and artistic values were the only parameters, other additional ones have now been added, the cultural value, its value of identity and the capacity of the object to interact with memory (Vecco 2010). From a purely normative approach, of an objective, mud houses have a certain value that represents the local identity of Bangladesh, which has been practiced from generation to generation. The organization of traditional houses in rural Bangladesh is formatted with respect to the functional uses and activities of various spaces. The traditional house form evolved rather slowly through centuries under the influence of many factors working simultaneously, e.g. the land, the culture, the climate, the available resources and so on (Hasan 1985). Shawpno Nagar village represents the rich culture and memory of mud houses of Bangladesh with the influencing factor of animal attack as well as available resources of construction material nearby. Houses of Shawpno Nagar village are usually made of mud, timber and bamboos. The roofs are covered by chan (straw). Only one house has tin (corrugated iron) roof. Villagers sleep on floor using mats made from jute and bamboo. Some use wooden bed which was abandoned by other people. Generally, they construct dual layer house enclosure. The outer layer is a service corridor which actually protects the inner core from direct elephant attack. Some house has upper storey provision which is used for storage and also for sleeping (Fig. 18.7). Courtyard formation is not very common in the locality. Elephant attack was the cause not to build courtyard which is a very common element for settlement in Bangladesh.

Court Pattern (Unique)

The layout of the homestead is defined by the courtyard, formed by the arrangement of dwelling units and ancillary building around an open space (Ahmed 2006). The courtyard is a common characteristic for a rural house in Bangladesh. But this courtyard pattern is not familiar in the locality though only one house follows the traditional way of living around the courtyard. Due to adequate space in the courts, the elephants may get a chance to take the position and destroy the valuables. Only 'Jibon Das' with his family chain is dwelling with this particular type (Fig. 18.7).

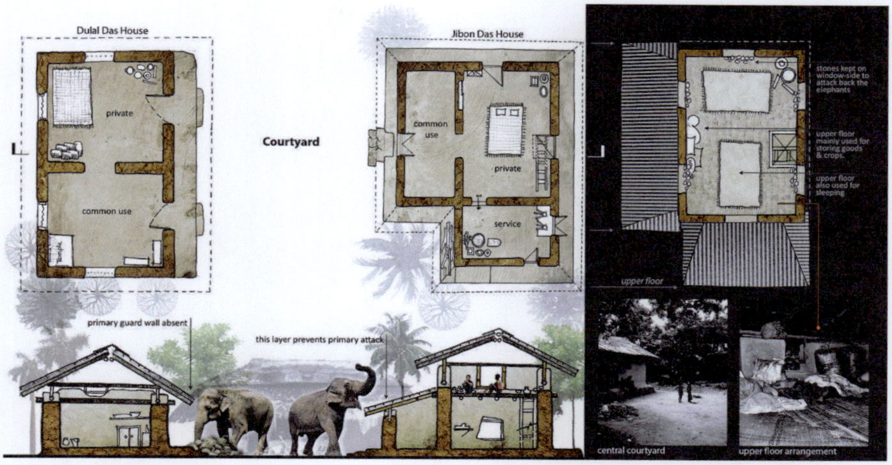

Fig. 18.7 Group settlement

Though guard wall is a prototype, some of the families still haven't yet adopted this precaution against direct elephant attack. Generally the lower storey is used for services and sleeping purpose.

Dulal Das House: Dulal Das's outer service layer was severely damaged by the 18 June incident when a troop of elephants broke through the primary guard wall. But hence the inner main house was protected since they adopted this building technique on an early stage. Furthermore, the backside protection wall encloses both the cooking and domestic area of this house.

Materials and Construction Technique

Their houses are usually made of mud, timber and bamboos. The roofs are covered either by CGI sheets or chon (straw) (Fig. 18.8). They sleep on floor using mats made from jute and bamboo. Some use wooden bed which is actually less frequent in the village. The basic structure is constructed with moulded mud and mud blocks. Due to frequent elephant attack and for structural stability of the mud houses, the windows are created as minimal as possible. However, though the basic material is mud, timber is also used to reinforce the whole structure, especially the upper roof, with the surrounding wall. Wood has been widely used in religious and civil architecture due to its easily accessible and renewable characteristics (Arun 2012; Irbe et al. 2012; Lebow and Anthony 2012). Also, wood has been chosen as one of the oldest traditional construction materials in many parts of the world, especially where dense forested areas exist (Yongwei 2006; Dăneasă 2013; Laurila 2015).

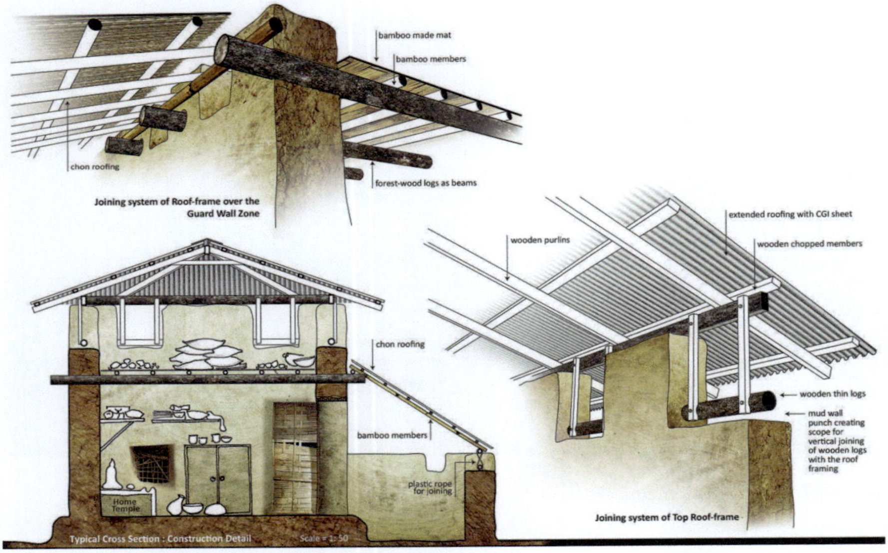

Fig. 18.8 Materials and construction technique

In this area, for instance, forest wood is chopped to use as horizontal beams to increase the bracing strength. The logs are punched through the mud wall at a level of 8′ (8 feet), and with the extension on the outside of mud wall, the frame for chon shading roof is generally bonded. And again the top roof frame is connected with the wooden members punched through mud walls by aid of some vertical wood joints. So basically no structural post is injected into the mud plinth. It is indeed a unique construction technique. Hence, documenting such wooden construction technique and detail of mud houses are necessary to nominate such traditional entity as cultural heritage, as well as determining and defining its architectural period for further conservation (Rochner et al. 2017).

Case Study

The Methodology of Dwelling Unit Survey

Habitation and its embedded architectural features have been analysed based on literature research, interviews and on-site studies. In the field study, detailed survey measurements, photographic documentation and oral interviews were performed with the inhabitants of the community and few local officials to know about their traditions and histories. Afterwards, mapping of the household's locations on Google Earth images based on the landscape photos was done. Furthermore, plan layouts, elevations, sections and 3D drawings were produced using the detailed survey measurements and photographic documentation to aid the research investigation. There were four types of house pattern. They are single storey dwelling, double storey dwelling, double storey dwelling without rear service zone/guard wall and triple layer house: house inside the Muslim community.

House Pattern No. 01: Single Storey Dwelling

This is the most simplistic type and mostly copied in the majority of the community. The guard wall encloses a service zone. At one side, it is used for cooking, and on the adjacent side, there is a domestic zone. The inner house is divided into two spaces. The latter one is more private.

The structural system is same as illustrated before. The mud wall bears the main vertical load as a shear support. The wooden beams and bamboo members run horizontally at a level of at least 8′ which pierce through the mud wall and promotes at least a feet of projection (Fig. 18.9). And with this extended hands, the roofing frame over the service zones are tied up. The top wooden roof frame is connected with a thin wooden log injected into mud punches. The top roofing is done with CGI sheet and the other with chon (straw).

House Pattern No. 02: Double Storey Dwelling

This is the first building in the territory which was constructed as a double storey residence. Their front primary guard wall is a bit high than the general types of the area (Fig. 18.10). They use this zone as a semiprivate sleeping zone since their family is extended. There are two more bedroom areas in the inner zone.

Fig. 18.9 Pictures and section of house pattern 01

Fig. 18.10 Picture and floor plan of house pattern 02

Fig. 18.11 Pictures and section of house pattern 02

The upper storey has no partition, mainly used for storage and also for sleeping purpose. They often keep some stone blocks at the upper storey window. During elephant attack they get to the upper floor and try to distract the troop from destroying their valuables by throwing stone blocks from the window apertures. In the interior mud walls, there are alcoved niches where they keep some regular goods. The protected back service zone is used for cooking and for domestics.

The mud structure is two-storey and separated by a bamboo-made mat, placed on evenly spaced bamboo members which actually create ceiling for the lower storey (Fig. 18.11). Both the top roofing and roofing over the semiprivate zone are done by CGI sheets, and roofing of the backyard service zone is covered with chon. The upper storey window apertures through the mud wall create scope for cross ventilation. Therefore, a system of passive cooling is evident. And in the same way, the upper storey acts as a heat trapper by creating a buffer zone.

House Pattern No. 03

Double Storey Dwelling: Without Rear Service Zone/Guard Wall

This house is a dual family double storey house. They use the upper storey as frequently as they do with the lower storey. The frontal zone, though semiprivate, is basically a guard wall zone which has been frequently attacked by elephants (Fig. 18.12). On the rear side, there is no service block. So it is often vulnerable since elephant troop has damaged the back wall several times to grab dry rice and stocked salt from the upper floor. On the left side, there is a small domestic zone enclosed by thatched sheets (Fig. 18.12).

House Pattern No. 04

Triple Layer House: House Inside the Muslim Community

In a Hindu-based settlement, there are only five Muslim families. Particularly this house has a unique layering system. They have been attacked by the wild troops so many times that they almost invented this triple layer building system. The first two layers are simply enclosed by CGI sheets and the third layer is by mud walls which also includes a service zone both for domestic use and cooking purposes (Fig. 18.13).

Fig. 18.12 Floor plan and elevations of house pattern 03

Fig. 18.13 Picture and drawing of house pattern 04

Conclusion

The 'Dherr' community of the Shawpno Nagar village exhibits a unique example of dwelling pattern and amplifies a state of co-existence with the given context they inhabit, however having no place, land and identity of their own. They are just caretakers and guards of assets of others that are worth millions. But what is praiseworthy is they not only guard assets but also guard our culture. Their settlement showcases ideal examples of mud house settlements in Bangladesh. The four types of house pattern studied and analysed here in this chapter through pictorial evidence and supporting drawings are testaments of cultural heritage for their historical value and authenticity. The detailed construction technique of the mud houses and the application of wood as an efficient structural combination is indeed a representation of traditional knowledge that has passed from generation to generation. However, we are living in an age when mud houses are going to be erased from our memory. Surprisingly, people of Shawpno Nagar village are still sustaining the tradition of mud houses as a form of co-existence with animal and nature. In this modern time, people here at Shawpno Nagar village are weaving the story of fighting with nature by adopting the cultural norms, traditions and building techniques for their habitation for sheer survival. They are not only adopting the culture and local customs but also transmitting the cultural knowledge to the future generation. In a sense, they are acting as a protector of cultural heritage. Their aspiration towards adopting mud house construction techniques and co-existence with nature is indeed a profound example for mud house preservation and also an attempt to sustain a rare tradition, which contributes to redefining the cultural heritage of southern Bangladesh. We believe this study shall uphold the importance of preserving traditional houses in rural societies and encourage to enable future avenues to research further in the possible process of conserving such cultural heritage of this region.

Acknowledgements We want to give special thanks to Mr. Dhrubajyoti Hore for helping us during the survey as well as providing us two photos of elephant attack. We also want to express special gratitude to Ar. Nabila Binte Nasir and Ar. Fahim Ul Alam for their support during field survey.

References

Ahmed KI (2006) The rural Bangladeshi courtyard. BRAC Univ J III(1):9–15

Anas M (1954) The pattern of the rural settlement in sub-himalayan region (east). Geogr 6(2):25

Arun G (2012) Evolution of timber construction in Turkey, archi-cultural translations through the silk road. In: Arun G (ed) 2nd international conference 2012 proceedings, Japan, pp 225–230

Bashem AL (1959) The wonder that was India, New York, p 11

Dăneasă CM (2013) The study of conservation state of a wooden church from Bozvillage Hunedoara county Romania. Int J Conserv Sci 4(1):53–64

Gude PH, Hansen AJ, Rasker R, Maxwell B (2006) Rates and drivers of rural residential development in the Greater Yellowstone. Landsc Urban Plan 77:131–151

Hasan DM (1985) A study of traditional house in rural Bangladesh, Master's thesis from Bangladesh University of Engineering and Technology, Dhaka, Bangladesh

ICOMOS (1964) International charter for the conservation and restoration of monuments and sites (The Venice Charter 1964). https://www.icomos.org/charters/venice e.pdf. Date accessed: 10 Aug 2018

ICOMOS (1975) The declaration of Amsterdam. http://www.icomos.org/en/charters-and-texts/179-articles-en-francais/ressources/charters-and-standards/169-the-declaration-of-amsterdam. Date accessed: 10 Aug 2018

ICOMOS (1999a) Charter on the built vernacular heritage. http://www.icomos.org/images/DOCUMENTS/Charters/vernacular e.pdf. Date accessed:10 Aug 2018

ICOMOS (1999b) Principles for the preservation of historic timber structures. http://www.icomos.org/charters/wood e.pdf. Date accessed: 10 Aug 2018

Irbe I, Karadelev M, Andersone I, Andersons B (2012) Biodeterioration of external wooden structures of the Latvian cultural heritage. J Cult Herit 13:79–84

Laurila A (2015) The old rauma and the old Church of Petajavesi – principles and practice on restoration and maintenance. In: Sürücü D (ed) Symposium onrestoration of timber structures 3, KUDEB, ˙Istanbul, pp 65–79

Lebow S, Anthony RW (2012) Guide for use of wood preservatives in historic structures. https://www.fs.usda.gov/treesearch-beta/pubs/42091. Date accessed: 10 Aug 2018

Li F, Zhang SW, Yang JC, Chang LP, Kun BU, Wang Q (2015) Spatial distribution of rural settlements in farming-pastoral zone of northern China and its impact on land use pattern: a case study of Korqin left wing middle banner. Sci Geogr Sin 5(3):328–333

Polat HE, Olgun M (2004) Analysis of the rural dwellings at new residential areas in the Southeastern Anatolia, Turkey. Build Environ 39:1505–1515

Rochner ML, Kelley HW, Wilson CS, Bennett T, Grissiono-Mayer HD (2017) Dendrochronological dating of the Warner House and barn (20LV334), Livingston County, Michigan, USA. Dendrochronologia 43:4–11

Sanjay KN (2007) Tourism and rural settlements Nepal's Annapurna region. Ann Tour Res 34(4):855–875

Segun O (2012) Behavioral outcomes of culture and socio-economic status on urban residential morphology: a case study of Lagos. Procedia Soc Behav Sci 50:294–306

Shmueli A (1980) The bedouin of the land of Israel—settlement and changes. Urban Ecol 4(4):253–286

Stone KH (1965) The development of a focus for the geography of settlement. Econ Geogr 41(4):346–355

Vecco M (2010) A definition of cultural heritage: From the tangible to the intangible. J Cult Herit 11:321–324

Yongwei W (2006) Restoration of ancient wood buildings. Int Hoizbau-Forum. http://www.forum-holzbau.com/pdf/ihf06WangYongwei.pdf. Date accessed 10 Aug 2018

Chapter 19
Altered Perception of Culture: Based on Features of Pedestrian Experience and Aesthetic Regeneration of Built Environment

Tanima Bhattacharya, Suparna Dasgupta, Tushar Kanti Saha, and Joy Sen

Abstract Culture and tradition are based on participation, engagement, and an idea that prioritizes equality for humanity. And "perception" toward tradition has created the bonding between human values, tradition, and technological progression. "Perception" is the identification, organization, and interpretation of sensory information into coherent unified meaning to self and to the world. Since perception is based on human interpretation and meaning; it varies with factors like culture, civilization, and age of development. Hence, perception forms the "imageability" of the space and altered with time and space.

The chapter would concentrate on the rapid urbanization that has led to the disparity of identity, imbalances of the indigenous knowledge system, deterioration of cultural and regional assets, and hence, a decline in individual empowerment. The chapter focused on the indigenous architectural forms and knowledge system as a part of everyday aesthetics. It attempts to grasp the physical, tangible architectural forms and patterns of space. Therefore, define and evaluate the form, style, and pattern of the architecture itself and space around by analyzing people's perception toward it. The study would culminate through assessment of architectural patterns and built environment in two different case studies, namely, vernacular architectures at Santiniketan in Birbhum and Jorasanko Thakur Bari and their alternative responses to different versions and priority of nature and ecology.

This chapter brings to light how a pre-transitional context of Jorasanko Thakur Bari grafted from a hardcore urban footprint in British Industrial Calcutta had contrasted with a posttransitional eco-friendly setup of Birbhum in Rural Bengal from which Santiniketan and its eco-exploratory setup was carved out. Thus there is a

T. Bhattacharya (✉) · S. Dasgupta · T. K. Saha
Ranbir and Chitra Gupta School of Infrastructure Design and Management, Indian Institute of Technology, Kharagpur, Kharagpur, West Bengal, India

J. Sen
Department of Architecture and Regional Planning &Ranbir and Chitra Gupta School of Infrastructure Design and Management, Indian Institute of Technology, Kharagpur, Kharagpur, West Bengal, India

© Springer Nature Switzerland AG 2020
M. Ghosh (ed.), *Perception, Design and Ecology of the Built Environment*, Springer Geography, https://doi.org/10.1007/978-3-030-25879-5_19

need of a search for qualities of regional vernacular architectural aesthetics, use of space, and people's perception toward it as against the more formal Bauhaus kind of design elements of formal urban spaces and its built environment. But the chapter does not end in contrasting and likening the two. Rather it takes another step to bring together the best of the two worlds. Therefore, integrating indigenous knowledge with modern technicalities may help to promote, sustain, and alter people's perception toward both vernacular and modern forms and patterns, and arrive at a design vocabulary of the third kind. Hence the chapter has been segmented into three sequences of discussion, i.e., firstly, indigenous architecture as explored through the analyses and discussions of vernacularism and regionalism, on the one hand, and contemporary formal architectural aesthetics, on the other hand; secondly, a composite focus on the range and gaps of the experiences in cultural and social contexts based on the twin cases and its allied people's perception; and thirdly, at a need to delineate an attempt to chart out recommendations on the basis of an innovative and creative approach to integrate vernacular into modern architecture and vice versa. With these three successions of discussions, the deliberation hopes to alter general perception toward the extremities, like "Indigenous" and "Modern," and portray a new design vocabulary of the third order.

Keywords Forms and patterns of perception · Vernacular architecture · Modern architecture · Perception · Pattern of space · Integration and a third order

Introduction

The recent pattern of urban and regional development shows a trend of rapid urbanization in the developing countries. It has caused a huge gap between the true shades of aspirations that human perception seeks and the imaginary propensities of material life that are featured by current notions of human comfort. The gap has further caused significant degradation in the livability conditions, both in the rural regions and in the urban environment, and it has also accelerated the loss of healthy linkages between variety in culture and habitation, people and their habit, and spaces and the appropriate placemaking. As the "sense of good placemaking" destabilizes, it triggers a certain shade of inconsistency in cultural continuity. It hinders emotional attachment between the stakeholders and their spaces. On the contrary, the trends of globalization and modernization have begun to facilitate with a whole new virtual world to explore and create perceived pleasure more in the unreal and less in the natural world. Hence, the global platform has now become the periphery to an understanding of a good interrelationship between a variety of community spaces, walked, lived, and perceived. On the one hand, the departure from the core to the periphery has set off a journey from an indigenous (introvert) community system to more and more objects, notions, and perceptions that are guided by the global (extrovert) phenomenon. The variety in the local and the regional are suppressed and altered in the battle of this perceived alienation. On the other hand, exposure to global phenomena has subtly altered people's perception toward their own culture. It promotes individualization, creativity, and diversity and, simultaneously, alters values, habits, customs, and preferences to shape the way of life and looking at things. Thus, there is a scope of transition and iteration between the two. And perceptions of culture may evolve, through one or many cycles of forward and backward linkages between the two.

The present note is an attempt to forward that "perception toward culture" and "experience of a space" are integrally related to each other. The former has the potential to alter the experience of place and the people or vice versa. It transforms the ways of apprehension. Perception forms a mental map to be imaginable based on what is imagery and being legible, audible too based on experiences of a space. As a person moves (walk), views vista and physical features changes and external stimulus accordingly, apprehend and forward the imageability of that particular space. Creating an adequate environment for walking further generates local economic activity Owen et al. (2004), Berrigan and Troiano (2002), Giles-Corti (2002) and is also ecologically benefitting. Hence, there is a need to understand a greater range of perception from a complex mosaic of economic activities that is battling between what is strictly urban and more urban to more interspersed, organic ribbons of the economies of rural settlements.

At present, overtly populated urban cores have lost its indigenous character, which had once steeped into history and heritage. People have lost their roots to the space that further shatters the emotional attachment and the sense of place. Thus, there is a need of transit, both in the mind (intangible) and the body (tangible). There is a further need of departure from the services that cater to physical needs only, to suffice the need of the mind. Incorporation of aesthetic elements in augmenting the features of physical, tangible heritage helps to regenerate the larger meaning of space. Therefore, this chapter concentrates on the betterment of experience based on aesthetic regeneration of the built environment based on the cited experience of a transition.

To best explore the ideas of aesthetic regeneration based on the perception of culture, two living distinctive spaces have been selected. The first one is Jorasanko Thakur Bari, located in the heart of Kolkata. It is a space where poet Rabindranath Tagore was born and brought up. The other is Santiniketan, a living site of experimentation on modern education and space design based on the recovery of India's natural heritage and values of the landscape by the poet himself. Accordingly, the chapter further attempts to tap Tagore's ideology of an ideal society based on the transition and the symbiosis of the two spaces posing to altered versions of humanity and nature. On the one hand, it is a reminiscence of the poet's upbringing against the British dominion in Calcutta, and on the other hand, it is a reflection of his journey from a city's core to a rural hinterland. To address this, the chapter brings to life and revisits a history which is about a hundred years back.

Today, Santiniketan is a site rich in cultural heritage; and it generates enough economy through national and international tourism to sustain itself and its surrounding areas. While Jorasanko area is steeped with influx of people who tend to move from fringes or peri-urban to the core city, it transformed to a space created or surrounded by public squalors, blighted areas, and degenerated spaces. On the other hand, the serenity of Santiniketan is also much hampered by a battle between an academic institution not able to preserve its original goals and governance considering increase in the footfalls. The simultaneous movement shows development in two directions and two histories of change, one following humanity, nature, and cultural orientation and, now, not able to preserve it, and the other facing cultural, infrastructural, and social challenges, which once was a citadel of Calcutta's renaissance. Hence, this difference of perception serves as an inspiration to study the two diametrically opposite spaces, both undergoing dialectical transitions and both connected to the life and celebration of the great visionary Rabindranath Tagore. To best forward the argument, discussions are initiated in the form of literature review inclusive of study approach and the cases, followed by one-to-one analyses of each, considering the view of the altered perceptions.

Review of Altered Perception of Culture and Built Environment

A brief literature review has been done to address the issues of altered perception toward culture and its impact in shaping the imageability, acceptability, and perception of the built environment and its adjunct spaces. The review is on the exploration of spaces from two different perspectives, i.e., (a) built environment and (b) perception of culture.

Built Environment

Human civilization is based on the extraction of resources for safety, survival, and comfort (Sarkar 2005) that guided human beings to design "structural and functional aspects of built forms." Therefore, ancient cities are developed organically, and these were in response to people's needs following a human scale, promoting the culture of "live-work" based on natural resources. But industrialization resulted in rapid urbanization and consequent congestion that begun to undermine the quality of health and well-being. Hence, there was a revival to include nature in the "park movement" of the late nineteenth century. Lewis Mumford opined an organic relationship between people and their spaces, while Sir Patrick Geddes postulated the idea of fundamental "holism" unifying the triad of "place-work-folk."

However, postindustrial and modern planning felt the urge to segregate industrial belts from the "livable" residential areas. The "Garden City Movement" (Eckert 2009) by Ebenezer Howard promoted the induction of nature in the form of a green belt to develop a self-sustained community system. On the other hand, the idea of "Broadacres city" by Frank Lloyd Wright introduced multicentred, low-density, auto-oriented suburbia. Finally, it had culminated to an urge to revive the live-work culture through "City Beautiful Movement" under the influence of eminent sociologist and urban planners like Daniel Burnham, Ebenezer Howard, and Frederick Olmsted. It was a revival of the "Agrarian," as it was once advanced by Thomas Jefferson.

At present, urban spaces and built environment are considered to be unique resources to promote culture and heritage (Landry and Bianchini 1995). It is expected to be diverse and tolerant (Florida 2003). Thus to design a successful place, it is important to explore its local culture and heritage. Addressing contemporary spaces in the urban and rural backdrop can help to better understand the scenarios.

Santiniketan, considering its rural backdrop, was one of the first ever experiments in space design by Rabindranath Tagore. He shared his initial and original views with the subsequent practices of "place-work-folk" triad of Patrick Geddes. Tagore considered built environment to manifest the relationship between nature and human beings. He has contextualized the design elements as "nature-tradition-originality" (Bhattacharya and Sen 2017) as per the rural settings of Bengal.

Therefore, if put in a nutshell, design-based regeneration of built environment has to consider three broad aspects, namely:

(i) Components
(ii) Activities or functions
(iii) Scale of interventions

The three aspects can be further categorized in terms of a detailed taxonomy (Table 19.1).

Table 19.2 represents the relative positions of Jorasanko and Santiniketan in terms of components, activities or functions, and scale of interventions in the built environment.

Table 19.1 Taxonomy of components, functions, and scale of intervention

Components	Activities or functions	Scale of interventions
Physical infrastructure	Urban context and functions like land-use and transport at city scale	City scale
Social infrastructure		The neighborhood scale
Economy	Pedestrian supportiveness and experience (at human scale)	People scale
Nature		

Table 19.2 Position of Jorasanko and Santiniketan in terms of their respective built environment

	Jorasanko	Santiniketan
Urban context and functions	Degenerated city core	A self-sustained community system rural hinterland based on education and tourism
Pedestrian supportiveness and experience of space	Unsafe and unfavorable for pedestrians due to excessive usage. No or negligible sense of space	Walkable neighborhoods but today, under the threat of degeneration due to excessive use of informal transports (Totos) increased tourism
Scale of intervention	No distinctly visible scale of intervention with respect to people and neighborhood level	Human scale, neighborhood scale design of spaces

Perception of Culture

Cultural value shapes society's way of life and has the potential to bring about the change of attitudes needed to ensure the achievement of sustainable development. Cultural sustainability involves efforts to preserve the tangible and intangible cultural elements of society in ways that promote environmental, economic, and social sustainability (NZMCH 2006). The quality of the built environment is the product of the sociocultural environment. Social needs guided the forms of the buildings according to its functionality, namely, social, political, economic, religious, and cultural. The structure and the physical forms are not only guided by the size, appearance, and location but by the society's perception, culture, and ideas and its economic and social organizations, its distribution of resources and authority, its activities, and its beliefs and values, which prevail at any one period of time (Ghinita 2016). By studying the connection between culture and the built environment, one can determine how architecture, as well as the larger built environment, creates better imageability of space.

Culture and values are significant components of society, but these values are dynamic; they evolve over time and with a sociopolitical alteration. Thus, culture first alters the ways of seeing and then creates paradigm shifts in the worldview on sustainability (Fithian and Powell 2009). Hence, along with economic, social, or environmental challenges, creativity, knowledge, diversity, aesthetics, and beauty have become the unavoidable basis for dialogue for peace and progress as these values are intrinsically connected to human development and freedom (UCLG 2013).

Of the two study areas, Jorasanko and Santiniketan, the former is facing the economic, social, and environmental challenges to retain its former glory. As can be seen in Table 19.3, the lack of aesthetics, sense of place, social solidarity, and rootedness of the stakeholders, Jorasanko area has caged in a state of urbanization and flux.

On the contrary, in Santiniketan, there has been a growing recognition that culture and the environment function as parts of a single system, and that they cannot

Table 19.3 Position of Jorasanko and Santiniketan in terms of perception of culture

	Jorasanko	Santiniketan
Learning about people's behavior	Lack of social solidarity, ethnicity, and sense of place	Sense of place and strong community bonding
Understanding and analyzing culture	Diverse ethnicity, social, and economic status quo	Promoting nature-human symbiosis in terms of built environment and presenting culture as a way of living
Defining context	Cradle of Bengal Renaissance	Hub of experiment with modern alternative education system, rural construction, and self-sustained community system
Maintaining the historic context that strengthens the sense of community	Lack of contextualization of the strong history	Harmonious relationship between nature, culture, and built heritage

be understood separately. Culture can be seen as the acting driver of sustainable development since culture-related initiatives have a direct impact in achieving sustainable development (NZMCH 2006; Fithian and Powell 2009).

Therefore, the objective of this chapter can be framed as follows:

1. To improve the human experience and enhance their perception or the imageability of the space
2. To consider human beings as an element of design that demands consideration of their perception, culture needs, and aspirations

The Study Approach

For aesthetic regeneration, one needs to consider the impact of altered perception on culture, built environment, and its adjunct spaces. This chapter probes into two spaces, which geographically, socially, and culturally belong to different contexts, yet manifest the vision of poet Rabindranath Tagore. The analysis consists of four major parts:

1. Firstly, a brief literature review has been done to examine the influence of culture on the built environment and adjunct spaces, dotted with heritage precincts. It also explores the link between culture and peoples' perception (imageability) about the built environment. This literature review aims to clearly identify the knowledge gap on the role of culture on the built environment and perceiving culture (Opoku 2015).
2. Secondly, a description of the past (use of the space in times of Tagore) and present usage of the space. A brief idea of both the sites and its structures.
3. Thirdly, an attempt to identify the available basic physical amenities and the significance and satisfaction of the elements which promote aesthetic regeneration.
4. And finally, a section has been carved out to forward a comparative study of both an urban (Jorasanko) and a rural (Santiniketan) neighborhood to find the design features that aid to regenerate cultural, social, and the built environment.

The Study Areas

To study and analyze the impact of the "perception of culture" on the built environment and the ambient space design, this paper explores two spaces, i.e., Jorasanko and Santiniketan, the flagbearer of two altered perceptions of culture, as perceived and extracted from the life of Rabindranath Tagore. The study is based on features of pedestrian experience and aesthetic regeneration of built environment.

Both Jorasanko and Santiniketan are deeply connected with the Bengal Renaissance. Jorasanko housed the famous Tagore family, who were the proponents of Bengal Renaissance. It is also the birthplace of Nobel laureate and poet

Rabindranath Tagore. The place at present is increasingly surrounded by dense neighborhoods and an overuse of physical infrastructure and sociocultural resources. On the other hand, Santiniketan is a living experiment of modern education drawing in inputs from the best of Western and ancient Eastern values. Santiniketan was mainly conceived and established after the concept of rural, serene hermitages of ancient India, called "The Ashrams." It is the first attempt to synthesize nature, tradition, and originality into a one and ever dynamic social life. Santiniketan propagates an alternative way of life and seeing of modern India. The former, situated in the densely populated core of the North (old) Kolkata, is suffused, degenerated, and lost amidst the engulfing business-driven city core, and the latter is situated in the lush-green scape of Birbhum. Tagore had made a transition from the first to the second, and the departure marked a revival of a lost philosophy of live, living, and livability, rooted in the Indian spiritual ethos.

For the sake of a brief discussion on the study of the two places, the subsequent portion of the chapter is broadly divided into two parts. Part 1 deals with the locational details of the places to understand the journey of Tagore moving out of the city into the abode of nature. And part 2 attempts to tap the present scenarios based on the altered perception of the two. The second part first concentrates on the availability of the elements of physical infrastructure, as only by securing the basic needs one can aspire for the fructification of cognitive and aesthetic needs. Then the part explores the presence, availability, and arrangements of the elements of aesthetic regeneration. Hence, the places are explored in terms of aesthetic sensibility, perception of culture, and users' experience and satisfaction, which the exploration intends to underscore.

Part 1

Location and Historical Significance of Jorasanko and Santiniketan

The first study area is Jorasanko. It is a heritage precinct caged in the overtly populated urban core of Kolkata. The area is currently subject to rapid degeneration under the pressure of encroachment, coupled with lack of basic infrastructure, and burgeoning population pressure due to proximity to the biggest economic hub, i.e., Burrabazar.

Location and History

Jorasanko is located in Rabindra Sarani (earlier Chitpore Road). In the early twentieth century, Rabindra Sarani was described by H.E.A. Cotton as:

"The great thoroughfare, which commencing in the extreme south, assumes the various names of Russa Road, Chowringhee Road, Bentinck Street, Chitpore Road, and Barrackpore Trunk Road, forms a continuation of the Dum Dum Road and was the old line of communication between Murshidabad and Kalighat. It is said to occupy the site of the old road made by the Sabarna Roy Choudhury, the old zamindars of Calcutta, from Barisha, where the junior branch resided, to Halisahar, beyond Barrackpore, which was the seat of the senior branch." (Cotton 1909)

From the aforesaid description, the importance of the location is evident. At present, the area is under Kolkata Municipal Corporation. The area stretches across wards 21, 22, and 23 (Fig. 19.1).

Tagore's family house is the most prominent landmark of the area, traditionally known as the Jorasanko Thakur Bari. Jorasanko also housed the home of the Singhas (including Kaliprasanna Singha), the Pals (including Krishnadas Pal), and the families of Dewan Banarasi Ghosh, Gokul Chandra Daw, Narsingha Chandra Daw, Prafulla Chandra Gain, and Chandramohan Chatterji. "The area thus became the cradle of Bengal Renaissance" (Nair 1991). It was earlier known as Mechuabazar (Deb 1991). Today, crossing a span of ten decades or so, the area has undergone a loss in maintenance of older building stock and adaptation of the existing cultural heritage by the new forces of gentrification and commercialization.

On the contrary, Santiniketan is a university town near Bolpur in the Birbhum district of West Bengal, India, situated approximately 180 km north of Kolkata (Jorasanko). The original Ashram around the core-green verge (Chatimtala) was established by Maharshi Devendranath Tagore. Later, his son Rabindranath Tagore transformed the Ashramic complex as an internationally acclaimed university, Visva-Bharati, which was also a nonconformist movement based on Ashramic patterns of imparting education. Figure 19.2 delineates locational map of Visva-Bharati and its adjunct land-use pattern. Santiniketan is a site of experimentation in terms of modern vision on education based on a balance of ancient and modern constructs. And it was parallel to the education system of a colonized, industrialized society. The area chosen is unique in terms of use of spaces and elements of design and use of traditional architecture and construction techniques. The order and a blend of both the ancient light and the new light of modernity rooted in ecological landscapes are even today way ahead of time.

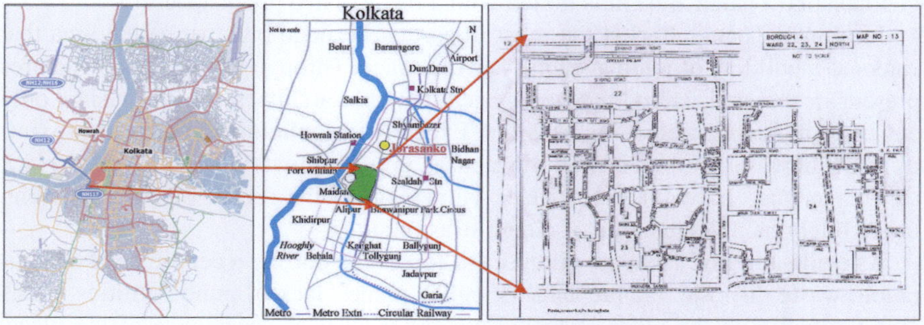

Fig. 19.1 Locational details and map (not to scale) of Jorasanko

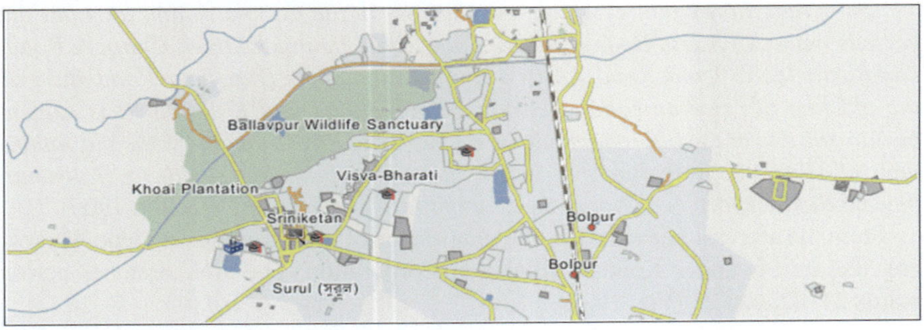

Fig. 19.2 Santiniketan town. (Source: Wikimapia 2007)

Part 2

Part 2 deals with present scenarios of the two study areas. It concentrates on the availability of the elements of physical infrastructure. It then explores the presence, availability, and arrangements of the elements of aesthetic regeneration.

Present Situation of Jorasanko

Following the land-use pattern, Jorasanko is predominantly a heritage precinct, intensely dotted with residential and commercial land use. The place has been an important node in Bengal Renaissance, and it is now buzzing with commercial activities and extends as a part of Burrabazar.

Condition of Physical Tangible Infrastructure

The land-use map of Jorasanko, as depicted in Fig. 19.3, shows the present knit of residential, commercial, and mixed land-use pattern along with the heritage precinct of Jorasanko Thakur Bari. It is dotted with a few isolated green spaces.

Jorasanko Thakur Bari is the most important heritage precinct of the area that caters to a building of an imageability of the space. Through the heritage structures present in the area, including the physical tangible architectures, the visitor only enhances his or her mental imageability. Excessive population pressure, its proximity to the biggest commercial hub Burrabazar, influx of floating population, and lack of maintenance have made the place vulnerable to the forces of encroachment, urban blight, and degradation of older infrastructure.

As evident in Fig. 19.4, the glimpses of Jorasanko portray a certain level of sanitation, waste disposal, maintenance of public places, lack of public utilities in this area undermining the older imageability, and a level of acceptance of the place, which is receding among the stakeholders and visitors.

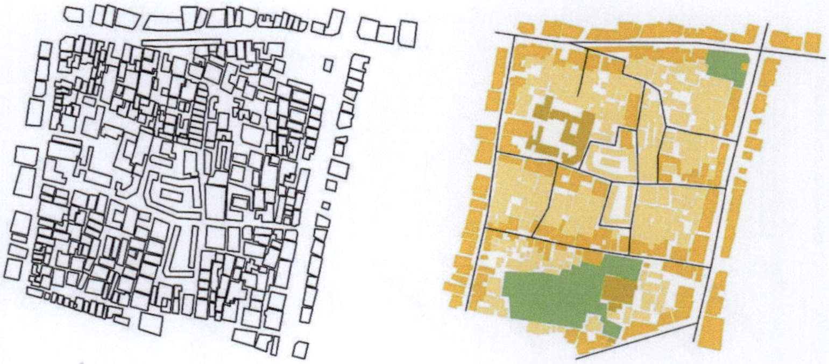

Fig. 19.3 Land-use map of Jorasanko area

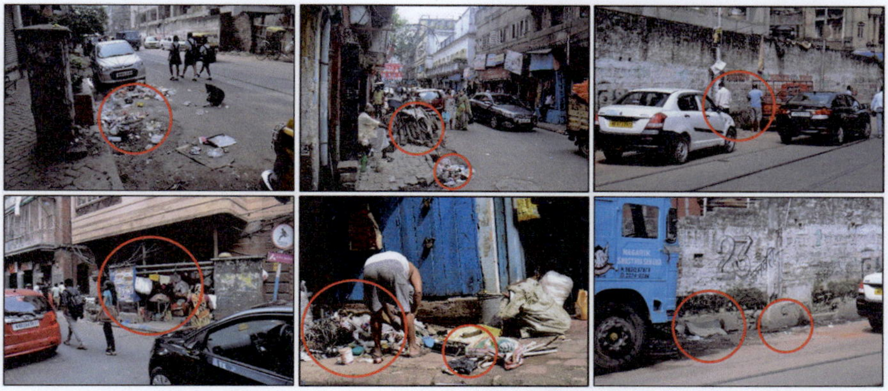

Fig. 19.4 Glimpses of degenerated spaces of Jorasanko

Table 19.4 Consolidated scoring of the significance and satisfaction level of availability of physical infrastructure

	Transportation facility	Drinking water	Sanitation facility	Waste disposal	Healthcare facility	Education facility	Electricity	Information center	Proper lighting	Pollution check	IOT
Significance	8.58	8.65	8.3	8.65	8.7	8.75	8.8	3.85	8.65	7.55	2.7
Satisfaction	7.85	6.3	4.65	4.5	5.9	6.75	8.3	1.8	6.65	5	1.15

Result and Discussion

To get the present imageability of the space, a survey has been done among 40 stakeholders. They were asked about the availability of basic physical infrastructure, their sociocultural life, the presence (significance) and effect (satisfaction) of tangible and intangible heritage on them, and their aspirations regarding the betterment of the place.

Table 19.4 shows the consolidated scoring of the significance and the satisfaction level of the stakeholders about different elements of physical infrastructure.

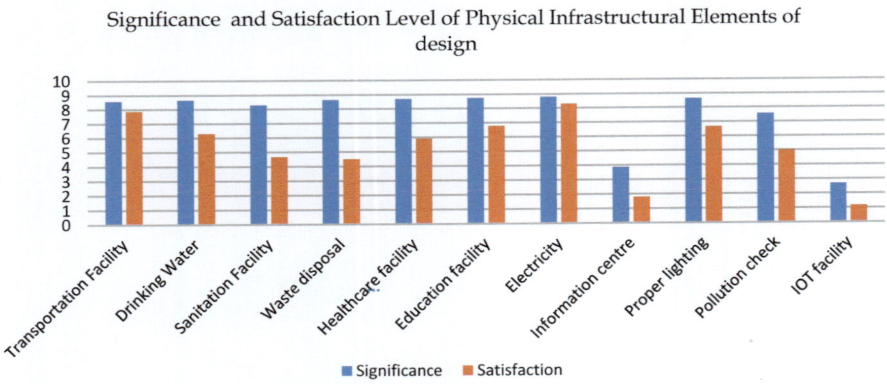

Fig. 19.5 Significance and satisfaction level of availability of basic physical infrastructure

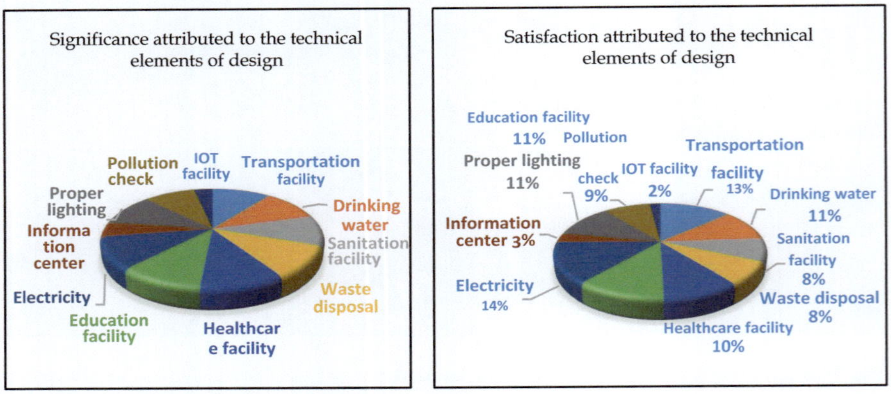

Fig. 19.6 Significance and satisfaction ratio of physical infrastructure

Fig. 19.5 shows some interesting trends while comparing the significance and the satisfaction level of the elements as transportation facility, drinking water, sanitation facility, waste disposal, healthcare facility, education facility, electricity, information center, proper lighting, pollution check, and IOT facility:

- In terms of significance, electricity, transportation, and education get the highest scores successively.
- The highest scoring element, i.e., electricity, has also attributed the highest satisfaction level. As can be seen in Table 19.4, electricity scored 8.3, transportation 7.85, and education 6.75.
- Significance of the presence of information center and IOT facility is as less as 3.85 and 2.7; therefore these elements can be regarded as negligible in terms of availability. The negligibility can be seen in Fig. 19.6.

Experience of Space

Experience of a space considers all the elements related to it, namely, physical (tangible) infrastructure, sociocultural scenarios, elements of space design, built, and semi-built structures, and the natural elements interspersed to it. Availability of infrastructural and aesthetic elements combined to create the imageability and aided to create the experience of a space.

Pedestrian Infrastructure Conditions at Jorasanko

Jorasanko, being in close proximity to Burrabazar, is a place for a large volume of pedestrian and non-motorized traffic (NMT). As Jorasanko falls in one of the oldest zones of Kolkata, addition and alteration of space in terms of pedestrian infrastructure has been done incrementally and thus reduces the coherence and rhythm of the place.

Haphazard arrangements of buildings and structures have led to inadequate articulation of spaces. The buildings with extended porticos are projected on pedestrian walks, encroached up to an extent of 90% at places; hence, pedestrians are forced to walk on the carriage way. Moreover, the newly constructed over-bridge further decreases the imageability of the place. Absences of adequate pedestrian infrastructure, like street furniture and street curbs, and adequate night illumination have reduced the space to a mere means of commute (see Fig. 19.7).

To understand the significance of different elements and services, a survey has been done among 40 stakeholders. Fig. 19.8 represents the significance and satisfaction on the available pedestrian infrastructure. The high level of significance of the elements shows their importance with respect to walking, while they score significantly low on availability and satisfaction. This situation calls for immediate action.

Fig. 19.7 Glimpses of the condition of pedestrian movement at Jorasanko

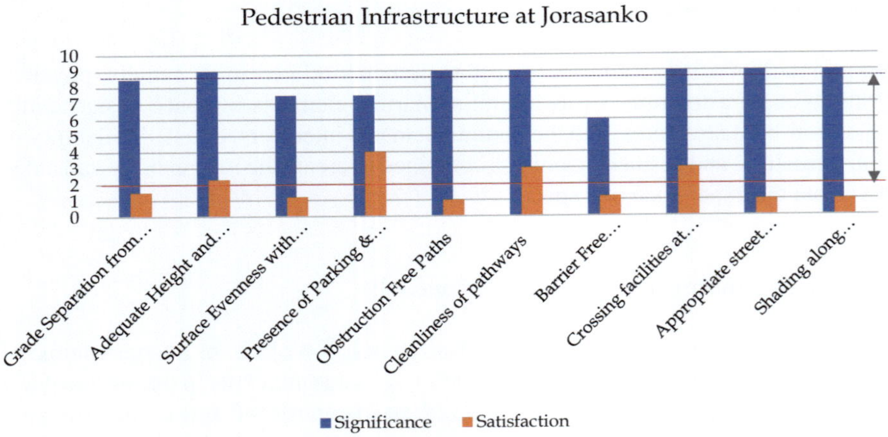

Fig. 19.8 Pedestrian infrastructure: significance and satisfaction

Table 19.5 Consolidated scoring of the significance and satisfaction level of availability of elements of aesthetic regeneration

	Green space	Open area	Common recreational space	Color scape	Murals	Graffiti	Paintings	Sculptures
Significance	8.25	6.75	7	6.1	5.15	4.8	6.2	4.85
Satisfaction	2	2.9	3.15	2.55	2.25	1.4	1.65	1.8

	Art installations	Topiary	Art performances	Performing art	Waterbody	Heritage structure	Economic regeneration	Community space
Significance	4.6	5.7	6.45	7.45	8.15	8.8	6.85	7.35
Satisfaction	1.1	2.55	2.3	3.9	3.45	5.55	2.75	3.95

Analyzing the Elements of Aesthetic Regeneration

After analyzing the tangible features of physical infrastructure available in Jorasanko, a study on the possibility of aesthetic regeneration has been done with 40 stakeholders. Table 19.5 shows the consolidated scoring of significance and satisfaction level of the stakeholders based on the elements of aesthetic regeneration. This chapter considers 16 aesthetic elements to tap on the perspective of the stakeholders and to analyze the subsequent possibilities. The elements are green space, open area (Kent 2008), common recreational space, color scape, murals, graffiti, paintings, sculptures, (Pollock and Sharp 2007) art installations, topiary, art performances (Remesar 1997), performing art, waterbody, heritage structure, economic regeneration, and community space.

Result and Discussion

Table 19.5 exemplifies the significance and satisfaction level of the 16 chosen elements of aesthetic regeneration. Consolidated scoring of all the stakeholders reveals general perception toward these elements (Fig. 19.9).

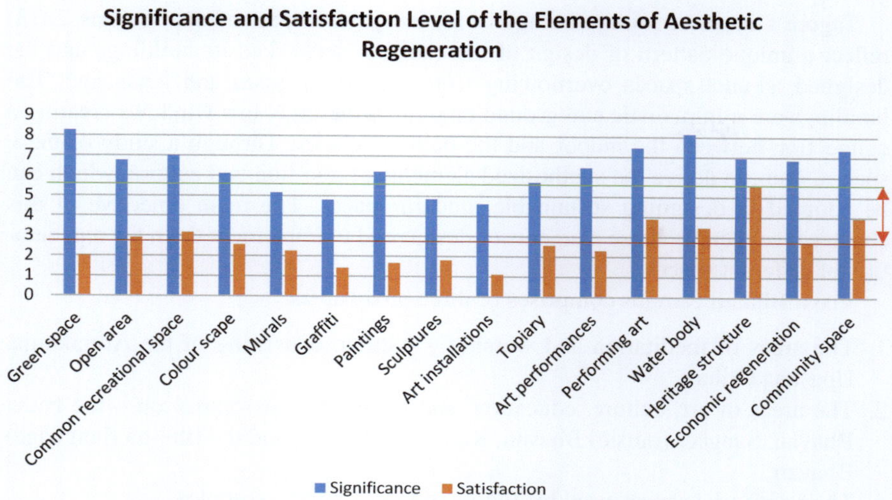

Fig. 19.9 Visual representation of the obtained scores by elements of aesthetic regeneration

1. Perception of the stakeholders reflects that few elements get a significant score of less than 5; therefore, those elements like graffiti, sculptures, and art installations show less possibility of regeneration. Thus, these elements can be regarded as negligible from stakeholder's perspective.
2. Heritage structures, availability of green space, waterbody, community space, common recreational space, and open area have scored more than 6 in terms of their significance. But the satisfaction scores of these elements are 2 (green space), 2.9 (open area), 3.15 (common recreational space), 3.45 (waterbody), and 3.95 (community space). This tendency of the score shows that the possibility of regeneration by implementing these above stated elements is quite high.
3. The gap between the average line of significance (6.52) and satisfaction (2.70) indicates the absence or degeneration of the aesthetic elements of regeneration (if present at all).
4. The difference between the scores (3.89) of average satisfaction and significance indicates the disparity between the present situation and stakeholder's aspirations. Hence, there is a dire need of aesthetic regeneration of Jorasanko area is suggested.

Present Situation of Santiniketan

Santiniketan is a living experiment in terms of education with nature, making of the built environment from nature, rural reconstruction along with modern ones, and a self-sustained community system that is rooted in nature. Within its own ambit, the campus is a cradle of creativity and innovation, which welcomes national and international students and tourists.

Tagore's ideas of human beings as an inseparable part of nature (Das 2013) reflect a unique pattern of design of the campus, networked its buildings and the designed adjunct spaces overflowing from the larger green and landscape. The buildings are minimalistic and ground hugging with large, low windows creating a connection between the indoor and the outdoor spaces. Through a study of these spaces, one can derive the synthesized elements of tradition and culture, which can be adopted in designing sustainable modern spaces. The main objective of this chapter is to identify these elements of design and to implement them for regeneration of different spaces.

Visva-Bharati campus comprises of three broad areas:

1. The areas of meditation and worship – Ashram consisting of the Amrakunja, Upasana Griha
2. The areas of art, culture, education, and knowledge dissemination – the Patha Bhavan, Sangeet (music) Bhavan, Kala (art) Bhavan, and the Bhasha (language) Bhavan
3. The residential spaces used by Tagore in the Uttarayan complex

This chapter primarily considers residential buildings, and spaces of Santiniketan (Uttarayan complex) designed by the poet as residence is a place under maximum personal control; it reflects the ideas and values of the dweller (Marcus 2006). Any additional alteration is very easily possible in this space (Fig. 19.10).

Uttarayan Complex at Santiniketan

The Uttarayan complex at Santiniketan was constructed in the last three decades (1919–1941) of Tagore's life. The premises are a combination of a set of aesthetic elements like buildings (presently converted to exhibits and museums), sculptures by eminent students of Kalavaban, gardens, open area, and sitting spaces designed with utmost anthropomorphic detail and human scale. This chapter is an attempt to analyze the elements of space design and how the stakeholders perceive and experience this space today.

The premises include building as Konark, Udayan, Udichi, Punaschya, and Shyamali along with the aesthetic elements of landscape like colonnades of trees, sculptures, gardens, and waterbodies, arranged after the principles of axis, line of vision, datum, symmetry, etc.

Pedestrian Experience of Space in the Uttarayan Complex

When a visitor actually enters the Uttarayan complex, he or she is not only attracted to a single element, an idea, or a vision but to the holistic simplicity and ambient design of the space. It is very difficult for visitors to identify the reasons for getting drawn to certain locations and liking the place. Since it is difficult to understand the reason behind selection of paths within the complex, observational inferences are depicted in the form of mapping the pedestrian behavior. The view shown in

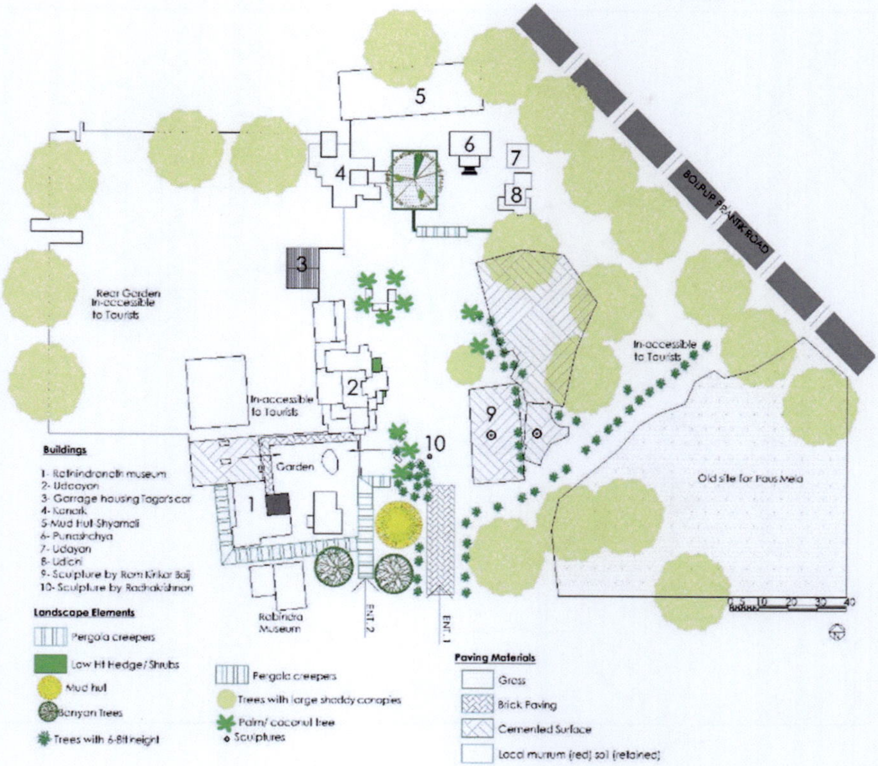

Fig. 19.10 Udayan complex plan

Fig. 19.11 (adopted from (Sanyal 2015)) is an attempt to reconstruct a visitor's pedestrian journey from the entry point.

Accessibility and Path of Navigation

Axis and Vista

Uttarayan is situated behind the Rabindra Museum. This has two distinct approaches from the museum. Both paths are predominantly used and provide different access to reach up to the focus of the journey. The series of images explain the nature of experience of the journey.

The path constitutes of a vista (refer to Fig. 19.12), and a person is attracted to the sculptures of Ramkinkar Baij (refer to Fig. 19.13). After having an interaction and experience, which in today's world can be noticed in terms of people taking digital photographs and selfies, people are attracted toward the architecture of Uttarayan and tend to move predominantly toward the building.

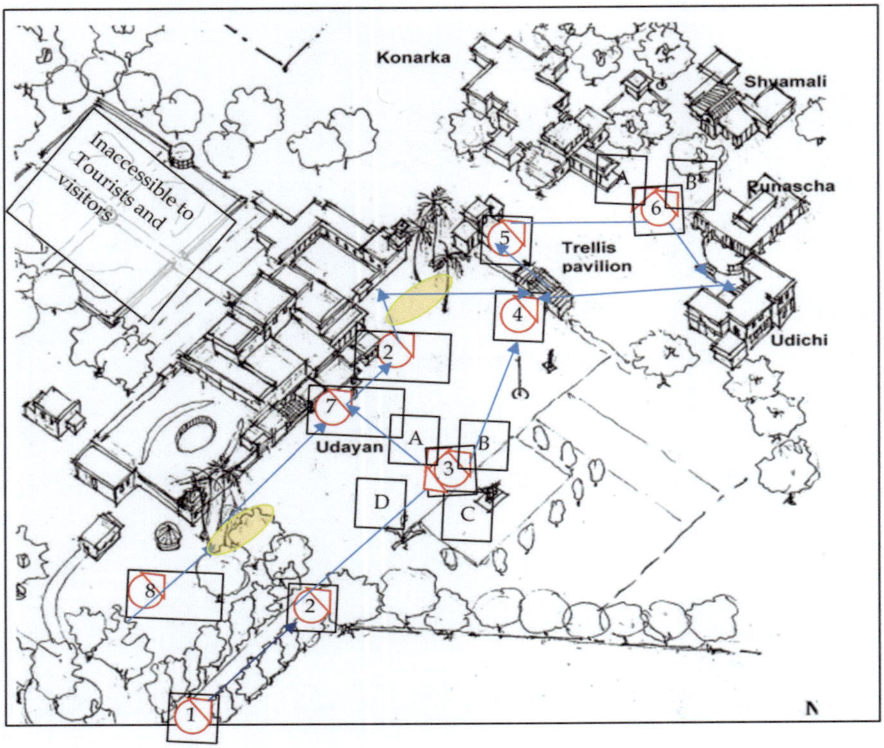

Fig. 19.11 Axonometric view of Santiniketan. (Source: Sanyal 2015)

Fig. 19.12 View from the entrance of Uttarayan complex

Existing and Emerging Views (Purview of "Here and There")

Natural elements like hedges and colonnades of trees are used to create screens around elements like waterbodies, sculptures to obstruct direct vision outside. This creates an element of surprise for the observer (outsider), thereby providing a personalized space for the observed space and its users (insiders).

Fig. 19.13 Explaining examples of existing and emerging views

Fig. 19.14 Way through the shaded pavilion (on the ramp)

Alternative Access

Barrier-free accessibility features have been added to the site in around 2013–2014. The Uttarayan complex is about 450 mm raised from, and the alternative universal accessibility is provided by a ramp covered with creepers and proproots of the banyan. The following set of images describes the journey (Fig. 19.14).

The Spaces

The Uttarayan complex is divided into rusty outer courtyards and the intricately planned inner gardens. Nowadays the inner gardens are not accessible to the visitors. Spaces can be broadly categorized into the following parts.

Outer Courtyards

The trellis pavilion divides the outer courtyard of the Uttarayan complex into two parts. The first one being the Udayan building and the sculptures of Ramkinkar Baij, one facing the other. Both places act as a vantage point to observe the overall site and are the places where people predominantly spent time. The other part consists of a cluster of smaller buildings (Konark, Udichi, Punaschya, and Shyamali) which act as an exhibit where people spent lesser time.

Gardens

Contrary to the outer courtyard, the inner gardens are greener. They also constitute spaces like galleries (vantage points), waterbodies with provisions for sitting, and paved pathway along with them. The following are the views and elements of the gardens (Figs. 19.15, 19.16, 19.17, and 19.18).

Effects of Spatial Forms on the Visitors

Santiniketan is inherently a tourist space, where vistors are exposed to the living philosophy of Tagore, feel his lifestyle, enjoy the built heritage, and finally attempt to derive meaning and relationship to Tagore and culture of Bengal. The site has juxtaposed different architectural elements from around the world, knitted in traditional styles. The design elements of the space respond to the micro-level user requirements. The "subtle" and simple designs coupled with natural elements of surprise, rhythm, and coherence compel the visitors to explore and discover a "place" and meaning of their own. It also serves as a brilliant example for the architects and designers to integrate human beings with nature through designs. Therefore, the holistic design enables people to associate themselves to the space.

Fig. 19.15 A clock at eye level

Fig. 19.16 Entrance to the inner courtyard (gardens)

Fig. 19.17 Gallery of Pratima Devi acts as a vantage point to have bird's-eye view of garden

Fig. 19.18 View of garden from gallery of Pratima Devi

People's Opinion on the Elements of Space Design

To understand the impact of the space, the design elements are analyzed from the point of view of the perceiver. In the present study, around 34 people were interviewed. None of the visitors realized the pedestrian flow pattern but are able to point out the elements of design. The survey reveals that people predominantly visit the place for its heritage value, gardens, and natural elements. The following pie charts represent the people's opinion regarding the design elements based on the following:

1. Age of visitors
2. Reasons for visiting
3. Spaces they liked
4. Things they felt the site was lacking
5. Remarks and suggestions (if any)

Though the site is one of the finest examples to showcase aesthetic sensibilities, the premises lack in terms of basic services like (well-maintained) toilets, when in fact the drinking water fountain is placed at the entrance. Among the remarks and suggestions, people (30%) pointed out that since most of the spaces are inaccessible, there should be detailing on family history, building architecture, its use, and history of spaces. And about 15% of the respondents have pointed out disliking of the tarring and artificial paving (i.e., betterment of infrastructure) toward destroying the natural "feeling" which

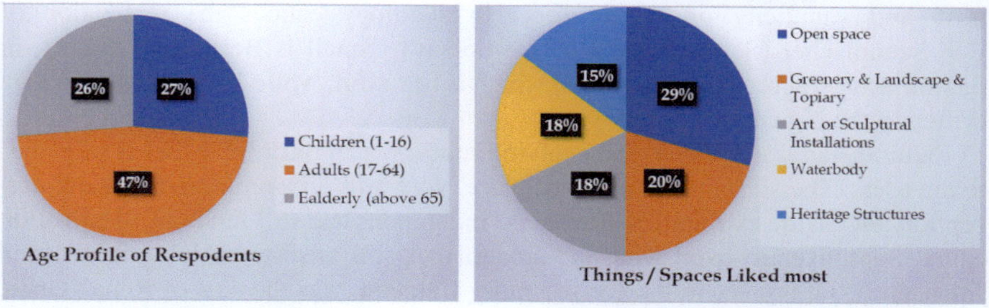

Fig. 19.19 Age of respondents and liking of a particular space

they expect from this space. Hence, the results suggest the need to sensitively combined elements that lead to aesthetic regeneration and basic infrastructure facilities that enhance the imageability without hindering the serenity of the place (Fig. 19.19).

Conclusion

The twin cases of Jorasanko and Santiniketan have formed a single life, the one life of Tagore. Yet, they present two poles of life and evolution of life itself. Tagore, however, makes a departure from one to the other, in search of his own growth and development. And this departure was the very core of his entire creative outburst. Literature, poetry, fine arts, educational pedagogy revival, rural development, the resurrection of local inhabitants, and creative economy are just, to name a few, his plethora of outbursts. The premises at Jorasanko definitely provided him with the first seed to format his multifaceted creativity. But the lush-green landscapes of Birbhum and, particularly, Bolpur provided him with the final earthing, sunlight, and water to germinate that very seed. So deep behind a structured separation of the two, there actually remain a tie and a linkage between the two. The present chapter has forwarded that thread as a deeper layer of understanding.

The observed pattern of innovation and creative approach embedded in the design of Santiniketan reflects the urge to integrate the nature, people, and space. It also reflects the ideology of Tagore toward making an aesthetically sensitive, people-centric, sustainable built environment. Conditions of the built environment as evident in Table 19.2, of the two spaces, i.e., Jorasanko and Santiniketan, portray both the dialectics, the need of a departure, and the required transition to ensure the evolution of human life and creativity. The former one exemplifies the lost heritage amidst the gradually degenerated city core, and in the latter one, a sense of promoting the built environment as an inseparable element of place along with nature, culture, and human being is still strong. The dire difference between the present situation of Jorasanko and Santiniketan has shown the journey and the poet's ideology regarding sustainable space design. Tagore realizes that beauty lies in the truth and simplicity of human life. His ideology echoes the realization of E.F. Schumacher (Schumacher 1973), who says "Small is Beautiful." Finally, in Santiniketan, Tagore found the small which is beautiful, while Jorasanko in Kolkata drifted to the big and the flamboyant.

Santiniketan is a dynamic and evolving space, which intrigues a "sense of place" to the users. The shift in the "sense of place" forwards an alteration of general perception toward the extremities of what is "indigenous and still modern" (destination called Santiniketan) from what is "modern but still unsustainable" origin called Jorasanko). Jorasanko, in the early century Calcutta, was the seat of Renaissance, not just of a family but of Bengal and, perhaps, India. But the growth of the city in later years was not commensurate with these urges. Hence, we experience an epic and biographic Tagore's departure from Jorasanko to Santiniketan. That departure is the evolution of a single lifetime called Rabindranath Tagore.

The flow of population from the village toward the city core saturates the capacity of the physical spaces and eventually triggers a lopsided urban growth causing further decay and degeneration. But such an arrow is an inversion of what Tagore had contemplated in his lifetime. Hence, to best understand the essence of the inversion, a reversal of the arrow is required. The present study of the altered perception of the two cases based on a pedestrian's viewpoint and the possibilities of regeneration of any environment accomplishes this requirement.

Only through a deep understanding of the alteration between the two contrasting case studies, enacted in a single life of Tagore, that shaping of a holistic design and its understanding is best assured. Only then a creative designer becomes holistic and is able to carve out a dynamic "sense of placemaking" that not only enhances uniqueness, creativity, and cultural continuity of the particular, but embraces the opposites leading to the integral perspective of space design. Dialectics in design transform into what is dialogical. The present study has threaded the two altered perceptions to reach this truth. Through one life of Tagore, the study has accommodated different users, who are also evolving over time, like Tagore. Only then a designer may reach the universality and vastness of pan-design semantics, which is embedded impersonally in human evolution itself. For the present chapter, universality and impersonality are reflected in the life of someone so personal and intimate like Tagore.

Bibliography

Berrigan D, Troiano RP (2002) The association between urban form and physical activity in U.S. adults. Am J Prev Med 23:74–79

Bhattacharya T, Sen J (2017) An exploration of design: based on Pan-Asian semiotics and semantics. Springer, pp 751–762

Cotton H (1909) Calcutta, old and new: a historical & descriptive handbook to the city. W. Newman, New York

Das S (2013) Tagore's concept of space. s.l.:s.n

Deb C (1991) Jorasanko and the Thakur family. In: Calcutta, the living city, vol I. Oxford University Press, Kolkata, pp 64–66. ISBN 0-19-563696-1

Doeksen GA, Johnson TG, Willoughby C (1997) Measuring the economic importance of the health sector on a local economy: A brief literature review and procedures to measure local impacts, s.l. The Southern Rural Development Center, Mississsippi
Eckert K (2009) Garden cities: lessons from Germany. Focus 6:52–58
Fithian C, Powell A (2009) Cultural aspects of sustainable development. University of Texas School of Architecture, Texas
Florida R (2003) Cities and the creative class. City Community 2:3–19
Ghinita AM (2016) How buildings influence society and how society is influenced by buildings – an introduction. [Online] Available at: https://www.linkedin.com/pulse/how-buildings-influence-society-influenced-ana-maria-ghinita
Giles-Corti R (2002) The relative influence of individual, social and physical environment determinants of physical activity. Am J Prev Med:1793–1812
Kent F (2008) Creating great urban parks. [Online]. Available at: https://www.pps.org/article/creating-great-urban-parks
Landry C, Bianchini F (1995) The creative city. Demos, London
Lynch K (1959) The image of the city. The M.I.T. Press, Cambridge, MA
Marcus C (2006) House as a mirror of self. Nicolas Hays, Maine, Berkeley
Nair PT (1991) The growth and development of old Calcutta. In: Calcutta, the living city. Oxford University Press, Kolkata, pp 15–17. ISBN 0-19-563696-1
NZMCH (2006) Cultural well-being and local government report 1: definition and context of cultural well-being. New Zealand Ministry for Culture and Heritage (NZMCH), Wellington
Opoku A (2015) The role of culture in a sustainable built environment. Springer, Cham
Owen N, Humpel N, Leslie E, Bauman A, Sallis JF (2004) Understanding environmental influences on walking: review and research agenda. Am J Prev Med 27:67–76
Pollock VL, Sharp JP (2007) Constellations of identity: place-ma(r)king beyond heritage. Environ Plan D Soc Space:1061–1078
Remesar A (1997) Urban regeneration: a challenge for public art. Edicions Universitat Barcelona, Barcelona
Sanyal S (2015) The Poet's home: architectural innovations in Rabindranath Tagore's dwellings in Santiniketan. Int J Arts Sci:593–612
Sarkar A (2005) Morphological approach towards socio-spatial design of creative city: context: post industrial economy. s.l., s.n., pp 1–14
Schumacher E (1973) Small is beautiful: economics as if people mattered. Harper and Row, New York
UCLG (2013) The role of culture in sustainable development to be explicitly recognised. UCLG Committee on culture world secretariat, Barcelona
Wikimapia (2007) Visva-Bharati. [Online]. Available at: http://wikimapia.org/1172957/Visva-Bharati. Accessed 27 Aug 2018

Chapter 20
Gerontology and Urban Public Spaces of Global South: Case of China

Saptarshi Kolay

Abstract Urban public space has multiple facets for successful placemaking. Appropriation of a space is a manifestation of privatization and globalization. Through these various aspects, a public space becomes inclusive. Here the aspect of inclusivity is perceived through the eyes of elderly population of an urban space. Gerontological aspects deal with the need of elderly population of the society, and an inclusive urban design should confront with the elderly-friendly design. A critical analysis of urban public space in South Asian country like China is done with respect to the sociocultural scenario of elderly living condition. Key principles of elderly inclusive urban space are discussed. These principles become the tool to evaluate the elderly-friendly inclusive urban public design.

The reality of an urban public space can be revealed by looking at it from both the aspects of design guidelines, rules, and regulations and the user's opinion about the space. The top-down approach of design guidelines is the depiction of what the city aspires to be, while bottom-up user-centric study reveals the effect of the guidelines and what the elderly people feel about the space. The article analyzes various strategies adopted by urban local body to create elderly-friendly urban public space in China. Study through Delphi method, post-occupancy survey, and ethnography reveals the reality of user's experience about the urban space. Correlation of these factors was drawn to through various literature reviews to shed light on the triadic interrelationship of design guidelines, urban public space, and elderly population.

Based on the discussions on various examples from the country of China, a correlation of an urban space and its elderly friendliness is collated. Possibilities of contextual applications and necessary deviations from the Chinese aspects are there to implement in other countries of South Asia, as they characterize sociocultural similarities as well as dissimilarities.

S. Kolay (✉)
Department of Architecture and Planning, Indian Institute of Technology Roorkee, Roorkee, Uttarakhand, India
e-mail: kolayfap@iitr.ac.in

© Springer Nature Switzerland AG 2020
M. Ghosh (ed.), *Perception, Design and Ecology of the Built Environment*, Springer Geography, https://doi.org/10.1007/978-3-030-25879-5_20

Keywords Gerontology · Elderly · Inclusive design · Age-friendly · China

The people of third age are gradually being isolated by the society due to the urbanization and modernization of family structure in Asian countries. Elderly people are the resource of our cultural roots. To increase interaction between the elderly population and the rest of the society and to initiate the exchange of dialog from them integrating them with the society through are a basic need. Design interventions of today are also not catering to elderly users (Moor 2012), which are acting as a catalyst to debar elderly people from the rest of the society. Social inclusion of elderly can be achieved by redesigning the urban public spaces with higher inclusiveness for elderly, as these spaces have a greater potentiality to be transformed into spaces for interaction among different people of a society. Despite the terminologies like "elderly" or "older people" being coined frequently in academic usage, the absolute definitions of these terms are not universally acclaimed, and the exact age of being "elderly" is subjective. The identified group is also not homogeneous in nature (Victor 2005). There are also differences between "third age" (people aged between 50 and 74) and "fourth age" (75 years old and above) (Laslet 1989).

According to URDPFI, any place or building which is open to the use of public, whether it is actually used or enjoyed by the public or not and whether the entry is regulated by any charge or not (URDPFI 2014), can be classified under public places. Urban public place has a major contribution in the inclusivity and elderly friendliness of a city. According to UTTIPE, there exists a need to redefine the street-design guidelines to increase inclusiveness and comfort (UTTIPE 2010).

The article attempts to decipher the theoretical construct of elderly friendliness in urban fabric. The strategies of inclusions have been discussed for generic concepts of elderly-friendly urban designs.

In the next stage, a case of Hong Kong has been discussed based on an ethnographic study done on this Chinese city.

Strategies for Inclusive Design for Elderly

As described by Elizabeth Berton and Lynne Mitchell in their book *Inclusive Urban Design: Streets for Life*, the inclusivity of public realm is dependent on six key design principles. These are familiarity, legibility, distinctiveness, accessibility, comfort, and safety. Among these principles the initial three factors are visual communication design attributes and directly related toward the aesthetic image of the public spaces. The latter three principles are functional and utilitarian attributes and are not directly linked with visual communication design rather the attributes of transportation, climatology, and system-level design issues.

Familiarity

Familiarity depicts the degree of acquaintance of a space for an elderly people. The more the place looks familiar, the better the inclusivity. More the place looks familiar better the inclusivity, because people feels rooted towards that place and generate a sense of belongingness for the place.

Design features of familiarity are:

- Streets, open spaces, and buildings are long established in terms of visual style.
- Incremental and gradual changes are preferred.
- Hierarchy of spaces has to be maintained.
- Architectural features and street furniture are in designs familiar to or easily understood by older people.

Users, requirements for familiarity of urban space:

- To design with contextual architectural style and space sensitivity. Only time sensitivity can respond to contemporary trends without acknowledging the cultural ethos and familiarity of the neighborhood.
- To achieve hierarchy in design of street scaping. Clarity of pedestrian, vehicular pathway. Clear distinctions between transportation and para transportation network.

Legibility

Legibility refers to the semiotic value of a space, by which people understand the meaning of the space. Legibility of a space can be further divided into three stages, urban design level legibility (macro), building level legibility, and visual communication design level legibility (micro). Urban design level legibility will refer to the clarity of road network. Clarity of entrance spaces of public and semipublic buildings adds to the legibility of architectural design. Well-designed signage design with higher value of understandability refers to visual communication design level legibility. A very contemporary signage design might be difficult to comprehend by elderly people, because of the lesser association of them with contemporary symbols. Universally accepted signage should be adopted by designers. In all these cases: "behavioral design" should be accurate—so in the system image, designer's conceptual model matches the user's mental model (Norman 2006). This attribute refers to the extent street helps older people, where they are and which way they want to go. According to Berton and Mitchell, the tourist information sign shown in Fig. 3 is not legible for elderly due to its extreme contemporary design style derived out of web-based design logo.

Design features of legibility are:

- Efficient road network planning
- Better semiotics for public and semipublic buildings
- Effective visual communication design for street signage

Urban elderly people's requirements for legibility of urban space:

- To establish well-connected road network following a single pattern. Grid-iron and radial cannot be mixed randomly.
- To achieve clarity of visibility of entrance spaces of public and semipublic buildings from access roads. Color coding for very important public buildings will help the users to identify them.
- To maintain universally accepted and traditionally followed logo design for signage.
- To follow the necessary color contrast in communication design.

Distinctiveness

For this character of a public realm, a space becomes memorable. Due to the anomaly of design features, the places can be more memorable and can have a more distinctive identity of its own. According to Kevin Lynch, "landmarks" add to the distinctiveness of an image of a city. The essential characteristics of a landmark are its singularity and uniqueness of architectural features and contrast of figure with respect to background. Visually distinctive utilitarian objects like clock tower and distinctive cycle parking spaces and street furniture can add to the distinctiveness of the urban public space.

Design features of legibility are:

- Context-sensitive anomalistic design in the places of public gathering
- Distinctive of important buildings and spaces

Users' Requirements for Distinctiveness are:

- To achieve distinctiveness by incorporating visual or utilitarian installations and landmarks in public realm design
- To design visually distinguishable features for important built forms and spaces to attract user's attention

Accessibility

The physical parameter for elderly inclusive neighborhood design is accessibility. It incorporates the materiality and design of public realm. It enables elderly to access various spaces, walk around, and provide them sense of independence. Street and walkway design, mode of level changing devices, and smaller design elements like hand rails, ramps, etc. can increase the inclusiveness of a neighborhood.

Design features of accessibility are:

- Comprehensive street layout.
- Segregation of vehicular had pedestrian pathways.
- Minimizing level changes.
- Public amenities should not have level changes.
- Gates/doors with no more than 2 kg pressure to open and levers rather than knobs.

Comfort

Comfortable streets facilitate elderly with the required amenities, which enhance psychophysical comfort. Public amenities like toilet, seating spaces, shades, vehicle-free zones, etc. can be welcoming parameters for elderly. Comfort also includes outdoor thermal comfort. Gust wind can create discomfort to elderly people. Other climatic conditions in severity can impact in discomfort for elderly.

Design features of accessibility are:

- Public seating every 100–125 m
- Ample shading devices for thermal comfort
- Noise barriers
- Some pedestrianized areas to offer protection from traffic

Safety

Safety is a psychological parameter, which enables users to move around the locality without fear. Public realm design can reduce fear of tripping, fear of being lost, and fear from heavy traffic. Sense of familiarity and belongingness can reduce fear to a great extent. Sense of safety can be achieved by micro-level inclusive visual communication design to much larger scale of inclusive urban design.

Design features of accessibility are:

- Mixed land use to provide eyes on the street
- Building doors and windows facing the street
- Signal-controlled pedestrian crossings with visual signals on both sides of the crossing and audible cues at a pitch and timing suitable for frail older people
- Wide, well-maintained, clean footways
- Flat, smooth, nonslip paving
- Grates and drains flush with paving with openings smaller than walking stick or shoe heel size

- Spaces and buildings designed and oriented to avoid areas of dark shadow or bright light
- Street lighting adequate for people with visual impairments

These design principles are the broad aspects for inclusion in urban spaces. These parameters can further be divided into micro parameters. In different countries, with the variance of the social structure and cultural aspects, there may be influence of certain parameters more than the other. For instance, if the city has more crime rate, safety issues will be more important parameter for elderly inclusion. Based on the geographic location, outdoor thermal comfort can be different in various cities. Similarly, topography plays an important role in accessibility, safety, and comfort. Many cities already have well-designed public space and streets, which enables elderly people to navigate freely within the urban fabric. Many of the cities follow urban design guidelines by incorporating proper seating spaces, public toilets, and other urban infrastructure and amenities in the public realm design. Many cities have a color specified for the exterior, which impacts the familiarity of the place. Well-designed public places and plazas add to the distinctiveness of the place, which helps to create an impactful image of the urban fabric.

Digital Technology in Elderly Inclusive Urban Spaces

In this digital era, cyber-physical system can improve all the aspects of inclusive design manifold. Many of the elderly people can be familiar with the usage of digital technology. Navigational web-based applications can help in accessibility. Other applications facilitating navigation, like car-pool apps, can also contribute to this. Weather-related applications can provide weather forecast and suggest precautionary measures. Other urban infrastructures and amenities can be accessible through mobile-based applications. Better accessibility toward information can improve quality of life. Language barrier and hearing impairment can also be minimized through the recent technological advancement.

Global South is an amalgamation of diverse countries; the cities of these countries are situated in different degrees of elderly inclusion. The cities of Global South need to assess the rate of elderly incisiveness of their urban spaces and intervene where it is necessary to achieve a symbiotic relationship between the elderly and society.

Symbiotic Relationship of the Elderly and Society

With the rapid urbanization, sociocultural paradigms are shifting. Interrelationship of social community and urban spaces is changing with the changes of urban fabric. To what extent communities are able to optimize their living within the changing spatial context varies from country to country. And as a manifestation of this optimization, elderly people struggle to be relevant and amalgamated within the urban community.

In 2018, a workshop has been conducted by Hong Kong Institute of Urban Design to explore what elderly people perceive to be, within an urban community, their psychosocial and physical needs, and what urban spaces offer to them. In reference to WHO Age-friendly Cities Project Methodology: Vancouver Protocol, they have conducted design evaluation by participatory approach, for an Age-Friendly City baseline assessment in two districts of Hong Kong, China. Participants were asked to share their views on their respective district of residence and identify aspects of the city they found unfriendly. Data generated from interviews were analyzed using thematic analysis.

The following factors were identified as key issues for elderly inclusive urban design. The failure of public transportation to cater to the needs of older adults is one of the primary factors, which connects to Berton and Mitchell's concept derived in the book *Inclusive Urban Design: Streets for Life,* Lack of public spaces for the elderly for their recreation and socialization can diminish elderly inclusion. Human interactions in welfare services are identified as a basic need, as solo living is increasing among elderly population. Finance, housing, and workplace-related issues have also been identified as important parameters for elderly inclusion.

The report of Hong Kong Institute of Urban Design workshop also highlights the challenges in fostering an inclusive urban community while ensuring efficiency and profit maximization. Apart from the urban design interventions and what community can offer to elderly, it is also important to identify what elderly people can offer to community, and how they can be an integral part of community by offering their service to society and stay relevant and valuable for the community. This helps them to enhance their sense of belongingness within the urban social community.

The concept of active aging and social relevance is also fostered by WHO. If the elderly, repositories of knowledge and cultural legacy, are alienated, balance of social sustainability would collapse. User-friendly urban fabric, a platform for sociocultural exchange, has the potential to augment mainstreaming and visibility of the elderly in public realm. increased mobility and independency of the aged may result in economic independency and strengthen inclusive urban social fabric. In Indian context, URDPFI (2015) defined smart urban spaces; UTTIPEC (2010) suggests use of trees (shade and reduction in solar gain), pavers to reduce UHI, buildings with overhangs, creating eyes on streets for security, crosswalks at grade, tactile paving, handicapped ramp, public toilets, dustbins, postboxes, signage, etc. to increase inclusiveness and comfort in general. Though identification of "Assistive and Enabling Devices" in the category of "Mobility" and "Recreation" for Elderly Indian is commenced (Kumar et al. 2009), third age-centric design is yet to come. Sample Registration System (SRS-Data-Census 2011) classifies three age compositions, 0–14, 15–59, and 60+; and the elderly have been recognized as huge untapped resource (National Policy on Older Persons 1999). West Bengal has the 2nd highest (with Tamil Nadu) urban Elderly population percentage among the larger states (Census of India 2011).

Active Aging and Social Relevance: Elderly to Society

Elderly inclusive urban space "encourages active aging by optimizing opportunities for health, participation and security in order to enhance quality of life as people age. In practical terms, an age friendly city adapts its structures and services to be accessible to and inclusive of older people with varying needs and capacities" (World Health Organisation 2007). In 2005 WHO has identified eight key aspects of age-friendly cities. Among the key parameters, "social participation" and "respect and social inclusion" were of the key aspect. Other parameters are—outdoor spaces and buildings, transportation, housing, civic participation and employment, communication and information, and community and health services. WHO Global Network of Age-friendly Cities and Communities consists of 287 cities and communities in 33 countries, covering 113 million people (Chui et al. 2018).

Today age-friendly communities are gaining importance among urban designers and government policies across the globe; in the context of Global South, countries are far behind with comparison to developed occidental urban communities. Global South has to cross many hurdles, including neoliberal policies and resultant welfare retrenchment (Phillipson 2013), obsolete land use regulations and policies (Lehning et al. 2007), a lack of opportunities or communication with potential organizations to volunteer, shortages of reliable, frequent and accessible transportation (Emlet and Moceri 2012), and underrepresentation of older adults in policy making processes (Alley et al. 2007).

Elderly can involve themselves into knowledge dissemination. Solo-living elderly can be attached with children day care system where a child and an old person get the benefit of each other's company.

Elderly Inclusion in Global South

As the hurdles of infusing elderly inclusion into urban spaces are more in Asian countries because of socioeconomic aspect, Global South struggles to implement them and as a result is far behind from western countries. Though generalization in Asia is not possible, as the context of countries like Japan, South Korea, Singapore, and Afghanistan, Pakistan, India, and Bangladesh are totally diverse in nature; the demand of elderly is increasing as their proportion in the population size increases. Government initiatives did not match the growth of elderly population (Aspalter 2006). Asia today consists of the largest population of elderly population in the world, and Hong Kong, Taiwan, Japan, and South Korea will convert into the world's oldest places by 2050 (United Nations 2015). In this context the urban local body of various Asian countries is developing planning guidelines to create elderly inclusive urban fabric. But the disconnect is there in the design guidelines to understand the real need of the elderly and translating the voice of the elderly into design guidelines (Chui et al. 2018). Though there exist a lot of sociocultural differences between oriental and occidental countries, we see a manifestation of planning guidelines of western into eastern countries. So there exists a disconnect and discord between the urban space and the elderly. For example, Kadoya (2013) found that solo living in elderly people in Japanese cities is common phenomenon. They are reluctant to involve into the community and participate in social activities despite the government's approach of inclusive elderly-friendly urban design. Studies in the context of Taiwan reveal that social relevance and active aging is not addressed holistically. There exist barriers of commination and delivering information to the elderly. Issues of transportation and physical connect of the elderly with the society exist (Chen and Truc 2013). For a significant period of time, Hong Kong also lacked long-term social welfare planning including elderly services (Chen and Truc 2013), though the number of older adults aged above 65 is expected to increase from 1.16 million (16.6% of the total population) in 2016 to approximately 2.37 million (31.1%) by 2036 (Census and Statistics Department 2017). The question arises, in what way global south can include elderly into society and make them relevant for the rest of the urban community?

Urban Design Interventions: Society to Elderly

Elderly-friendly transportation system becomes one of the major parameters of elderly inclusion into urban fabric. Selection of relevant mode of transportation for the elderly based on their economic and physiological need should be catered to. In the case of western countries, though most of the urban elderly population is dependent on private transport, in Global South, public transportation system should be elderly inclusive. Transportation can be the prime parameter to augment their quality of life (Paez et al. 2006). They prefer less physically and cognitively demanding transport systems. A study by Paez reveals, elderly may prefer a slower and cheaper aboveground public transport system based on their physical need. They will avoid level changes and fast moving transport like mass rapid railways and underground metro rails. River-based water transport system in Bangkok works well for the elderly as it is cheaper and safe. Integration of multimodal transport system can also be looked upon for elderly inclusion.

Housings designed for the elderly can induce feelings of safety and belongingness and increase quality of living and are often connected with place appropriation among older adults (Brown et al. 2003; Buffel et al. 2014) and ability to live independently and be relevant for the society. Failing this can negatively impact elderly adults' psychological well-being. The housing developments and the growth of cities also need to cater to elderly need (HelpAge International 2000). Affordability of elderly and low-income group might match in terms of income, but their living style might be different. A solo-living elderly person might need a smaller space but might not be comfortable within the economically weaker section. The housing for the elderly must inculcate the social belongingness and community living. Amenities and infrastructure, crucial for the elderly, must be available within the vicinity of residential spaces.

In cyber-physical era, digital infrastructure can enhance the quality of life for the elderly. User-friendly mobile applications integrated with urban infrastructure can help the elderly to communicate with society and avail amenities and infrastructure with ease (Goyal and Dixit 2008).

Moreover, the social inclusion will come from the awareness of society and involvement of the elderly into society. Club and park dedicated to the elderly and designed for them are few to mention. In the case of Hong Kong, the average amount of open space available per person is 2.7–2.8 m^2, which is lower than other Asian cities like Tokyo (5.8 m^2), Seoul (6.1 m^2), Shanghai (7.6 m^2), and Singapore (7.4 m^2) (Lai et al. 2014). Due to the unavailability of adequate public space to enjoy their leisure time, many elderly people confined themselves into isolation. In many of the cities of Global South, development heavily favored the economic growth rather than the social need of an urban community, which impacted adversely on elderly inclusion. Government need to take initiatives and optimum decision where the conflicting interest of public realm development occurs. Participatory design approach and community involvement may be few options to resolve this problem. In cases of Hong Kong, older residential buildings had better communal areas for social interactions, whereas with increase of land price, real estate developers are decreasing the shared facilities in housing complexes. Thus facilities for sociocultural interactions are decreasing (Lai 2018).

Urban Public Space for Elderly

An optimized urban public space for the elderly needs a two-way interaction between the urban fabric and elderly people. The elderly people should be involved with tangible and intangible aspects of urban fabric—the urban space and the urban community. The space should be inculcating the inclusive urban design parameters, whereas the elderly people should be an integral part of urban community.

While integrating the elderly into urban fabric, urban local body plays a pivotal role, by virtue of holding the power of deploying guidelines, rules, and regulations. They can monitor and take feedback from the residents and implement the guidelines accordingly.

Here a model of this three-way connection between urban local body, elderly population, and urban design interventions is demonstrated to maintain a homeostasis of inclusive elderly-friendly urban fabric.

Maintaining Homeostasis of Inclusive Elderly-Friendly Urban Fabric

Hong Kong Institute of Urban Design has conducted a study on the elderly peoples' perception of design interventions in the city fabric (Chan 2018). The perception of happiness and possibility of social inclusion have been delineated through several

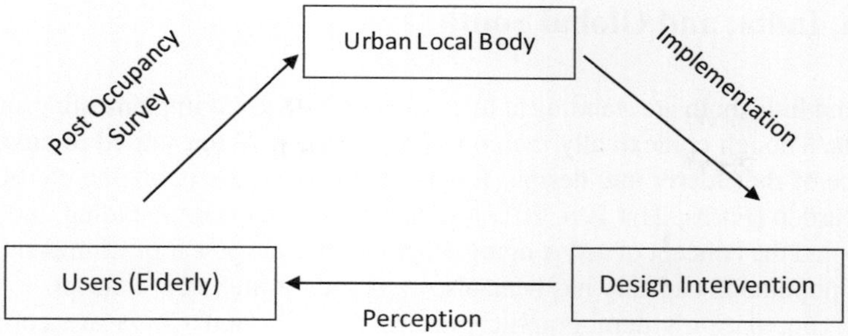

Fig. 20.1 Inductive approach of design feedback. (Source: Author)

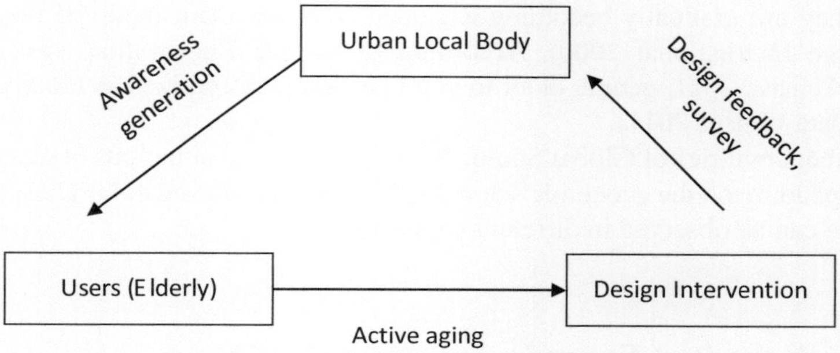

Fig. 20.2 Deductive approach of system design. (Source: Author)

interviews and questionnaire survey. Evaluation process with the triadic node of urban local body, elderly users, and design interventions is established to incorporate a participatory approach for inclusive design.

Through participatory approach a feedback loop can be created as shown in Fig. 20.1. Elderly people will generate their perception of intangible and tangible aspects of urban spaces, while urban local body must conduct post-occupancy survey of urban spaces, if interventions are done. This feedback must reflect into urban design guidelines, and voice of elderly needs to be addressed during design implementations. Figure 20.1 is a bottom-up or inductive approach of design where user feedback is the key element.

In Fig. 20.2, it is shown how urban local body can adopt a top-down or deductive approach of system design to achieve elderly inclusive urban space. Urban local body can have awareness generation campaigns to make elderly people familiar with the urban space and amenities; they can connect elderly people with the society through infrastructure. By adopting active living, elderly people can impact on design intervention, as they become integral part of society. Again urban local bodies can get the feedback of design interventions through surveys of urban spaces and monitor how they function.

China, India, and Global South

China and India both are standing at their inception stages of implementing smart city concepts. Though contextually the concepts will vary, China started to amalgamate the voice of the elderly into design. Elderly populations are not being considered as social burden (Forbs, 21st Feb, 2017); rather through capacity building, the country has adopted the concept of active aging. Population is the power of Chinese economy. Elderly population capacity has been also explored through new concept.

Elder population's demographic profiles of India and China are comparable (Acoca et al. 2014). 7.4% of Indian population is above 60, and the number is growing rapidly (Das 2011). The elderly people are gradually isolated with urbanization and change of sociocultural attributes of urban family structure. Though the third age people are the source of our country's cultural root and repository of knowledge, they are gradually becoming excluded from the mainstream of the society (HelpAge International 2000). According to Sample Registration System data, Indian Census, 2011, people older than 60 years are categorized as elderly people (SRS-Data-Census 2011).

In other countries of Global South, there will be varied situations of elderly people. Depending on the economic scenario, social structure, and cultural aspects, the variance can be observed in different countries.

China: Not a Best Example, but Has a Head Start

Though examples of urban design policies for elderly inclusion in Chinese cities are not the best example and still in a nascent conceptual level, but the genesis of empathy toward the elderly is there. Cities like Hong Kong and Beijing are trying to explore the capacity of huge number of elderly population by providing livelihood opportunities to them and creating urban infrastructure development to enhance independency of elderly population. New smart cities of China are being conceptualized including the voice of the elderly into the city fabric.

On the other hand, India is still to think from this aspect of elderly inclusion. smart-city policies are still very fluid in India. But in this stage, if India can explore an inclusive design strategy, the growing elderly population will not remain as a burden society and be excluded from the rest.

Conclusion

In conclusion it can be stated that a communal integration is required for elderly inclusive urban design across the Global South. Only safe, accessible, and comfortable public realm design cannot achieve the urban inclusion of the elderly; there has to be social inclusion as well. Concepts of active aging should be deployed to make the elderly people socially relevant and connected with more people. This will also help them in livelihood after retirement and strengthen their economy and make them more independent. Inclusive urban design also makes the elderly physically independent. As solo living is becoming a trend, especially, at urban areas, independency will provide elderly people psychological comfort.

Society, elderly people, and urban local body need to be integrated into an ecosystem of design. These three facets have to be correlated to create an elderly inclusive urban space. Here an inductive and a deductive mode of design approach have been demonstrated. The flow of feedback loop is dependent on the interrelationship between these facets. Participatory approach of co-design process, where all the stakeholders come together in the design process, can enhance the probability of social inclusion of elderly people.

Though the article proposes the process of elderly inclusion, contextual variable should be taken into consideration. As the context and nature of cities varies, every country has its own challenges and advantages. There exists a divergence between the cities of western countries and Global South, but each city of global south has its own variances. While adopting the inductive or deductive approach design, local context should be considered, and design should be a manifestation of the voice of the elderly to shape an elderly inclusive urban fabric.

This article provides a broad scenario of Global South in the context of elderly people and social inclusion in urban space. Each city tells its own story and has its own shortcoming. In Salt Lake, a starlight township of Kolkata, India, problem faced by the elderly is the similarity of urban parameters. A study conducted by the author reveals that extreme similarity of built form, and pathways does not lead to a positive parameter of "familiarity" as described by Berton and Mitchell (Berton and Mitchell 2006); it creates more confusion in navigation. The sense of orientation and direction can be affected by similar built form, and absence of distinctive nodes. The grid-iron pattern of pathways adds to the confusion. But the satellite township provides a well-delineated residential neighborhoods; the segregation of various hierarchy of roads provides safe navigation option for the elderly. Public spaces, parks, and community halls are accessible for the elderly. Salt Lake community also includes the elderly, as the area is highly populated by elderly people, and there are social gatherings for elderly people. In the case of Singapore, the scenario might be very different. Urban spaces are much more legible; street canyons are properly designed, though community inclusion might not be much strong, if compared with the Salt Lake township of Kolkata. Japan faces more difficulties in community inclusion, though from design aspect, public realm is much more inclusive in physical design aspect. So, the degree of inclusiveness differs from places to places, and the parameters of inclusion also change. The interventions have to be made keeping in mind the contextual relevance of the city.

Bibliography

Acoca A, Chattaraj S, Watchter S (2014) Urban governance and development of informality in China and India. In: Informal Real Estate Markets. University of Pennsylvania Press
Alley D, Liebig P, Pynoos J, Banerjee T, Choi IH (2007) Creating elder-friendly communities: preparations for an aging society. J Gerontol Soc Work 49:1–18
Aspalter C (2006) The East Asian welfare model. Int J Soc Welf 15:290–301
Berton E, Mitchell L (2006) Inclusive urban design: streets for life. In: ElsevierBuffel T, Phillipson C (eds) Can global cities be 'age-friendly cities'? Urban development and ageing populations. Cities, p 55, 94–100. (2016)
Brown B, Perkins DD, Brown G (2003) Place attachment in a revitalizing neighborhood: individual and block levels of analysis. J Environ Psychol 23:259–271
Buffel T, De Donder L, Phillipson C, De Witte N, Dury S, Verté D (2014) Place attachment among older adults living in four communities in Flanders, Belgium. Housing Studies 29:800–822
Census Report of India (2011)
Census and Statistics Department (2017) Hong Kong 2016 population by-census — thematic report: household income distribution in Hong Kong. Census and Statistics Department, Hong Kong
Chan J (2018) Insight from the HKIUD experimental workshop on active aging. HKIUD Press
Chen PH, Truc TN (2013) Computer-aided visual communication for way-finding in emergency in indoor environment. ISARC
Chui CHK et al (2018) Older adults' perceptions of age-friendliness in Hong Kong. Gerontological Society of America
Das SK (2011) Situation analysis of the elderly in India. Central Statistics Office, Ministry of Statistics & Programme Implementation. Government of India, New Delhi
Emlet CA, Moceri JT (2012) The importance of social connectedness in building age-friendly communities. J Aging Res 201
Goyal VC, Dixit U (2008) Technology interventions for elderly people. HelpAge India
HelpAge International (2000) The ageing and development report: poverty, independence and the world's older people. HelpAge International
Hunter S (2010) Architectural wayfinding. Design Resources
Kadoya Y (2013) Toward an age-friendly city: the constraints preventing the elderly's participation in community programs in Akita city, Working with older people, vol 17, pp 101–108
Kumar P, Dixit U, Goyal VC (2009) Assistive and enabling technology needs of elderly people in India: issues and initial results., Associated Book Service
Lai C (2018) Open space opinion survey. Summary report. Civic Exchange. The Hong Kong Jockey Club Charities Trust
Lai D, Guo D, Hou Y, Lin C, Chen Q (2014) Studies of outdoor thermal comfort in northern China. Build Environ 77:110–118
Laslet P (1989) A fresh map of life: the emergence of the third age. Weidenfeld and Nicolson
Lehning A, Chun Y, Scharlach A (2007) Structural barriers to developing 'aging-friendly' communities. Public Policy Aging Rep 17:15–20
Moor P (2012) Quality of life for all ages, by design
National Policy on Older People (1999)
Norman D (2006) The design of everyday things. Basic Books
Paez A, Scott D, Potoglou D, Kanaroglou P, Newbold KB (2006) Elderly mobility: demographic and spatial analysis of trip making in the Hamilton CMA, Canada. Urban Stud 44:123–146
Phillipson C (2013) Ageing. Polity Press, Cambridge
SRS-Data-Census (2011) SRS statistical report 2011. Office of the Registrar General & Census Commissioner, India Ministry of Home Affairs, Government of India
United Nations (2015) The world population prospects. Department of Economic and Social Affairs, New York
URDPFI (2015) Urban and regional development plans formulation and implementation guidelines. MOUD, 2015
UTTIPEC (2010) Street design guidelines. Delhi Development Authority
Victor C (2005) The social context of aging: a textbook of gerontology. Taylor and Francis
World Health Organisation (2007) Global age-friendly cities: a guide

Chapter 21
Living in Alleys: A Story of Kampung Kota

Achmad Syaiful Lathif

Abstract This chapter discusses about local urban phenomenon in Indonesia called kampung kota (urban village), based on a premise that living space is a manifestation of contextual urban situation into form. From which, its form is actually a dynamic embedded object evolving with the city.

One of its examples is kampung kota in Indonesian cities. It has been common to view urban kampung phenomenon as problem about agglomeration in a city. Alternatively, it is part of the dynamic life and a learning place for surviving in urban space. One of kampung kota spatial elements is alley. Alley is having a function more than only giving a better circulation and connectivity between houses or streets. This chapter discusses about other (social, economic) functions and meaning of alley which factually contribute to maintain stability of connectivity and communication. The discussion assumes dwelling in a settlement as a process of evolvement which contain a transformational capacity and which arbitrarily possible to produce a stable dwelling environment or settling.

In one hand kampung kota can be considered as a "trans-form" within space and like what is underlined by Jan Newberry (Back door Java: Negara, Rumah Tangga dan Kampung di Keluarga Jawa. KITLV-Jakarta & Yayasan Pustaka Obor Indonesia, Jakarta, 2013) that kampung in Indonesia cannot be concluded into fixed forms. On the other hand, kampung has a form, a being which is able to adapt quickly to its surroundings and can be read as a fixed or permanent form through its physical or may be narrative way of living. Eventually, it is better for us to look urban kampung phenomenon not as a problem but an ideal way of urbanizing in the future, which requires more exploration and understanding.

Keywords Alley · Kampung kota · Memetic · Strategy · Tactic · Indonesia · Dwelling

---

A. S. Lathif (✉)
Department of Architecture, Institut Teknologi Bandung, Bandung, Indonesia

© Springer Nature Switzerland AG 2020

M. Ghosh (ed.), *Perception, Design and Ecology of the Built Environment*, Springer Geography, https://doi.org/10.1007/978-3-030-25879-5_21

City Growth

A city grows and expands along the time. It grows in organically or planned way, which is often blurred, so we cannot classify every city only in one type. An organic city is usually more workable, more efficient, more equitable, and more democratic than a planned city. Organic is symmetric involving both growth and decline, while planned change is more asymmetric, frequently embodying growth but rarely dealing with decline (Batty and Longley 1994, p. 8). These two characters of processes help us in the development of our urban area through their works of combination that is interwoven both in space and time.

People develop a city through their work both organic and planned. Organic is usually more humane because it involves a myriad of individual decisions. On the other hand, planned embodies the action of larger agencies (Batty and Longley 1994, p. 8), reflecting the will of the majority through their representatives which often does not represent individual people's will. The myriad of individual decisions in organic city, especially people and its dwelling process, create a more alive built environment. The process is in line with Certau's statement about the beginning of city history at ground level, with footsteps (Certeau 1985, p. 129). Footsteps can be seen not only as steps but also as stages, a process, a slowly growing and shaping phases from nowhere to somewhere, with a set of individual decisions which adapt space or place into their daily habit. Even walking between one site to the other will make a spatial creation (Certeau 1985, p. 129).

Pedestrian network and its walk activity, though has no physical receivability (Alexander 1967, as cited in Certeau 1985), are the true system whose existence connects the city sites. Its true system is effective because a pedestrian cannot be localized; they spatialize (Certeau 1985, p. 129). It is an interesting perspective to understand the process of settling of a dynamic living space through movement and motion. A city is a growing organism, an entity that always changes and makes a solution to survive its problem.

In Indonesia context, how does that happen? Can it be understandable through a discussion about kampung disparity with the city between formal and informality? By looking into this fundamental question, we will disclose a more humane process of inhabiting city especially in Indonesia.

Dynamic Organism

As we know, our cosmic space contains energy and forces that generate human involvement through them. The attachment between space and man is an inseparable relation. "Where there is space there is being" (Lefebvre 1991, p. 22).

Whether we realize it or not, settlement itself is a life energy that modifies and transforms the place. It is a process of continuous survival to settling and dwelling. It is explained further by Doxiadis (1968) in the relationship between spatial forces and settlements, which comprises directional and nondirectional forces. Doxiadis (1968, p. 22) in his narration also explained how the flow of force is shaping our settlement by using five elements consist of nature, man, networks, society, and shells.

> Chronologically speaking, the settlement can be presented as follows: Nature is the container, Man arrives in it, and forms social groups which function as a society. The social group, in its need for protection, finally creates the Shells and then, when it becomes larger and more complex, Networks.

Doxiadis's description of human settlement is built in a model of ongoing or growing development. He calls it morphogenesis, from the Greek *morphê* shape and *genesis* creation, literally "beginning of the shape," a biological process that causes an organism to develop its form. The process itself is working together with control of cell growth and cellular differentiation in evolutionary developmental biology framework.

The growth and development of settlement, organic, or planned, can be referred as the process that is happening in biology (Alexander et al. 1977; Batty and Longley 1994) because the built environment resembles an organism more than an artifact (Habraken 1998, p. 6). The growth and decline within the city, as a result of the development and evolution of human settlement, is happening slowly, just like in biology (in a cell, gametes, organism, etc.). According to Rapoport, their replacement of old forms is often due to prestige value of novelty rather than lack of utility or even not satisfied with their way of life (Rapoport 1969, p. 78).

Alexander explained further about this interesting insight, from which we can learn and give a new perspective about space itself and its natural character as living system. The mysteries of appearance of any spatial system will be more easily understandable when we assume the process which works on the centers, strengthens some of the centers through preserving and enhancing as much of the structure, and destroys as little as possible (Alexander 2002a). The stability within the wholeness (Alexander 2002a) will ensure the sustainability or survival of the system.

If we look around, see the built environment and look at the landscape, we see that people tend to make changes and adapt. If it works well, we see it being replicated in another place. This process helps to keep the wholeness, the living structure which arises from the context (Alexander 2002a). According to D'Arcy Thompson (1917, 1961), the form of an object is a "diagram of forces," and in this sense, the study of form without the processes is meaningless (Batty and Longley 1994). Alexander (2002b) also underlines about failures of twentieth-century architecture for finding the living structure, form, and language because they do not look into the trace of a smoothness of process as the result of continuous stepwise adaptations.

> It is rather as if someone gave you a ruler and T-square and said, "Use these drawing tools to draw a human face." You would say, "But that is almost impossible: the ruler and the T-square create the wrong geometry. A human face is made of different shapes and different shapes and different relationship than can be drawn with these tools.

In this process through time, the built environment shows us their transformation. They also represent values shared with ancestor and passed down to descendants, uniting past and future, and in that similar continuity exists (Habraken 1998). The process underlies in two kinds of activities, namely, "making do" and "mimesis." The first, "making do" forms the city with the work of everyday decision or practice that has structure and system (Certeau 1985). And the second activity, "mimesis" has its role by imitating the nature in the process of space making (Lefevbre 1991, p. 376). By collaborations of these two, living structure in the city is being created and developed.

Kampung Kota

In Indonesia, people view kampung kota as a severe problem. Kampung becomes an agglomeration problem even though it is an essential element that forms Indonesia city. Kampung becomes something contradictory with the city, especially in its visual appearance, and it changes our city into a big village (Widjaja 2013, p. 17).

If we look into Jan Newberry's writing (Newberry 2013, pp. 51, 211–212), there is a consideration of kampung kota like an illusion; I choose the word trans-form to describe what Newberry wrote. She mentioned the recognition of social life in kampung kota cannot be concluded into fixed forms. It is also supported by William's writing about common social structure, "fixed forms can be social consciousness if we live it actively in the relationship (between people) that is real." It is important to remember about the feeling structure in kampung that is defined as a social experience that can be differentiated from other social meanings which have been underway, clearer, and faster formed (William 1977, pp. 133–134).

Even though the term of kampung is used to differentiate between people who live in small and narrow lane in the city and people who live in modern housing complexes or street-siders, which is also known as *gedongan* (the affluent) people, the notion of kampung is signified with the particular lifestyle of urban inhabitant. Silas described kampung as an "indigenous concept of housing and community which caters for mostly low and middle to lower income families" (Silas 1993 as cited in Ida 2017). According to Sullivan (1992), kampung people tend to view themselves as "not-rich," but "not-exactly poor," as the word "poor" has an unpleasant connotation in the kampung (Ida 2017). Kampung is also meant for a settlement which is limited to a group of houses, usually populated by low-income people and united under business and ethnicity. In Javanese and Sundanese, a kampung is a fenced yard or residential area led by one head (Veth 2003, p. 142; Jansz 1906, p. 348, as cited in Raap 2015, p. 139). In the past, at first, kampung is the original settlement entity within Indonesia landscape. But then in 1930, every Indonesia city disbands the kampung autonomy (Faber 1935, p. 44, as cited in Raap 2015, p. 140), and kampung territory falls directly under the government or municipal. This change may be one of the backgrounds that bring contradiction between planned city and kampung kota.

The contradiction makes us see kampung only as a problem, without considering it as a natural way or system of Indonesia dwelling culture. People's values and beliefs, often conveyed symbols, are, and always, expressed in a multitude of ways. It makes built environment, forms, and patterns, that is created by societies extend into, and the extensions of, human psychological realms (Roberts 1996). And maybe this is one of the factors that makes it difficult to define every settlement in the world with precision especially as words like "town," "village," or "hamlet" (Roberts 1996). It is also happening in kampung kota case where we have difficulties to describe and explain how it is formed or the reason why it is not visually and looked good as an idea of habitation or settlement.

Therefore, interpreting the daily language of an alley is important to reveal how is Indonesia natural system of settlement. The formed language within the alley becomes important to be unfolded and disclosed to have a better understanding of how living structure is formed, grown, and developed. "How" becomes crucial for our study because "how" is a process of how to negotiate with the condition, to formulate an identity, or to modulate the character in a continuous stepwise process of adapting (Mohamad 2011) especially in built environment.

Alley in Kampung Kota

As a system, kampung kota concise a network of alleys that make the space. Alley is not only working as a physical element of kampung kota but also working as social and economic investment for ensuring community survival through good connection and communication. Furthermore, it helps kampung to keep attached to the city as a bigger settlement system. Grady Clay defines alley as a path used to conceal things that do not want to be seen from the front road or area (Clay 1978). This definition is different from kampung kota alley that works as a unifying tool of its surrounding entity (Jelinek 1991 as cited in Hans Dieter Evers and Rudiger Korf 2000) which is a consequence of its inhabitant adaptation to survive. They survive by bond of togetherness along the alley. It can be seen clearly from the nexus of neighborhood in the form of houses, windows facing the same alley, so they can meet regularly and help each other lend or loan household goods to each other (Bremm 1988 as cited in Hans Dieter Evers and Rudiger Korf 2000).

To live in an environment like an alley, people usually need a simplified model of the real situation from points, routes, or region (Simon 1957), and this will help the primary function of the brain to create a spatial abstraction. They learned from trial and error to make a guide using a generalized image of the environment (Porteous 1977).

This process helps a person make everyday decisions easily in kampung kota and adapt to changes or modification. Habraken explains it further with the correlation of human control and form of the environment which comprise other mechanisms: physical/form, territorial/place, and understanding (Habraken 1998). These mechanisms shape the habit of people in the alley.

The dynamic structure of spatial changes that Habraken explain is also in line with Santoso's writings about how dynamic life in kampung make an alley becomes a learning place to survive (Santoso 2009). A network of alleys is important for this learning place so people can learn through society in the city, an educational system which is congruent with the urban structure itself, a concept of living and learning as one. The learning system is built from wandering around the city, freedom of choosing whatever you want to learn and what to teach, a decentralized and "enriched" method. This condition ensures the sustainability of human life through better connectivity between a semiprivate space and public space, between houses and streets. Of course, this kind of space will bring a lot of interconnected situation which according to Alexander will emphasize learning network and comprise the city's "curriculum" (Alexander et al. 1977). From this point of view, we may conclude that the alley works as a reservoir for keeping the wholeness and the structure that keep the settlement alive by working as a network that is spatialized.

The Act of Making a Place

Kampung kota especially in its alley is a good sample of how unpredictable condition always happens even in planned area. There are "sentences" that remain unpredictable, even though space ordered by organizing techniques of systems (Certeau 1985). Space itself is a dialectical dynamic (Lefevbre 1991) that respond to particular action like strategy and tactic.

Certeau explains these dynamics as sentences that are becoming a vocabulary of their language, composing a new story through strategy and tactic for the very activity of "making do" (Certeau 1985, p. 48). The "making do" can be described as the practice of surviving, it must be done for keeping our life continuity.

Certeau explains this "making do" in two words, strategy and tactic. Strategy is defined as the calculation (or manipulation) of power relationship that becomes possible as soon as a subject with will and power (a business, an army, a city, a scientific institution) can be isolated. And tactic is a calculated action determined by the absence of proper locus. In short, a tactic is an art of the weak and strategy is the other (Certeau 1985, pp. 53–54).

Unwittingly, "making do" forms our living space with the work of everyday decision or practice which works through its structure or system. To exist in the built environment, a system must be appropriate with the social body and depend on the nature of the system and the circumstances in which it is manifested. This condition put the man who modifies and transforms space as an important agent because built environment is created by virtue of human intervention, imbued with life and spirit of the place (Habraken 1998).

Mimesis

Similarities often found in the kampung, one of the examples is guardhouse or gate which is always present in kampung boundary. Even though this phenomenon arose because of political influence and adapted in city context differently (Kusno 2007, pp. 2–6), this is an artifact that is significant to identify and then reproduced by others. This kind of guardhouse or gate is being replicated in every place, and over time it becomes a distinct character of the kampung. The replicating process itself involves two steps: interpret and then replicate. Dawkins referred this as mimesis or *meme/memetic*, he use *meme* to have the same syllable with the word *gene*.

The theory develops an idea about how culture can propagate successfully in human. Memetic, is the process of mime, copying and making the same behavior with its ancestor (Dawkins 1989). It keeps human survival by transferring the knowledge to the next generation just like gene/DNA in organism. Organisms that assimilate knowledge for increasing their fitness are more likely to survive. Transferred knowledge should be easy to learn because ideas that are easy to express in a particular language or medium will be propagated more easily (Heylighen 1997) and this makes human sustain their survival.

Language of Alley

Arrangement presents in the alley can be used as a form language (see Fig. 21.1). Bhatt and Rybczynski write out about spatial recognition for informal activity: house extensions, workplaces, small shops, trees, public structures, vehicles, and access street (Bhatt et al. 1990). But this is only a few compared with what is actually used by people to communicate, for instance, street vendor cart or mobile stands, motorcycles, plant pot, chair, cooler box, clothesline, etc.

Even though these forms have its regular function (to sit and put something, as storage, etc.), it usually helps to inform others about things happening in the place, it also can mark the area or territory, and sometimes it is being used as a tool for arguing or confronting, or else it becomes a symbol of how the community connects each other. In Jamika (Bandung), we find a row of rented house which is inhabited

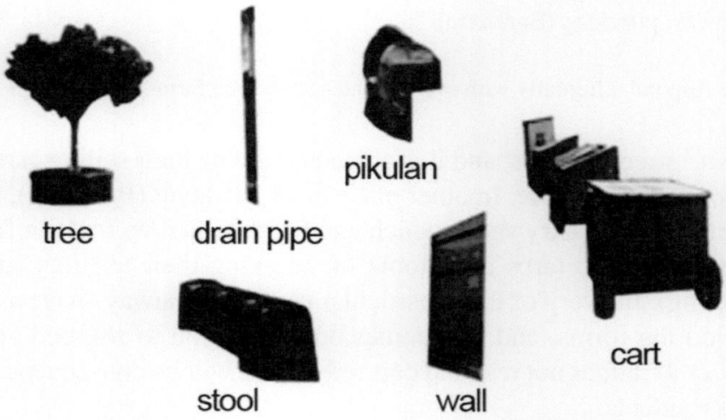

Fig. 21.1 Examples of physical elements that can be used as daily language in alley

Kampung Behind Stasiun Jakarta Kota (Jakarta)

Kampung Behind Stasiun Jakarta Kota (Jakarta)

RT 3 RW 5 Kelurahan Jamika (Bandung)

Gang Cikapundung (Bandung)

Gang Cikapundung (Bandung)

Fig. 21.2 We respond differently with every "sentence" within form

by *bakso tahu*[1] street vendors, and it is characterized by their selling carts which are placed in front of their house. In other place like Cibaduyut (Bandung), there is also a network of home industry shoes which can be identified by tools in front of their house. People also use tarps and stools for marking their territory in Cibaduyut Alley (Bandung). In every of these particular cases, they always have a social norm hidden behind the forms, and it becomes our obligation to respond appropriately (see Fig. 21.2). If it does not respond correctly, there will be consequences like quarrel between residents.

[1] A traditional food that is made as a combination of meat and tofu.

Living in Alley

Unconsciously, people in kampung always do their "making do," by strategy and tactic. It makes a similar language, with the help of memetic. People create similarity to avoid the wrong word in communication. One of the examples is how people build kampung gate in the city (see Fig. 21.3), it still has similar resemblance with village gate, albeit it is being modified according to context, especially considering its durability and local worker in the kampung. The form of the gate is made very clear and distinct with surroundings, and sometimes it has a label so that everyone can recognize it easily. It is a type of strategy to make everyone know that they are entering a kampung.

Placing seats in the alley is another example of a strategy to communicate that this is a place for gathering or meeting (see Fig. 21.4). Location of the seat is usually formed in step by step process following the needs and agreements between people in the neighborhood. According to Alexander, no social group can survive without constant informal contact among its members, and this brings form and location of the communal areas so critical and important. The balanced situation is the one where a common path, which people use every day, runs tangent to the common areas and is open to them in passing. In process, if they want to, they can keep going, come right in, and settle down (Alexander et al. 1977). The position of the

Kampung Behind Stasiun Jakarta Kota (Jakarta)

Kampung Behind Stasiun Jakarta Kota (Jakarta)

RT 3 RW 5 Kelurahan Jamika (Bandung)

Gang Cikapundung (Bandung)

Fig. 21.3 How kampung gate is created in kampung kota

Fig. 21.4 How people arrange their seat placing in strategic location within the alley

seat gives awareness to passers by that they are being watched. This kind of feeling usually makes people walk slower.

Positioning a small shop (*warung*) in the alley is also interesting strategy (see Fig. 21.5). These small shops always place in the intersection – crossroads or T-junction – of an alley. This arrangement makes customers reach small shop easily. In several cases, there are conflicts between shops if one shop has more customer than the other. To avoid this kind of conflict, people usually negotiate their selling goods or maneuvering their place. They negotiate so in the end they can have a good economic fairness.

In kampung kota, we often find people placing something above the alley (see Fig. 21.6). There are many types, namely, tarps, clothesline, and birdcage. We can see it clearly in the case of Cikapundung Alley where every small shop place things and put something above to mark a spatial arrangement in the alley. These imaginary territories are made by placing signage or things, making a shade, or making an enclosure. They are placing things as a strategy, a semi-permanently signs which mark a spatial arrangement, create a comfortable space for activity and adapt to a limited space. On the other hand, people also apply tactic like put a birdcage which usually put at night and then change it again in the next morning.

People also build a bridge between the houses as a solution in some places and below it is still access for public. A clear sign is also added to the place to notify every passers by (see Fig. 21.7). People usually build this bridge after living in the neighborhood for years and have a good bond with the surroundings. The increasing

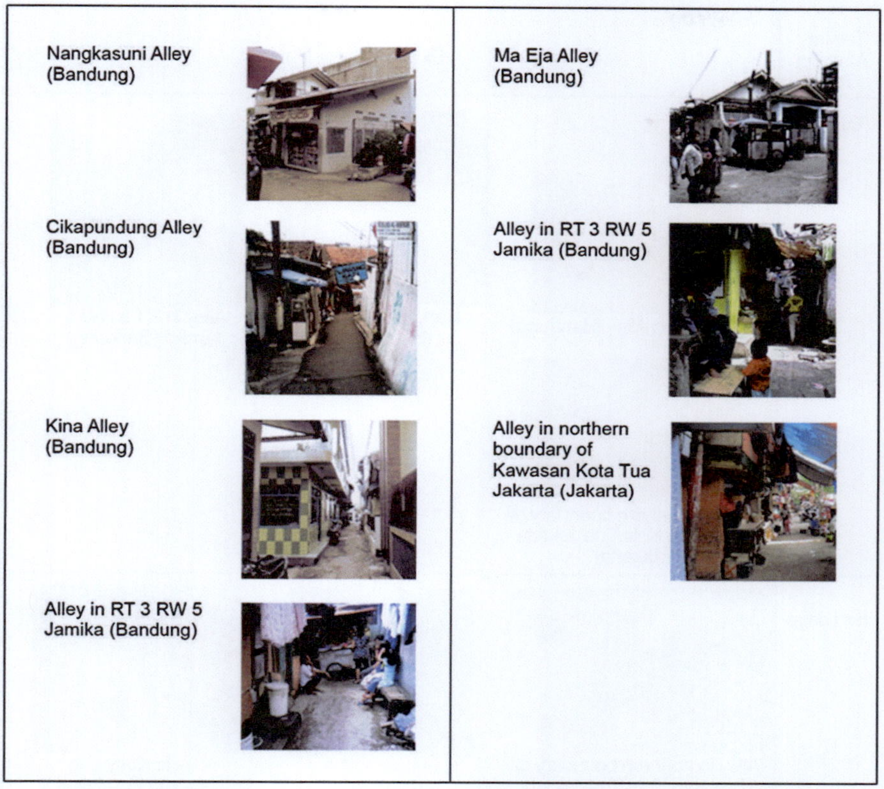

Fig. 21.5 Composition of small shop with the main street in the alley

needs and their investment capital will make someone to make this kind of space modification as long as it will not interfere with other people.

Limited space has caused several practical solutions like bridge between houses. Put one part of their house outside their land or house is one of their practicality, this makes alley a space reservoir. There is another example, the placement of the stairs. Stairs can be placed in the alley, in the public area as long as there is still space for people to walk, even though it is only a small space. They prefer it because it is difficult to make it inside the house and alley has become one part with their house, as a home, so it has become a legit way for people to use alley to place their stairs. But this method also used for personal gain by taking public area secretly with stairs and later on, these kind of phenomenon becomes usual conflict in kampung (Fig. 21.8).

These everyday actions and artifacts like bridge or stairs play a great deal for defining the life of kampung. Their limited resource and capacity make them adopt a practical strategy and tactic based on the environment and condition. Alley at first, in concept, is a public infrastructure for circulation, then it develops into reservoir space for social and economy needs. People keep the alley as public, but in some cases,

Territory Marking Tools	Examples		
Tarpaulin	 Cikapundung Alley (Bandung)	 Nangkasuni Alley (Bandung)	 Alley in RT 3 RW 5 Jamika (Bandung)
	 Alley in northern boundary of Kawasan Kota Tua Jakarta (Jakarta)		
Bird cage	 Alley in northern boundary of Kawasan Kota Tua Jakarta (Jakarta)	 Nangkasuni Alley (Bandung)	 Alley in Kampung Babakan Surabaya (Bandung)
Clothes line	 Cikapundung Alley (Bandung)	 Alley in RT 3 RW 5 Jamika (Bandung)	 Alley in RT 3 RW 5 Jamika (Bandung)

Fig. 21.6 People mark their territory in the alley by putting something

there are people who take some part of the alley as their property. This unconscious action can be understood because house is more than a living unit, it is a structure or area in which human invests emotion, an individual or a group (Porteous 1977). This condition made social control dependable. If someone makes mistake or fault in the alley (like taking someone else's or public property), people will warn or scold the wrongdoer. This normally happens, but in several cases it does not work as usual,

Fig. 21.7 A bridge is built above a public alley as a solution

Fig. 21.8 Stairs placement becomes strategy in the alley

people will ignore it. The ignorance is comes from the high pressure of living conditions and thus prevents them from doing the right thing. Therefore, it usually takes a third person or higher authority to intervene and settle through negotiation if not it will generate conflicts, internal or external. So, in the end, people are usually insisted on their mutual understanding. The kampung society writes a mutual understanding through a verbal agreement which works on fluidity or flexibility of their settlement. It will change to fit with the condition that exists. People keep this flexibility with the same language which works fast and is known by every people in kampung. If there is a new word, people must reach a common agreement or understanding again.

Inside the "making do" in kampung alley, we can see John Habraken's concept of "open building." His concern about how the natural relationship of human and environment in a dwelling (Habraken 1972) can be seen entirely from the alley. The "making do" makes urban space more convivial, and from this continuous stepwise adapting, we can feel the natural living form. In the same time, it can interweave happiness and human dignity emphasizing the fact that it is far more than a mere process of dwelling. Unfortunately, until now we have still not achieved a useful and coherent geometry which helps us laying the stepping stones to create a more livable city. Kampung has a potential of becoming an intermediate agent with how it grows and adapts, and it can make a city more humane and natural, not only a mechanical product that is precisely happening without any souls imbued in it.

Conclusion

In settling or dwelling, people do something to meet their needs. Alley in its context answers as a spatial formation to answer people's needs. Alley not only has its relation to a person but also entangled with a network of people and forms an entity as one kampung. Wholeness becomes important because alley connects a person's identity to kampung and someone is still a member of a kampung if it linked to the kampung through another alley.

Alley works in a way not only accommodating physical needs, but it also supports mentally and can be considered as a life reservoir. As a reservoir, the alley has a homeostasis capability of keeping balance within kampung. Homeostasis, a concept coined by James Lovelock in 1979, tells us about earth's capability of maintaining balance if there is an imbalance or disequilibrium (Dewi 2015, p. 2). This capability can be seen in how fluidly alley gives space to the "making do" activity, and how the modification can still be changed according to context and adapts with the changing condition. It is like the cell wall. The cell wall of an organic cell is, in most cases, as larger as, or larger, than the cell interior. It is not a surface which divides inside from outside, but a coherent entity in its own right, which preserves the functional integrity of the cell and also provide a multitude of transactions between the cell interior and the ambient fluids (Alexander et al. 1977, p. 87).

The presence of alley in kampung is important because it is able to maintain the wholeness or integrity of a kampung. A life reservoir that makes us self-sufficient by providing us a temporary space. This new kind of urban collectivity should present in the city, to fill the gap between asymmetric changes that happens in our fast development progress. We must realize that even though we encourage development by avoiding decay, the system always works as an organic system. It will put the decay somehow in our city. Just like one of Garrett Hardin's paraphrase of the laws of thermodynamics, "We can't get out of the game." This particular thinking makes us realize that between the development areas there should be a room for "escape," to let the aftereffect, force of a development, to find its balance, and to restore its condition. This is why the presence of kampung is so important, because it gives a room between buildings in our city, a space to feel the natural life, to give a little breathing space and then it can bring our humane back. It will be the space where people can live in a relaxed and undemanding way.

Bibliography

Alexander C (1967) La Cité semi-treillis, mais non arbre. Architecture, Mouvement, Continuité

Alexander C (2002a) The nature of order, book one: the phenomenon of life. The Center for Environmental Structure, Berkeley

Alexander C (2002b) The nature of order, book two: the process of creating life. The Center for Environmental Structure, Berkeley

Alexander C, Ishikawa S, Silverstein M (1977) Pattern language: towns, buildings, constructions. Oxford University Press, New York

Batty M, Longley P (1994) Fractal city. Academic Press Limited, London

Bhatt V, Rybczynski W, Barquin C et al (1990) How the other half builds. A three-volume study, December 1984–March 1990. McGill University, Montreal

Clay G (1978) Alleys: a hidden resource. Clay Grady and Company, Louisville

Dawkins R (1989) The selfish gene. Oxford University Press, Oxford

de Certeau M (1984) The practice of everyday life. University of California Press, Berkeley

de Certeau M (1985) The practice of space. In: Blonsky M (ed) In signs. The John Hopkins University Press, Baltimore

Dewi S (2015) Ekofenomenologi: Mengurai Disekuilibrium Relasi Manusia dengan Alam. Marjin Kiri, Serpong

Doxiadis CA (1968) Ekistics: an introduction to the science of human settlement. Oxford University Press, New York

Ever H-D, Korff R (2000) Urbanisme di Asia Tenggara: Makna dan Kekuasaan dalam Ruang-ruang Sosial. Yayasan Pustaka Obor Indonesia, Jakarta

Habraken NJ (1972) Supports. Architectural Press, London

Habraken NJ (1998) The structure of the ordinary: form and control in the built environment. MIT Press, London

Heylighen F (1997) Objective, subjective and intersubjective selectors of knowledge. Evol Cognit 3:63–67

Ida R (2017) The Kampung women's space and place: the look of urban Kampung women in cosmopolitan Surabaya. In: Novianto MC (ed) Surabaya: city within Kampung universe. The Urban Laboratory of Surabaya, Jakarta

Jansz P (1906) Practisch Javaansch-Nederlandsch woordenboek. G.C.T. van Dorp, Semarang/Soerabaja/Bandoeng/'s-Gravenhage

Kusno A (2007) Penjaga Memori: Gardu di Perkotaan Jawa. Penerbit Ombak, Yogyakarta

Lefebvre H (1991) The production of space. Blackwell Publishers Ltd, Oxford

Lovelock J (2000) Gaia: a new look at life on earth. Oxford University Press, Oxford

Mohamad G (2011) Indonesia/proses. Tempo & Pt. Graffiti, Jakarta

Newberry J (2013) Back door Java: Negara, Rumah Tangga dan Kampung di Keluarga Jawa. KITLV-Jakarta & Yayasan Pustaka Obor Indonesia, Jakarta

Porteous JD (1977) Environment & behavior: planning and everyday urban life. Addison-Wesley, Reading

Raap OJ (2015) Kota di Djawa tempo doeloe. Kepustakaan Populer Gramedia, Jakarta

Rapoport A (1969) House form and culture. Prentice-Hall International Inc., London

Roberts BK (1996) Landscape of settlement: prehistory to the present. Routledge, London

Santoso J (2006) Kota Tanpa Warga. Kepustakaan Populer Gramedia, Jakarta

Santoso J (2009) The fifth layer of Jakarta. Jakarta, Centropolis

Siahaan H, Purnomo C (eds) (1996) Johan Silas: Kampung Surabaya menuju metropolitan. Yayasan Keluarga Bhakti dan Surabaya Post, Surabaya

Simon HA (1957) Models of man: social and rational; mathematical essays on rational human behavior in society setting. Wiley, New York

Sullivan N (1992) Local government and community in Java: an urban case study. Oxford University Press, Singapore

Thompson D'AW (1917, 1961) On growth and form. Cambridge University Press, Cambridge
Veth PJ (2003) Uit Oost en West, Verklaring van 1000 woorden uit Nederlands-Indie. L. J. Veen, Amsterdam/Antwerpen
von Faber GH (1935) Nieuw Soerabaia. H. van Ingen, Soerabaia
Widjaja P (2013) Kampung-Kota Bandung. Graha Ilmu, Yogyakarta
William R (1977) Marxism and literature. Oxford University Press, Oxford

Chapter 22
Connecting the Past and the Present for a Better Future of Historic City of Developing Country: Case of Heritage Walks of Hyderabad

Madhu Vottery

Abstract This chapter presents the journey of the author about understanding the city one loves the most. She takes pride in showing its historic and modern pockets to the world. The chapter is also about the challenges the people like her face in the city as they love the local heritage, and they also attempt to resolve the conflict of choosing future over past. In this chapter we shall understand and appreciate how a city in history lives in harmony with its futuristic counterpart. There is a discussion about the solutions to work with a balance between the old and the new. The Heritage, which means 'inherit', should be discussed, but not in close rooms. In open historic areas and with the people who make the true meaning of the word 'heritage', that is, the with younger generation the discussion must go on.

Keywords Hyderabad · Walled city · Golconda · Charminar · Qutb Shahi · Asaf Jahi

A City and Its Being

Indian subcontinent like many other Southeast Asian countries is right in history and cultural background. The fertile lands have given way to growth of old civilizations and important settlements had emerged throughout Asia. Some of the world's oldest cities are still functional in these areas. Hyderabad is one among many historic cities of India. These cities, look quite complex at present, are product of changing culture, art and architecture. The cities that have witnessed many well-known political decisions, wars and victories, today stand unchallenged with its architectural icons – built by patrons under different dynasties. Indian historic cities have beautifully emerged and evolved from prehistoric period to modern period.

M. Vottery (✉)
Telangana Tourism, NSC Cultural Tourism, ICOMOS India, Hyderabad, India

© Springer Nature Switzerland AG 2020

M. Ghosh (ed.), *Perception, Design and Ecology of the Built Environment*, Springer Geography, https://doi.org/10.1007/978-3-030-25879-5_22

Hyderabad is nestled among one of the oldest rock systems in the world which is 2.5 billion years old, a result of volcanic eruption. The Deccan rock, had millions of years of weathering, that has given them various shapes. 'Deccan' comes from the word *Dakshina,*[1] means the south. Some of the unique rock formations can still be seen in the city, fondly called as tortoise rock, mushroom rock, bear nose rock, obelisk rock and many more. Today's Hyderabad city, and a princely state of the Nizams till 1948, is known to have a very rich history.

Presently a well-known tourist spot of the city, the Golconda Fort was originally a Kakatiyan outpost named as Mankal. The Kakatiya rulers (1000–1321 CE) chose this strategic location as it was beautifully covered with different natural barriers. Golconda fort is a unique blend of various fort typologies, namely, *Giri Durg* (Hill Fort), *Jal Durg* (Fort surrounded by a water body), *Vann Durg* (Fort surrounded by a forest), all together make it one of the most impregnable fort. Not only the fort and the security systems, the water systems of Kakatiya rulers are also holding good till date. Hundreds of the villages across present Telangana State are getting the water for their drinking and agriculture needs from the water structures built by Kakatiya rulers, who worked with the concept of tank, temple and town.

Rudramma Devi, the courageous queen of Kakatiya dynasty, ruled for 34 years (1262–96 CE) from the capital city Oragallu (now Warangal, 130 km from Hyderabad to the north-west). She withstood the combined attacks of the Pandya, the Yadava and the Hoysala rulers with the help of her grandson Prataprudra (1296–1323), who succeeded her. During the rule of Prataprudra there were multiple attacks by Delhi Sultans. Finally, Ulugh Khan imprisoned Prataprudra and became the Sultan of Hindustan (Indian subcontinent). Ulugh Khan is better known as Mohammed Tughlaq in the annals of history.

Deccan region that was rich in mineral wealth attracted many evil eyes. After multiple attacks and wars, this region went into the hands of the Bahmani rulers who appointed their *Subedar* (governor). In this period this fort was called Mohammed Nagar. The important phase of history of Hyderabad starts from the time when the Bahmani Governor Sultan Quli came from Bidar (via Patancheruvu, or Potala Cheruvu as named originally) to the Golconda Fort that stood at a commanding location. Quli was said to have shown this site by a shepherd boy hence it was called 'Gollakonda'. 'Golla' means a shepherd and 'Konda' refers to a hillock.

Sultan Quli Qutb-ul-Mulk, founded the Qutb Shahi dynasty after the fall of Bahmani kingdom. One by one all the Bahmani governors became independent, namely, Bidar, Berar, Ahmednagar and Bijapur. Located on Goa – Masulipatnam (later called Machlipatnam) trade route, Golconda – the Qutb Shahi capital, got to be known as a major business centre.

The Qutb Shahi rulers, who ruled from Golconda, added many more sites into this Kakatiyan fort and extended it too. While ruling from Golconda Fort the third Qutb Shah,[2] Ibrahim Quli, planned a bridge Pul-e-Narva, now called

[1] The area of Deccan in the maps have been marked as *Dakshinapatha*, means 'the path to south'.

[2] The next Qutb Shah, Subhan Quli ruled for a very short period and was a minor hence we consider his next ruler as third Qutb Shah-Ibrahim Quli.

Puranapul, 8 km to the south-east of Golconda. He also planned a new township named Ibrahimpatnam. Ibrahim is remembered to have some gardens laid out, one is still surviving called Ibrahim Bagh,[3] near Golconda. Fourth Qutb Shah Mohammed Quli, planned a new capital city, connected with the bridge, later named it Hyderabad; the fortress town of Golconda was becoming overpopulated and plague had spread there.

Saara Sheher logon se mammor kar
Rakhiya jo Dariya mei too meen ya saami.

This couplet was read by the ruler, as he founded the new Qutb Shahi capital.[4]

The couplet meant 'O lord fill my city with people the way you have filled fish in the ocean'.

Charminar at the centre of the new city was ready in late sixteenth century. The new capital city had various quarters, planned by Qutb Shahi Prime Minister Mir Momin to serve different purposes. The four majestic gateways named Kali Kaman (the Black Arch), Machli Kaman (the Fish Arch), Sehr-e-Batil ki Kaman (the Arch of the Magic Breaker) and Charminar ki Kaman (the gateway facing Charminar) were built. These gateways faced the four cardinal directions, built around an octagonal fountain named Gulzar Houz (also called *Char-su-ka-hauz*). It is said that along these four roads were water channels symbolizing the heavenly canals of water, milk, honey and wine.

The Shia mourning place Badshahi Ashurkhana[5] and famous hospital Dar-ush-Shifa[6] were built near the north bank of Musi River at the same point of time. Dar-ush-Shifa (means the House of Health) was a famous medical college and hospital, where people from all over Asia came for treatment under the system of Unani medicine. A congregational mosque-Jama Masjid was built in front of Charminar that had a beautiful *Hamam* (bath) and a caravanserai too. The top floor of Charminar too has a mosque, built in the 1590s. There were two big caravanserais built on the trade route in Qutb Shahi period. One on the west side, close to Golconda, named Shekhpet Sarai and the other on the east side named Hayat Bakhsh Begum[7] Sarai along with a grand mosque (Fig. 22.1).

[3] The sixth Qutb Shah Abdulla built a Baradari here for courtesans Taramati and a mosque was built by another courtesan named Pemamati on a nearby hillock.

[4] The capital was named Hyderabad. It was also called Baghnagar (the city of gardens). It might have been Bhaganagar, as the scholars believe that Mohammed Quli name the new city after Bhagmati, whom he married.

[5] Ashurkhana is a place where *Alams* are kept. The mourning in the month of Moharrum is observed here.

[6] This hospital had inspired the Asaf Jah rulers to upgrade medical facilities in their dominion by introducing modern medicines in the early part of the nineteenth century. The Hyderabad Medical School was set up in 1846, before the modern Osmania Medical College. The city also witnessed one of the major medical discoveries – malarial parasite 'Plasmodium' that is transmitted by mosquitoes. This discovery was made by Dr Ronald Ross (1857–1932); his laboratory in Begumpet is a Heritage site now.

[7] Hayat Baksh was an illustrious woman in the Qutb Shahi history. She was the mother of sixth Qutb Shah, wife of fifth Qutb Shah and daughter of fourth ruler Mohammed Quli.

Fig. 22.1 Mohammed Quli Qutb Shah, who built Charminar

In the period of fifth Qutb Shah, Sultan Mohammed Quli, the son-in-law of Mohammed Quli, the construction of Mecca Masjid started, which is now the second biggest mosque of India. He could not complete it in his lifetime, even his successors could not do it. After the Mughal Attack of 1687, Aurangazeb the Mughal ruler was approached for the funds to complete the mosque construction. He sanctioned the amount but not with a lavish budget. Hence the minarets of the mosque, which were planned to be quite tall, proportionately close to the ones of Charminar, look stunted.

Close to Golconda Fort was planned the Qutb Shahi graveyard, by the first ruler Sultan Quli who built his tomb there. Mohammed Quli built his tomb, with a mix of Hindu and Persian architecture elements. The tombs were built with the best of the Qutb Shahi design and ornamentation skills crafted in Deccani lime plasterwork. The Murda Darwaza connected the tombs complex with the fort to carry out the dead bodies. The monumental domes were built with the squinches on a square base. The calligraphy work done on the graves is still at its best here.

Deccan region, rich in mineral wealth and other natural resources, had good trade links from other continents. It became quite famous during this period. The traders from different parts of north India, mainly chose to set up their trade on caravan route, in the area which is till date called Karvan – from Puranapul Gate (Walled City entrance) to Golconda. Here one still finds the historic localities named Rangrezpura (colony of dyers), Kulsumpura (named after Qutb Shahi noble women Kulsum Begum), Sabzi Mandi, Mustaidpura and Karvan Sahu (the area belonging to the Hindu traders community). The Hindu traders' houses that came up on this route are worth a mention here. They also built walled temple complexes, a unique typology in the historic architecture of Hyderabad, named Raja Bhagwan Das Pavillion,

Lakshman Bagh (on Old Bombay Road) and Sitaram Bagh (in Nampally). The mosques and graveyards were built along with other religious sites, on this stretch.

We also find the Kos Minars (the milestones) on this route, which was part of the East West coast trade route. This road from Charminar to Golconda is 9 km long which has historic water bodies, other than numerous built heritage sites. Starting from Charminar via Laad Bazar, Shahi Jilukhana and the secondary entrance gateway to Mecca Masjid to the left, it reaches Mehboob Chowk. Mehboob Chowk has a historic metal and mechanization market and also a clock tower (built by Sir Asman Jah Bahadur). Around the clock tower is a park, where till date one can see the cast iron ornamental railing on all four-sided compound wall. Further on this road one witnesses a lot of religious structures. This road leads to *Puranapul*, the bridge that was built in 1578, by Ibrahim Quli, more than a decade before the construction of the new Qutb Shahi capital by his son Mohammed Quli. Along with the bridge, here stands a huge gateway now known as Puaranapul Darwaza (gate) that was also built at the same point of time. It served as the main entry point to the people who came to Walled City from north side.

The prosperity of the Qutb Shahi kingdom grew and reached the Mughal ruler. The attempts were made during the period of Shahjahan. After multiple trials, finally the next Mughal ruler, Aurangazeb attacked the city in 1687. Golconda, the most impregnable fort of southern India, and the newly built Qutb Shahi capital city, Hyderabad, were ransacked and slowly fell in partial ruins. Many Qutb Shahi palaces[8] were destroyed. The only remain of these palaces in the Walled City is a Qutb Shahi fountain that can still be seen in a historic complex near the Musi River. The Mughal army left the plundered city with loads of valuables. The last Qutb Shahi ruler Abul Hasan Tana Shah was taken prisoner to Daulatabad Fort (earlier a Hindu fort named Devagiri). He died there in captivity after 13 years. Mir Qamaruddin Asaf Jah I was appointed the *Subedar* (governor) of Deccan. After the death of Aurangazeb, Asaf Jah became independent and founded the Asaf Jahi dynasty that ruled for 228 years.

The new ruler started building the City Wall to protect it from Maratha attacks. The Walled City had 13 gateways and 13 posterns; two of these gateways still stand. Salabat Jung[9] (1751–62) started the construction of the famed Asaf Jahi palace Chowmahalla in the Walled City. *Chow* means four and *Mahalla* means a group of palaces. The main structure called '*Khilwat*' in Chowmahalla functioned as the Durbar Hall where the public audience took place. After the advent of British, in the nineteenth to twentieth century, the Nizams held state receptions here and entertained the British Governor Generals. In 1762, the then ruler Nizam Ali Khan (1762–1803) acquired a palace and added more building to it. He used it as his residence and the next ruler was born here. This palace is presently known as Purani Haveli (Fig. 22.2).

The city of Hyderabad slowly grew with more buildings and recovered the business links. The trader families of Bohra Muslims lived in Sarai Bohra Kalan or Bohron ki Sarai. The Hindu jeweller families lived near Sehr-e-Batil ki Kaman.

[8] Most of these palaces are mentioned to be of three to four storeyed high.

[9] He was not granted the title of Nizam.

Fig. 22.2 Charminar

Near Chatta Bazar was the palace of Diwan Salarjung.[10] Near Chowmahalla, lived the Paigah nobles in a street, called Amir-e-Kabir Galli. The area was also called Shahgunj. *Amir-e-Kabir* was the title given to a Paigah noble.

Paigah means the high rank, and pomp. As per some scholars '*Pai*' refers to foot, and '*Gah*' means the place to rest on. Hence Paigah were the army chieftains and the nobles closest to Nizam. They also had marital alliance with the Asaf Jahi family; Paigah boys married with the girls from Nizam's family. One can still see the Asman Jah Bahadur Devdi,[11] Khursheed Jah Bahadur Devdi and Iqbal-ud-Daula Devdi in the Amir-e-Kabir Street. A mural of sun is noticed in the *devdi* of Asman Jah. The Paigah nobles built their graveyard at the Paigah Tombs, in another locality, now near Santosh Nagar, Phisal Banda. Hyderabad has a French connection too. A young man named Monsieur Raymond reached Pondicherry, India in 1775, as a trader. Sometime later he met French General Bussy and served under him. In 1786, Raymond joined the service of the then Nizam; he began as any other soldier. He was given 300 soldiers to command over a period of time. In 1796. Raymond was appointed *Amir-e-Jhinsi*, Controller of Ordnance. He established several cannon and cannonball manufacturing units. Raymond, fondly called as Musa Ram and Musa Rahim, amongst his army. He had set up more such factories in the erstwhile state of Hyderabad, Raymond died an early death and his tomb is in new Malakpet, Hyderabad. One of these cannon manufacturing units in the city still stands with its

[10] It later on was occupied to be used as the Salarjung Museum.

[11] *Devdi* or *deorhi* is a palatial residence.

Fig. 22.3 One of the gateways of the British residency

tall, massive brick walls and the cubicles supported by brick and lime mortar arches at ground level. They still have brick kiln-like spaces underneath. The area surrounding this site is called Gunfoundry.

Another major political event happened in 1798 that changed the map of Hyderabad and India too. It was known as the Subsidiary Alliance, a treaty between the British East India Company and the Indian princely states. By virtue of this, the Indian kingdoms lost their sovereignty to the English. It also was a major process that led to the building of the British Empire in India. A new cantonment came up in the north part of the city, called Sikanderabad (also Secunderabad) named after the Nizam Sikander Jah. The Nizams paid a huge sum for the maintenance of the British Army. The Subsidiary Alliance also got a major shift in city's architectural style.

British Residency was built in the early 1800s, away from the Walled City, towards the north of the Musi River. Other than the impressive European style of the Residency Complex, it also housed a small Rangmahal in local style with multi cusped arches and *Jharokha* windows. Residency, built by Major Kirkpatrick was considered the most important building in India next only to the Governor General's House in Kolkata (Fig. 22.3).

In the north and away from the Walled City, United Service Club was also built during the same period; it was later called the Secunderabad Club. The club became a hub of cross culture activities. The nobility of Hyderabad moved towards north part of Musi River, the people who were greatly influenced by the British. House of the famous businessman Dewan Bahadur Seth Ramgopaldas was built in James Street. A clock tower was fixed over the structure in 1900 by him.

In the Walled City, the Prime Minister Salarjung built two gateways to his residence, Dewan Devdi. The front/western gate was built in European style where the side/north gate was built in local style. Salarjung family purchased many artefacts, from European dealers and displayed them in different rooms of this house. These articles are now a major part of the collection of the world famous Salarjung Museum. In Devdis of Paigahs in Shahgunj, British architectural façade and other features were added, yet retaining the Deccani planning (courtyard type) in their pre-existing houses. House of Parsi Banker Pestonji[12] close to British Residency (presently ENT Hospital) is worth a mention here. The houses of these eminent people were built on a monumental scale with huge doorways, windows with fan-light on top that was operated with tower bolts and brass hinges. The window glass panels were etched with monograms of the initials of the name of the owner. The residences of the Paigah in the Walled City saw lots of extensions coming up to suit to the British lifestyle, adopted by the nobles.

In 1892, the Falaknuma Palace was built in the south of Charminar on the hillock named Koh-e-Tur, by Paigah Noble Sir Vikhar-ul-Umra (also called Nawab Iqbal-ud-Daula). Influenced by the design of Palace of Versailles, the construction of Falaknuma became another major architectural event in the sequence. Falaknuma means 'the mirror of the sky'. It indeed is the most famous and grandest palace of Hyderabad till date. Nawab Sahib planned for the construction of this palace after his long trip to Europe. Influenced by European art, architecture and culture, he built it with lots of passion.

Later in the period of seventh Nizam, a palace complex was constructed by one Kamal Khan, It has a mix of architectural styles starting from European to Indian. Bought for the Seventh Nizam Mir Osman Ali Khan in 1911 (when he was a prince), the royal palace complex was named as King Kothi (means the King's Mansion). It happened to be his last palace.

Hyderabad also has its own race course. It is said that one of the wives of the sixth Nizam was fond of horses. For her, a palace named Mehboob Mansion and the race course were built in Malakpet in late nineteenth century. During mid-nineteenth century the race course was in Maula Ali.

Since Secunderabad was associated with Europeans, Paigah nobles also moved towards this part of the town in late nineteenth to early twentieth century and built these palaces in Begumpet[13] along with a school. The school earlier named 'Jaagirdar

[12] Pestonji also contributed to the construction of a Parsi Fire Temple (presently located at Tilak Nagar), thereby giving a very secular and cosmopolitan character to the city.

[13] The area came to be known as Begumpet as Basher-un-Nisa, wife of a Paigah Noble got this land in her marriage. The village after the '*begum*', was called Begumpet. The school was built for the children of Jagirdars now known as Hyderabad Public School.

College' (now Hyderabad Public School) was meant for the sons of the nobles. Devdi Nazir Nawaz Jung (presently the Chiran Fort Club building) was built in the year 1890. Paigah Palace, where the present office of the US Consulate Hyderabad is functioning from, happens to be one another magnificent structure of a part of huge palace complex, built by Nawab Vikhar-ul-Umra. Spanish Mosque, a mosque influenced by Spanish architecture, was built within this complex. Now different palaces stand as independent sites and also being put to appropriate reuse.

Another major event was floods in Musi River in the year 1908. A good number of structures were lost due to the floods. The Zenana Hospital on the north bank of Musi River suffered major damages. People climbed on the tamarind tree in the courtyard of the river and saved their lives. The benevolent and secular ruler Mir Mehboob Ali Khan, fondly remembered as Mehboob Ali Pasha, is said to have worshipped the river and prayed for the safety of his people. After the floods the Asaf Jahi ruler understood the need of redevelopment of the city. Some of the city gates were also demolished at this point of time in the process of decongesting the city.

To work on the water systems of Hyderabad Sir Vishveshwaraiah was called from Mysore. The work of City Improvement Board resulted into expansion of the Walled City. The walls were pulled down, new reservoirs like Himayatsagar and Osmansagar were planned. In this period a new architectural style, 'Indo-Saracenic', came up for the public buildings of Hyderabad. Vincent Esch designed the High Court, Unani Hospital, City College, etc. Pathargatti Market was built by the City Improvement Board to the north of Charminar. On the south of the river bank the new hospital was constructed; it was called the Osmania General Hospital. Unani Hospital was built in front of Mecca Masjid. The story of Hyderabad Heritage gets over at 1948, when the Nizam's state was merged into Indian Union after the efforts of the Indian leaders. With the reorganization of states, Andhra Pradesh was formed in 1956 and Hyderabad became the capital of Andhra Pradesh. In 1967, the last Asaf Jahi ruler Mir Osman Ali Khan passed away. In the year 2014, the Andhra Pradesh was bifurcated. Telangana became the 29th state of the Union of India with Hyderabad as its capital. Both states will share the capital for 10 years until Seemandhra (the Present Andhra Pradesh) can establish its own.

Hyderabad has a wonderful mix of art, architecture, language and culture with respect to its people, place and time. It is about the people, who built the spaces, and the period they built it and the architectural impression of the place where they migrated from.

Dakkan hai nagina angoothi hai jag
Angoothi ko humrat Nagine hi lag.

To explain the uniqueness of Hyderabad (Deccan) this couplet is sufficient.

If the world is the ring, Hyderabad (Deccan) is the jewel. The beauty of the ring cannot be described without appreciating the diamond.

Heritage Interpretation: For the Next Generation

Heritage Walk

Hyderabad Heritage Walks have become the identity of the city which is in it's the 10th year now. Under the aegis of Telangana Tourism officers, with the help of a Heritage Education consultant, walk guides and Tourist Police, the Heritage Walks are a happy morning affair at the Walled City. There are four different walk routes

1. **Charminar to Chowmahalla** – South-western quarter, residential area belonging to the Nizam and the Paigah nobles.
2. **Charminar to Badshahi Ashurkhana** – Nucleus of the city about the main market stretch
3. **Charminar to Purani Haveli** – the Spice Walk through a historic market with many religious sites like Alava, Bargah, Naqarkhana are on the way etc.
4. **State Central Library to City College** – the River Walk from north side of the city entering the Walled City

Each route covers almost a dozen of heritage components across different historic building typologies, mapped on the historic route. The walks help people to understand and appreciate the city's culture starting early morning by 7:30 hrs and getting over by 9 or 9:30 hrs.

Priced at Rs 50/- including breakfast, and a walking distance of a kilometre and half, the walks have been loved by locals and also by international tourist. At any given Sunday 15–20% international visitors are part of the crowd. We have conducted almost 700 walks in last 9 years attended to around 25,000 visitors. Walks are designed keeping different values in mind. These values are getting reflected here (Fig. 22.4).

Academic Value

During the period of last 9 years the walks have made place into the outdoor classroom sessions of many institutions. The primary, middle and senior schools book a session at Heritage Walks to learn more about the local history, culture and the historic architecture. The government schools are supported by individuals and by the state government too. The architecture, history, tourism and media colleges bring their students to learn different facets of local art, culture and architecture. The MCRHRD Institute sends the administrative officers for appreciating heritage. Forest Academy trainees join it regularly.

Most of the time when the walk gets over in a museum, (walk 1 and 3) the guide tends to help the visitors and take them inside the Museum too. The school teachers send the works of their students back to the Walk coordinator and take guidance for their class projects in the long run. They also connect with the Heritage Walks for their regular Tourism Club activity.

Fig. 22.4 Heritage Walk I closing at Chowmahalla

Nature Walk

Nature Walk is conducted at Telangana State Forest Academy (TSFA). The Forest Academy has been conducting many public outreach programs to create awareness about the forests and environment, under CAFE program – ‘Children And Forest Education’ from past 15 years. Launched in 242,004, under CAFE, Nature Trails, Eco Sensitization Meets (activities with 3Rs) in schools and road shows have been consistently conducted by the forest officials. A Nature Walk starts by 10 AM and gets over around 2 PM. The school kids are given a brief introductory session where the value of Forest is told to them and basic statistics about the forest cover of India and Telangana State is discussed. Many a times students are also gifted the saplings to be taken to their homes from the Academy and the government run nurseries. The kids walk almost one and a half kilometre inside a man-made forest. Here kids get to see some rare varieties of plants. They also visit a herbal garden. Children also learn to segregate the waste in to various categories. They upcycle the newspaper; using organic glue and jute strings they make paper bags. Nature walk, priced at Rs 1000/- for a group of 100 students, is an all-time favourite of all the schools and colleges. Government schools are admitted for free (Fig. 22.5).

Fig. 22.5 Kids at the check dam, during a Nature Walk at TSFA

Academic Value

The Nature Walk is specially designed with lot of fun learning ideas. Forest at TSFA is home to rocks, man-made water bodies, a check dam, trees, plants, shrubs and many animals. TSFA has 75-hectare campus where they train forest staff at different levels. Some of them also guide the students for this walk, connecting their school/college curriculum with the species from the plant and animal kingdom they witness in the forest. In this forest, students find animals of various sizes starting from the tiny creatures – from avian to aquatic animals also herbivores to carnivores. They sometimes also spot our State Bird '*Paala Pitta*', the Blue Jay or Indian Roller here. This jungle has more than 70 types of trees. Many of them have medicinal properties, some of them are good for timber and all of them together give us so much of oxygen. These trees mainly have single trunk with hard and woody branches like mango, teak, rosewood, sandalwood, etc. We have a rich collection of medicinal variety of trees. To name a few *Vepa*/Neem, *Nilagiri*/Eucalyptus, *Tani* and *Tuniki*. This forest has shrubs like *Danti*, *Ulinda*, *Kodisha*, *Vaka*/*Karonda*, *Nakkera*.

Visit to the 'Herbal Garden' really surprises the visitors and as many of them end up tickling their taste buds with some of the rare Indian herbs. The forest cover of the state has grown in last 3 years due to the combined efforts of the Government, individuals and the Non-Government Organizations (NGO). They target to increase the state's forest cover from 24% to 33%.

Craft Walk

This is a craft awareness program conducted in Shilparamam, the village of craftsmen at the Hitech City, modern part of the city to the west of Hyderabad. Designed by following the design methods of rural areas of state, Shilparamam is frequently visited by thousands of people. The craftsmen from different parts of India come here to demonstrate their regional art forms: visual and performing arts both. The students come here by 10:30 hrs and given a guided tour of all the stalls. After their 2 km visit inside the craft village, visitors attend the live demonstration of the art and crafts by the craft persons. They all assemble in an indoor space which is built with natural materials and climatically designed well with traditional methods. Craft village is also home to some water bodies and rock formations. A sculpture park adds to the beauty of Shilparamam. Priced at Rs 25/- per student, this Craft Awareness program connects the children to the variety of craft traditions of India. The Madhubani artist and Pattachitra artist tell the children about making colours from flowers and leaves. Use of Lamp black as black colour in these painting astonishes the children.

The performing arts like puppetry, folk dance, classical dance are a major attraction of Shilparamam in the evening.

Museum Walk

This walk discusses the historicity of area in and around the Centenary Museum that houses the objects focussing on the prehistory of cultural regions of Telangana. The physical growth of the state and city of Hyderabad from Qutb Shahi period till the British developments also find mention here.

This walk opens doors to many other such walks and talks across the state of Telangana for the people interested in art, architecture and archaeology. A visit to the 225 years old Gunfoundry across the road builds great excitement among the visitors. This walk is a must for parents and teachers – a lesson in prehistory.

Centenary Museum

In this museum the visitors experience the Krishna River valley civilization, which in some aspects comparable to the great civilizations of Indus and Nile rivers. This valley was threatened with submergence after the construction of high dam across the river Krishna. A pavilion was constructed by the Irrigation and Projects Department in the year 1976 and handed over to the Department of Archaeology and Museums in 1990 with a view to display antiquities and art objects recovered from submergence areas of Yeleswaram and Srisailam Reservoirs of Krishna River Valley. This also displays other collections excavated in the archaeological sites of the state.

The Centenary Heritage Museum showcases terracotta, stucco, prehistoric, megalithic and iron artefacts, early historic objects, temple sculptures and door-jambs collected from the submerged area of the Srisailam Project.

Ground Floor Gallery: Art Through the Ages

The ground floor displays objects and antiquities such as Terracotta figurines from nearby cities and villages. It also includes display of stone, wood, ivory, porcelain objects, etc., in the remaining sections.

First Floor:

(i) **Prehistoric Gallery:** In this gallery the Palaeolithic tools are displayed which explaining the pre-historic times. The tool kits used by early humans are organized according to the era such as lower Paleolithic, Middle Paleolithic, Upper Paleolithic and Mesolithic times.

(ii) **Proto-historic Gallery:** Proto-history is a period between prehistory and history.

A good number of stone and copper items such as polished edges, axes, rubbers, sling ball, Saddle querns etc., and pottery recovered from various parts of state shows the day-to-day life of Neolithic man in terms of tool making technology, domestication of animals, hunting, agriculture, art of pottery making and paintings on the ceilings of caves and caverns. Further the department has conducted excavations at several megalithic burial sites in state such as Chinnamarru, Peddamarru and Uppaladu which are on display. These show the burial practices, religious customs and beliefs of the Megalithic folk during the period.

(iii) **Early Historic Gallery:** In the Early Historic Gallery, priceless antiquities recovered from excavations are on display. The most important objects are beads from Kondapur excavations (Medak district). The other antiquities and art objects include iron implements, beads, ivory combs, dice, terracotta moulds, coins, semi-precious beads, silver beaded necklace, shell bangles, antimony rods, copper foils and pottery etc., from excavated sites such as Kotilingala, Pedabankur and Dulikatta (Karimnagar District) and Neelakondapalli (Khammam District).

(iv) **Scale Models Temple Gallery:** The committee of Srisailam Power Project had selected nearly 44 temples for their relocation to the higher contours of nearby rehabilitated villages in the Srisailam submergence areas of the mid Krishna river valley. The gallery includes scale models of the selected temples for the benefit of scholars and visitors. The department has acquired such scale models of temples from the endowment board and is displayed here.

(v) **Rock Art Gallery:** The top floor of Museum building has a wonderful display of rock sites with the prehistoric artwork of Telangana Karimnagar, Warangal and Rangareddy district. Most interesting one is of Pandavalugutta.

(vi) **Objects displayed outdoors:** Temple pillars, sculptures, ceiling panels, murals, miniature temples and cannons belonging to different periods of history are displayed in the Museum Complex, in covered. Semi-covered and areas open to sky (Fig. 22.6).

Other Cultural Facets of Hyderabad

Not only the historic buildings, the city is famous also known for its ethnic ensembles, pearls, Bidri art form, Deccani miniature paintings (mainly seen in the museums), *Daastangoi* (the art of storytelling) and Hyderabadi food. Tantalize your taste

Fig. 22.6 Megalithic tombs from Tadvai forest described in the museum

buds with Biryani (rice cooked with oriental spices and meat/vegetables), Haleem (wheat cooked with spices and meat), variety of pickles, soups, curries, *Nahari Kulcha (slow cooked meat with flat bread to be eaten in breakfast), Paye ka shorba* (leg soup or bone broth), Badam ki Jaali (sweet made with almonds), Irani Chai and Osmania biscuits; the list is actually very long. Visitors from across the world come to Hyderabad for shopping and learning the Hyderabadi traditional knowledge system. The city is becoming more famous after entering into the UNESCO Creative Cities Network (UCCN) since October 2019 in gastronomy category.

Conservation Efforts

Hyderabad home to hundreds of natural and built heritage components has created some good example of adaptive reuse the historic sites. The examples are Taj Falaknuma Palace Hotel; Falaknuma Palace was built by Paigah noble Sir Vikhar-ul-Umra (mentioned above) and has been topping into the categories of best luxury hotel. Chowmahalla Palace, an erstwhile palace of Nizam was awarded by UNESCO for the best upkeeping and maintenance. It has been successfully functioning as a Museum cum convention centre and on the must-see list of all the local and international visitors. Present US Consulate office is in a historic complex known as Paigah Palace. Country Club and Chiran Fort Club are functioning in well-known historic properties and have been doing well.

The present local and state government have been working for saving the historicity of the city. Some historic clock towers are currently being spruced up by the municipal corporation and many more sites are there in the list. The former British Residency Complex is also under restoration. This site has a scaled model, which is being restored recently and standing at the site itself. A British cemetery is also part of this complex and getting restored with the joint efforts of WMF and state government. A 10-year-long very well-planned conservation scheme is going on in phased manner by the joint efforts of the Aga Khan Trust and the state government, at the Qutb Shahi Walled Tomb Garden. A lot of hidden heritage components are being discovered over a period of time. The large *baolis* (wells), part of Qutb Shahi groundwater network, have been revived, and the historic complex is setting a great example of sustainable methods. A state level study on ‘Step wells of Telangana’ is also being supported by the Heritage Telangana (earlier known as the State Archaeology and Museums).

Hyderabad a Futuristic City

From the being the cultural capital of southern India, presently Hyderabad, is strengthening its status as a new global tech capital – Microsoft, Amazon, Google, Facebook and many other international companies have set up offices in the south-central Indian city. A good number of national and international research and training institutes have their campuses in Hyderabad. The city is getting ready with a cutting-edge metro system. The International Airport and the Hyderabad International Convention Centre (HICC), definitely makes our city an international destination for many more global events. Yet, the challenge is to maintain the balance of the modern and maiden. The joint efforts of people and the government will definitely take Hyderabad into the right direction.

Bibliography

Banagiri V (ed) (2008) Hyderabad hazir hai. Rupa and Co, New Delhi
Luther N (2006) Hyderabad a biography. Oxford University Press, New Delhi
Sajjad S (2016) The Times of India, Hyderabad, India, newspaper articles
Vottery M (2010) A guide to the heritage of Hyderabad: the natural and the built. Rupa and Co, New Delhi
Vottery M (2014) Heritage of Hyderabad: from children for children. Blue Jay Books, New Delhi
Yajnik BR (2018) Kakatiya dynasty – the golden age of Telugu civilization. Visual Quest Books, Hyderabad

Chapter 23
Understanding 'Rural' and Village Society

Abhishek Bhutoria

Abstract Nepal is predominantly characterised and perceived by her rural and village societies. It is only about one-fifth of the total population that forms the city and urban spaces. Despite the dominance of rural and its associated historical manifestation in the Global South, we still lack a comprehensive understanding of what it is or of what it constitutes or how it is perceived or lived. The rural of South Asia hasn't been defined of what it is but of what it is not and the phenomena of ruralism constructed on an interlinked social, cultural, political, economic and environmental matrix; same as of urbanism, still, it finds itself with a reputation of unestablished discourse. It is certain that urban spaces play a critical role in the contemporary world and especially in the underdeveloped and developing nations; however, the environment, perception and design of the 'rural' and village society which is a continuum of a different space that we have termed as 'urban' is yet to be explored with comparable intensity. In the meantime, it is certain that such a complex subject cannot be understood without a framework; hence, this study attempts to understand 'rural' as a nested arrangement of space, people, a social system (not separated from economic and political system), production, development and a way of living where a village society breathes and sustains itself. In that attempt, the study also aims to comprehend the methodology of understanding the rural system and the village society, as this would determine the framework on which the understanding will be constructed. Along with this, the study presents a case methodology for understanding Nepali village from the lens of architecture in order to comprehend the household, environment, social processes and perception of the villagers about their belief, design, production, continuity and change. This study therefore is an introductory understanding of rural and village society and offers a way forward for a more detailed, comprehensive and interlinked understanding of the frameworks discussed here.

Keywords Rural · Village · Society · Nepal · Space · Theoretical framework · Methodology

---

A. Bhutoria (✉)
Independent Researcher, Kathmandu, Nepal

© Springer Nature Switzerland AG 2020

M. Ghosh (ed.), *Perception, Design and Ecology of the Built Environment*, Springer Geography, https://doi.org/10.1007/978-3-030-25879-5_23

Understanding Rural in the Context of Space

'Rural' and 'urban' are the characteristics of the space that humans through reductionist logic have categorised. The association to a particular or multiple abstract logic becomes a belief system on which science is built. This abstract understanding from a singular perspective to converge the concept of rural and present as a hegemonic explanation is questioned by the variance of other abstract logic. In order to comprehend the multiple abstract logic framed under certain parameters or measurements or phenomenon, it seems crucial to understand the social system of the village through a deductive framework. The first layer on which the multiple layers act and then command as a cyclic relation is space. Hence, it is important to understand the space in order to understand 'rural'. Space is an everyday term yet one of the most complex words to understand. The word with its constant evolution, definition and practice cannot be understood without muddling in some labyrinth of complexity. The diversity in the association of the word space (with personal, social, public, psychic, liminal, metaphorical and more) gives it a varied multiple contextualities. Hence, the generalisation of the word would serve no purpose in its understanding as the meaning of it is exclusively conditional upon the context.

A philosophical discussion of space in *Social Justice and the City* (Harvey 2009, pp. 13–14) explains space in a tripartite division. The *absolute* space is a permanent and immovable entity, which is independent in nature and is a 'thing in itself'. It then acquires a structure, which can be understood as a relationship between objects that relate to each other building a *relative* space. The final layer of space is the *relational* space where 'processes do not occur in space but define their own spatial frame' and internal relations and external influences are the integrals of the specific process of a time frame. The tripartite division of the space seems to have a hierarchy, the relational space can comprise the relative and the absolute, the relative space can comprise the absolute, but the absolute space is just absolute. For instance, a house is a physical and legal unit that is built on absolute space; however, the position of that absolute space given its location is dependent on the relative space with respect to services, livelihood, activities, recreation and people. However, it is further associated with relational space that is the social, cultural and cosmic relations, memories, desires and sentiments and more. So, to comprehensively understand the house or for say other space or form, it is crucial to understand the three forms of spatiotemporality simultaneously.

In this understanding, Halfacree (1993, p. 26) debates that the absolute form of space has 'no inherent powers', whereby spaces do possess causal power, and hence are not absolute. A spatial establishment may generate social practices, but it cannot be reduced to the 'sum of relationships' between the objects and hence is not relative. However, space and spatial relations are both languages of fundamental structures and a means of creation; hence, space is produced and space is a resource (Smith 1984). Space as per Lefebvre's theory (Lefebvre 1992) is also divided into three – material space (experienced), representations of space (conceptualised) and spaces of representation (lived). The material space for humans is the space of experience and perception of physical and sensual interaction with matter. The representations of space are the appropriate abstract representation of the material realities conceived physically and sensually. And the spaces of representation are the lived spaces of sensations, the imagination and emotions of the way we live in the world. This complex understanding of space invites us to consider the ways we organically, perceptually and symbolically shape our environment and the ways in which we both represent and get to live it. Hence, it is critical to understand the production of space, which Molotch explains in reviewing *The production of space* by Lefebvre –

> Humans create the space in which they make their lives; it is a project shaped by interests of classes, experts, the grassroots, and other contending forces. Space is not simply inherited from nature, or passed on by the dead hand of the past, or autonomously determined by 'laws' of spatial geometry.... Space is produced and reproduced through human intentions, even if unanticipated consequences also develop, and even as space constraints and influences those producing it. (Molotch 1993, p. 887)

The production also indicates that space should be considered comparable to another economic good. The space as goods and services is capable of generating the economy, and the economy, in turn, is capable of generating spaces. Space contains more than we ordinarily appreciate – it is neither merely a medium nor a list of ingredients but an interlinkage of geographic form, built environment, symbolic meanings and routines of life. Space is produced for two purposes, domination and appropriation. The space produced for domination is put to the service of some abstract purpose to express power and strength, whereas space produced for appropriation is carved into forms of exchange in the market so as to be interchangeable as commodities. The resultant space represents the "triumph of homogeneity" as Lefebvre mentions it and stands, both in its totality and in its constituent part as a product.

Marx's later writings reflect his belief in a tripartite basis of capitalism: the earth, along with labour and capital, replacing the binary opposition of labour and capital. Lefebvre states that 'Marxism should be treated as one moment in the development of theory and not, dogmatically, as a definitive theory' (Lefebvre 1992, p. 321). Lefebvre argues this on a number of grounds. Capitalism now has laid "its imprint upon the total occupation of all pre-existing space and upon the production of new space" (Lefebvre 1992, p. 326). Capitalism has spread so globally, so deep into the earth and so far into outer space that it threatens nature itself. Lefebvre exclaims that 'along with God, nature is dying', and he asks whether 'by destroying nature and nature's time, (is) there not a danger that the economic sphere, fetishized as the world market, along with the space that it determines, and the political sphere made absolute, might destroy their own foundation – namely land, space, town and country – and thus in effect self-destruct?' (Lefebvre 1992, p. 326).

The space as a resource for the production of social, physical or mental space is dependent upon the capacity, tendencies and constraints of humans. These spaces generate the possibilities for variation in the human action, and the action of humans on the spaces facilitates or restricts the reproduction or transformation of the spaces. This cycle in the continuity creates diversity and complex spaces. However, in the mode of understanding rural on the framework of space as product and resource, the existence of human agency is unduly overlooked (Sayer 1989, p. 267).

As there are multiple abstract sets of logic to understand rural, the logic of understanding it as space has also been criticised while also being celebrated. In the course of understanding rural through a framework or a concept, it is important to understand the limitation and criticism of it. The generalisation of rural solely through this concept condenses the richness of complexity and diversity of a society (Gould and Olsson 1982). Also, this conceptualisation fails to appreciate the dynamism of space (Halfacree 1993) and understanding and treating the society and space as a dualism rather than as duality (Giddens 1986). In the process of production and as the state of resource, it is important to explore the ways in which rurality is constructed and deployed in a variety of contexts (Murdoch and Pratt 1993, p. 423). On the same principle, the concept of space raises the question of hegemony and power and how the power relation conditions the production, transferring and receiving capacity of the tangible and non-tangible space and "… which or whose spaces matter and what power relations are embedded in these particular discourses of space and locality" (Pratt 1991, p. 264).

Understanding Rural in the Context of 'Social'

Nepal is principally a rural nation, where most of its population live in villages and sustain their lives on agriculture. For the majority of Nepalese, villages are home, community, the very circumstances of daily life and a social centre to which they are intimately connected (Pigg 1992, p. 491). So, what does the village represent to, in contemporary Nepalese environment? In the social science discourse, the term rural in a paradigmatic sense linked with empirico-positivism is conceptually a subordinate to the definition of urban; it is whatever is non-urban in character. Along with that the statistical variable of socio-spatial characteristics that represents the essence of rural through the population, land use, employment, housing condition and migration helps in defining and understanding rural (Pandey 2003, p. 34).

On these principles, villages of Nepal are neither bazaars (local trading centres), district centres (regional centre for government offices and agencies) nor towns (population strength much higher than either). This classification of space allows a statistical understanding, for instance, the rural population of Nepal in 2018 accounted to 80.26% as compared to 97% in 1960 (The World Bank 2018), and the level of urbanisation was 18.2% in 2014 (UN DESA 2014). The 2011 census indicates that Nepal's demographic profile is changing rapidly, and the rural-urban classification is coded into the census figures. This coding leads to a reasonably obvious conclusion that what isn't urban is rural. This approach in an attempt to create a geographical and populace distinction between urban and rural further questions '..at precisely what population does a rural area become urban?' (Woods 2005, p. 5) and does the geographical urban starts immediately after geographical rural, or is there a transition or a continuum? For Nepal, there isn't a clear distinction; one finds a strong rural character (geographical and people) in the urban fabric and vice versa. There are numerous examples to quote for such scenarios in one of the fastest urbanising nations, yet quite rural in their character (Bhaktapur, Kirtipur, Namche Bazar, Jiri). However, this coexistence exists in the form of transition, continuum and relation.

In the rural versus urban dualism (Pandey 2003) and the idea of the country and the city (Williams 1973), both authors discuss urban and rural as a continuum and relational. On a sociological construct, the understanding of villages in Nepal through universal or self-evident rural-urban dichotomy can be a precarious method. The census distinction of rural and urban limits the understanding to figures, and hence the dynamics of the relationship, transition and perception are crucial to understand their physical and social forms.

Rural as socially constructed brings attention towards the phenomenological understanding. The subjective construction of the reality of rural diverging from the determinism of structure-functionalism emphasise on the way people perceive and react to the experiences of the physical and social settings in which they live. The experiences of urban as dynamic, unstable and rational and it's opposite rural as stable, integrated and traditional question the oversimplification of urbanity and rurality; in rural areas, urban societies can exist and the other way around (Young and Wilmott 1957, in Woods 2005). It is in the sense of what people think is rural; it exists because it is an establishment of communal experiences, meanings and culture (Pandey 2003).

However, in the course of understanding rural, it is also understood as social relations of production, as villages are often the spring that feeds the urban areas. On the contrary, both doctrines of rural-urban classification and the subjective construction of rural, this approach, with an orthodox Marxist construct, stresses on the structural process associated to capital use and power distribution in agriculture (Pandey 2003; Davidson 2014). The spirit of rural is agriculture, and the production and capital associated with it determines the differences in geographical areas, populace and living and hence allowing to understand rural from another perspective.

Understanding Rural in the Context of Development

In Nepal, the understanding of the village doesn't advance on the ground of comparison with the city but in relation to development (or *bikas* as termed in Nepali). The term *bikas*, historically and socially, is comprehended in the context of Nepal and should not be confused with the theories and framework of development discussed by international institutions (Pigg 1992, p. 495). However, both the terms differ in its meaning because of the context; they are not completely distinct. The influence of the philosophy of the international form of development has contested the concepts of the villages of Nepal. For Nepalese, the ideology of development in the village comes through social relationship, the relationship between local communities and other places to progress. The progress and development has a magnitude and depending on that, the understanding of the social territories, transitions and relationships can be established. Development can vary from new varieties of seeds and breeds of goats to hydroelectricity plants and highway roads. It is the affiliation to the type of ideology and degree of development that brings the distinction and image of a place.

> For me, India begins and ends in the villages (Gandhi 1979 in Jodhka 2002, p. 3343). The old Indian social structure which, has so powerfully influenced our people, was based on three concepts: the autonomous village community; caste and the joint family system (Nehru 1946 in Jodhka 2002, p. 3343). The Hindu village is the working plant of the Hindu social order. One can see there the Hindu social order in operation in full swing (Ambedkar, in Moon 1979). For Gramsci, the peasantry would remain in the foreseeable future as a significant political force; their ethos and values would, therefore, have to be incorporated in any Marxist theory of transition and socialism; and developmentalism or modernisation would have to be re-examined as a core part of the Marxist strategy. (Davidson 2014, p. 139)

All four discussed village in the context of the civilisation and development, but the way they understood and presented village was on different principles. Gandhi celebrated Indian village life and saw the village as a political symbol of the republic. He perceived village life as the essence of India. The reformist vision of Gandhi advocated the need for reconstructing and reviving the 'essential spirit' of village life, which is harmonious and self-sufficient, uncontaminated by the modernisation of the city and technology. Nehru perceived Indian village as 'a new picture of India, which is naked, starving, crushed and utterly miserable. He believed "village republic" had long degenerated due to various ills, which is a consequence of internal differences in the rural society challenging its unity and equality'. Unlike Gandhi, Nehru saw no merits in reconstructing and reviving the traditional social order and the essence of the village, but his modernist vision wanted to transform the village social and economic structure by using new technology and innovative mechanisms for development. His aim for transforming the village was built on social security, stability and continuance of the rural. Ambedkar criticised the idea of the village and stated that village republics have been the ruination of India. He perceived village as a sink of localism; a den of ignorance, narrow-mindedness and communalist; and a model of Hindu social organisation. He strongly condemned the thought of recognising the village republic as a legal structure of India as it would bring a great calamity. He believed that the emergence of Dalit from its present condition didn't lie in the revival of the village or for a say in the development of the village. He imagined a 'subaltern view' of the village, which escapes the caste structure of the village.

Gramsci, from a developmentalist position, prior to the rise of fascism believed that as a prerequisite to socialism, there should be the alliance of the workers of the factory and countryside (village) through self-establishment of the organisation, which is the approach for socialist civilisation. However, with the rise of fascism, Gramsci realised the peasants won't disappear with the development of capitalism, and there would be no initiation for transformation without them. This type of development brought advancement and backwardness simultaneously, and the benefits of such advancement mostly rested with the power of the industrial capitalist. For a progressive collective conscious development of the peasants, their folklore should be studied as a 'view of the world', as an impulse of the life which they have been living and will continue to live and as something outside 'modern currents of thought' (Gramsci in Davidson 2014, p. 38). The obvious truth of modernity penetrating every stratum of the society would have infiltrated the traditional wisdom and conservative qualities. However, this modernity in the traditional system could also be a current of innovations and should be conditioned for the development of the space. For Gramsci and Gandhi, the future didn't lie in the modern cities, but on the popular town and country (village).

The world as a course of history is built on the concept of development, production, reproduction and survival; and it was dominantly the country (village) that saw the process. Though it is now the cities and urban societies that are more associated with the concept of development because of its magnitude, it has not disappeared from the village. The village crystallises into a different social category in the context of the concept of development. The idea of understanding the village from the concept of development is to interlink the construct of society, culture, economics and politics. These aspects as an interlinked system from the lens of development determine the life for the villagers who seek to grow and progress. The magnitude, approach and understanding of development or progress, in turn, mark the social identities and differences, cultural practices, economic aspirations, political affiliation and in short, the way of living of village society.

Understanding Rural as a Way of Life

Despite the preponderant significance of the villages in Nepal, our knowledge and understanding of rural and village society are meager. There have been attempts to understand the characteristics of rural life and village society. However, researchers from different disciplines and agenda offer understanding in view of their respective discipline. The characterisation of society as rural solely on the basis of size or density or the classification of it under an administrative structure is certainly arbitrary. The association of space with rural characteristics based on the physical entity and development also limits the understanding to mere rigid tangible parameters, which doesn't offer an adequate conception of rural as a way of life. However, with no intent to supersede, these and many other approaches, the formulation of understanding rural and village society comprehensively through a sociological approach that attempts to emphasise on the interrelations between different approaches, can offer a better understanding of it as a social entity.

The fundamental concern of the sociologist of the village is to discover the forms of social action, customs, habits, relations and beliefs that typically emerge in the society and settlements of heterogeneous individuals and sometimes a close-knit homogenous society. These aspects determine the way the villagers live their life, and hence understanding those aspects will enable to understand the rural and village society. Hence, this section will attempt to present an understanding of rural and village society as a way of life by interrelating space, society and development. Although it is certain that this will present only the essential characteristics that different villages of Nepal have in common as social entities, the specific characteristics can be further developed as a way forward in this study.

Nepalese villages can be located in the splendid peaks of pure white Himalayas or totally immersed in the green of the serene rolling hills or in the plain beauty of plains. Shaping themselves on the steep of mountains, dense jungle, trails of cultivated steps, banks of rapid flowing white-water river or on the flats of the plains, the villages possess a power of its own kind from the absoluteness of its geography which to an extent determines the way of life. However, regardless of this power of geography, in the close-knit communities of the villages, the relationships emerge through its architecture and material through peoples' association with public and private spaces; through the emotions, sensations and attachment associated with space; and through the heterogeneous mix of the ethnicity/caste of the villagers. These relationships are the binding factor for any space but especially for the villages. The relationship of villagers with the external environment was mostly reflected through agriculture or related activity, while with human beings, it was facilitated through caste, language and cultural practices.

Nepali villages are generally self-dependent, and there is little evidence of other world. The meaning of development and growth is very different for the villagers. There are limited shops or none at all; electricity and running water are still a dream for many, and there is little or no sign of the government or police or health workers. If a Nepali villager wants something, they generally get or do it themselves, to the extent that they cut mountains in a traditional way and build connectivity; education comes at the cost of traveling 1–3 h across mountains and rivers, and hospitals are hours apart, and many die on the way. However, there are certainly some exceptions where the influence of other world due to tourism has changed the dynamics of a village, and these villages sometimes happen to be the face of all the villages due to publicity.

Among such perfections and imperfections, what remains common in most villages is the preoccupation with agriculture, land and other objects or occasions either incidental to or derived from agriculture (Srinivas 1976). For most villagers, agriculture is the way of life; it is the source of living, social and economic relations and daily activities. There is a hierarchy in the system of agriculture where there are landowners, tenants, sharecroppers, labourers and servants. However, this hierarchy is more prevalent in the Terai villages of Nepal. In the villages of hills and valley, it's generally the family members who take the responsibility of the land and production. This can be possible because of the less rigid occupation-based caste structure in the villages of hills and valley, which however were quite pertinent decades back.

The process of production isn't an easy task; in the majority of the villages, there isn't a shimmering and shiny John Deere or Mahindra tractor pulling an equally modern plough. It is the water buffalo that pulls a wooden plough, which is controlled by the males of the house, while the females work in the rice paddies that is more exhaustive and labour-intensive. This is the result of cultural practices where much of the labour-intensive task like padding the rice field, taking the produce from the fields which are near to the river banks to the houses in the hills, collecting firewood and water, washing clothes, cooking and pretty much everything related to household are done by the women in the household. The males are generally artisans, construction worker, service providers and labourers and are involved in politics of village. The understanding of gender and the differences between the role of male and female in the rural communities will bring the understanding of social order, social structure and functioning of the village.

The economic system like anywhere else, but exclusively in the village, cannot be seen in isolation from the social and cultural system of the village. "If the principles of economics are to have a predictive value, social institutions of which economic activities are a part should form the background of economic studies" (Aiyappan 1945, p. 86). The economic system is a construct of space, social relations, production and development. In the economic system of a village, the land occupies the key position, though there can be a group which is land deprived, who either work as agricultural labourers or have a small 'non-farm' occupation for livelihood (such as small retail shops, repair centres, tailoring store, etc.). However, the land with its monetary value does not limit itself purely to economic frontiers but reacts to social relations, especially one connected to the power of production and ultimately status. The status and power of the villagers, to a great extent, is defined by the size and quality of landholdings. The use of this status and power for the development/growth of the village is very contextual from village to village and is at the level of individuals or a few collective people. However, the discussion on the development of village highlights and questions the role of government and its governance system.

The question raised is the consequences of the political instability within Nepal. This instability, especially in the last three decades, has been a result of the insurgency of the Maoist movement for the economically backward parts of the country (dominantly villages) where the worst legacy of the caste system and prolonged ignorance of the government changed the dynamics of the village in their social relationship and order, functioning and development. In Nepal, democracy and its benefits came to people in district centres and towns and to the people with ostentatious voices. Village people had no idea what democracy is or feels like, and eventually the apparent democratic state was viewed as system alienated from the aspirations and needs of rural individuals (Shneiderman 2009). However, this political conflict has plunged much of the villager's life into chaos and difficulties, be it prior to the insurgency or after. The political instability, economic distress, social conflict and development crisis in totality has affected the villagers and their way of living.

Methodology of Understanding a Village/Rural Society

In an attempt to comprehend the methodology of understanding a village/rural society or as conventionally called 'village studies', this section aims to identify how agenda determines the methods, which methods are adopted for it and how it should be shepherded. The history of village studies has brought varied theoretical impulses, methodological approaches and eccentric and conventional stories of the life of people in the village. Depending upon the purpose of research, a village study is either descriptive, exploratory or explanatory, and these studies were carried out as part of duties of the government for the census and administration, to understand village economy and to understand traditional social life and culture, village structure and social relations, for political agendas and for development projects. Each method determined by the agenda is further controlled by the position we take as researchers. Hence, the position, the agenda and the methods together formulate the research design.

The research is a process of acquiring knowledge using a ‘scientific’ method. The association to a specific scientific thought or school of thought determines the type of research conducted and methods adopted. For a researcher, it is crucial to be aware of their association with the school of thought. The school of thought can vary from *rationalism* (systematic logical reasoning, philosophy of Plato, Aristotle, and Socrates), *empiricism* (empirical activity rather than reasoning activity, Baconian method), *natural philosophy* (the philosophy of nature and physical universe, Galileo Galilei and Sir Isaac Newton), *idealism* (experience is purely subjective, Kant’s philosophy), *positivism* (blend of rationalism and empiricism, Auguste Comte), *anti-positivism* (qualitative method, interpretive sociologists), *post-positivism* (knowledge based on tentative conjectures which can never be proven conclusively, but only disproven, Karl Popper), and *critical research* (criticism to antipositivism, critiquing and not just understanding, Karl Marx and Friedrich Engels).

The research design is the process through which research is conducted and in turn is fashioned by our mental constructs or belief systems, which we use to systematise our reasoning and observations (Bhattacherjee 2012). These mental constructs or belief systems are called paradigms, the term popularized by Kuhn (1962). The term ‘social reality’ is perceived and further comprehended by different people in different ways based on both their thinking and reasoning (or say their paradigm), and this association to one paradigm limits the knowledge of that particular school. The judgement and ability to understand is based on the ‘lens worn’ or the position taken by the researcher and most importantly as to what the individual infers from what he sees in the world. The research design that builds on multiple paradigms for holistic learning is a very idealistic scenario; however, one should be aware of different schools of thought and paradigms while conducting a study.

In social science research, the two popular paradigms are positivism and post-positivism. The principle of positivism is that the construction of science and knowledge should be limited to what can be observed, measured and tested. The logic of it is that the research must be able to prove or disapprove and anything that is extended or reasoned beyond observable and measurable facts doesn’t qualify for scientific research. The strictness and the empiricist nature of positivist philosophy led to the emergence of post-positivism. The principle on which post-positivism builds its argument is, science and knowledge can be constructed on logical reasoning, criticism and subjectivity. The phenomenon can be studied and understood by combining empirical observations with logical reasoning built on the subjective construct.

Burrell and Morgan (1979) suggested that there are two fundamental sets of philosophical assumptions that shape the study of social phenomena, which is ontology and epistemology. Ontology refers to our assumptions about how we see the world and knowledge; for instance, is the world or knowledge a construct of social order or constant change? While epistemology refers to our assumptions about how do we study the world and obtain knowledge, is it through a subjective or an objective approach to studying social phenomena? The model of social science research of Burrell and Morgan (1979, p. 22) suggests four paradigms based on the ontological and epistemological position of the researcher: functionalism, radical structuralism, radical humanism and interpretivism. The paradigm of functionalism is adopted if the world and knowledge are perceived as social order and the knowledge of patterns of social order, events or behaviours are obtained using objective approach, whereas interpretivism is a construct of the same worldview but with a subjective approach. The paradigm of radical structuralism is a construct of radical change and objective approach; however if the social order is understood using the subjective perspective, it is a radical humanism paradigm.

The study of village/rural society can be constructed on any of the mentioned paradigms. However, in the context of such studies, the paradigms of functionalism and radical structuralism or jointly called as positivism were more prevalent prior to the 1950s. The research rested content with the observation, measurements and calculations of censuses, reports and administrators survey of villages. Later to that period, the discourse underwent a change; there seemed a paradigm shift from the domination of structural-functionalism to a combination of structural-functionalism and interpretative-humanism jointly called as post-positivism. The paradigm of post-positivism accepted the relation of the theory and the description and the subjective interpretation in research. However, training in research methodology was still focused on survey techniques, elementary statistics, questionnaires and scaling. Many of the research then and now still rely on data collected by various agencies, organisations and committees of government, the UN and scholars. What this results into is the rise of a school of thought where there are groups of manual labourers (data gatherers), administrators (number crunchers) and potentates (elite theoreticians, practitioners, officials and intellectuals) who with an approach of problem-solving negate the understanding of the village and people by coming in close contact with the space and the people of concern.

> The lack of a field-work tradition in the social sciences, excluding social anthropology and sociology, has had adverse results on their growth and development. Most important, it has alienated them from grassroots reality and led to fanciful assumptions about the behaviour of ordinary people. It has realized in woeful ignorance about the complex interaction of economic, political and social forces at local levels. (Srinivas 1975, p. 1389)

However, even anthropologists and sociologists depending more on the field-work tradition(s) shy away from empiricism of the positivist school establishing the validity of the case studies and qualitative narratives. Hence, it is perhaps thus not the education but the dominance of certain methods that subverts other methods. In the meantime, the methods applied to understand the village underplay the importance of understanding the villagers and their life and community. The village communities have their own logic, resources, intellectuality and power; and hence rather than applying methods just to build our understanding of the village, it is also crucial that methods allow to understand how villagers perceive their life and community. This will allow looking at the research from the plural lens rather than a singular lens of the researcher. Also, scientific reductionism that tends to close the scope of complex understanding and yet celebrate objectivity will be challenged by the complexity of multiple understanding and layers of the village society rather than getting stuck with objectivity, linearity and rigidity.

The image of the village and villagers is only a caricature and untrue until they are first-hand observed, recorded, lived and understood. The disconnect between field research and theory has led to the misunderstanding of our villages. Village studies have a strong relation with the social environment, unlike natural science. When the study comprises of people, culture, society and their function, it is inevitable to work and develop a comprehensive understanding without a combination of field research and theory, interaction/observation and interpretation, and also, functionalism and structuralism play a critical role in substantiating the understanding.

A Case of Methodology for Understanding Nepali Village from Its Built Form

In contemporary sociology, the concept of structure and agency aims to progress from the idea of 'dualism of structure and agency' to the 'duality of structure', in which social structure is both the instrument and the product of social action, and the mutual construct of these two independent entities must comprehend time-space relation for all social interaction (Giddens 1986). The interaction of the agent and the structure is not a succession of separate acts but a continuum of social actions in a social system, which is located in time-space. The social action, which is dependent on spatial and temporal configuration, produces and reproduces under the structure that is employed in the social system. Under the same principle, the research aims to study the social structure and social action that is the household and social life through the structure that is the built form. The social structure that is the household is composite of two components, which are not two separate entities but relational – the 'pattern of interaction' (relation between members of the household and the relation between the households) and the 'continuity of interaction' in time that results as social life.

"… while the household is conceived of as a specific combination of social relations – a kinship morphology and several distinct spheres of practice (production, distribution, transmission, reproduction and coresidence)", the understanding of this social structure through the combination of the social ingredients mentioned is not holistic in its approach to understand the social life of the village (Gray 1995, p. 21). To identify the holism and particularity of the social experiences of the village, the structure, the agency and the social action of individuals will need to be comprehended from the conception of the relation between parts and whole rather than blocks which minimises the importance of wider structure and social system of society as a whole. This complexity of inside-outside approach in relating the internal structure of the household with the external structure of the village is precisely the opportunity such study should engage with.

In order to attempt this ethnographic study, the research from an ethnographic strategy aims to build from villagers' experience and understanding of their household and life in it and consequently move out to the other territories of their society and social life. This account from the 'inside-out' approach uses a family of methods under the shade of ethnography methodology, which involves direct and sustained contact with the villagers within the context of their daily lives and culture. However, in order to carry out the 'inside-out' method, a preliminary research needs to be conducted in order to find a representative example(s) that will voice for larger faction. The method for such preliminary architectural investigation has been outlined by Carter and Cromley (2014) as preliminary survey, architectural documentation, taking photographs, measured drawings, recording a building and reading the physical fabric.

Once the representative is established through observing what happens and what is present, recording and documenting of the form, listening to what is said and asking questions to determine the accountability of the representative(s), there will be three objectives of the research. First one is to observe and understand the belief and practice of the household in their house and how their life is shaped by it. Second, is to establish and analyse the rationale of architecture practices and processes that determine certain form and function and in turn a certain way of a social and cultural system. The third is to critically examine the change and continuity in the process of production of the built environment and its impact on the household and social and cultural dynamics of the village.

To observe, understand, analyse, establish and critically examine the same, this ethnographic study will request mixed methods of data collection and interpretive techniques (participant observation, unstructured and in-depth interviews, still, audio, written and visual documentation) to get maximum of villager's experiences for the mentioned objectives. People with a historical or traditional continuity of resources and lifestyle often hold an indigenous knowledge system. This knowledge system has been built through a long series of observations and practices transferred from generation to generation. Such 'diachronic' knowledge is of great value and in the meantime difficult to extract, as it is well embedded in the lives and experiences of the people. Hence, a considerable time in the context with the people needs to be devoted in order to develop a reliable medium for disclosure of the real world of the villagers, which would manifest the understanding of the human society under study.

In the meantime, the critical aspect of this and as a matter of fact for any ethnographic research is the process of writing down and writing up, which is equally and importantly a part of the methodology. The phases of writing down, analysis and writing up are distinct phases of the research process that are indistinguishably interlinked (O'Reilly 2005; Atkinson 2007). The process of writing down will incorporate taking notes of setting, observed practices and activities, what villagers have said and overheard conversations among the villagers, transcribing the interviews conducted and translating the recordings (audio, still and video). The mass of data that will be collected from the methods applied and the written down notes will be summarised and categorised to tell a story of what was seen and heard. But, in the meantime the study aims to go further than the description and analyse the data that will try to reflect upon the how and the why of the occurrences, practices and phenomenon of the human society under study. The framework that's been discussed in detail in previous sections will offer the guidance for the analysis and also the conceptualisation and formalisation of the summarised data as a narrative. However, these predetermined frameworks are by no means complete in sense as the research is open to add/develop new concepts in order to incorporate the findings that emerge out of the study, which has not been considered. The process of writing up will be in the form of narration, interpretation and presentation of the findings and analysis. The writing technique/style will reflect on accounts that are to be delivered. When describing the findings of the everyday practices, actions and the setting, which happens in social time that includes past and present, past tense will be used. However, when a reflexive stance will be taken while interpreting and presenting, present tense will be used to reflect the interpretive voice and the knowledge that will emerge in the process of understanding the human society while writing up this ethnography. The entire presentation of the research will be supported by photographs and sketches which will allow presenting the human society 'as life as it is lived'.

The everyday world of practices, that is motivated action, takes place in the context of space, social time that includes past, present and future which results as a way of life in the understanding of life progressing. And this everyday world of practices is everyday life, and it is the domestic life for the villagers. The establishment of understanding such a complex relation rejects the reductionism of domestic life into parts, which are then studied as simple building blocks of a complex whole. It rather proposes that the parts which build up a whole are just as complex as the whole domestic life itself. In order to appreciate that complexity, a methodology (ethnography) that constitutes a family of methods (participant observation; unstructured and in-depth interviews; focus group discussion; written, audio and video documentation) allows to comprehend the multifaceted subject of understanding the society. The framework established can be applied at a range of multiple hierarchies – from an individual to household, to cluster to a village(s), and it is those methods mentioned that will come in practice when dealing with those multiple scales.

Conclusion

The tendencies to discuss 'rural' as known review only the concept and neglect any questions of its reality and complexity (Falk and Pinhey 1978). It is only so far as an introductory study; the understanding of rural and village society has built a foundation to understand the village society as a compound of multiple theories, concept and realities. These few theories, concepts and realities are further perceived and presented from different positions, which gives an opportunity to engage with the multiple realities of the village in multiple possibilities. There is a realization that quest for any single, one comprehensive definition, which embraces all the aspects of rural and village society, is neither feasible nor desirable or appropriate. In no point in time, space is fixed, and knowledge is universally neutral. Hence, the dynamism of space and different logics of abstraction and power bring the rural as space, representation of space, a social representation, a subjective construction of experiences, development (bikas), and way of living, and certainly there are more multiple frameworks on which it could be comprehended. Rural is a complex abstract, and village societies are an amalgamation of multiple spatial layers, temporal evolution, and emotions and experiences. Thus, the breadth and depth of complexity of the subject cannot be captured in one such brief study. However, this introductory study to understand rural and village society and the methodological co-ordinates of the conventional village studies tradition enables to build a foundation for a future study that aims to understand the historical evolution, cultural matrix and social structure which gives the shape to the village society of present.

Bibliography

Aiyappan A (1945) Iravas and culture change. Superintendent Government Press, Madras

Atkinson P (2007) Handbook of ethnography. SAGE, Los Angeles

Bhattacherjee A (2012) Social science research: principles, methods and practices. University of South Florida, Tampa

Burrell G, Morgan G (1979) Sociological paradigms and organisational analysis. Heinemann, London

Carter T, Cromley EC (2014) Invitation to vernacular architecture. The University of Tennessee Press, Knoxville

Davidson A (2014) Gramsci, the peasantry and popular culture. J Peasant Stud 11(4):139–154

Falk W, Pinhey T (1978) Making sense of the concept rural and doing rural sociology: an interpretive perspective. Rural Sociol 43(4):547–558

Giddens A (1986) The constitution of society. Polity Press, Cambridge

Gould P, Olsson G (1982) A search for common ground. Pion, London

Gray J (1995) The household's world: purity, power and dominance in a Nepali village. Oxford Press University, Delhi

Halfacree K (1993) Locality and social representation: space, discourse and alternative definitions of the rural. J Rural Stud 9(1):23–37

Harvey D (2009) Social justice and the city. University of Georgia Press, Athens

Jodhka S (2002) Nation and village. Econ Polit Wkly 37(32):3343–3353

Kuhn T (1962) The structure of scientific revolutions. University of Chicago Press, Chicago

Lefebvre H (1992) The production of space. Blackwell, Oxford, UK

Molotch H (1993) The space of Lefebvre. Theory Soc 22(6):887–895

Moon V (1979) Dr. Babasaheb Ambedkar, writings and speeches. Education Department, Government of Maharashtra, Bombay

Murdoch J, Pratt A (1993) Rural studies: modernism, postmodernism and the post-rural. J Rural Stud 9(4):411–427

O'Reilly K (2005) Ethnographic method. Routledge, New York

Pandey V (2003) Re-presenting rural: from definition to discourse. Sociol Bull 52(1):32–52

Pigg S (1992) Inventing social categories through place: social representations and development in Nepal. Comp Stud Soc Hist 34(3):491–513

Pratt A (1991) Discourses of locality. Environ Plan 23(2):257–266

Sayer A (1989) The 'new' regional geography and problems of narrative. Environ Plann D 7(3):253–276
Shneiderman S (2009) The formation of political consciousness in rural Nepal. Dialect Anthropol 33(3–4):287–308
Smith N (1984) Uneven development. Blackwell, New York
Srinivas M (1975) Village studies, participant observation and social science research in India. Econ Polit Wkly 10(33/35):1387–1394
Srinivas M (1976) The remembered village. University of California Press, Berkeley
The World Bank (2018) Rural population (% of total population). The World Bank, viewed 28 July 2019. https://data.worldbank.org/indicator/SP.RUR.TOTL.ZS?locations=NP
UN DESA (2014) World urbanisation prospects: 2014 revision. United Nations Department of Economic and Social Affairs, New York
Williams R (1973) The country and the city. Oxford University Press, New York
Woods M (2005) Rural geography. SAGE, London

Chapter 24
Lively Urban Spaces

Mariana de Cillo Malufe Spignardi

Abstract The investigation aims to illustrate how design strategies, analyzed through a graphic method, can contribute in creating a lively and attractive urban space. It compares two parks, one located in the Global North and the other in the Global South. Parc La Villette, in Paris, and Parque da Juventude, in São Paulo, have similarities. Both are located within a dense urban fabric and can be considered a reference of open and green space in the neighborhood.

The concept about the features that contribute to a lively and attractive urban space is not easy to be defined, the investigation is about how design strategies contribute to more lively and attractive urban spaces and focuses on interrelated elements and the relation of the space with the city.

It considers the organization and division of functions of these public and open areas, and the location of points of orientation inside and outside the park, as anchors. The investigation aims to understand into the main and side entrances of the parks, and how they are opened to the city the spatial and visual continuity of the parks, and its relationship with the urban fabric are part of the analyses.

The two examples are located in the city fringe and must deal with homeless and drug user issues. Social issues should also be considered. The design of urban areas focused on pedestrians and the connectivity between cities' uses and activities, instead of that prioritizing traffic flows, seems to contribute to it.

Keywords Urban spaces · Design strategies · Urban parks · Global South

Manuscript

This is an investigation of the features that contribute to a lively and attractive urban space, but it must be considered that this concept is not easy to be defined. This characteristic is a result of distinct aspects and design strategies, and it is not possible to point the isolated reasons and elements responsible for it. There are distinct

M. de Cillo Malufe Spignardi (✉)
Faculty of Architecture and Urbanism, University of Sao Paulo-USP, São Paulo, Brazil

© Springer Nature Switzerland AG 2020
M. Ghosh (ed.), *Perception, Design and Ecology of the Built Environment*, Springer Geography, https://doi.org/10.1007/978-3-030-25879-5_24

elements that can contribute in some cases and have no impact in others, as the resultant situation is always a product of interrelated elements and the relation of the space with the city.

Two examples will be used to illustrate what is being considered as a successful urban space, and they will be analyzed through a graphic method aiming to find general aspects and design strategies that contribute in creating and emphasizing this characteristic.

There is an increasing lack of lively spaces in our cities nowadays, especially in cities where security is an issue to be considered, as it is in São Paulo, Brazil. Assuming that it decreases the quality of the urban spaces, this investigation is made necessary in order to understand this process and how design strategies can deal with it. The number of empty or poorly used urban spaces seems to increase as time goes by. The reason can be the changes in the society that are not correctly addressed in urban projects or a lack of investigation about design strategies and how these spaces perform. As other "car-oriented" cities around the world, the design of urban areas focused on pedestrians and the connectivity between cities' uses and activities gives rise to larger streets and viaducts, which prioritizes traffic flows.

Social issues should also be considered. In São Paulo, urban parks and squares are gated to avoid hosting homeless and drug users. There is also a lack of maintenance of public spaces in general, which leads to its reduced use, specially by families, elderly people, and children, instead of reinforcing its role as facilitator of social interaction.

A few children are concentrated in front of their house building a sandcastle, as their mother is complaining to a neighbor about the rising prices of the food markets in the surrounding area and wondering what could be done against it. A tall and elegant man, who works at the town house and would later help the ladies with the question of the prices, stop by the children and explain how a piece of wood inside the castle would help them to build it higher. The children run after a big piece of wood, while a dog goes in the direction of the castle.

The dog's owner, who is playing chess a few meters from there with a group of friends, doesn't notice the dog; he is teaching the game to a young girl who stopped by and get amazed with the chessboard she haven't seen before. Her father, while waiting for her, recognizes one of the players as an old friend from school and starts an excited conversation. The old lady, who sits in the same bench every day to feed the birds, adverts the dog and impedes the castle of being destroyed. She joins the ladies in the conversation about risen prices and introduces to them the elegant man, who lives next door to her.

The children cross the street running when another person, who was late for a meeting, adverts them to be careful when crossing the streets. One of the children looking back at him shouted: "The bus doesn't run here on evenings, what are you doing at the bus stop?" The man, who has lost almost 20 min waiting there, goes to the corner to wait for the bus where it actually passes. Really close to them, a painter is observing the movement and starting some sketches of a painting he would later call "our city."

The scene described above would be common in an urban environment if lively streets and city spaces weren't in decline. It seems that high-quality urban spaces that contribute to the interaction between people are not in a sufficient number comparing with the cities' demand for them. The less successful are the design strategies and urban projects, the less people use these spaces when they have other options. As a consequence, the cities become, among several serious consequences, less lively. There is a lack of high-quality urban spaces nowadays, and more successful examples of them are needed.

Added to that, the sustainability challenges we face nowadays must be considered. Cities occupy less than 2% of the Earth's surface but account for more than 70% of the planet's environmental impact. Besides that, the urbanization rates are increasing dramatically and tend to occur in developing countries. Those countries concentrate the highest rates of lack of infrastructure, increasing the challenge of designing cities and its public spaces. Cities concentrate more that 80% of the world's population, is where we must concentrate our efforts to provide a "better life" for all. As Ban Ki-moon, ex-Secretary General of United Nations, would say: "Our struggle for global sustainability will be won or lost in cities."

But what's the role of the design? How can urban design projects face these challenges? And how could the design initiatives create more opportunities? The creation of social interaction is a powerful tool to increase social opportunities, to generate safer urban spaces, healthier spaces, and more meaningful lives.

Maybe the little girl watching the chess game would start playing it after that evening, and the ladies of the neighborhood, with the help of the elegant man, would organize themselves to become a powerful force against the high prices in the area. Maybe one of the children would get more interested in constructing and projecting as they realize how different materials perform and can be used as powerful structures, or maybe not. Simple actions and meetings can create opportunities and generate diversity. That's what cities are about.

Some spaces within the city seem dead, while others, with similar physical characteristics, are full of people and life. To investigate lively spaces, many aspects must be considered, as innumerous and interrelated reasons contribute to it. The social and physical characteristics, the role of it within the city, and the functions and activities that take place are some of these reasons. Lively spaces translate the very special reason why we live in cities: to exchange, to meet people, to have opportunities and on time information, and to learn how to live in society.

Nowadays, the way lives are being organized, over the communication revolution and transportation technologies, the role of urban spaces as a place of market, trade, and business is shrinking. If meeting was an essential and automatic part of daily life, as these activities were made in public spaces, nowadays this relationship has changed. The search for optimizing our routine and daily life reduces casual encounters. Technology also plays an important role in reducing social diversity in our lives. This virtual dimension contributes to change the role of streets, squares, parks, and urban spaces in general within people's lives. As a consequence there is a tendency of urban spaces to be less essential as a place of meeting, which makes them being even less used.

The changes in living standards define new roles for public lives and consequently for urban space. According to Gehl et al. (2006), changes in living standards, working life, and economy have contributed in various ways to the new functions of city space over the past century. During the last 100 years, the amount of free time is increasing. The students are staying longer at school, and the longevity index shows more years after retirement, which means that there is more free time. Together with the greater resources for consumption and pleasure proportioned by production conditions and economic changes, it contributes to a new configuration of urban needs and spaces. Among that, the mobility that the new systems of communication proportionate redefines the role of urban spaces. As affirmed by Barzilay et al. (1984), the city of the end of this century is becoming increasingly utilitarian and inhospitable, responding poorly to the needs of its inhabitants.

The kinds of urban space that interest me are the ones that respond to citizen's needs and encourage social activities and where the interaction between people flourishes. It might address the interests of diverse kinds of people, attracting groups with diverse purposes. In spaces where it successfully happens, there is an interesting mix of people.

I would argue that lively and successful spaces are accessible by all layers of society and are where different groups feel they belong to the place; these spaces have an identity and are easy to reach, with good access to public transportation and connections to the surrounding areas. It attracts both local residents and people of different ages and social conditions. People with distinct purposes want to use the same place, and it can occur at the same time. It is, after all, and as said before, where people's activities flourish. All these elements can be important components but are not compulsory. If they are not combined within each other, they have a relative effectiveness.

The concept is complex enough to include or exclude examples depending on different points of view. I've chosen an example of South America, Parque da Juventude, in São Paulo and compared it with Parc La Villette, in Paris. The analysis will investigate how these parks are organized and relate to the city. There is no recipe or list of combined elements that can guarantee liveliness in an urban space, especially because it can be a park, an area in between buildings, a street, or something else. Some of them have similarities, but others are strongly different from each other. It is not possible to specify the success factors or point the reasons why they are strongly used and how decisive are the design strategies.

Parks are interesting examples as they have a direct relation with the city and have a flexible and adaptable form. They can reflect the society as their function and the role they play adapted over the years. It can accumulate distinct functions and act in numerous scales within the urban fabric. Their role within the city can achieve bigger proportions if they accommodate important cultural, social, or sportive elements.

Park La Villette in Paris and Parque da Juventude in São Paulo, the two successful examples chosen to be analyzed, have some similarities. Both are urban parks, proposed by a government initiative of placing a park as a tool to revitalize a degraded and suburban area of the city. The projects, as winners of a public competition, have been successful in providing an interesting and lively urban environment. It is important to notice that they are multifunctional and strongly integrated with the surrounding area. All the functions provided are essential to the role they play within the city, improving the quality of the surrounding area and, consequently, the quality of citizen's life.

According to Sassen, S. (1961), parks affect the part of the city they touch and are also directly and drastically affected by the way neighborhood acts upon them. Although one could argue that a park is able to revitalize a degraded area, improving the quality of the urban space, it definitely cannot do it disconnected from its surroundings.

She also affirms that it is not possible to explain what constitute a lively park, successful in attracting people, but it is possible to generalize about a few principles that deeply affect them. The principles will be investigated aiming to understand the influences that act upon a park and how design strategies can be effective in creating popular urban spaces. These principles, described below, together with a tool of graphic analysis, will be used to investigate the parks through their physical characteristics.

The park should not constitute a physical barrier or interrupt the continuity of the urban fabric. Instead, it must connect and provide continuity between the diverse neighborhoods. The accessibility and permeability of the park's boundaries can be used to emphasize these characteristics and integrate with the city.

Having functional diversity and attracting different visitors make the park being used during different times of the day and different periods of the year. This diversity attracts more people and does not permit a lack of liveliness in certain periods, as the park is being used for diverse purposes. The design strategies might accommodate and emphasize this diversity to guarantee a high-quality space. If the park is located in an already lively area of the city, or with a potential to become, with distinct uses and full of people, it has a bigger tendency for success.

If the park adds a new element to the diversity of the area, it increases its value and consolidates the integration with the city. Points of orientation within the park can help in creating a sense of identity, a positive characteristic in the view of community life.

It is essential considering the urban fabric where the park is settled. If the park is a unique open space in the surrounding area, it will perform differently than if people have many choices of open and green spaces. If there are parks both close to each other and similar in purpose, there would be an inevitable competition between them, and the diversity might be dissipated. If a park is a unique choice of open space in the area, it stimulates different groups of people to use the same space.

As affirmed by Sassen, S. (1961), certain qualities, such as intricacy, centering, and sun-enclosure relation, seem to make a difference in attracting people to a park. Intricacy is related with the reasons why people go to the park, the diversity of uses. The complexity of the plan, defined by the hierarchy of functions and spaces and how they are related, creates different scales, stimulating the visitors and giving them reasons to return for different reasons in different times of the day. An important element of intricacy is centering and how the elements are articulated, as it gives diverse significances to the spaces, and people can easily identify themselves with the park. Sun is an important factor in attracting people, especially in winter. It is strongly related to the relation of volumes, voids, and shadows created by buildings and other elements. A shadow of a building cutting the field can kill its vitality and popularity. The buildings around the park, if they don't cut the sun, can enclose it and emphasize the openness of the park, which might be a positive characteristic.

Regarding the principles pointed by Saskia Sassen (1961) and based in the method of Balion, L. in *designing parks*, a graphic analysis of Parc La Villette and Parque da Juventude will be made. The idea is using these two successful examples of urban spaces to investigate how the design strategies, the organization of functions, and the relation to the surrounding area have contributed to build a lively environment.

The graphic composition and spatial organization of the parks will be analyzed through "components of a plan's coherence" such as access, organization, anchoring, articulation, and openness regarding the principles pointed above. It investigates how the elements and functions are organized, the relation of the parks with the city, and if a coherent organization has been achieved. The components are related to each other, and the concepts overlap. This coherence is given by the designer's choices and has an essential role in the quality of the space. Achieving a high-quality organization is essential to enable the visitors to identify and orient themselves within the park.

The initiatives to create the parks were, in both cases, a political decision. In 1982, François Mitterrand and a group of advisors elaborated the park's program for the competition to Parc La Villette, and in 1999 Mario Covas, the governor of São Paulo, made the competition to transform the Carandiru prison into an urban park. Both projects are located in the suburban area of the city and were conceived as a tool to revitalize a degraded area. They have a multiple-use program, which includes cultural and sportive facilities, learning centers, and leisure activities.

The analysis of the access will investigate the accessibility of the parks in different scales. Firstly it will analyze the approachability of public transports, such as metro lines, trains, bus lines, and avenues, and in which scale these elements which constitute barriers or connections. Secondly it will examine the links between inside and outside the park, and finally it will concern about the connectivity of functions and activities within the park and the accessibility to them.

La Villette is provided with a good access by metro lines and buses in the north and south areas, the main entrances. On the one hand, it has avenues that provide easy access by cars on the north and east sides and has a parking close to the south entrance; on the other hand, the avenues constitute barriers for pedestrians and cyclists. There is a link between inside and outside especially through the main entrances (north and south) where the boundary of the park is permeable, and there is an area of reception which invites the visitors to go inside the park.

The park can be considered a link between the opposite surrounding areas with the well-defined direct connections between the main entrances. It is crossed by two main paths, north-south and east-west, which are along the canal. The visitors can easily identify the routes and orient themselves, which contributes to be a space of passage. Within the park, beyond the direct routes, there are pathways to go for a walk or connect the gardens.

Parque da Juventude is well served by public transportation, including metro lines and buses. It is also surrounded by avenues in the northwest and east direction, where the main entrances are located inviting the visitors to the park. The avenues that connect the park with the rest of the city are barriers to pedestrians and cyclists, but the parking is a facility for those who reach the park by car.

It has two main routes crossing the area, a north-south and an east-west link, which allow the visitors to easily recognize the path and use it as a connection between the opposite neighbors. These connections also link the main functions and activities, and besides them there are alternative routes for other activities including relaxation.

The analysis of the organization examines how the functions are distributed, combined, and linked over the site. It examines the system of paths and connections and their relation with the buildings. The quality and coherence of the park's organization are related with giving the users the necessary information about its functioning and the capacity to accommodate the number of visitors expected to each activity that takes place.

Parc La Villette has different functions addressed to diverse groups of people, including children of all ages. The main functions are distributed along the axis of the park (north-south and east-west) and the amenities located within a grid of small buildings, the Folies. The elements and the connections of the park are easily recognized by the users. The system of points (Folies), lines, and planes developed on the park clarify and orient the visitors of the direct path, main functions, and park amenities.

There are direct connections for the users passing by the park or going from one main attraction to the other but also connections for the purpose of walking, wandering, or relaxing. The hierarchy of spaces and functions is clear and allows the users to choose between different scales of spaces and between more or less lively areas, once the main connections have more visitors than the other spaces of the park. Together, the system of points, lines, and planes cancel the notion of unique centrality.

The organization of the Parque da Juventude divides its functions into three main areas: the institutional park where the cultural and educational activities take place, the central park where the significant gardens and landscape spaces are, and the sportive park where we can find the sports facilities. These areas are not divided physically and are connected with direct and indirect paths. The activities are carefully organized according to their function, and the concentration of similar activities helps the visitors to orient themselves (Images 24.1 and 24.2).

The anchoring analysis considers the position of the park within the city and the relation between points of orientation inside and outside the park. It relates spatial situations and the role of points, buildings, and edges in orienting the visitors.

The north and south entrances are the main connections between the Parc La Villette and its surrounding, and the entrance areas that opened to the city consolidate the inside-outside connections. These spaces, where the boundary of the park

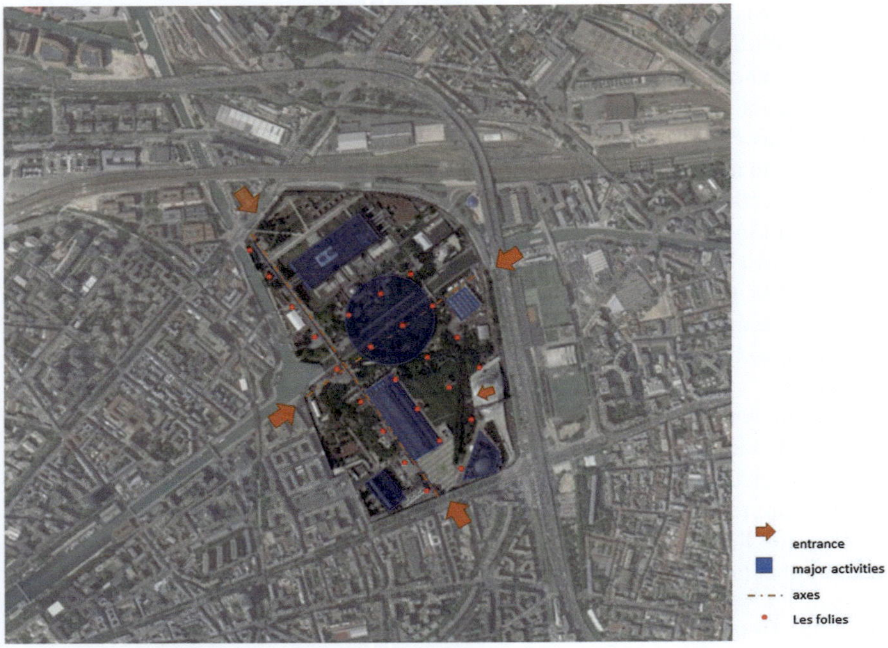

Image 24.1 Park La Villette: access and organization

is permeable, welcomes the visitors. There are internal and external orientation points. The Cite Des Sciences et de Líndustrie and the Cite de la Musique are the most visible elements inside the park, and there are elements outside that enclosure it, such as the large avenues and the train line. The regular grids of Folies and paths (system of points and lines) are elements of reference which help the visitors to orient themselves. The canal that crosses the site is another important line of reference. Together, these elements and paths make the park easily readable for who is there for the first time.

In Parque da Juventude, the buildings are the strongest visual elements, and together with an active prison next to the park, it constitutes strong visual elements. The elevated metro line is an external influence on the west face of the park.

The analysis of openness investigates the spatial and visual continuity of the park. It investigates which elements are landmarks and which constitute visual barriers. Openness is an important quality of a park, especially within a dense urban fabric; it allows the visitors to experience the continuity of the space and contributes to the coherence of the park. The manipulation of views can give different characters to certain spaces.

Parc La Villette is located within a dense urban fabric with a strong presence of transportation infrastructure. There are no parks with the same purpose and size in the area, contributing to the park to be considered a reference of open and green space. The density of the surrounding areas emphasizes the feeling of openness of

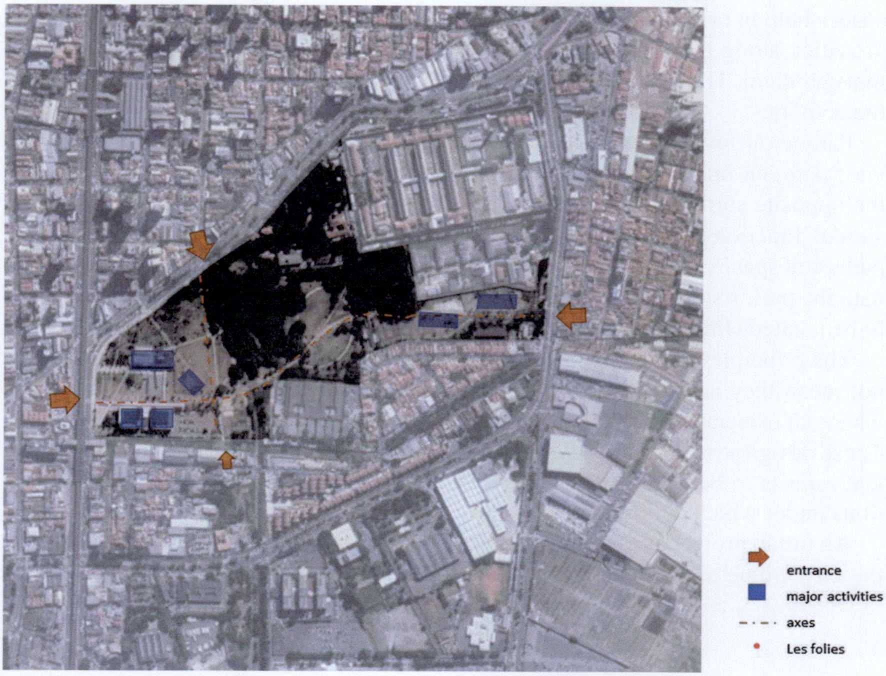

Image 24.2 Parque da Juventude: access and organization

the park. The main buildings constitute visual barriers but are equilibrated with the visual continuity in the connecting pathways (the north-south with the canopy and the east-west, along the canal) and help the visitors to move around. There is coherence and continuity of the visual images along the paths and out of them.

Parque da Juventude has a high-density surrounding and is the only significant open and green space in the area. The other green space that can be seen in the map is part of another prison, which will be soon deactivated and incorporated to the existing park. It has the presence of significant transport infrastructure. The visual barriers are the buildings, and the relation of volumes and voids in the rest of the park is made within the landscape.

The analysis of the articulation considers the interaction between spaces and volumes and how it contributes to the organization and coherence of the park. It is related to the hierarchy of the different components and the dominance of certain buildings, axes, or spaces. The articulation can be related to both visual and functional aspects, in view of the visitors' orientation.

Parc La Villette has a clear visual hierarchy of the buildings, paths, and spaces. The main buildings, located close to the bigger entrances, are visual orientation to the visitors and where most of the indoor activities take place. They are articulated with direct connections that get stronger when combined with the different landscapes. The activities and function of the park are easy to identify, and the paths and spaces help in orientating the visitors. The articulation of the buildings and outdoor activities along the paths contributes to the high intensity of use as visitors pass through them. The visual orientation is strong and contributes to the articulation of the activities.

Parque da Juventude has its principal activities articulated along the main east-west path, an important visual and functional connection of the park and between the opposite surroundings. The park has distinct centralities, as the activities with similar functions are combined and located close to each other. The hierarchy of paths and spaces is an easy orientation for the visitors. The buildings visually dominate the park's spatiality as they are all in the west entrance, where the prison used to be located (Images 24.3 and 24.4).

The principles pointed by Sassen, S. (1961) can be found in both projects. It does not mean they are essential, but it indicates they can help in constructing a place where an exchange between different social groups is possible and actually occurs. These design strategies must be combined with each other and with social and political aspects to be successful but are essential in providing opportunity and conditions under which cultural and social exchange can flourish.

An urban project can provide the possibility to become lively and integrated with the city. Some alternatives that could emphasize these characteristics are, among

Image 24.3 Park La Villette: anchoring, openness, and articulation

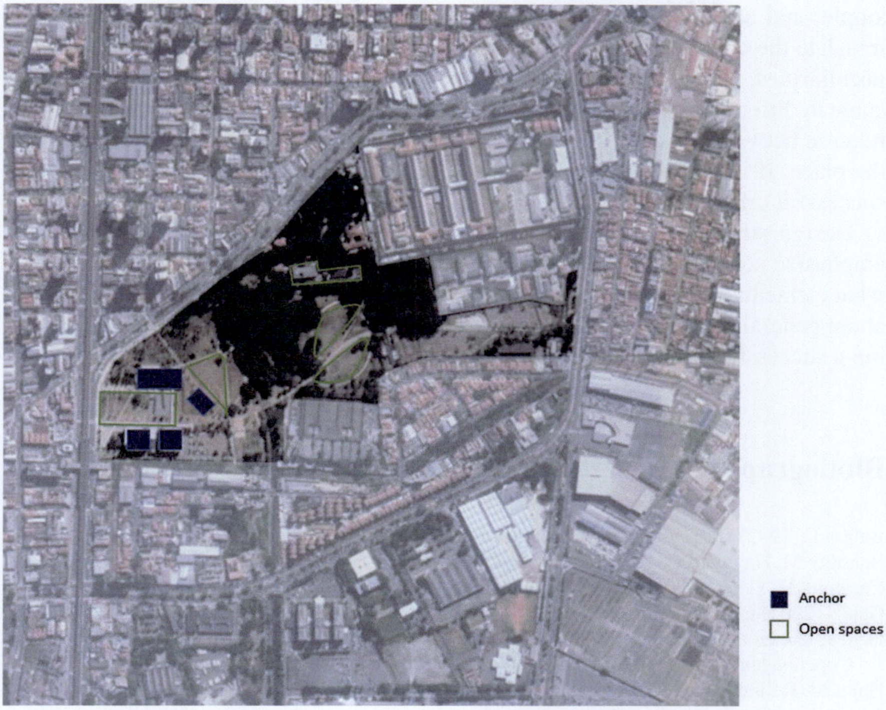

Image 24.4 Parque da Juventude Park: anchoring, openness and articulation

others, the multifunctional character, the plurality of interests addressed, and the connectivity and accessibility to the surrounding.

Parks, between other urban projects, can be used to revitalize a degraded area, if carefully designed and strongly related to the city. It can be used as an object capable of transforming the area and reverse the decline of its quality, as shown by the two examples. In both cases, the diverse character of the surrounding, composed by residents, commerce, hotels, and entertainment centers among others, bring people with diverse interests and during different periods of the day to the parks, which is essential to their vitality.

There are social, political, and security aspects not mentioned here which are significant. For instance, Parque da Juventude has a strong sense of identity with the local residents, especially considering the transformation of the prison, scenery of a huge tragedy with a negative impact in the neighborhood and strong significance, into an urban park. It contributed in the way that citizens have sense of belonging to the place and fill the park with life, especially in a country where there is a lack of successful urban interventions.

It is not possible to explain the exact reasons of the success or failure of an urban project or intervention. What the analysis pointed was some general characteristics that contribute to its vitality. The concept of lively and successful urban space is ample, and an effective way to explain it is giving examples of real spaces that translate the characteristic. Each case must be analyzed separately as it might have peculiarities, but the interesting fact is the general principles that seem to be applicable in different cases. These principles might be a useful tool in understanding the relation between physical characteristics (design decisions) and the performance of the place. If its effectiveness is proved, it might help in designing more lively and successful urban spaces.

Design strategies cannot guarantee liveliness to an urban space, but they can emphasize some aspects that contribute to it. Although it is not possible to affirm what elements are essential and how they should be organized, it is possible to agree about general concepts and organization strategies that might act positively in building a successful urban space.

Bibliography

Baljon L (1992) Designing parks. Architectura & Natura Press, Amsterdam
Barzilay M, Hayward C, Valentino L (1984) L'Invention Du Parc Graphite Editions/E.P.P.V
Carmona M, Heath T et al (2004) Public places – urban spaces. Architectural Press, Oxford
Gehl J (2001) Life between buildings using public space. Danish Architectural Press, Copenhagen
Gehl J, Gemzoe L, Kirknaes S, Sondergaard B (2006) New city life. Danish Architectural Press, Copenhagen
Hajer M, Reijndorp A (2001) In search of new public domain. NAi, Rottherdam
Jacobs J (1961) The death and life of great American cities. Random House, New York
Macedo S, Sakata F (2002) Parques Urbanos no Brasil Imprensa Oficial do Estado de São Paulo
Rowe P (1997) Civic realism. MIT Press, Cambridge, MA
Sassen, Saskia. Cities in a World Economy. 1961

Chapter 25
The Smart City in Relation to Its Environment, Perception, and Urban Planning Process: Lessons for Developing Countries

Vasiliki Geropanta

Abstract This research presents some of the key smart city strategies in Spain and the Middle East with a dual objective. Firstly, it aims to underline how the integration of information technologies in city management in the two cases altered the perception and design of the built. Secondly, it argues that they might be a useful example for the urban policy discourse in the developing countries. The two cases represent a different conception of the smart city project. The first one describes an intervention within the consolidated urban fabric of Barcelona in Spain. The second case takes place in the outskirts of Doha in Qatar. Each one was examined from the perspectives of architecture, technology, and communication and the analysis included diverse media, such as textual material, drawings, review of literature, empirical observation, and a critical elaboration of the results with the use of matrixes. Within this variety, it became easier to discuss the processes that are latent when juxtaposing studies on physical space, social practices, and real-time data with urban fabrics, culturally and historically charged. The results revealed the spatial implications that new technologies would have in diverse settings. They also hinted on the way sustainability has become part of the project for the city. Latent under this integration of space and ICT, there are user-driven actions and human interaction with artificial intelligence.

Keywords Smart city · ICT in developing countries · Sustainability · Smart policy · Economic welfare · Urban and spatial planning

---

V. Geropanta (✉)
School of Architecture, Technical University of Crete, Crete, Greece
e-mail: vgeropanta@arch.tuc.gr

© Springer Nature Switzerland AG 2020

M. Ghosh (ed.), *Perception, Design and Ecology of the Built Environment*, Springer Geography, https://doi.org/10.1007/978-3-030-25879-5_25

Introduction

Urban configuration and urban planning are strongly affected by the technologies of transport, communications, and production already from the second part of the twentieth century. Initially, a number of interwoven lattices of utility networks made possible the functioning of the modern "networked" city and their integrated urban system (Tarr and Dupuy 1988; Tarr 1984). In some cases, they affected urban planning to change its entire landscape; in others, they improved the quality of urban life (Serag El Din et al. 2013). However, all these, strongly connected with the societal and economic changes of this century, stressed dramatically the natural environment, contributed to the fragmentation of cities, splintering of neighborhoods and as a consequence to marginalization (Graham and Marvin 2001).

The last decade, urban planning acquired a new type of infrastructure, the one that provides the transport of data and information related to the city functioning in real time (Komninos 2015). The idea was that by introducing information and communication technologies (ICT) in urban planning, the problematic processes of the previous practices would resolve radically, transforming it into an urban settlement smartly connected and therefore holistically managed with optimized resources. The difference here was that ICT affects urban design through the processes of economic development and citizens' participation without compromising the ability of future generations to meet their own needs. The immaterial nature of ICT, and the pervasiveness in all layers of human life, became a catalyst for spatial planning redevelopment, land use transformations, and therefore a new conception, use, plan, and control of the built space in cities that did not stress anymore the natural resources (Gunder and Hillier 2009; Firmino et al. 2008).

Among the various city paradigms that emerged through the use of ICT, the smart city dominated and already counts for the majority of the venture capitals investments worldwide (Larson 2017). In its full completion, it is an urban settlement, constructed by "top down or bottom up 'educational' processes that act as the strategic device to properly connect, smartly manage and optimize the city resources" (Komninos 2015). This urban settlement emerges as a complex mixture of networks, places, flows, and information, in which multiple relations and activities have the possibility to generate creative synergies and respond to the previous mentioned urban and natural demands (Healey 2007). ICT, space, and users become the paramount means of affecting the urban, allowing planners to work with spaces a unified "whole." Completely new districts are created to serve the smart city, and new means of digital "zoning" are changing the conception of the traditional urban tissue (Geropanta 2016). Financially at least, these evolutions seem to have opened a new world of possibilities, collaborations, and problem-solving strategies towards the twenty-first century problematic of cities.

In this framework, the objective of this paper is to evaluate how the implementation process of the smart city influences the outcomes of such strategies in context to its environment, perception, and design. Latent behind this target is the belief that while more and more cities view such strategies as a possible way to enhance urban competitiveness and solve existing urban issues, it is important to establish an analytical framework to explore the relation between the means of implementation and the outcomes of such strategies. For this reason, the paper will try to identify the main factors and their interrelationships via two case studies (Spain, and Qatar) that follow two different approaches: the top-down process and the bottom-up as policy-making logic in the two urban environments. The author believes that the findings of these two cases offer insights into parts of the world where these transformations are yet to start and can become therefore a good resource of information for the developing world. These two realities are treated therefore as a major and unexploited deposit of ideas for innovation, for the developing world.

Spain and Qatar Embracing Smart Urban Transformations

In Spain, the smart city projects of Barcelona delivered real-time information and recreated any environment, enhanced, virtual, or augmented in a historical, stratified, and culturally rich city. As a result, new layers of meanings and new potentialities for the enhancement of the city performance revealed and altered the perception of the existing built environment (Bakici et al. 2012). Specifically, with a core at the smart district @22 in Poblenou, the smart city strategy of Barcelona regards the collection of small-scale "smart" pieces, all spread in the existing urban fabric, generated and motivated by the transformation of @22 ex's industrial district to a new innovative environment (Pyzyk 2018).

In Qatar, the smart city appeared as a possibility to discuss city growth in a wide range of dimensions spanning among the social, economic, cultural, political, physical understanding of ICT. In this case, the smart city became the central topic to speak always about the city in architectural terms (Harrison 2015) and in the transformation of the future image of the capital city. In fact, the project for the construction of the smart city in the outskirts of Doha regards the construction of a new large city of 38 km^2 of waterfront land sub, home to 250,000 people (Harrison 2015). Both cases were designed following objectives for regional sustainable growth, and both cases introduced new concepts of city perception.

The smart city of Barcelona represents the way the deployment of broadband networks, smart spaces, web applications, and e-services helped the city to become more innovative and efficient in resources and environmental sustainability. Specifically, the city's objective is to use ICT in order to transform business processes and the public administration, both internally and externally (Bakici et al. 2012). This vision tried also to generate a sustainable city growth, by encouraging initiatives related to smart lighting, mobility (e.g., e-vehicle) or residual energy (e.g., heating and cooling networks) but also related to social innovation. Among others, the main principles of the smart city ecosystems included (1) the creation of a backbone of broadband networks, (2) the creation and agglomeration of digital applications in every domain of the city life and activity that bring the city closer to the citizens, and (3) the renewal of city-districts through smart systems and innovative communities. Strong reinforcement of alliances and synergies among public and private partners including also institutions, the European Commission, and the United Nations helped the city to engage and keep many stakeholders connected and, in this way, ensured strong support from anywhere (Ferrer 2017). During the creation of the hundreds of the smart city projects in the city, an ecosystem of IT businesses in the @22 district was accumulated, transforming the ex-industrial area to an IT hub. This district is located in the industrial area of Poblenou, spanning 115 urban blocks in 1.9826 km^2 and is considered today the core of the smart city of Barcelona (Pyzyk 2018).

Qatar's smart city is a case study that uses new technologies to create risk control, efficient, and sustainable services to its citizens and therefore offer an advanced quality of life (Harrison 2015). By the official website, it seems that the smart ecosystems are monitored by a central command and control center. A common network infrastructure integrates the various services and systems, which are grouped in telecom services (residential and business services), non-telecom services (ITS, Public Car Parks, LCCC, and ITS Information system), and value-added services (Vision Presentation). Briefly, these categories cover different services for security, mobility, traffic management and parking systems, municipality administration, advertisement, access control, billing, smart homes, light control, GSAS monitoring, energy management, and GIS, among others (Harrison 2015).

ICT Implementation on Two Large Territories - Two smart Cities

Qatar's smart city is one of the emerging cities in the Gulf, a new fragment in Doha's urban development. It is located in the municipality of Al Daayen, in the northern end of West Bay Lagoon of Doha (Alhorr 2016). The schematic design reveals that the new urban environment covers 38 km^2 of waterfront land sub and consists of 5175 plots that are under construction. Specifically, the city is planned to provide accommodation for 250,000 people permanently living there, while it is expected that 170,000 more people will be entering the city on daily basis for work (Vision Presentation). The total estimated capacity of this territory is considered to reach 450,000 people including tourists (Harrison 2015). The city will offer all community facilities such as schools, mosques, community retail, cultural facilities, petrol stations, medical care, special accommodations, and diverse strategies for public, massive football stadiums, hotels, residences, mass rapid transit, and a marina, all to debut when the country hosts the 2022 FIFA World Cup (Pyzyk 2018). It is a result of more than 2500 architects involved in the design process, while world famous architects such as Foster and Partners and many others are delivering the most central parts of the cities (www.lusail.com, 2018). The master plan has been divided in 19 mixed-use districts, which have all completely different design styles, 3 smaller islands, 3 ports, and a large number of network infrastructures (Alhorr 2016).

The districts in the schematic plan are divided by function and are designed in diverse architectural forms. The energy city, for example, has a circular form, almost amp theatrical however plane, while the Marina District is developed along a parallel axis to the sea. A number of smaller buildings are spread all over the city devoted respectively to the utilities channels and the spaces of the IT and engineering. The main blocks are conceived as a sequence of rectangular blocks without inner courtyards that were placed at a distance from one another and were linked by built mass only on one side. The new districts include the new ports, new marinas, residential areas, island resorts, commercial districts, and luxury shopping and leisure facilities. Its master plan depends on the design of an open space network, a pedestrian and bicycle network, with special focus on pedestrian infrastructure (promenades, connection bridges, squares). The infrastructure that bares the hardware for the support of the smart city is dug into the ground, creating an additional floor underneath the level of the ground. Such gesture is unique remodeling the underground desert landscape underneath the city. It is based on a canonical figure-ground relationship through control centers and the employment of new technologies all connected underground, but installed differently with every district.

The city has been therefore conceived as a part of a simultaneous smart settlement in which the majority of the built masses that support the existence of the smart city are sunken into 30.5 km wide over ground by 28 km, with 36 passenger stations. The overall length of the city exceeds 60 km, with its southern edge located almost inside to the built fabric of Doha.

Conversely, in Barcelona, the project for the smart city relied on a more traditional understanding of the contrasting relationship between the smart as a built object and the ground onto which it was placed. Where the implementation of Qatar's smart city emerged out of the morphology and design of a place, the scattered pieces for Barcelona were superimposed from the top over the existing landscape. Then, these pieces derived from a zoning of the programmatic requirements gathered in the project for @22 that were 3.2 million sq. used for productive activities and 800.00 for a residential center. Today Barcelona has 83,640 business premises and a relatively large number of the new companies are technology or knowledge based.

Poblenou was a point of focus in these transformations. Covering 2 km^2 of industrial land and considering the needs both of the Olympic Games and the preservation of the economic productivity of the former industrial district, along with the housing needs of the immigrants arriving in the area, the district was the perfect field condition for a strategy to alleviate the housing shortage in the city. The result of this intervention led to the creation of @22 districts, a central business district, spanning 115 urban blocks in 1.9826 km^2 (Pareja-Eastaway and Pique 2011). It promotes a dual vision of smart city: an area that is marked both from the use of ICT as a means for offering better services in a city scale and also by the existence of IT and research institutes that promote knowledge and innovation. As such, it has attracted attention both of the academic world, but also of international stakeholders, policy-makers, and large corporations. Examples of these interests can be found to the research by Pareja-Eastaway and Pique, Casellas and Pallares, Urenio, and Engel but also to the International Association of Science Parks that consider this district as a reference source for policy transferability and experienced based knowledge (Pareja-Eastaway and Pique 2011; Pallares-Barbera 2015; Engel 2014).

Proposals for the regeneration of Poblenu respect completely the preexisting structures and maintain the streets and historic features and elements as the base of the new organization. In this way, this part of the city became more attractive, fighting against the low density that characterizes traditional industrial zones. The overall project opts for a dense and complex urban space, allowing for more efficient land use and creating an environment where different urban agents can interact and exchange information to develop an agglomerative economy (Barcelona City Council 2000). Architecturally, it was divided into 15 main blocks and arranged in an overall "cerda" shape. A first block was composed of two parallel bends devoted, respectively, to teaching and research spaces for the departments of the IT and engineering. The teaching part was conceived as a sequence of rectangular blocks with inner courtyards that were placed at a distance from one another. They were linked by built mass only on one side. This resulted in an alternation of open and closed courtyards of which the former were extended into the smaller courtyards of the adjacent research block. The spaces that were provided in this first research-teaching block included departmental libraries, classrooms of different sizes (between 100 and 250 seats), teaching laboratories, and offices for the professors. A second block was placed parallel to the first, and the two were connected only at the southern edge by a third block that housed the main common facilities of the university – including the central library, a music and art center, administrative offices, and some communal facilities like dining halls, clubs, meeting rooms, bars, etc. The third block was meant for social purposes and was provided for office space work. Finally, a fourth block for research was deployed to house the scientific departments. This block was composed of two parallel linear structures of which one replicated that for the research spaces of the first block, while the other was provided for larger floor plates for laboratories. The residential area was split into two sites that were divided by traffic infrastructure and were intended to be developed in different stages. It was meant to provide housing for 1500 people that included 600 student residences and 800 apartments for professors and staff. Student residences were designed as linear structures on two levels, with a portico and common spaces at ground floor and bedrooms on the first floor. They were conceived as continuous structures of varying length that bended at right angles, thus creating inner courtyards that, however, were never completely enclosed.

Through a system of incentives that allow for better exploitation of this sector's urban areas, the urban regeneration processes contributed to the re-urbanization of all streets. In this way, they would generate land that was free of compensation to the community – from a plot that was 100% private, in which 30% became public – in order to create new green areas, amenities, and subsidized housing (Pareja-Eastaway et al. 2007). Therefore, productive activities would coexist with dwellings, commerce, and research, continued training, and technology transfer centers in an area that could balance high density with open spaces and amenities. Low density and functional obsolesce could be solved through the rehabilitation of the urban fabric. Mix of uses and coexistence between preexistence and transformation were part of the design process. New green spaces, new advanced infrastructures, industrial heritage protection, new public amenities, and new social housing were only some of the main axes of the urban renewal.

Synergies between urban agents led to including a building index, which attracts real estate developers and allows them to contribute financially to the sector's new infrastructures. This brought about an increase in the private sector building index (from 2 to 2.7), and the re-designation of land are linked to certain obligations: payment of urbanization fees, obligatory presence of e-activities, land concessions, etc., which establish a system of incentives for transformation (Engel 2014).

Novel Design Concepts in the City, District, and Building Scale Because of ICT

In Qatar's smart city, there was a number of "smart" infrastructures as the result of the installation of ICT and the integration of telecommunications, computers, as well as necessary enterprise, software, middle ware, storage, and audio-visual systems. The system of work seems to follow the IoT logic, since it consists of hundreds of mini networks that support the whole city. However, the strategy for the smart implementation seems to be divided into four main structural concepts.

a) One is related to the creation of an open space (outdoors) network pedestrian street system. This network aims at becoming a breaking point in the traditional use of the interior of the buildings in the Middle East because of the desert's high temperature. A large number of technologies and sustainable environmental design gestures manage to create a minor climate similar to the South European one. This is found in different ways such as ventilation systems in the main facades of the buildings, in the cooling underground system, and in the hydrating underground system. For example, the creation of an open space network obviously refers to traditional planning concepts of European climates. This strategy has been largely concerned the development of the Marina District. This district is a part of the city strictly connected with the waterfront, mixed use that is serviced by a light rail transit (LRT) network and water transport system that connects to the rest of the smart city. Underground stations, all linked directly to underground car parking facilities, aim at dissolving the traffic and monitoring the mobility.
b) The second is related to sustainable utilities on behalf of the city, such as district cooling, pneumatic waste collection and the creation of a gas network, and as such a large number of hardware and built infrastructure.
c) Intensive use of ICT in the level of the building led to a complete different design phase for every property, since diverse traditional practices, now will be atomized.

d) Additional to these visions, there are also a number of smart city services offered to the local citizens with numerous social implications. Mainly, an amount of daily actions will be facilitated through a free access internet system, some related to communication, some to low energy consumption (public lighting, smart metering etc.), some to environment (air and noise monitoring, early warning systems), and some related to the lifestyle (Tele assistance, virtual meetings, emergency responsive systems).
e) Lastly, very important sector in the development of the smart city strategy is the creation of the integrated management system, where an amount of sensors and nanotechnologies aim at developing methods for traffic tracking, incident detection with vehicle count and speed. Electronic messages will be informing citizens in real time about the weather, the traffic situation, and so on.

It is clear that these actions all allow users to access, store, transmit, and use information. This refers to a collection of practices guided by the "Operating Command and Control Center LCCC." The Command and Control Center is the human interface to all the intelligent smart systems on the city. The main tasks of this control center are network management system, call center services, enterprise Resource Planning ERP, safety and security systems monitoring, facility management system, utility telemetry and billing, traffic management, and parking control systems.

Barcelona and City, District, Building Scale Observations

All over, the city has been filled with new broadband infrastructure and e-services that are available to the users. Among the new infrastructures are as follows:

a) A corporate fiber-optic network of 325 km that connects all municipal buildings and other complementary infrastructure.
b) A corporate network that is a closed broadband infrastructure, most with 1 Gbps links, which covers council-related groups.
c) A Wi-Fi network that provides wireless connection to those municipal services at the level of the street municipal Wi-Fi network: The municipal Wi-Fi network complements the corporate network's connectivity, and provides workers and services throughout the city. The council Wi-Fi network consists of 469 active nodes, reaching 30% of the city.
d) Multivendor and multipurpose sensors.
e) Public Wi-Fi that offers open broadband access.

Also at the district scale, the ICT cluster brings together about 60 ICT firms mainly from the metropolitan area of Barcelona (companies, associations, chambers of commerce, and technology centers). It is made up of totally 120 ICT companies (products, integrators, VAR, projects, and retailers) and institutions (universities, technology centers, business organizations, support organizations) involved in the field of information security. This cluster aims to boost the competitiveness of Catalonia's ICT sector, promoting new business culture based on collaboration and open innovation that can be solved as we saw earlier through the corporations' help. In the sector of mobility, there was created a project named Clúster6m, which contains more than 100 companies (navigation and positioning solutions, vehicle communications, mobile devices and applications, integrators) and institutions (research centers, universities, and government) in order to carry out projects that improve mobility, business competitiveness, and the welfare of citizens (Barceló Rota 2005).

In the field of knowledge production and diffuse of information, a world of sensors is being developed which currently includes about 100 companies (sensors, communications networks and applications, street furniture and equipment, and services) and institutions (local administrations, universities and technology centers, support organizations). The initiative is being promoted by i2cat Foundation, with support from the Generalitat of Catalonia. In addition, Education (LogosNet) is an initiative promoted by CETEI-Fundació Joan XXIII (Ituarte Technologies Centre) Bdigital Technology Centre with the support of ACC1Ó, the Ministry of Telecommunications, and the Information Society at the Generalitat of Catalonia and Barcelona City Council (2016). In this way, more than 200 companies (consulting, computer services, medical equipment, communications, application developers, etc.) and institutions (groups of hospitals, health centers, technology centers, public administrations) generated promotion strategies in the various sectors and networking across the health field.

One of the key factors of Catalonia's ICT development to ensure the success of a cluster policy is its long-standing associative tradition, collaboration, and networking capacity. Two out of three companies are involved in a Catalan ICT association 22. These traits have helped Barcelona, for example, become a leader regarding the number of collaborative projects carried out as part of the Framework Program in Spain and in Southern Europe. Some projects executed under the Living Lab concept:

- The SIIUR Living Lab project is an urban pioneer and innovator in the intelligent management and efficient use of public lighting and the integration of municipal services via this infrastructure. The city of Eindhoven chose the SIIUR project as one of five finalists from over 245 entries in the open Living Labs Global competition in order to develop and implement innovative intelligent lighting and sustainability solutions in one of the city's main districts.
- The REMEI (Real-time Medical reporting service) project, an i2cat sponsored Living Labs project, became the implementation of shared medical report solutions in real time between specialists from different medical centers solving emergencies.
- The Edulab initiative, led by Citilab Cornella, is a laboratory for the educational community. It aims to foster innovation in teaching and learning with information technologies and specific communication methodologies.
- Project ICING: Eight cameras using artificial vision systems were installed as part of the Barcelona Urban Lab to count the number of vehicles travelling on

city streets. This pilot is part of the ICING European project involving 10 companies.
- The Project MOVELE Barcelona: As part of the MOVELE project promoted by the Institute for Diversification and Energy Saving (IDAE), this project involves deploying a network of 190 re-charging points for electric vehicles for users in underground and surface car parks across the city and which are run via different promoters yet interoperable.
- Barcelona City Council's Open Data Project: This project is about making public data that the city has – at any aggregation level – available to the public to increase the transparency of council management, to encourage the social use of public data, and to boost business and innovation.

Complementing Living Lab initiatives and located in the 22@Barcelona district, in Media-TIC building, which was designed using the values of sustainability and the digital world created a home and a transparent citizen's platform to all the above projects. The building is an excellent networking space, a laboratory-come-showroom that allows IT companies to show and demonstrate their most innovative products and services and citizens to participate. The Open University of Catalonia (UOC), Barcelona Digital Technology Centre, the new headquarters of Cibernàrium, and other IT companies and professionals currently occupy the Media-TIC building.

The Physical Presence of the Smart City: Confirmed

The construction of brand new IT poles in Barcelona, and the creation of an intelligent high-tech city as Qatar's smart city, activated a complete relocation or construction of buildings and lead to clustering, concentration, agglomeration, and proximity of companies and IT institutions. Qatar is becoming the new tertian settlement for smart society in the Middle East. @22 have metabolized in time attracting attention for being excellent settlements for innovation, reactivating the urban fabric, and pushing forward architectural reflections on continuity, collective spaces, and integration of expat communities in the city functioning. The leading companies created synergies in large scale, beyond national borders, reaching at minor building scale. These strategies, as it is the case with most smart cities, are scattered throughout the urban fabric of the city centers or are recreating city centers by concentrating business ecologies and research institutions. In Qatar's smart city, the IT poles are expressed through a series of buildings on the desertous new grid, stating the technological manifestation of the new city. These new poles are located inside the clusters even if their borders and limits stretch in the borders and limits of the potential networking reach out of them. The plan of Barcelona, for example, roughly indicated an area for the smart city headquarters on the western periphery of Barcelona. However, that was only a minor part of a much larger area that had been allocated for innovation related uses in Poblenou. Architects were asked to take into consideration the respective urban master plan with the idea that a number of spin-offs and occupation by more innovative businesses would be brought about. Something similar happened in Qatar's smart city. The project's master plan indicated an area in the outskirts of Doha, and participants could not rely on any existing intermunicipal plan, but the requirements were new settlements that could respond to issues of innovation and welfare. In other words, from a series of vague definitions of the digital settlements, in both cases the smart city made a substantial leap forward. It declared itself as a large-scale urban reconstruction element.

The dependence of cities on innovation ecosystems requested that architects would take into consideration the possible interactions among the new districts or buildings with the presence and use of people. The @22 district is actually a case to provide practical examples of how this engagement can be realized. Softening of the building edges, zoning, permeability in buildings, and new urban spaces are only some of the outcomes of the Spanish smart city. The smart city thus asked for a multiscale thinking to keep together the large territorial scale of a metropolitan area, the scale of the district itself, and a more detailed architectural scale of a typological and technical study. Given this requirement for multi-scalarity automatically became a pretext for architects to develop wider thinking on the spatial organization of an urban territory. In this sense, the term "smart city" does not refer any more only to something "invisible" or something that could not find a spatial essence. Rather it took a larger meaning in relation to an increasingly complex urbanized territory.

In fact, @22 proposed a way of re-conceiving the overall spatial organization of the city restoring a central urban role of the district. The Living Lab type sets a spatial principle in this sense and added within the new organizational system an idea of a place in which each piece is found its main role: Media Tic became the main center for tertiary services, leisure, and business activities; the former industrial block gathered all IT activities and so on. A large number of housing were also offered, and this was the main isotopy with Qatar: the intention to turn the new smart city into an ideal residential community, that is, into a city in miniature. Services for the smart infrastructure are not isolated but are designed as an integrated service core that includes every possible city function. Both cases actually revealed a large working and residential area where governmental activities merge with learning, research, and leisure. The new cities are completely hybrid in this sense, reviving the sensation of the center of older historical cities, where a monument (church for Rossi and IT headquarters for the smart city) reinvented a human collective way of living. For example, among the various strategies, @22 proposed also to use the existing buildings for student housing and claimed the importance of keeping the research departments inside the same blocks. However, the project also argued for the need to create a mixture of activities in the urban fabric, stating the way the new knowledge activities are inserted in the city, by mixing. This created a difference also from the trend towards privatization that existed up to now to relevant compounds. It was thus intended as a place free occupation and continuous reconfiguration around which the residential and proper working areas for the smart cluster were housed inside converted existing buildings. In other words, rather than a project for a large city project, a duplicate of the city center is conceived as a place for the encounter of people that would act as the smart community. This is exactly the point where the organization of space alone would be enough to affect real and social change and an enacted utopia.

In Qatar's smart city, a similar trend is followed: the possibility of defining the future organization of a large area by means of clear architectural drawing that depicts and mirrors its social heritage. Planning in this sense can be interpreted as a process in which drawings and technicalities come together, something that goes against an increasingly bureaucratized version of urban planning. Both cases encompass an idea of realizing a smart city settlement as a major node along which the new axis for urban development and urban restructuring are realized. While sharing a similar argument towards this restructuring of the whole, they also differed in the decisions at the architectural scale. Barcelona conceived the possibility of regeneration as a composition that respects the traditional buildings and therefore took action in the interiority of the buildings or in the urban scale creating "open spaces." Qatar's smart city craved for a free composition of basic architectural forms scattered in the urban scale. In opposition to the open plan of @22, Qatar proposed a clearly bounded area that followed a pre designed grid.

An important aspect of Qatar is that rather than defining the space according to a given institutional organization (either by faculties, or institutes, or departments), it followed a functional logic in which different spatial configurations corresponded to the main activities of a city regardless of any possible hierarchical organization of the institution. By translating those activities into a diagram made of juxtaposed linear bands, which alternated indoor and outdoor spaces, the project stated that the functioning of a smart city ultimately depended on the coexistence of various activities happening simultaneously. While deploying a level of internal flexibility through a juxtaposition of generic but differentiated spaces, the project reaffirmed a cornerstone concept of archipelago, namely, the necessity of a clear figure-ground relation between pockets of built-form and the rural-park land. In opposition to Barcelona, the project proposed is to consider the smart city as a possible "matrix" capable of shaping the city in which continuity in the urban environment is ensured through the digital net. The attempt of grounding such idea of continuum on the analysis of the smart city as something that permeates all heritage, built and non-built, can be seen as the reaction of architects towards this new city paradigm. Continuity was found in Archizoom and the Non Stop City, or Zenetos and Electronic Urbanism. Architecture is more absorbed within the production of new services chain, and it makes no sense to associate a specific form to the smart city.

Conclusions and a Glance to the Developing World

This paper has proposed a contribution to the field by analyzing two case studies from the combined perspectives of architecture and strategic implementation of the smart city. The new spaces created by the application of this strategy bring different experiences of the city and confirms the existence of a smart city settlement in both cases, something that certainly will have also implications for daily social practices. In this way, the article suggests that there is a strong connection between the process of creating policies making to apply a smart city project and their urban regeneration and city growth. In both cases, the smart city was not an alternative idea to sustainable development but a strategic imperative that boosted or transformed the economic future of both areas.

Considering the current predictions on the worlds' growth in developing countries as well as the strains that this growth will bring on local economies to produce adequate jobs, infrastructure, and public services for its citizens (United Nations 2014), the smart city seems a promising scenario. However, despite widespread adoption of the urban technologies, there is still a lot of consideration on the appropriation of the smart city as project to solve the challenges of the global south. This happens because the focus of most companies in smart city technologies, then, is on a narrow selection of between 300 and 600 developed cities driving the global economy (WEF 2018).

For this reason, the analysis of the two cases studies serves here also to highlight the potentialities and the strengths of such a strategy in the developing world.

The specificities of the smart city implementation projects in both cases highlighted that real-time monitoring through IoT devices brings rapid action to manage risk control and to minimize the impacts of any security, leakage, and contamination issue. The synergies and collaborations that stand in the creation and support of each smart city strategy can empower these actions and augment the opportunities to success. Since it is widely considered that the cost of smart monitoring is low regarding the infrastructure cost, then updating the infrastructures through smart city implementations and smart design reduces the overall cost of construction as well as the running and maintenance costs. In other words, the research suggests that using the digital revolution to conduct efficient transformation at lower cost might be an opportunity in the suggested areas.

Furthermore, the common strategies in both case studies revealed solutions that are extremely challenging in the developing world. For instance, security seems to be highly established through the use if CCT cameras in both cases, allowing for better experience of the cities. Another example is the wastewater efficiency that was brought about in the desertous environment of Qatar by the use of smart water grids, or SWGs, which ensure the security of water quantity and the safety of consumption. The ICT in this case allowed the cities to accurately monitor the quantity and quality of water being transported. Moreover, the smart pumps and valves in Qatar can adjust or block flow in water pipes depending on what is necessary. This diminishes the amount of water and energy wasted in each process and increases efficiency at the same time.

Furthermore, there are three components for a smart city as highlighted in the case studies: urban space, ICTs, and skilled users (Komninos 2014). In the case of the skilled users, it seems that in developing countries the mobile is highly penetrating and social media are highly popular. This brings good skills in information technologies and therefore can be easily used by the citizens. From the two case studies (the bottom-up and the top-down), it is not clear what strategy should be followed when it comes to people engagement, but it became clear that users are not any more objects of observation but rather active participants in the creation of a new human spatiality. The resulting schematic nature of the place where users interact is a cluster or an enclave, and this is a fact that cannot be considered in negative terms. In fact, the potentiality that stands in people's participation and the production of knowledge in cities allowed more generic but useful arguments to be developed. Reading the two as part of a structure where the spatial practices, the representations of space, and representational space revive opened an opportunity for various observations, through which people and human touch came in the forefront. Mostly, it offered the reading of a micro sensor not as a miniature in space, but in fact, it refused it. The smart city or the smart technologies were not to replicate the spaces of the city but to advance novel spatial configurations in the service of people. The district configuration (Fox Hills and @22) hardly re-proposed an existing typology but opened possibilities of inventions and experimental spaces produced by social interactions. If previously similar city paradigms around in the world were limited as to be conceived as self-contained compounds, as a device a new spatial organization within their boundaries, in Barcelona human participation led to a total reconception of the area.

References

Alhorr Y (2016) Development of sustainable cities in GCC: Lusail case study, Jeddah Retrieved by: https://www.google.com/search?q=lusail+smart+city+research&newwindow=1&client=firefox-b d&ei=XxahXMezFoyWkwWlori4CA&start=10&sa=N&ved=0ahUKEwiH34L9j63hAhUMy6QKHSURDocQ8NMDCJ8B&biw=1280&bih=585&dpr=1.5#

Bakici TY, Almirall E, Wareham J (2012) A smart city initiative: the case of Barcelona. J Know Econ 4. https://doi.org/10.1007/s13132-012-0084-9

Barceló Rota, M (2005) 22@Barcelona: a new district for the creative economy from making spaces for the creative economy

Barcelona City Council (2000) Modificacion del Plan General Metropolitano de Barcelona (MPGM). Normativa vigente. Barcelona, Spain: Ajuntament de Barcelona. Retrieved from http://www.22barcelona.com/component/option,com_remository/Itemid,750/func,select/id,16/orderby,1/lang,es/

Barcelona City Council (2016) 22@Barcelona, 2000–2015. Eldistricte de la innovaciondeBarcelona. Barcelona, Spain: Ajuntament de Barcelona. Retrieved from https://dades.naciodigital.cat/redaccio/arxius/documents/148534772422@_15_ANYS_CATALA_p.pdf

Case Study: 22@ Barcelona Innovation District | Smart Cities Dive, www.smartcitiesdive.com. Retrieved 13-03-2019

Engel JS (2014) Global Clusters of Innovation: Entrepreneurial Engines of Eco-nomic Growth around the world. Edward Elgar Publishing, Cheltenham

Ferrer, Josep-Ramon (2017) Barcelona's smart city vision: an opportunity for transformation, Field Actions Science Reports, Special Issue 16, 70–75 https://journals.openedition.org/factsreports/4367. Retrieved 14-03-2019

Firmino RJ, Duarte F, Moreira T (2008) Pervasive technologies and urban planning in the augmented city. J Urban Technol 15:77–93

Foster + Partners designs golden stadium for Qatar World Cup final, https://www.dezeen.com/2018/12/19/lusail-stadium-foster-partners-qatar-world-cup-2022-final/ Retrieved 14-03-2019

Geropanta V (2016) PhD thesis

Graham S, Marvin S (2001) SPLINTERING URBANISM networked infrastructures, technological mobilities and the urban condition by Routledge

Gunder M, Hillier J (2009) Planning in ten words or less: A Lacanian entanglement with spatial planning. Ashgate Publishing Ltd, Farnham

Harrison D (2015) Handbook of research on digital media and creative technologies. Idea Group, U.S, Harrisburg. https://doi.org/10.4018/978-1-4666-8205-4

Healey P (2007) Urban complexity and sapatial strategies towards a relational planning for our times, The RTPILibrary Series. Routledge Barcelona, Healey

Komninos N (2014) The age of intelligent cities: smart environments and innovation-for-all strategies. Routledge, London

Komninos N (2015) The age of intelligent cities: smart environments and innovation-for-all strategies. Routledge, London/New York

Larson K (2017) Barcelona smart city expo world congress. Retrieved from https://www.youtube.com/watch?v=aXFwZZ9cXD0

Lusail Smart City vision, http://www.lusail.com/the-project/smart-city/. Retrieved 14-03-2019a

Lusail Smart City vision, http://www.lusail.com/wp-content/File%20Store/DOC/Lusail%20Smart%20City%20Vision.pdf. Retrieved 14-03-2019b

Lusail Towers, https://www.protenders.com/companies/foster-and-partners/projects/lusail-towers. Retrieved 14-03-2019

Pallares-Barbera, M. (2015). Gentrificación productiva en barcelona: efectos del nuevo espacio económico. https://doi.org/10.13140/RG.2.1.4809.1041. https://www.urenio.org/

Pareja-Eastaway M, Pique JM (2011) Urban regeneration and the creative knowledge economy: the case of 22@ in barcelona, 4:319–327.

Pareja-Eastaway M, Turmo J, Pradel M, Simó-Solsona M, Garcia-Ferrando L (2007) The city of marvels? Multiple endeavours towards competitiveness in Barcelona, ACRE report WP2.2. AMIDSt, University of Amsterdam, Amsterdam

Pyzyk K (2018) Case study: 22@ Barcelona innovation district | smart cities dive. www.smartcitiesdive.com. Retrieved 13-03-2019

Qatar National Vision 2030, http://www.lusail.com/who-we-are/the-foundations-to-qatar-2030/. Retrieved 14-03-2019

Serag El Din H, Shalaby A, Farouh HE, Elariane SA (2013) Principles of urban quality of life for a neighborhood. HBRC J 9(1):86–92. ISSN 1687-4048

Smart city experts should be looking to emerging markets. Here's why, World Economic Forum, https://www.weforum.org/agenda/2018/10/how-the-developing-world-can-kickstart-the-smart-cities-revolution/. Retrieved 14-03-2019

Tarr JA (1984) The evolution of urban infrastructure in the nineteenth and twentieth centuries. In: Hanson R (ed) Perspectives on urban infrastructure. National Academy Press, Washington, DC, pp 4–66

Tarr JA, Dupuy G (1988) Technology and the rise of the networked city in Europe and America. Temple University Press, Philadelphia

United Nations Department of Economic and Social Affairs Population Division World Urbanization prospects: 2014 Revision (New York 2014)

WEF (2018) https://www.weforum.org/agenda/2018/10/how-the-developing-world-can-kickstart-the-smart-cities-revolution/

Chapter 26
The Road Ahead

Uttaran Dutta and Mainak Ghosh

Abstract Scholars and international institutions have noted that toward making the world more habitable and humane, we need to be mindful about anthropocentric activities and be proactive to raise ecological/environmental consciousness. In other words, we need to diligently engage and invest in designing meaningful solutions for all to co-create contextually appropriate outcomes and sustained access to resources and services, particularly when the Global South is experiencing severe disparities (including multidimensional poverty) and gaps in terms of power, structure, and resources. The impacts of these inadvertently shape up the forms that are designed or built by the people, be it product, environment, policy, or any artifact to run the society. Another delinquency is the perceptual, conceptual, and cognitive gaps, which often (influenced by Eurocentric conceptualizations) portray the knowledge and knowledge production processes of the Global South as inferior (if not pseudoscientific). It is therefore important for the design scholarship to recognize and understand various gaps that need to be bridged; they include—value-/ideology-related gaps; priority-, preference, and intention-related gaps, as well as cognitive-, thought process- and perception-related gaps. The issue of lack of access (to the environmental resources, to income and employment, and to basic infrastructures, among others), unfairness, and illegitimacy is quadrupled in the developing world due to illiteracy (including functional, computational, ecological, and cultural illiteracy) and lack of plurality of language, which divide the Global North and the Global South. Moreover, intercultural incompetence, lack of critical listening, and lack of reflexive dialogic engagement farther worsen the scenario. This chapter envisions a few approaches and calls for prudent use of these approaches or principles to make this world a better place to live in, including co-design and co-creation; community-centered an agency-centric approach; de-colonial and de-westernization approach; contextually/locally meaningful solutions; communicative engagements; intercultural competence; unlearning, relearning,

U. Dutta (✉)
Hugh Downs School of Human Communication, Arizona State University, Tempe, AZ, USA
e-mail: Uttaran.Dutta@asu.edu

M. Ghosh
Department of Architecture, Jadavpur University, Kolkata, West Bengal, India

© Springer Nature Switzerland AG 2020
M. Ghosh (ed.), *Perception, Design and Ecology of the Built Environment*,
Springer Geography, https://doi.org/10.1007/978-3-030-25879-5_26

and co-learning; transdisciplinary and multisensory approach; and ethics and reflexivity.

Keywords Co-creation · De-colonial · Intercultural communication · Reflexivity · Transdisciplinary · Social transformation

Sustainable development goals (SDG) of the United Nations as well as other international missions are diligently working toward making the world more habitable and humane. In doing so, on one hand, we need to be mindful about anthropocentric activities and be proactive to raise ecological/environmental consciousness. On the other hand, we need to actively engage and invest in designing meaningful solutions for all to co-create contextually appropriate outcomes, as well as to ensure adequate and sustained access to resources and services especially in the spaces of the Global South. Engaged involvement in such inclusive and socially embedded missions is a need of the hour, particularly when the Global South is experiencing severe disparities and gaps in terms of power, structure, and resources. Some of such obstacles and/or absences have been identified as discussed later, and some still remains obscure. The impacts of these inadvertently shape up the forms that are designed or built by the people, be it product, environment, policy, or any artifact to run the society. These get apparent in visible and usable forms and thus start creating a marked difference in terms of the Global North and Global South or any other terminology which refers to this phenomenon of perceptible difference. For example, the cars designed by the local companies of the developing countries are often questioned for its environmental efficiency, safety, and comfort. The compromise of these factors is significantly stark in comparison to their Global North counterparts. These reconciliations are often based on the issues, discussed further, which characterize the Global South. The way forward for the design and development requires careful attention and meaningful interpretation of the same.

Many spaces of the Global South are experiencing scarcity of resources and material disparities, which are potential precursors to multidimensional poverty (a measure to understand overlapping human deprivations in education, health, and standard of living) particularly at the margins. In this twenty-first century, many people of the Global south are still fighting for survival, particularly to overcome abject poverty and hunger. Poverty intertwined with disparities in the distribution of wealth, taxation, and policies related to poverty alleviation and related governmental or nongovernmental mechanisms constantly influences the design and development of the developing countries. The built environment takes up the largest share of this biased growth; a walk around the city reveals the signs of poverty. One such common view is presence of slums, shabby temporary shelters of poor people encroaching the streets, and vacant pockets of land. Such formations are unheard of in the Global North. Multidimensional poverty shapes up such built-up environment in the Global South.

Along with poverty, another delinquency is the perceptual, conceptual, and cognitive gaps. Eurocentric conceptualizations often portray the knowledge and knowledge production processes of the Global South as inferior (if not pseudoscientific). It is therefore important for the design scholarship to recognize and understand various gaps that need to be bridged; they include—value-/ideology-related gaps; priority-, preference, and intention-related gaps; as well as cognitive-, thought process, and perception-related gaps. This particular topic has been discussed in-depth in the book. The perceptual issues related to the Global South need exposure to wider global audience for its deeper understanding and thus putting "label" to any of the facts and figures of the Global South is rather approximate and relative. The very derivation of the words like Global South or Global North has been the fruit of these perceptual gaps. Today, the perception of a place in comparison to another is perhaps not the best way to evaluate anymore. For example, a meager vernacular building of a tropical region in the developing country should not be judged for its energy efficiency in contrast to or in a scale, which relates to the mechanized active climate-controlled building of the "west." The perception of that dwelling is unique and self-contained in its own way; it should not call for semantic elucidation based on perception, such as "traditional," "redundant," "inefficient," etc.

SDGs and other international initiatives show that lack of access to structure, facilities, and services is one of the key concerns of this hour; they include access to environmental resources, access to income and employment, and access to basic infrastructures, among others. These issues are intimately intertwined to governance and transparency; for example, increasing social injustice, inequality, and conflicts across the globe are some of the deeper concerns of this contemporary era. For example, the housing shortage is one of the burning issues in the population-heavy Global South. The lack of a proper shelter for living and surviving is predominant; however the available provisions seem to be neither adequate nor affordable or accommodative. In other words, they are provided to people based on some other criteria than the need itself. The transparency is likely to be at stake, consequently rendering the society unsolicited ground of inequality and prejudice.

The issue of unfairness and illegitimacy is quadrupled in the developing world due to illiteracy and lack of plurality of language. Functional illiteracy, computational illiteracy, ecological illiteracy, and cultural illiteracy are some of the major challenges of this era, which divide the Global North and the Global South. Moreover, dominance of mainstream languages (such as English) and rapid disappearance of endangered languages not only hinders the plurality of languages and diversity of cultural expressions but also makes the marginalized population vulnerable to the oppressions and exploitations. The sublime effects of language, as a part of cultural expressions, proliferate into other forms. The vernacular starts appearing useless and inferior as against prevalent forms of ascendency. The urbanization needs to meet this cruel transformation every day. In the quest to a create a fabric of widespread very average permanent houses made out of concrete, steel, and glass, they are advertised profusely, for commercial benefit only. This ingrains the idea into the minds of the people that other forms of buildings are perhaps not adequate and effectual enough. Of course, here the issue is not just about the invading propaganda, and the parties preaching, it but also with the takers of it, blinded by the layer of illiteracy.

Power-disparity and power-distances are also higher in many places of the Global South (in comparison to the Global North), particularly in those societies, which are collectivistic and vertical in nature. Such contextual realities along with skepticism and mistrust in various discursive spaces potentially pose additional challenges to design research interventions and implementations.

A variety aforementioned gaps, including lack of access to structural and communicative resources along with linguistic and cognitive gaps, essentially constitute (and aggravate) interactive barriers and silences and even cause communicative impossibilities that divide the societies. Moreover, intercultural incompetence, lack of critical listening, and lack of reflexive dialogic engagement farther worsen the scenario. All of these together or in separate parcels start affecting the appearance and formation of the Global South. This assumes a distinct character, but at the same time always gets compared for its inferiority compared to the Global North. The difference would continue to remain and perhaps widen with changing political, economic, and social epicenters. The chapters have highlighted issues, suggested recommendations, and described situations which could be rethought or worth considering in a different approach than the way it is dealt today. In order to overcome the aforementioned absences and disparities, future scholarship might want to pay attention to the following to bring about transformation, emancipation, and empowerment in the Global South.

Co-design and Co-creation As opposed to modernist and universal approaches to design, scholars are increasingly arguing in favor of participatory and co-creating design interventions, where the designers and community members work hand in hand as co-researchers (Kapuire et al. 2010; Nieusma 2004). In contrast to dominant and trickle-down approaches, co-designing interventions seek to reduce dependency on external resources or expertise toward bringing about empowerment and change in the societal spaces (Braund and Schwittay 2006).

Community-Centered and Agency-Centric Approach Challenging the dominant/hegemonic approaches to design, scholars envision cultural participants as the central organizing elements and active guiding force for community mobilization/organization. Furthermore, in legitimizing and developing local agentic capacities, they have argued in favor of critically examining and de-centering the exogenous experts to bring about transformations in the spaces of the Global South.

De-colonial and De-westernization Approach De-colonial and de-westernization approaches essentially call for questioning and problematizing Eurocentric and hegemonic assumptions by including values, philosophies, practices, experiences, and evidences from the Global South. Moreover, such approaches legitimize local knowledge as wealthy sources of resources and prioritize local autonomy to foster multicultural interactions that are socially embedded and culturally responsive.

Contextually/Locally Meaningful Solutions Inclusive approach such as this pays attention to local knowledge, resources, and skills as well as to situated needs, practices, norms, and cultural participants' aspirations and worldviews. This approach also seeks co-create cost-effective, customized, and sustainable solutions and that too by utilizing locally available social and cultural resources (Galdeano-Gómez et al. 2011).

Communicative Engagements Critical listening and mindful dialogue are key components of meaningful communicative engagements. While the process of listening questions societal power dynamics and taken-for-granted assumptions that foreclose community voices, a dialogic process seeks to ensure respectful and impactful interactive environment that is essential for creating transformation (Ganesh and Zoller 2012; Sorrells 2015). Moreover, such engagements are crucial for bridging communicative and discursive gaps to make design research initiatives open, unobtrusive, and conducive.

Intercultural Competence In the Global South (particularly in the underserved contexts), unforeseen and less known contextual complexities pose challenges to design research. In addressing/overcoming the same, it is often essential to develop intercultural competences and sensitivities. Inculcating and nurturing the qualities like empathy and compassion are crucial to make design approaches and/or interventions inclusive, respectful, and emancipatory.

Unlearning, Relearning, and Co-learning As future researchers and designers, we need to unlearn and introspectively evaluate our embodied ethnocentrism, unearned privileges, as well as unexamined practices that constitute formidable barriers to knowledge production. Such engagements not only would help us in exploring new learning avenues as well as guide us in developing newer theoretical and methodological insights. In addition, future scholarship needs to explore mutual learning endeavors to open up innovative and creative knowledge production/problem-solving avenues.

Transdisciplinary and Multisensory Approach Increasingly professional and academic research interactions are becoming transdisciplinary, where inputs and inferences from multiple paradigms, lenses, and domains intermingle and enrich design outcomes. In addition, it is also important to explore the complexities and nuances of multisensory interactions and alternative design avenues for the future scholarships.

Ethics and Reflexivity Reflexivity and ethics are two key quality designers needed to inculcate and nurture in bringing about responsible and responsive transformations in the Global South. By being reflexive, we question our unexplored privileges and sub-/unconscious senses of superiority as well as use self-introspections to act mindfully and ethically. Being cognizant about our own ethnocentric assumptions potentially prepares ourselves to conduct compassionate, inclusive, and socially embedded research.

Though much of the preceding discussion revolves around design, design should be transcended to a more holistic approach centered around man's encounter with earth for a humanitarian cause. The domain of design is thus vast and is definitely going to play a great role in the twenty-first century (Dunne and Martin 2006; Razzouk and Shute 2012). While negotiating with economic, social, political, and ecological complexities, the challenges of the Global South are pivoted around our ability to meaningfully design the solutions. The built environment is no exception. This particular book looks into different facets of the built, in the Global South, in terms its designed forms related to the environment and perception of people. Many of the chapters hint lacunas, which relate to lack of these approaches, indirectly or try to provide a solution based on these holistic approaches, without necessarily naming them. A way forward requires prudent use of these approaches or principles to make this world a better place to live in. And this is a stepping-stone to perhaps change the discourses on which this book revolves around, the Global South. An earth with no divide is possible only if it is consciously built that way.

Bibliography

Braund P, Schwittay A (2006) The missing piece: human-driven design and research in ICT and development. In: 2006 international conference on information and communication technologies and development. IEEE, pp 2–10

Dunne D, Martin R (2006) Design thinking and how it will change management education: an interview and discussion. Acad Manag Learn Edu 5(4):512–523. https://doi.org/10.5465/AMLE.2006.23473212

Galdeano-Gómez E, Aznar-Sánchez JA, Pérez-Mesa JC (2011) The complexity of theories on rural development in Europe: an analysis of the paradigmatic case of Almeria (South-East Spain). Sociol Rural 51(1):54–78

Ganesh S, Zoller HM (2012) Dialogue, activism, and democratic social change. Commun Theory 22(1):66–91

Kapuire GK, Winschiers-Theophilus H, Chivuno-Kurio S, Bidwell NJ, Blake E (2010) Revolution in ICT, the last hope for African rural communities' technology appropriation

Nieusma D (2004) Alternative design scholarship: working toward appropriate design. Des Issues 20(3):13–24

Razzouk R, Shute V (2012) What is design thinking and why is it important? Rev Educ Res 82(3):330–348

Sorrells K (2015) Intercultural communication: Globalization and social justice. Sage, Los Angeles

Correction to: From Grey to Green: Rethinking Setback and MGC Rules as a Sustainable Growth Strategy of Residential Areas – A Case Study of Anannya Residential Area of Chittagong, Bangladesh

Rezuana Islam, Kanu Kumar Das, and Samira Binte Bashar

Correction to:
Chapter 6 in: M. Ghosh (ed.),
***Perception, Design and Ecology of the Built Environment*, Springer Geography,**
https://doi.org/10.1007/978-3-030-25879-5_6

The original version of this chapter was published with some citations missing. The revised version has the missed citations added.

The updated version of this chapter can be found at
https://doi.org/10.1007/978-3-030-25879-5_6

© Springer Nature Switzerland AG 2020

M. Ghosh (ed.), *Perception, Design and Ecology of the Built Environment*,
Springer Geography, https://doi.org/10.1007/978-3-030-25879-5_27

Index

© Springer Nature Switzerland AG 2020

M. Ghosh (ed.), *Perception, Design and Ecology of the Built Environment*, Springer Geography, https://doi.org/10.1007/978-3-030-25879-5

Q

R

U

V

MIX
Papier aus verantwortungsvollen Quellen
Paper from responsible sources
FSC® C105338

FSC
www.fsc.org

If you have any concerns about our products,
you can contact us on
ProductSafety@springernature.com

In case Publisher is established outside the EU,
the EU authorized representative is:
Springer Nature Customer Service Center GmbH
Europaplatz 3, 69115 Heidelberg, Germany

Printed by Libri Plureos GmbH
in Hamburg, Germany